Handbook of Dielectric Materials

Handbook of Dielectric Materials

Editor: Miley Davis

NY RESEARCH PRESS

New York

Published by NY Research Press
118-35 Queens Blvd., Suite 400,
Forest Hills, NY 11375, USA
www.nyresearchpress.com

Handbook of Dielectric Materials
Edited by Miley Davis

International Standard Book Number: 978-1-63238-599-4 (Hardback)

Cataloging-in-Publication Data

Handbook of dielectric materials / edited by Miley Davis.
p. cm.
Includes bibliographical references and index.
ISBN 978-1-63238-599-4
1. Dielectrics. 2. Dielectric devices. 3. Electric insulators and insulation. I. Davis, Miley.
QC585 .H36 2018
537.24--dc23

Contents

Preface

This book aims to highlight the current researches and provides a platform to further the scope of innovations in this area. This book is a product of the combined efforts of many researchers and scientists, after going through thorough studies and analysis from different parts of the world. The objective of this book is to provide the readers with the latest information of the field.

Dielectric materials are materials that can store electrostatic charge. They are important materials that are used in the production of capacitors as well as industrial coatings, the electrophorus and the Van de Graaff generator. The electronics industry has greatly benefitted from the manufacture of such materials. Most dielectric materials are solids such as porcelain, plastics and mica but they can also include gases such as dry air and distilled water. Through this book, we attempt to further enlighten the readers about the new concepts in this field. It aims to serve as a resource guide for students and experts alike and contribute to the growth of the discipline.

I would like to express my sincere thanks to the authors for their dedicated efforts in the completion of this book. I acknowledge the efforts of the publisher for providing constant support. Lastly, I would like to thank my family for their support in all academic endeavors.

Editor

The Effect of Macromolecular Crowding on the Electrostatic Component of Barnase–Barstar Binding: A Computational, Implicit Solvent-Based Study

Helena W. Qi, Priyanka Nakka, Connie Chen, Mala L. Radhakrishnan*

Department of Chemistry, Wellesley College, Wellesley, Massachusetts, United States of America

Abstract

Macromolecular crowding within the cell can impact both protein folding and binding. Earlier models of cellular crowding focused on the excluded volume, entropic effect of crowding agents, which generally favors compact protein states. Recently, other effects of crowding have been explored, including enthalpically-related crowder–protein interactions and changes in solvation properties. In this work, we explore the effects of macromolecular crowding on the electrostatic desolvation and solvent-screened interaction components of protein–protein binding. Our simple model enables us to focus exclusively on the electrostatic effects of water depletion on protein binding due to crowding, providing us with the ability to systematically analyze and quantify these potentially intuitive effects. We use the barnase–barstar complex as a model system and randomly placed, uncharged spheres within implicit solvent to model crowding in an aqueous environment. On average, we find that the desolvation free energy penalties incurred by partners upon binding are lowered in a crowded environment and solvent-screened interactions are amplified. At a constant crowder density (fraction of total available volume occupied by crowders), this effect generally increases as the radius of model crowders decreases, but the strength and nature of this trend can depend on the water probe radius used to generate the molecular surface in the continuum model. In general, there is huge variation in desolvation penalties as a function of the random crowder positions. Results with explicit model crowders can be qualitatively similar to those using a lowered "effective" solvent dielectric to account for crowding, although the "best" effective dielectric constant will likely depend on multiple system properties. Taken together, this work systematically demonstrates, quantifies, and analyzes qualitative intuition-based insights into the effects of water depletion due to crowding on the electrostatic component of protein binding, and it provides an initial framework for future analyses.

Editor: Claudio M. Soares, Instituto de Tecnologica Química e Biológica, UNL, Portugal

Funding: This work was funded by a Wellesley College Staley Fellowship. The funders had no role in study design, data collection and analysis, decision to publish, or preparation of the manuscript.

Competing Interests: The authors have declared that no competing interests exist.

* E-mail: mradhakr@wellesley.edu

Introduction

It is believed that up to 40% of the cellular volume is occupied by macromolecules [1], making the cell a crowded place. Nevertheless, many *in vitro* experiments and computational studies model protein processes in a vast "sea" of aqueous solvent. To build better models of such processes, it is crucial to better understand the effect of cellular crowding on the physical determinants of protein folding and binding. While more attention has been given to these effects in recent years, reviews of crowding effects span multiple decades [2–9]. Experimental work has shown that crowding can cause a thermodynamic favoring of compact states – folded, bound, or aggregated states of proteins [10–13] – and could favor compaction of unfolded states as well [14,15], although sometimes certain effects were found to be small or even reversed [16,17], likely because of enthalpic interactions between crowding agents and the proteins being studied [18]. Nevertheless, even small, subtle effects could have important implications for aggregation associated with neurodegenerative diseases [10,19]. Crowding has also been experimentally shown to change the preferred conformations of protein and DNA systems [20–25] and

to alter drug–target interactions or affinities [26–28]. Finally, macromolecular crowding may slightly [16,29] or more greatly affect association rate kinetics [30] and reaction mechanisms [31,32].

Theoretical and computational studies have provided great insight into the physical bases for observed effects due to macromolecular crowding. Many thermodynamic studies to date have focused on the entropic "excluded volume" effect, in which crowding lowers the available cellular volume, thus lowering the entropy of noncompact states more than that of compact states, leading to a relative free energy stabilization of compact states. This effect was shown to have measurable consequences in theoretical and computational studies [33–36]. More recently, it was shown that favorable interactions between less compact states and the crowders could cancel out this effect or dominate over it [37–39], demonstrating not only that the physical properties of the crowders are important, but also that crowding could significantly affect the enthalpic component of the binding free energy in addition to the entropic component. The subtle interplay between multiple energetic components as well as dynamical effects have been considered via molecular dynamics simulations of proteins within a crowded environment [37,38,40,41]. These and other time-dependent simulations [42,43] have also provided insight into the association rates of proteins within the cellular milieu.

There have been relatively few studies that focus on how crowding affects the *electrostatic* component of protein–protein interactions and their solvation energetics. As a reasonable hypothesis, crowding can both affect the hydration dynamics of water [44] and deplete the number of polarizable water molecules surrounding the proteins, thereby potentially descreening their electrostatic interactions relative to the infinite dilution limit (i.e., the uncrowded case). While crowding has been incorporated into electrostatic models via a screened Coulomb potential-based implicit solvent model [45] and a lowered effective solvent dielectric constant [46], to our knowledge, only very recent work has probed more specifically to study how crowding affects electrostatic interactions within a solvated medium [47,48]. Such work demonstrated that it may be possible to capture certain electrostatic effects of crowding by a lowered solvent dielectric constant, a result that supports other work suggesting that the observed dielectric constants within cellular environments may be quite lower than that of water [49–53]. Specifically, Harada *et al.* [47] found via explicit solvent molecular dynamics simulations that water mobility was hindered in a crowded environment, providing one physical mechanism for this lowered dielectric constant. However, as they note, another mechanism for a lowered dielectric constant may stem from the fact that crowding depletes bulk water from around molecules, an idea that was explored further in an implicit model study [48]. It is this latter mechanism that provides the focus of the current study, although here, we extend this idea to study protein–protein binding.

This work uses simplified models to study how water depletion due to crowders can alter electrostatic binding free energies between proteins. We use the barnase–barstar protein complex as a model system, as it has been shown previously [54,55] that electrostatic interactions play a crucial role in their interaction, and it has also been used in previous studies investigating crowding or similar phenomena [35,45]. While a more realistic model may use explicit solvent and actual proteins as crowding agents, we wished to separate out electrostatic effects due to water depletion from other electrostatic effects, such as loss of mobility of individual water molecules or electrostatic interactions with crowder molecules. To that end, our study uses spherical, uncharged model crowders within an implicit solvent, and electrostatic free energies are computed through obtaining potentials via the Poisson Equation (or the Linearized Poisson-Boltzmann equation, if applicable). To again focus on the water depletion effect in a controlled manner, we assume rigid binding, although we recognize that crowding may affect protein conformations [48]. Our thermodynamic cycle allows us to separately quantify the effects of crowding on desolvation and on solvent-screened interaction. The use of simple model crowders enables us to systematically study these effects as a function of crowder density and size. Adequately sampling crowder locations to get proper Boltzmann-weighted distributions of states would be computationally infeasible, and so we limited our results to simple averages over 50 randomly-generated crowder placements in the bound and unbound states per data point, especially since Boltzmann-weighting based only on electrostatic solvation energies may be less realistic than assuming that other factors can also contribute to crowder placement.

We find that on average, crowding lowers desolvation penalties and amplifies solvent-screened interactions, stabilizing favorable interactions and destabilizing unfavorable ones. This effect is more pronounced when crowder size is reduced, assuming a standard-size water probe radius within the continuum model. The mean stabilization or destabilization of solvent-screened interactions was robust to the specific placement of the random crowders, but the average desolvation effects were not, with very large standard error values. While an overall reduced dielectric constant may capture average water depletion effects, there may be system specific conditions that lead to uncertainty in the mean effect of crowder placement as a simple function of crowder density and size. Finally, we show that crowding can differentially affect the electrostatic contributions of individual protein residue side chains toward binding, with the relative effects on desolvation and interaction depending on the residue's environment. This suggests that crowding could affect the consequences of specific mutations on binding, as well as the role that certain residues or binding "hot spots" play in varied cellular environments. While these results may qualitatively agree with intuition, our goal is to provide a systematic, controlled demonstration and quantitative analysis of these effects. Moreover, the methods used here provide experimentally testable hypotheses and an initial framework for understanding the role of crowding in modulating electrostatic interactions in protein–protein binding that can be built upon in future work.

Materials and Methods

Structure Preparation

Studies used a 2.0 Å resolution crystal structure of barnase complexed with a Cys -> Ala (40,82) double mutant of barstar (PDB ID 1BRS) [56]. The asymmetric unit consisted of 3 model complexes; the complex corresponding to chains A and D were used in this study. Crystallographic water molecules greater than 3.3 Å from either binding partner or with fewer than three potential hydrogen-bonding interactions with protein were removed. The remaining 17 water molecules were assigned to either protein partner based on proximity and hydrogen-bonding contacts. The amide groups of asparagine and glutamine and the imidazole group of histidine were flipped as necessary and the tautomerization states of histidine were assigned based on manual inspection of possible hydrogen bonding with surrounding residues. The two N-terminal residues of barnase and residues 64 and 65 of barstar were not resolved in the crystallographic experiment, and neighboring residues were patched with acetyl or N-methylamide groups. Hydrogens were modeled onto the structure with the HBUILD [57] functionality in CHARMM [58], using the CHARMM22 force field [59] and the TIP3P water model [60]. Patches and missing side chain density were added via CHARMM and were energy minimized.

Crowder Placement

Bound and unbound states in each binding free energy calculation were crowded separately. A box was created to contain both the protein complex (or each unbound state) and the model crowders, such that the box "walls" were each 70 Å from the most extreme (i.e., maximal and minimal) x, y, and z protein coordinates. The dimensions of the box were approximately 190×190×190 Å. Spherical crowders of either specified or random radii (up to 25 Å, roughly the size of the barnase–barstar complex) were added sequentially, and each potentially new crowder was accepted if it did not (1) overlap in space with any existing crowder or protein molecule, (2) partially or totally fall outside the total box volume, or (3) cause the volume density of crowders to be higher than the desired value. The volume density of crowders was calculated as the ratio of the total volume of the crowders to the originally available volume (i.e., volume not taken up by the protein(s)). Fig. 1 shows sample, random crowder placements around the bound state at denoted specifications. Preliminary analyses showed that one consequence of our crowder placement method is a depletion of crowder density at the system's extreme edges; future efforts to place crowders could adopt a strategy leading to more even placement throughout the entire system volume.

Continuum Electrostatics Calculations

A single-grid red-black successive over-relaxation finite-difference solver (M.D. Altman and B. Tidor, unpublished) [61] of the Poisson/Linearized Poisson Boltzmann Equation, distributed with the Integrated Continuum Electrostatics (ICE) software package (D.F. Green, E. Kangas, Z.S. Hendsch, and B. Tidor, Massachusetts Institute of Technology Technology Licensing Office), was used to solve for the electrostatic potentials of both crowded and uncrowded systems. Unless otherwise noted, a probe radius of 1.4 Å was used to define the molecular surface for the dielectric boundaries. Likewise, unless otherwise noted, a dielectric constant of 4 was used for all spherical crowders and protein atoms, and the solvent was modeled using a dielectric constant of 80. Potentials were solved on a 491×491×491 grid. A three-tiered focusing procedure was used, in which the system (the complex and all crowders) occupied 23%, 92%, and 184% of the grid. At the lowest focusing, the regions beyond the entire system were modeled as dielectric 80 and screened Coulombic (or Debye-Huckel, in cases of non-zero ionic strength) boundary conditions were used. Zero-radius dummy atoms were placed at identical extreme points of every run to maintain equal grid resolution for all states. At the highest focusing, this grid spacing yielded a resolution of approximately 4.6 grids/Å, and the grid was centered on barstar within the large system (for a small subset of runs, the grid was centered on a particular atom within the interfacial barstar Asp39 residue). PARSE radii and charges [62] were used. The ionic strength was set to zero except when implicit salt was modeled at a concentration of 0.145 M and a Stern layer of 2 Å was used. Due to memory limitations, runs with nonzero ionic strength were solved on a 401×401×401 grid, and to assess the effect of ionic strength, were compared only to other runs at the same grid resolution.

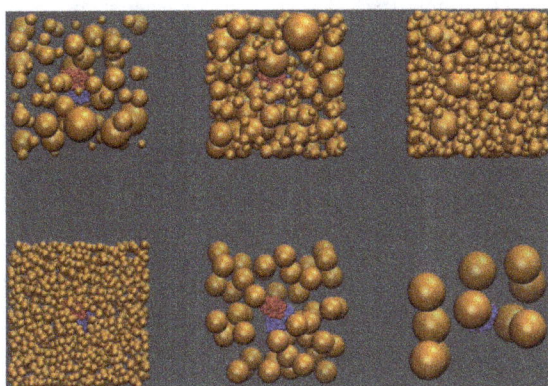

Figure 1. Sample simulated crowded environments. Here, the bound state barnase–barstar complex (red and blue) is surrounded by randomly-placed crowders (orange); the top row depicts environments in which the radius of crowders varied within a system (from 5–25 Å), at increasing crowder volume densities (left to right). The bottom row depicts environments at a constant crowder volume density, but with increasing crowder radius (left to right).

Potentials were solved for both the bound and unbound dielectric boundaries upon charging up one binding partner at a time. By multiplying (one-half) the potential differences due to charges on a given partner by the charges on that partner, desolvation penalties were obtained, and by multiplying the potentials due to charges on one partner by the charges on the other partner, solvent screened interactions were obtained [63] (Fig. 2).

Model Charge Variation

The monopole on each binding partner was changed by adding or subtracting random charge values of maximum magnitude 0.1 e to randomly selected atoms within the partner until the desired overall monopole was reached. No single atom was allowed to have an overall charge magnitude greater than 0.85 e. To test the robustness of the results, monopoles were changed by starting both with the original charge distribution and from a structure in which all the charges were set to zero. Here we show only the results produced by starting with the original barnase-barstar charge distribution.

Component Analyses

To quantify the contributions of selected residues toward the electrostatic component of binding in the presence and absence of model crowders, the partial atomic charges on the side chain of a given residue were all set to zero and the binding free energy re-evaluated, in a similar manner to component analyses in previous work on both protein and small molecule systems [55,64–68]. The effect of zeroing out the side chain was then computed via:

$$\Delta\Delta G_{res} = \Delta G_{zeroed} - \Delta G_{orig}$$

A positive value of $\Delta\Delta G_{res}$ implies that a residue's side chain contributes favorably toward the electrostatic component of binding, as zeroing out its charges worsens binding. The desolvation and interaction components of $\Delta\Delta G_{res}$ were computed by directly subtracting the desolvation and interaction components of the binding free energies between the system with zeroed charges and the original system, respectively.

Component Analyses of Residue Groups within Barstar

For analyses in which charges of groups of residues were zeroed, groups were determined by calculating the solvent accessible surface area (SASA) of residues within each partner (assuming

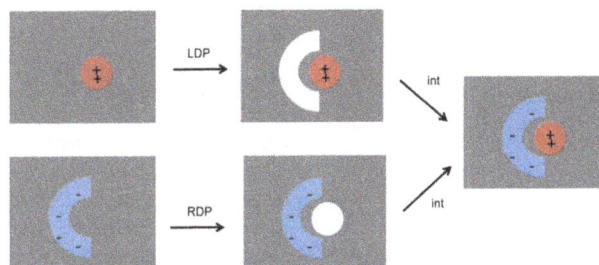

Figure 2. Schematic defining physically relevant components of the electrostatic binding free energy. Pictorially represented are the ligand (barstar) desolvation penalty (LDP), the receptor (barnase) desolvation penalty, (RDP) and the complex solvent-screened interaction (int). Gray regions denote solvent, and white regions denote low-dielectric cavities in the shape of a given partner, but without charges modeled. The total electrostatic binding free energy is LDP + RDP + int.

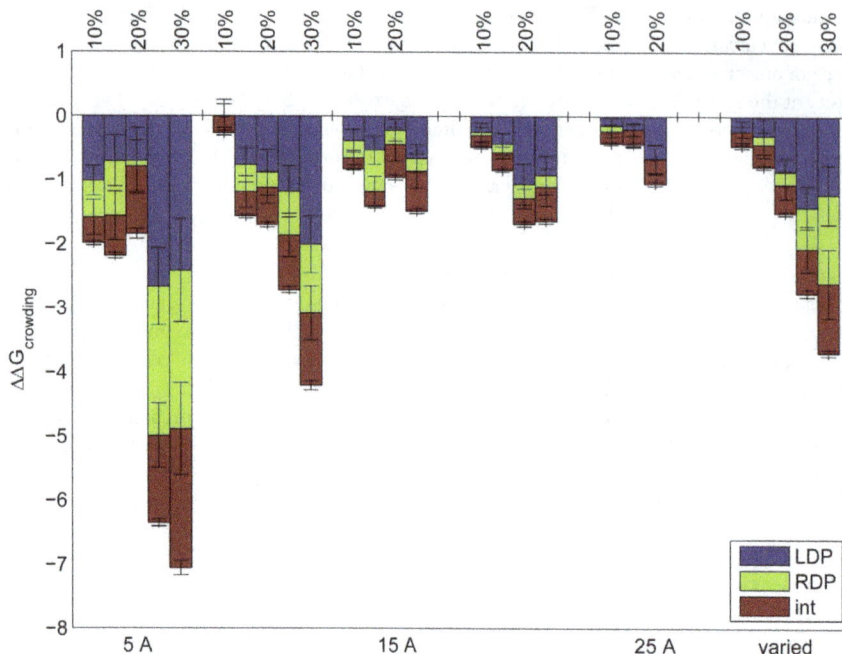

Figure 3. $\Delta\Delta G_{crowding}$, **in kcal/mol, for barnase-barstar vs. crowder volume density (top axis) and radius (bottom axis).** The bars at right ("varied") are for systems in which the crowder radius varies within each trial. Each bar is the average of 50 trials and is shown as a composite of its contributions of barstar desolvation penalty (LDP, blue), barnase desolvation penalty (RDP, green), and solvent-screened interaction (int, red). Error bars on each contribution represent +/−1 standard error. Missing bars are a result of unsatisfiable geometric constraints (see Results).

associated water molecules are considered residues and not bulk solvent) in the bound and unbound states. CHARMM was used to calculate SASA, using a 1.4 Å -radius probe and the CHARMM22 force field. Residues with non-zero burial upon binding were classified as either highly buried or peripheral depending on whether more or less than 50% of their unbound SASA remained in the bound state. Non-core residues were classified as either surface exposed or partially exposed depending on whether they have more or less than 50 Å^2 SASA in the unbound state. Here, the charges of both side chain and backbone atoms were set to zero so that the union of all atoms considered was the entire barstar protein (and associated explicit water molecules).

Data Analysis and Visualization

Figures of protein molecules and model crowder systems were generated using VMD [69]. All plots and data analyses were performed using Matlab (The Mathworks, Inc. Natick, MA).

Results

To assess the effect of water depletion due to crowding on the electrostatic component of protein–protein binding, binding free energies were computed in the presence and absence of model crowders. To model the crowded states in a controlled fashion and focus on water depletion, spherical, uncharged "crowders" were randomly placed around the bound and unbound state proteins at specified densities (Fig. 1). The effect of crowding on the electrostatic component of the binding free energy was quantified as the difference between the electrostatic binding free energies in the presence and absence of crowders:

$$\Delta\Delta G_{crowding} = \Delta G_{bind,crowded} - \Delta G_{bind,uncrowded}$$

A negative $\Delta\Delta G_{crowding}$ means that crowding lowers the electrostatic binding free energy (i.e., favors binding, all other components equal). With our model, $\Delta G_{bind,elec,uncrowded}$ was found to be 0.5 kcal/mol, suggesting that the electrostatic component of binding in this system (in pure aqueous solvent) is neither strongly favorable nor unfavorable, in qualitative agreement with previous work using quantitatively different parameters [70]. Given that the electrostatic binding free energies between proteins are generally quite unfavorable with models using an internal dielectric constant of 4 [71], our value supports the accepted view that electrostatics play an important role in this system.

Binding free energy contributions were broken into desolvation and interaction components (Fig. 2). The free energy cost upon binding to remove solvent interactions with barstar (considered the "ligand") is denoted the ligand desolvation penalty (LDP), and was found to be 41.7 kcal/mol for the uncrowded system. The energetic cost upon binding to remove solvent around barnase (the "receptor") is termed the receptor desolvation penalty (RDP, 37.2 kcal/mol when uncrowded). Finally, the solvent-screened interaction between the partners (int) was also quantified (−78.4 kcal/mol when uncrowded).

On Average, Crowding Lowers Desolvation Penalties and Amplifies Interactions

Figure 3 is a graph of $\Delta\Delta G_{crowding}$ as a function of crowder radius (bars grouped by bottom axis) and crowder volume density (top axis). In the rightmost set of bars, crowder radii vary within each system from 5–25 Å (the largest spheres were therefore approximately the size of the protein complex). Total $\Delta\Delta G_{crowding}$ values are broken up into contributions due to changes in barstar's desolvation penalty (LDP, blue), barnase's desolvation penalty (RDP, green), and solvent-screened interaction (int, red). Each bar is the result of 50 random trials, with average values +/− standard error (not standard deviation) shown for each contribution.

Figure 3 shows that on average, $\Delta\Delta G_{crowding}$ was negative for all crowder densities and radii, although generally, the effects were more pronounced at higher crowder densities and smaller crowder radii. Moreover, the changes in all contributions (LDP, RDP, and int) were generally negative on average, in this system. This result makes intuitive sense – in a crowded environment, each unbound state is already partially desolvated by crowders, with some crowders potentially occupying the same space in the unbound state as the binding partner does in the bound state. Hence, there may be less solvent displaced near the binding interface upon binding in the crowded system when compared to an uncrowded one, resulting in a reduced desolvation penalty on average. Moreover, the bound state is also partially desolvated due to the crowding, resulting in less solvent screening and more amplified interactions between the two partners. Because the interactions in this complex are favorable in general, amplifying them would increase their favorability.

The average effects seen in Fig. 3 are qualitatively similar to what one might obtain using a lower solvent dielectric constant. Previous work has modeled aspects of crowding via the use of a lower "effective" solvent dielectric constant [37,38,46,48], and experimental evidence suggests that a dielectric constant can be characterized for the cytoplasm [51,53] through measuring shifts in emission wavelength maxima of fluorescent probes due to the polarity of the microenvironment. This observed constant likely is a macroscopic average accounting for both the loss of water mobility and water depletion (and potentially other effects), the first of which is not accounted for in the present study. Nevertheless, it is instructive to measure the effects of a lowered, effective solvent dielectric on protein–protein binding. Figure S1 shows $\Delta\Delta G$ values (relative to a solvent dielectric constant of 80) for the desolvation and interaction components of barnase-barstar binding as a function of solvent dielectric constant. In addition, Table 1 shows numerical data using two potential values of solvent dielectric constant – an experimentally obtained value of 21.9[53] and the value of 55, similar to values found from explicit simulations at 30% crowder volume density, to model solely the effects of hindered water mobility [47]. A dielectric constant of 21.9 produced $\Delta\Delta G$ values that were several times more pronounced (Table 1) than the results obtained using explicit crowders (Fig. 3), but this may be because the experimentally-obtained constant would account for not only water depletion, but also hindered water mobility and other possible effects of crowding. A dielectric constant of 55 again produced more pronounced results than using explicit crowders within a dielectric 80 medium, although the effects were more quantitatively similar to our explicit crowding simulations ($\sim$1 kcal/mol difference in $\Delta\Delta G$ for desolvation components and $\sim$5 kcal/mol difference in $\Delta\Delta G$ for interaction, at a 30% crowding density and varied radius, Table 1). Again, differences could be due to the fact that this value was found to account for hindered water mobility and not water depletion.

The qualitative trends seen with lowered dielectric constants (Fig. S1) were similar to the trends found in this work for either increasing crowder volume density or decreasing radius, although for a given crowder radius and volume density, there may not exist an effective dielectric constant that provides quantitative agreement. Perhaps a "long-range" dielectric constant cannot model the full effect of hydration immediately surrounding each macromolecule; in a heterogeneous environment, the dampening of the electric fields due to a small amount of highly polar water might not be captured by an average, low macroscopic dielectric constant and therefore, effects of crowding may be overestimated. Nevertheless, one potential solution, similar to what was done in

work by Harada et al. [38], is to use a slightly lower dielectric constant to account for the loss of water mobility and explicitly model crowders to account for water depletion. Future work could also involve effective medium theory approaches to estimate effective dielectric constants of this composite environment as a function of crowder size and shape [72].

The relatively small standard error for interaction indicates that the mean stabilization due to the further descreening of interactions relative to infinite dilution is fairly robust to the ensemble of states sampled; there is little uncertainty in the mean effect. However, the large standard error for both desolvation contributions in all ensembles indicates great uncertainty in the mean reduction of desolvation penalties due to random crowder placement. As desolvation penalties depend strongly on the level of direct solvent exposure of charged or polar interfacial groups, it makes sense that they will be very sensitive to precise crowder placement. Interaction energies, on the other hand, are more long-ranged, except for interfacial interactions (and these are fairly unaffected by crowders in the bound state anyhow), and are therefore far less sensitive. The large standard error due to desolvation, by definition, implies an even larger standard deviation and therefore a huge amount of variability between trials, which suggests the necessity of thorough sampling. Currently, it is computationally infeasible to thoroughly sample all relevant crowder configurations. Preliminary attempts to use Boltzmann-weighting to more heavily account for lower-energy states by obtaining partition functions from each set of 50 sampled configurations resulted in similar qualitative trends to those shown in Fig. 3 (data not shown).

Our results suggest that the effects of crowding on water depletion are most pronounced at a given crowder volume density when the crowders are small, although large standard errors confound the robustness of this result, especially for desolvation. Presumably, very small molecules can more closely approach the irregular surface of a protein, more substantially desolvating it in its unbound state and more effectively descreening its interactions with a partner in the bound state relative to infinite dilution. Analyses of our model crowded systems showed that the minimum distance of approach between any one crowder and the proteins increases on average as the crowder radius increases (Figure S2), in support of this hypothesis.

It is plausible that aspects of this observed trend could be dependent on the use of a standard, nonzero-sized (here, 1.4 Å) "probe" used to generate the molecular surface in continuum models. The water-sized probe is intended (as standard practice) to approximately account for the nonzero size of discrete water molecules and the inability of "actual" water molecules to occupy cavities and crevices smaller than their size. A consequence of this model feature is that low-dielectric regions will be larger than the actual volume occupied by model crowders and protein, and this difference will likely be greater for systems with smaller-radius crowders due to the likelihood that they often closely approach each other and the protein.

To test this hypothesis, we redid a subset of the calculations shown in Figure 3 using a probe radius of zero to generate the molecular surface. The results are shown in Figure S3. Desolvation penalties were still reduced on average and interactions amplified, but as expected, the quantitative effects were now often $\sim$50–75% less pronounced ($\Delta\Delta G_{crowding} = \sim$2 kcal/mol or less). Additionally, the dependence of the desolvation effects on radius was not apparent (although they did not appear to be statistically significant even with a standard probe radius). However, the average effect on the interaction component still strengthened overall as the crowder radius decreased, suggesting some

Table 1. $\Delta\Delta G_{crowding}$ values for selected model systems described in the text.

$\Delta\Delta G_{crowding}$	LDP	RDP	int	TOT
$\varepsilon_{out} = 55$	−0.1	−1.1	−4.8	−6.0
$\varepsilon_{out} = 21.9$	−3.1	−6.6	−21.7	−31.4
$\varepsilon_{in} = 4$, control run	−1.2±0.5	−1.4±0.5	−1.08±0.05	−3.7±0.7
$\varepsilon_{in} = 1$, same	−3±2	−5±2	−1.4±0.1	−9±3
$\varepsilon_{in} = 1$, random	0±1	−5±1	−1.4±0.1	−6±2
0 M ions, same, lower grid	−1.2±0.5	−1.4±0.5	−1.08±0.05	−3.7±0.7
0 M ions, random, lower grid	−1.0±0.4	−0.1±0.3	−0.97±0.04	−2.1±0.5
0.145 M ions, same, lower grid	−0.9±0.5	−1.3±0.5	−0.48±0.05	−2.8±0.7
0.145 M ions, random, lower grid	−0.7±0.4	−0.2±0.3	−0.51±0.05	−1.4±0.5

$\Delta\Delta G_{crowding}$ values broken into components (LDP, RDP, int, and total) for systems not shown in Fig. 3. In the first two rows, the outer dielectric constant is varied as a substitute for explicitly modeling crowders. In the next set of rows ($\varepsilon_{in} = 1$, $\varepsilon_{in} = 4$), the internal dielectric constant was changed to 1 and compared with the control value of the reference system ($\varepsilon_{in} = 4$, also the rightmost bar in Fig. 3). The last four rows show the effect of nonzero ionic strength. For maximal control, all components were re-evaluated at a slightly lowered grid both without ions ("0 M ions, same, lower grid") and with ions ("0.145 ions, same, lower grid"). Additionally, crowders were either kept the same as they were in the 50 trials of the reference system ("same") or were randomly varied ("random").

robustness to the observation that smaller crowders may have greater impact. While it is standard practice to use a probe radius of 1.4 Å [73,74], results using a continuum model can be sensitive to this feature [74,75]. Our results demonstrate this limitation, specifically when modeling crowding effects using a continuum approach.

Even with the "standard" probe radius of 1.4 Å, at radii that more accurately model small proteins (20–25 Å), the mean effects on electrostatic interaction were found to be modest, but still significant on average, especially at higher crowding densities. These data suggest that the effects of crowding on electrostatics could be sensitive to the precise distribution of molecular sizes within the cell, and that it might be not be crowding due to proteins but rather, due to smaller metabolites and peptides that most greatly affects the electrostatic component of binding. We note that the trends for radii are curtailed here due to missing data at higher crowder densities and larger radii. Because of our purely random, sequential crowder placement, it became geometrically impossible to satisfy all constraints noted in the Methods when both crowder size and desired volume density were large. Future work can attempt to explore this region of property space while still maintaining a purely random crowder placement within the noted constraints.

Taken together, these results show that on average, the effects of crowding on electrostatic interactions can vary as a function of both crowder volume density and size, but desolvation effects are highly sensitive to crowder placement. To qualitatively account for crowding effects due to water depletion, therefore, it may be expedient to use an effective lowered solvent dielectric constant. Our work supports the idea that such a constant is likely to be specific to crowding volume fraction [47] and the distribution of crowder radii, and additional parameters may be needed to capture system-specific variations due to various arrangements of crowders.

In addition to the varied probe radius size discussed above, a subset of data was obtained under other different model conditions, to gauge the robustness of our results to parameters and physical conditions. First, we varied the internal dielectric constant used for both protein and model crowders. For maximal control, the precise locations of crowders in the bound and unbound states of the 50 trials were maintained in calculations

with different dielectric constants in one set of runs, and allowed to vary in another set. Results here used a varied crowder radius at a volume density of 30%. With an internal dielectric constant of 1, results were qualitatively similar to those with an internal dielectric constant of 4 when controlling for crowder placement and quantitatively more pronounced on average, especially for desolvation penalties (Table 1, $\varepsilon_{in} = 1$, same). However, standard errors were much larger, which may explain the difference in $\Delta\Delta LDP_{crowding}$ between trials in which the same crowders were used and when random crowders were used (Table 1, "$\varepsilon_{in} = 1$, random").

To understand how the presence of electrolytes could modulate the effect of crowding, data were gathered including implicit mobile ions at a concentration of 0.145 M through obtaining potentials via the linearized Poisson-Boltzmann equation. Again, we used a crowder volume density of 30% and randomly varied crowder radii, although all relevant runs with and without mobile ions were done at a somewhat lower grid resolution due to memory limitations when modeling salt (see Methods). We obtained qualitatively similar results when the solvent contained implicit, mobile ions, although the average lowering of the LDP, RDP, and especially int, were not as pronounced (Table 1).

If crowders descreen interactions relative to infinite dilution, they should amplify both attractive and repulsive interactions. To show this, we computationally modified the charge distributions on both barstar and barnase to vary their monopoles (see Methods). Of course, such charge distributions are not realistic, but they allow for a controlled, systematic study on how a system's charge distribution may affect its molecular recognition profile in a crowded environment. Figure 4 shows the average change in LDP, RDP, and int for three modeled pairs of monopoles – in which the partners either had opposite, large-magnitude monopoles (+/− 10 e), no net monopole, or the same, large-magnitude monopole (+10 e). Each bar is the average of 50 trials in which crowders of varied (5–25 Å) radius were used at a 30% volume density. The average effect of crowding on desolvation penalties was similarly stabilizing in all three cases, but the average effect on interactions is markedly different in the three cases. As expected, crowding greatly destabilized the (+10/+10) interaction and greatly stabilized the (+10/−10) one. This suggests that binding partners' overall monopoles can affect how they interact with partners in a

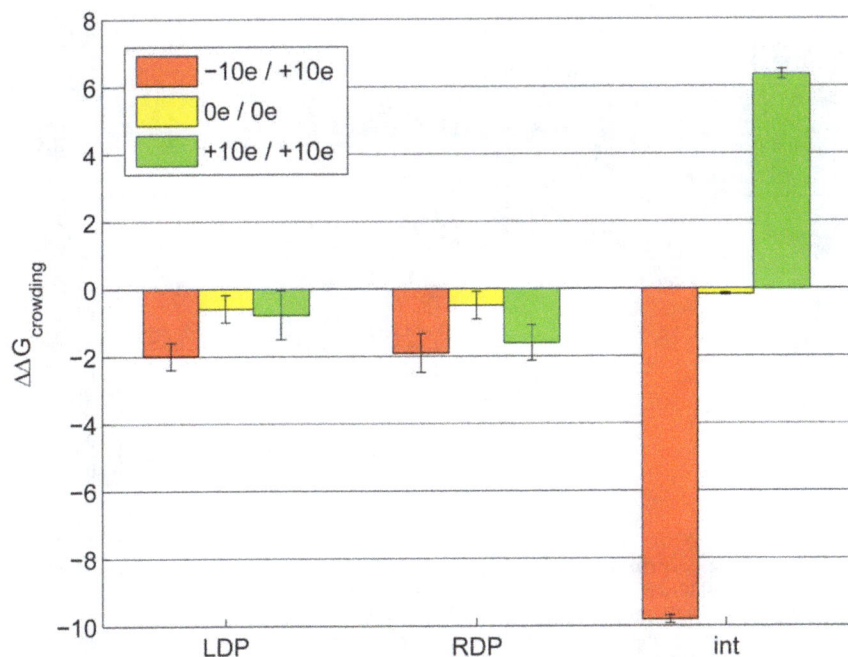

Figure 4. Effect of partner monopole on $\Delta\Delta G_{\text{crowding}}$. $\Delta\Delta G_{\text{crowding}}$, broken into barstar desolvation penalty (LDP), barnase desolvation penalty (RDP), and solvent-screened interaction (int) components, is shown in kcal/mol for the binding free energy of hypothetical proteins generated by randomly altering the charges of randomly selected atoms on the barnase–barstar complex until a desired overall monopole on each partner is reached (see legend). Each bar shows the average of 50 trials in which the bound and unbound states were crowded with spheres of random, varied radii (5–25 Å) to 30% crowder volume density. Error bars indicate +/−1 standard error.

crowded environment, although this effect is mediated more by interactions rather than the desolvation component.

Crowding can Differentially Affect Electrostatic Contributions of Side Chains toward Binding

Many protein–protein interactions have been shown to be mediated by one or more polar or charged residues or "hot-spots" [76–79]; such residues can be elucidated by experimental mutagenesis studies (e.g., alanine scanning) or through computational analyses. Presumably, if the overall electrostatic binding free energy can be modulated by the level of environmental crowding, as the model above suggests, then this implies that the specific contributions of individual residues toward that interaction can also be altered, but the nature of the alteration may depend on the properties of each residue.

To explicitly demonstrate, quantify, and better understand this intuitive idea, we began with the original (unaltered) charge distribution of the complex and quantified the electrostatic contribution of selected barstar residues toward the binding free energy by computationally setting the original partial atomic charges on a given side chain to zero and re-evaluating the binding free energy to obtain a $\Delta\Delta G_{\text{res}}$ (see Methods); this procedure was done both in the presence of crowders (the 50 trials used in the original analyses were used to obtain an average $\Delta\Delta G_{\text{res}}$) and in the absence of crowders. Consequently, we can define a $\Delta\Delta\Delta G_{\text{res,crowding}}$ that quantifies the effect of crowding on a residue's contribution toward the binding free energy:

$$\Delta\Delta\Delta G_{\text{res,crowding}} = \Delta\Delta G_{\text{res,crowded}} - \Delta\Delta G_{\text{res,uncrowded}}$$

A positive $\Delta\Delta\Delta G_{\text{res,crowding}}$ means that a residue contributes *more*

favorably (or less unfavorably) toward binding in the presence of crowding than in its absence.

In this study, we chose to calculate $\Delta\Delta\Delta G_{\text{res,crowding}}$ for five barstar residues whose side chains were previously shown to contribute significantly toward the electrostatic component of binding free energy [55]: Tyr29, Asp35, Asp39, Thr42, and Glu76. Figure 5a is a graph of the $\Delta\Delta\Delta G_{\text{res,crowding}}$ for each of these residues, broken up into barstar desolvation (LDP) and interaction (int) components (there is no change in the desolvation of barnase, RDP, as only charges on barstar were changed to zero). On average, the charged side chains contributed even more favorably in the presence of crowding, although the effect was quite small, with an average $\Delta\Delta\Delta G_{\text{res,crowding}}$ of only tenths of a kcal/mol. The contributions were not significantly changed on average for the two polar side chains studied.

Interestingly, the desolvation component of $\Delta\Delta\Delta G_{\text{res,crowding}}$ was altered more on average for Asp35 and Asp39, whereas the interaction component was altered more on average for Glu76. We hypothesize that the different mechanisms of altering $\Delta\Delta\Delta G_{\text{res,crowding}}$ is due to where these residues lie relative to the binding interface (Fig. 6). Both Asp35 and Asp39 are interfacial and highly buried upon binding, and so crowding may more greatly affect their desolvation penalties, by partially desolvating them already in the unbound state. Glu76, however, is more peripheral to the interface and so it remains more solvent exposed upon binding–this implies that crowding could more greatly impact the solvent-screening of its interactions in the bound state.

To further explore the idea that crowding might affect residue-based contributions differently, we grouped barstar residues based on both level of surface exposure and degree of burial upon binding (see Methods). Then, we zeroed out the charges simultaneously on all residues in each group (including both side chain and backbone) to determine $\Delta\Delta G_{\text{res}}$ for that group. This was

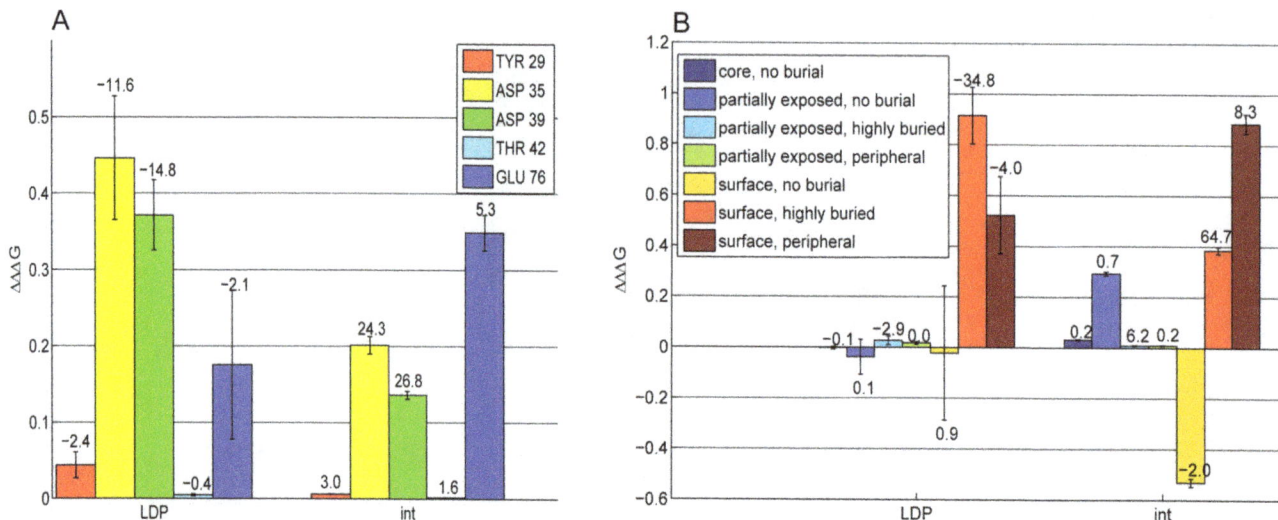

Figure 5. Effect of crowding on residue-based electrostatic contributions. $\Delta\Delta\Delta G_{res,crowding}$, broken into barstar desolvation penalty (LDP) and interaction (int), in kcal/mol, is shown for (a) selected barstar residues (see legend) and for (b) groups of barstar residues based on level of surface exposure and degree of burial (see Methods); The number above each bar indicates the actual magnitude of the selected component of $\Delta\Delta G_{res}$ without crowding present. Each bar indicates an average of 50 trials in which each crowded bound and unbound state is crowded with spheres of random, varied radii between 5 and 25 Å to 30% crowder volume density. Error bars indicate +/−1 standard error.

done both in the presence and absence of crowding to obtain a $\Delta\Delta\Delta G_{res,crowding}$ (using the 50 trials used in the original analyses). Indeed, surface residues that are highly buried upon binding showed the largest desolvation component of $\Delta\Delta\Delta G_{res,crowding}$ values (Fig. 5b), while surface residues that are peripheral to the interface (i.e., only partially buried upon binding) showed the largest interaction component of $\Delta\Delta\Delta G_{res, crowding}$. Interestingly, $\Delta\Delta\Delta G_{res,crowding}$ of surface residues with no burial upon binding (i.e., distal from the interface) was negative; here, crowding makes

these residues contribute *more unfavorably* toward binding. This result may be due to the dominating effect of the monopoles of distal groups; the monopoles on our model of barnase (+1) and the collection of distal, surface exposed residues on barstar (+3) have the same sign. The same trends are found when one controls for the number of residues in each group by finding the average $\Delta\Delta\Delta G_{res,crowding}$ per residue in each group (Fig. S4). These results explicitly demonstrate that electrostatic contributions–and therefore perhaps mutational energies–can be predictably altered in an

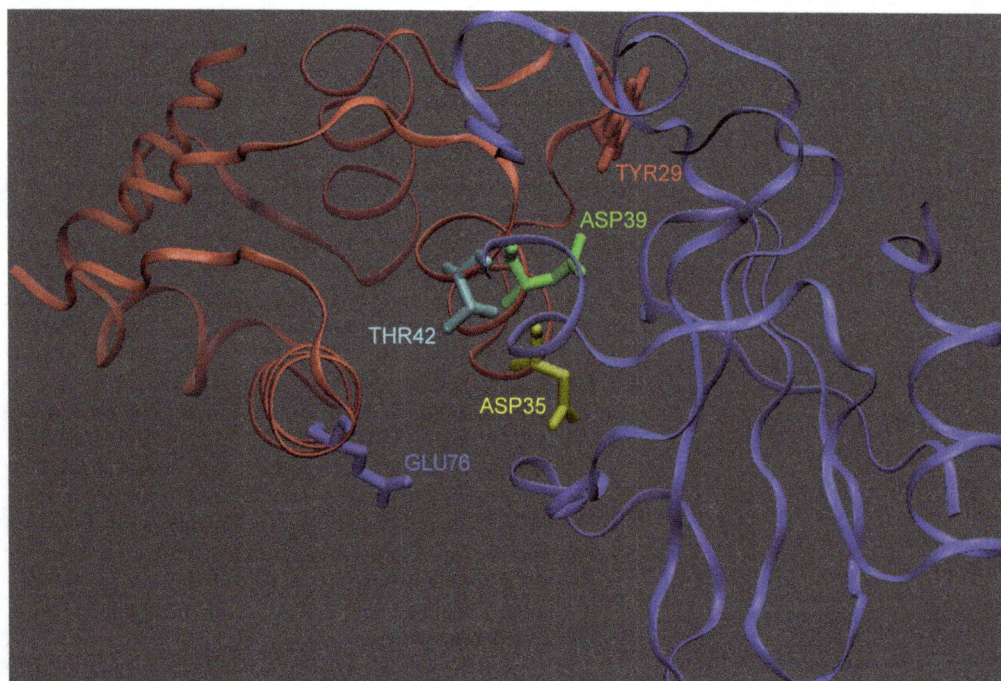

Figure 6. Location of the 5 barstar residues studied via component analysis within the barnase(blue)/barstar(red) complex.

environmentally-dependent way for residues in a crowded environment.

Discussion

In this work, we used simplified models to investigate the effect of macromolecular crowding on the electrostatic component of protein–protein binding free energy via water depletion. We found that for proteins with favorable electrostatic interactions, crowding can enhance the relative favoring of the bound state due to lowered desolvation penalties and enhanced interactions. For proteins with potentially unfavorable interactions, there may be opposing effects. The effects of crowding on desolvation were highly sensitive to crowder placement–yielding far more uncertainty in the mean effect on desolvation than in the mean effect on the interaction component.

Our results can potentially provide experimentally-testable hypotheses. For example, one could experimentally study the effect of monopole-changing yet relatively isosteric (e.g., Asn-→Asp) interfacial and peripheral mutations on protein–protein binding in crowded and uncrowded environments to see if crowding affects their relative contributions as predicted.; these experiments can be bolstered by varying ionic strength to highlight the interaction component of binding over desolvation components. Experimental tests would likely combine the effects of crowding due to both water depletion and lowered solvent mobility, so experimental results should reflect the predictions in this work in combination with other computational predictions [47].

The importance of crowder size was studied in a previous computational study that focused on the excluded volume effect of crowding on the binding of the barnase–barstar complex [35]. Like our study, it was also found that smaller crowders had a larger effect, but for a different reason–at a given volume density, smaller crowders left smaller voids for the proteins to occupy, lowering the available volume. This effect was confirmed in another study, and it was also shown that the ratio between crowder size and protein size is important [12]. Thus, smaller crowders may have a bigger impact for multiple reasons – by their excluding more volume and by their ability to more closely approach proteins to desolvate them and descreen their electrostatic interactions relative to infinite dilution.

We also demonstrated that crowding can differentially affect the relative contributions of residues toward binding. That these changes can be dominated by different phenomena (desolvation vs. interaction) could provide avenues for rational, environmentally-dependent design tasks.

This study provides a useful framework on which to build in future studies. With adequate computational resources, larger-sized model crowders and overall crowded volumes could be explored. Elements of "reality" can be added individually, in turn, to understand the effect of each on the binding free energy. Such elements include using actual protein shapes for the crowders (crowder shape has been shown to affect changes in folding and binding free energies [12,80]) as well as protein charge distributions to include direct enthalpic crowder interactions, which have been shown to be important for protein stability and conformation [18,48]; it would be interesting to quantify their precise effects on protein–protein binding. Another future goal is to increase the sampling of crowder configurations and potentially the conformational states of the binding partners, to allow for Boltzmann-weighted averages through Monte Carlo or dynamic simulations. In this study, the costs of Poisson-based models on such large systems prohibited exhaustive sampling (each binding free energy

calculation took ~0.5 day of CPU time and>1GB RAM with current resources).

To also account for the altered mobility of water molecules due to crowding, explicit solvent simulations are necessary, and have been previously attempted [38,47], although rigorously analyzing such effects on the energetics of specific protein–protein binding has yet to be done, to our knowledge. Given the potential computational cost of such studies, alchemical transitions [81,82] of individual residues (i.e., component analysis) or small molecule–protein binding systems may be good starting points.

In this study, we demonstrated and systematically explored the idea that macromolecular crowding can affect the electrostatic component of the free energy of binding between proteins through depleting regions of high dielectric water. Our results highlight yet another example of how environmental effects can have a quantitative and potentially qualitative impact on molecular recognition and should therefore be considered in both the analysis and the rational design of biomolecular systems.

Supporting Information

Figure S1 $\Delta\Delta G_{elec}$ vs. solvent dielectric (relative to a solvent dielectric constant of 80), without explicit crowders. A lowering of the external dielectric constant produces a similar qualitative trend as increasing the volume density or decreasing the radius of explicit crowders.

Figure S2 Average minimum distance of approach between crowders and protein vs. crowder radius. The minimum distance of approach is the shortest distance between the protein and crowder in each state, accounting for their radii. Data are shown for both 15% crowder volume density (data for 20% crowder density show a similar trend, not shown). Data are averaged over bound and unbound states for all 50 trials conducted for each radius and volume density. Error bars are +/− one standard deviation.

Figure S3 Effect on $\Delta\Delta G_{crowding}$ of using a zero-radius probe to generate the molecular surface. A subset of runs shown in Fig. 3 were redone using a zero-radius probe sphere to generate the molecular surface instead of the standard 1.4-Å probe. Identical crowder placements were used for each bar shown here and the bar corresponding to the same crowder density and radii in Fig. 3; the only different is in the size of the probe sphere.

Figure S4 Per residue $\Delta\Delta\Delta G$ for sets of residues on barstar. Residues were grouped by degree of burial and solvent exposure and values were *normalized by dividing by the number of residues in each group* (Figure 5b in the main text does not normalize per residue). Similar overall qualitative trends are seen in this Figure and in Figure 5b in the main text. The number above each bar indicates the per-residue value of the selected component of $\Delta\Delta G_{res}$.

File S1 Contains sample pdb files used in our analyses; each contains the barnase/barstar complex with a specific random placement of crowders of varying radii at 30% crowding density. Also included are files specifying, for each pdb file, the radii of each atom and model crowder.

Acknowledgments

The authors thank Michael Altman, Jaydeep Bardhan, David Green, and Bruce Tidor for helpful discussions and/or software.

Author Contributions

Conceived and designed the experiments: HWQ PN MLR. Performed the experiments: HWQ PN CC. Analyzed the data: HWQ PN CC MLR. Contributed reagents/materials/analysis tools: MLR. Wrote the paper: HWQ MLR. Reviewed and approved the manuscript: HWQ PN CC MLR.

References

1. Ellis RJ, Minton AP (2003) Cell biology-Join the crowd. Nature 425: 27–28.
2. Zimmerman SB, Minton AP (1993) Macromolecular crowding-Biochemical, biophysical, and physiological consequences. Annual Review of Biophysics and Biomolecular Structure 22: 27–65.
3. Ellis RJ (2001) Macromolecular crowding: Obvious but underappreciated. Trends in Biochemical Sciences 26: 597–604.
4. Hall D, Minton AP (2003) Macromolecular crowding: Qualitative and semiquantitative successes, quantitative challenges. Biochimica Et Biophysica Acta-Proteins and Proteomics 1649: 127–139.
5. Zhou HX, Rivas GN, Minton AP (2008) Macromolecular crowding and confinement: Biochemical, biophysical, and potential physiological consequences. Annual Review of Biophysics. 375–397.
6. Elcock AH (2010) Models of macromolecular crowding effects and the need for quantitative comparisons with experiment. Current Opinion in Structural Biology 20: 196–206.
7. Zhou HX (2013) Influence of crowded cellular environments on protein folding, binding, and oligomerization: Biological consequences and potentials of atomistic modeling. FEBS Letters 587: 1053–1061.
8. Phillip Y, Schreiber G (2013) Formation of protein complexes in crowded environments-From in vitro to in vivo. FEBS Letters 587: 1046–1052.
9. Minton AP (2005) Influence of macromolecular crowding upon the stability and state of association of proteins: Predictions and observations. Journal of Pharmaceutical Sciences 94: 1668–1675.
10. Batra J, Xu K, Qin SB, Zhou HX (2009) Effect of macromolecular crowding on protein binding stability: Modest stabilization and significant biological consequences. Biophysical Journal 97: 906–911.
11. Miklos AC, Li CG, Sharaf NG, Pielak GJ (2010) Volume exclusion and soft interaction effects on protein stability under crowded conditions. Biochemistry 49: 6984–6991.
12. Christiansen A, Wang Q, Samiotakis A, Cheung MS, Wittung-Stafshede P (2010) Factors defining effects of macromolecular crowding on protein stability: An in vitro/in silico case study using cytochrome c. Biochemistry 49: 6519–6530.
13. Sanfelice D, Politou A, Martin SR, De Los Rios P, Temussi P, et al. (2013) The effect of crowding and confinement: A comparison of Yfh1 stability in different environments. Physical Biology 10: 7.
14. Hong JA, Gierasch LM (2010) Macromolecular crowding remodels the energy landscape of a protein by favoring a more compact unfolded state. Journal of the American Chemical Society 132: 10445–10452.
15. Mikaelsson T, Aden J, Johansson LBA, Wittung-Stafshede P (2013) Direct observation of protein unfolded state compaction in the presence of macromolecular crowding. Biophysical Journal 104: 694–704.
16. Phillip Y, Sherman E, Haran G, Schreiber G (2009) Common crowding agents have only a small effect on protein-protein interactions. Biophysical Journal 97: 875–885.
17. Miklos AC, Sarkar M, Wang YQ, Pieak GJ (2011) Protein crowding tunes protein stability. Journal of the American Chemical Society 133: 7116–7120.
18. Wang YQ, Sarkar M, Smith AE, Krois AS, Pielak GJ (2012) Macromolecular crowding and protein stability. Journal of the American Chemical Society 134: 16614–16618.
19. Ma Q, Fan JB, Zhou Z, Zhou BR, Meng SR, et al. (2012) The contrasting effect of macromolecular crowding on amyloid fibril formation. Plos One 7.
20. Dhar A, Samiotakis A, Ebbinghaus S, Nienhaus L, Homouz D, et al. (2010) Structure, function, and folding of phosphoglycerate kinase are strongly perturbed by macromolecular crowding. Proceedings of the National Academy of Sciences of the United States of America 107: 17586–17591.
21. Heddi B, Phan AT (2011) Structure of human telomeric DNA in crowded solution. Journal of the American Chemical Society 133: 9824–9833.
22. Xue Y, Kan ZY, Wang Q, Yao Y, Liu J, et al. (2007) Human telomeric DNA forms parallel-stranded intramolecular G-quadruplex in K+ solution under molecular crowding condition. Journal of the American Chemical Society 129: 11185–11191.
23. Zheng KW, Chen Z, Hao YH, Tan Z (2010) Molecular crowding creates an essential environment for the formation of stable G-quadruplexes in long double-stranded DNA. Nucleic Acids Research 38: 327–338.
24. Homouz D, Sanabria H, Waxham MN, Cheung MS (2009) Modulation of calmodulin plasticity by the effect of macromolecular crowding. Journal of Molecular Biology 391: 933–943.
25. Homouz D, Perham M, Samiotakis A, Cheung MS, Wittung-Stafshede P (2008) Crowded, cell-like environment induces shape changes in aspherical protein. Proceedings of the National Academy of Sciences of the United States of America 105: 11754–11759.
26. Li W, Zhang M, Zhang JL, Li HQ, Zhang XC, et al. (2006) Interactions of daidzin with intramolecular G-quadruplex. FEBS Letters 580: 4905–4910.
27. Chen Z, Zheng KW, Hao YH, Tan Z (2009) Reduced or diminished stabilization of the telomere G-quadruplex and inhibition of telomerase by small chemical ligands under molecular crowding condition. Journal of the American Chemical Society 131: 10430–10438.
28. Rincon V, Bocanegra R, Rodriguez-Huete A, Rivas G, Mateu MG (2011) Effects of macromolecular crowding on the inhibition of virus assembly and virus-cell receptor recognition. Biophysical Journal 100: 738–746.
29. Phillip Y, Harel M, Khait R, Qin SB, Zhou HX, et al. (2012) Contrasting factors on the kinetic path to protein complex formation diminish the effects of crowding agents. Biophysical Journal 103: 1011–1019.
30. Wang YQ, Li CG, Pielak GJ (2010) Effects of proteins on protein diffusion. Journal of the American Chemical Society 132: 9392–9397.
31. Homouz D, Stagg L, Wittung-Stafshede P, Cheung MS (2009) Macromolecular Crowding Modulates Folding Mechanism of alpha/beta Protein Apoflavodoxin. Biophysical Journal 96: 671–680.
32. Chen E, Christiansen A, Wang Q, Cheung MS, Kliger DS, et al. (2012) Effects of Macromolecular Crowding on Burst Phase Kinetics of Cytochrome c Folding. Biochemistry 51: 9836–9845.
33. Kim YC, Best RB, Mittal J (2010) Macromolecular crowding effects on protein-protein binding affinity and specificity. Journal of Chemical Physics 133: 7.
34. Cheung MS, Klimov D, Thirumalai D (2005) Molecular crowding enhances native state stability and refolding rates of globular proteins. Proceedings of the National Academy of Sciences of the United States of America 102: 4753–4758.
35. Qin SB, Zhou HX (2009) Atomistic Modeling of Macromolecular Crowding Predicts Modest Increases in Protein Folding and Binding Stability. Biophysical Journal 97: 12–19.
36. Mittal J, Best RB (2010) Dependence of Protein Folding Stability and Dynamics on the Density and Composition of Macromolecular Crowders. Biophysical Journal 98: 315–320.
37. Feig M, Sugita Y (2011) Variable Interactions between Protein Crowders and Biomolecular Solutes Are Important in Understanding Cellular Crowding. Journal of Physical Chemistry B 116: 599–605.
38. Harada R, Tochio N, Kigawa T, Sugita Y, Feig M (2013) Reduced Native State Stability in Crowded Cellular Environment Due to Protein-Protein Interactions. Journal of the American Chemical Society 135: 3696–3701.
39. Rosen J, Kim YC, Mittal J (2011) Modest Protein-Crowder Attractive Interactions Can Counteract Enhancement of Protein Association by Intermolecular Excluded Volume Interactions. Journal of Physical Chemistry B 115: 2683–2689.
40. McGuffee SR, Elcock AH (2010) Diffusion, Crowding & Protein Stability in a Dynamic Molecular Model of the Bacterial Cytoplasm. PLoS Computational Biology 6.
41. Abriata LA, Spiga E, Dal Peraro M (2013) All-atom simulations of crowding effects on ubiquitin dynamics. Physical Biology 10: 8.
42. Kim JS, Yethiraj A (2009) Effect of Macromolecular Crowding on Reaction Rates: A Computational and Theoretical Study. Biophysical Journal 96: 1333–1340.
43. Wieczorek G, Zielenkiewicz P (2008) Influence of Macromolecular Crowding on Protein-Protein Association Rates-a Brownian Dynamics Study. Biophysical Journal 95: 5030–5036.
44. Verma PK, Rakshit S, Mitra RK, Pal SK (2011) Role of hydration on the functionality of a proteolytic enzyme alpha-chymotrypsin under crowded environment. Biochimie 93: 1424–1433.
45. Hassan SA, Steinbach PJ (2011) Water-Exclusion and Liquid-Structure Forces in Implicit Solvation. Journal of Physical Chemistry B 115: 14668–14682.
46. Tanizaki S, Clifford J, Connelly BD, Feig M (2008) Conformational sampling of peptides in cellular environments. Biophysical Journal 94: 747–759.
47. Harada R, Sugita Y, Feig M (2012) Protein Crowding Affects Hydration Structure and Dynamics. Journal of the American Chemical Society 134: 4842–4849.
48. Predeus AV, Gul S, Gopal SM, Feig M (2012) Conformational Sampling of Peptides in the Presence of Protein Crowders from AA/CG-Multiscale Simulations. Journal of Physical Chemistry B 116: 8610–8620.
49. Despa F, Fernandez A, Berry RS (2004) Dielectric modulation of biological water. Physical Review Letters 93.
50. Despa F, Fernandez A, Berry RS (2004) Dielectric modulation of biological water (vol 93, art no 228104, 2004). Physical Review Letters 93.
51. Sasmal DK, Ghosh S, Das AK, Bhattacharyya K (2013) Solvation Dynamics of Biological Water in a Single Live Cell under a Confocal Microscope. Langmuir 29: 2289–2298.

52. Tjong H, Zhou HX (2008) Prediction of protein solubility from calculation of transfer free energy. Biophysical Journal 95: 2601–2609.
53. Ghosh S, Chattoraj S, Mondal T, Bhattacharyya K (2013) Dynamics in Cytoplasm, Nucleus, and Lipid Droplet of a Live CHO Cell: Time-Resolved Confocal Microscopy. Langmuir in press.
54. Lee LP, Tidor B (2001) Barstar is electrostatically optimized for tight binding to barnase. Nature Structural Biology 8: 73–76.
55. Lee LP, Tidor B (2001) Optimization of binding electrostatics: Charge complementarity in the barnase-barstar protein complex. Protein Science 10: 362–377.
56. Buckle AM, Schreiber G, Fersht AR (1994) Protein-protein recognition-Crystal structural-analysis of a barnase barstar complex at 2.0-Angstrom resolution. Biochemistry 33: 8878–8889.
57. Brunger AT, Karplus M (1988) Polar hydrogen positions in proteins-Empirical energy placement and neutron-diffraction comparison. Proteins-Structure Function and Genetics 4: 148–156.
58. Brooks BR, Bruccoleri RE, Olafson BD, States DJ, Swaminathan S, et al. (1983) CHARMM-A program for macromolecular energy, minimization, and dynamics calculations. Journal of Computational Chemistry 4: 187–217.
59. MacKerell AD, Bashford D, Bellott M, Dunbrack RL, Evanseck JD, et al. (1998) All-atom empirical potential for molecular modeling and dynamics studies of proteins. Journal of Physical Chemistry B 102: 3586–3616.
60. Jorgensen WL, Chandrasekhar J, Madura JD, Impey RW, Klein ML (1983) Comparison of simple potential functions for simulating liquid water. Journal of Chemical Physics 79: 926–935.
61. Altman MD (2006) Computational ligand design and analysis in protein complexes using inverse methods, combinatorial search, and accurate solvation modeling [Ph.D. Thesis]. Cambridge: Massachusetts Institute of Technology.
62. Sitkoff D, Sharp KA, Honig B (1994) Accurate Calculation of Hydration Free-Energies Using Macroscopic Solvent Models. Journal of Physical Chemistry 98: 1978–1988.
63. Radhakrishnan ML (2012) Designing electrostatic interactions in biological systems via charge optimization or combinatorial approaches: insights and challenges with a continuum electrostatic framework. Theoretical Chemistry Accounts 131.
64. Hendsch ZS, Tidor B (1999) Electrostatic interactions in the GCN4 leucine zipper: Substantial contributions arise from intramolecular interactions enhanced on binding. Protein Science 8: 1381–1392.
65. Midelfort KS, Hernandez HH, Lippow SM, Tidor B, Drennan CL, et al. (2004) Substantial energetic improvement with minimal structural perturbation in a high affinity mutant antibody. Journal of Molecular Biology 343: 685–701.
66. Green DF, Tidor B (2005) Design of improved protein inhibitors of HIV-1 cell entry: Optimization of electrostatic interactions at the binding interface. Proteins-Structure Function and Bioinformatics 60: 644–657.
67. Carrascal N, Green DF (2010) Energetic Decomposition with the Generalized-Born and Poisson-Boltzmann Solvent Models: Lessons from Association of G-Protein Components. Journal of Physical Chemistry B 114: 5096–5116.
68. Minkara MS, Davis PH, Radhakrishnan ML (2012) Multiple drugs and multiple targets: An analysis of the electrostatic determinants of binding between non-nucleoside HIV-1 reverse transcriptase inhibitors and variants of HIV-1 RT. Proteins-Structure Function and Bioinformatics 80: 573–590.
69. Humphrey W, Dalke A, Schulten K (1996) VMD: Visual molecular dynamics. Journal of Molecular Graphics 14: 33–38.
70. Sheinerman FB, Honig B (2002) On the role of electrostatic interactions in the design of protein-protein interfaces. Journal of Molecular Biology 318: 161–177.
71. Talley K, Ng C, Shoppell M, Kundrotas P, Alexov E (2008) On the electrostatic component of protein-protein binding free energy. PMC Biophysics 1: 2.
72. Giordano S (2003) Effective medium theory for dispersions of dielectric ellipsoids. Journal of Electrostatics 58: 59–76.
73. Gerstein M, Tsai J, Levitt M (1995) The volume of atoms on the protein surface-calculated from simulation, using Voronoi polyhedra. Journal of Molecular Biology 249: 955–966.
74. Bhat S, Purisima EO (2006) Molecular surface generation using a variable-radius solvent probe. Proteins-Structure Function and Bioinformatics 62: 244–261.
75. Li AJ, Nussinov R (1998) A set of van der Waals and Coulombic radii of protein atoms for molecular and solvent-accessible surface calculation, packing evaluation, and docking. Proteins-Structure Function and Genetics 32: 111–127.
76. Cortines JR, Weigele PR, Gilcrease EB, Casjens SR, Teschke CM (2011) Decoding bacteriophage P22 assembly: Identification of two charged residues in scaffolding protein responsible for coat protein interaction. Virology 421: 1–11.
77. Brown CJ, Srinivasan D, Jun LH, Coomber D, Verma CS, et al. (2008) The electrostatic surface of MDM2 modulates the specificity of its interaction with phosphorylated and unphosphorylated p53 peptides. Cell Cycle 7: 608–610.
78. Busby B, Oashi T, Willis CD, Ackermann MA, Kontrogianni-Konstantopoulos A, et al. (2011) Electrostatic Interactions Mediate Binding of Obscurin to Small Ankyrin 1: Biochemical and Molecular Modeling Studies. Journal of Molecular Biology 408: 321–334.
79. Hu ZJ, Ma BY, Wolfson H, Nussinov R (2000) Conservation of polar residues as hot spots at protein interfaces. Proteins-Structure Function and Genetics 39: 331–342.
80. Qin SB, Zhou HX (2013) FFT-Based Method for Modeling Protein Folding and Binding under Crowding: Benchmarking on Ellipsoidal and All-Atom Crowders. Journal of Chemical Theory and Computation 9: 4633–4643.
81. Christ CD, Mark AE, van Gunsteren WF (2010) Feature Article Basic Ingredients of Free Energy Calculations: A Review. Journal of Computational Chemistry 31: 1569–1582.
82. de Ruiter A, Oostenbrink C (2011) Free energy calculations of protein-ligand interactions. Current Opinion in Chemical Biology 15: 547–552.

2

Broadband Transmission EPR Spectroscopy

Wilfred R. Hagen*

Department of Biotechnology, Delft University of Technology, Delft, The Netherlands

Abstract

EPR spectroscopy employs a resonator operating at a single microwave frequency and phase-sensitive detection using modulation of the magnetic field. The X-band spectrometer is the general standard with a frequency in the 9–10 GHz range. Most (bio)molecular EPR spectra are determined by a combination of the frequency-dependent electronic Zeeman interaction and a number of frequency-independent interactions, notably, electron spin – nuclear spin interactions and electron spin – electron spin interactions, and unambiguous analysis requires data collection at different frequencies. Extant and long-standing practice is to use a different spectrometer for each frequency. We explore the alternative of replacing the narrow-band source plus single-mode resonator with a continuously tunable microwave source plus a non-resonant coaxial transmission cell in an unmodulated external field. Our source is an arbitrary wave digital signal generator producing an amplitude-modulated sinusoidal microwave in combination with a broadband amplifier for 0.8–2.7 GHz. Theory is developed for coaxial transmission with EPR detection as a function of cell dimensions and materials. We explore examples of a doublet system, a high-spin system, and an integer-spin system. Long, straigth, helical, and helico-toroidal cells are developed and tested with dilute aqueous solutions of spin label hydroxy-tempo. A detection limit of circa 5 μM HO-tempo in water at 800 MHz is obtained for the present setup, and possibilities for future improvement are discussed.

Editor: Claudio M. Soares, Instituto de Tecnologica Química e Biológica, UNL, Portugal

Funding: This work was supported by a research grant from the Dutch National Research School Combination-Catalysis Controlled by Chemical Design (www.nrsc-catalysis.nl). The funder had no role in study design, data collection and analysis, decision to publish, or preparation of the manuscript.

Competing Interests: The author has declared that no competing interests exist.

* E-mail: w.r.hagen@tudelft.nl

Introduction

The phenomenon of electron paramagnetic resonance (EPR), or electron spin resonance (ESR), is at the basis of a well established spectroscopy widely used in multiple disciplines including, e.g., physics, chemistry, biology, for the characterization of electronic structure of open-shell systems with electron spin S ≠ 0 [1–3]. The word 'resonance' in EPR refers to the quantum-mechanical phenomenon of transition between molecular spin energy levels induced by microwave radiation (typically: gigahertz frequency or centimeter wavelength). The actual spectroscopic experiment also involves a classical, macroscopic resonance phenomenon because the sample is held inside a cavity which is constructed as a single-frequency, fundamental-mode resonator cell whose quality factor insures high radiation energy density at the sample to overcome the intrinsically low concentration sensitivity associated with the detection of molecular energy differences of the order of the thermal energy kT.

With very few exeptions the resonator has always been a key component of the EPR experiment ever since its inception nearly seven decades ago [4]. Also the modern EPR spectrometer is a single-frequency instrument, and its ability to produce spectra depends on the possibility of tuning molecular energy level differences by means of an external static magnetic field of variable strenght. However, microwave-frequency variation is an important and often essential strategy in EPR analysis for the disentanglement of multiple molecular magnetic interactions some of which are independent of the external magnetic field B, while others are linearly dependent on B. This has led to the cumbersome and costly practice of multi-frequency EPR as the serial deployment of a small number of physically separate spectrometers each one with its own operating frequency. Replacement by a single, frequency-tunable spectrometer would not only be very convenient, but it would also allow for the collection of two-dimensional frequency-field data at a much higher frequency-axis density.

Molecular EPR spectroscopy without a resonator has been occasionally explored previously. In the early '80 s Bramley and Strach reported on a zero-field (i.e. without a magnet) reflection spectrometer working with a sweep oscillator based on a backward wave tube with a typical tuning range of 1–8 GHz, and employing a non-resonant coaxial sample cell (of unspecified dimensions) with the sample in lieu of the dielectric, which gave good room-temperature frequency-swept spectra of 1.5 mol% Mn(II) in magnesiumammoniumsulfate powder [5]. This corresponds to a concentration of circa 50 mM high-spin paramagnet, and so for sensitivity reasons in a later version of the zero-field spectrometer the non-resonant coaxial setup was replaced with a loop-gap resonator [6]. A broadband, field-scanning, non-resonant reflection spectrometer was described in 1989 by Rubinson, who used a plate-ended brass coaxial tube and a 5.4–12.5 GHz signal generator [7]. The instrument afforded only a very weak signal at 9 GHz from 770 mM Mn(II) in pure hydrated manganese chloride. In later work the cell was modified into a quarter-wavelength truncated-line probe (a resonator) for work at <1 GHz [8], although a non-resonant version has also been used for FT-EPR at 0.2–0.4 GHz [9]. All in all it appears that the historical verdict on non-resonant sample cells for cw (continuous wave)-EPR has been to drop the subject from further consideration in view of the poor signal-to-noise values attained. Transmission cells have been used in high-frequency (≥

A

B

C

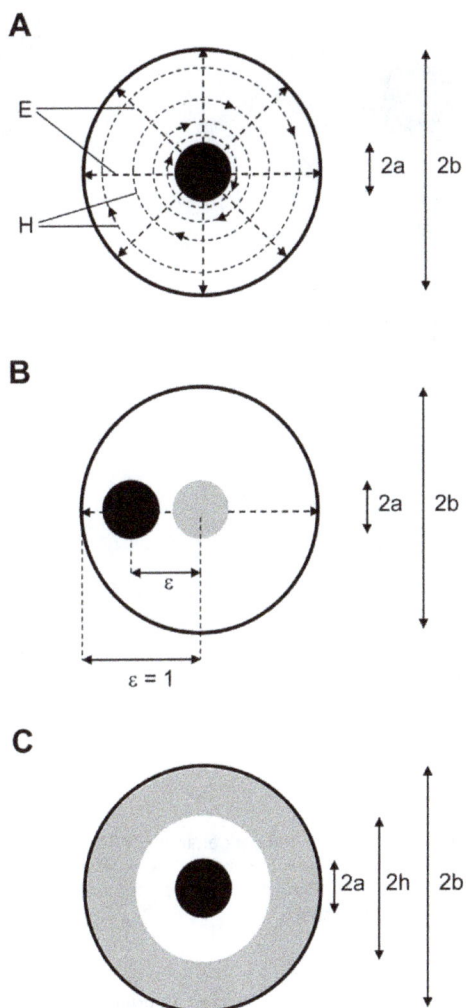

Figure 1. Cross sections of coaxial line. (A) Outline of electric (E) and magnetic (H) field lines; note radial the decrease in H-density. The outer diameter of the innner conductor is 2a and the inner diameter of the outer conductor is 2b. (B) Definition of eccentricity of an off-axis inner conductor. (C) Illustration of a coaxial cell in which a paramagnetic sample (white area) is contained in a diamagnetic holder (grey area).

$\approx$100 GHz) high-field EPR with low concentration sensitivity; these are oversized waveguide structures which sustain a multiplicity of excited modes [10].

The present work describes the development of theory, basic design principles, and applications for field-swept transmission EPR spectroscopy based on an instrument in which the single-frequency reflection resonator has been replaced by a continuously frequency-tunable, broad-band, fundamental mode transmission cell. A central design goal of the present work is to combine the added versatility of microwave source tunability over a broad frequency range with a concentration sensitivity comparable to that of extant conventional single-frequency spectrometers.

Results

Theory of Transmission EPR Spectroscopy

The magnetic susceptibility tensor of a compound is a complex quantity

$$\chi = \chi' - i\chi'' \tag{1}$$

when measured in response to a field with time-varying components. In magnetic resonance spectroscopy the field B is the vector sum of a static field B_0 from the external magnet and a perpendicular time-varying field B' from the RF source. At the resonance frequency $\nu = c/\lambda = \omega/2\pi$ the dynamic susceptibility is detected as an average power dissipated in the spins

$$P = \omega\chi'' B'^2 \tag{2}$$

(e.g., [11]). Conventional EPR spectrometers with field modulation produce spectra as graphs of $\delta\chi''/\delta B$ (arbitrary units) versus B (gauss or tesla).

The magnetic susceptibility is related to the complex magnetic permeability $\mu = \mu'\text{-}i\mu''$ and the relative magnetic permeability μ_R as

$$\frac{\mu}{\mu_0} = \mu_R = 1 + \chi_m \tag{3}$$

with complex components

$$\mu'_R = \chi' + 1 \tag{4}$$

$$\mu_R'' = \chi'' \tag{5}$$

in which

$$\mu_0 = 4\pi \cdot 10^{-7} \quad (henry/meter) \tag{6}$$

is the permeability of free space.

In field theory [12] the response of a compound to a transverse electromagnetic (TEM) wave is defined by a plane wave propagation constant γ and a characteristic impedance Z_0. The propagation constant

$$\gamma = i\omega\sqrt{\varepsilon\mu} = \alpha + i\beta \tag{7}$$

in which $\varepsilon = \varepsilon' - i\varepsilon'' = \varepsilon_R\varepsilon_0$ is the complex electric permittivity of the medium,

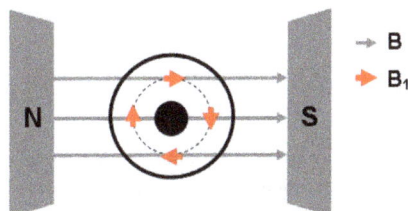

Figure 2. A coaxial structure with its TEM propagation axis perpendicular to the B-axis of an external magnetic field. The microwave magnetic-field component B_1 in the sample compartment runs through all possible angles with respect to the external field vector B.

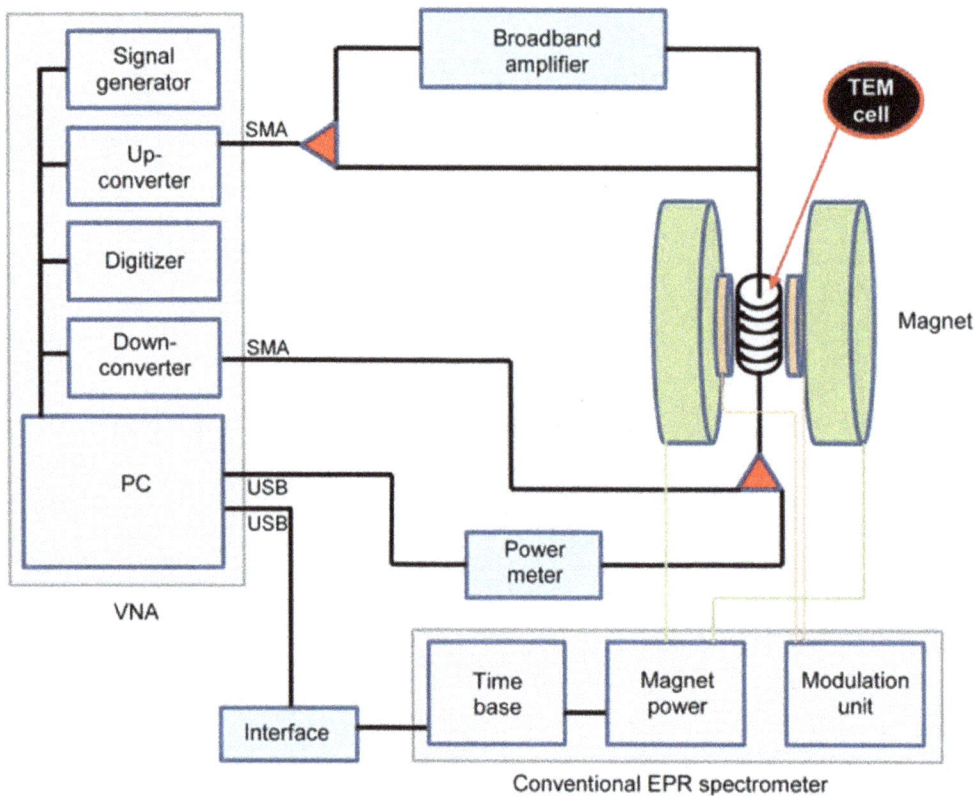

Figure 3. Outline of the transmission EPR spectrometer. An arbitrary microwave produced in the vector network analyzer (VNA) is send – either directly or via a broadband amplifier – to a transmission cell placed in the electromagnet of a conventional EPR spectrometer. The sample-attenuated output of the cell is either detected in a broadband power meter or as a carrier amplitude by the VNA. The time base of the magnet is used to trigger the onset of signal detection. The modulation coils of the magnet are employed in rapid-scan EPR.

$$\varepsilon_0 = \frac{1}{36\pi} \cdot 10^{-9} \quad (farad/meter) \qquad (8)$$

is the permittivity of free space, α is the attenuation factor, and β is the phase factor of the wave. The characteristic impedance, or the ratio of the electric field E and magnetic field H, is

$$\frac{E}{H} = Z_0 = \sqrt{\frac{\mu''\omega + i\omega\mu'}{\varepsilon''\omega + i\omega\varepsilon'}} \qquad (9)$$

The equivalent definition of characteristic impedance Z_0 in distributed transmission-line theory [12] is in terms of a ratio of voltage across a line segment over current through the segment

$$\frac{V}{I} = Z_0 = \sqrt{\frac{R + i\omega L}{G + i\omega C}} \qquad (10)$$

with series resistance R, series inductance L, shunt capacitance C, and shunt conductance G. The propagation constant becomes

$$\gamma = \sqrt{(R + i\omega L)(G + i\omega C)} \qquad (11)$$

The equivalence implies that the EPR effect in transmission spectroscopy ($\mu'' \neq 0$ for the paramagnet) will be detected as a

resistive loss serial to the resistance of the line conductors; it also announces a key experimental problem of non-resonant power dissipation by conductance of lossy diamagnetic hosts ($\varepsilon'' \gg 0$ for the diamagnetic host and/or the paramagnet) in particular in aqueous solutions.

For a particular transfer-line geometry the propagation constant and the characteristic impedance provide the design parameters in terms of mode sustainability, attenuation, and power handling capacity. We will work this out to obtain expressions for EPR absorption in a coaxial or quasi-coaxial cell in which part of the

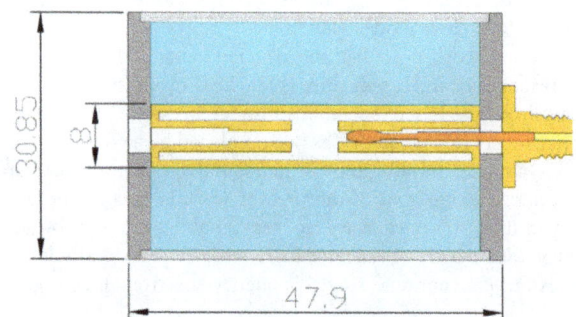

Figure 4. Drawing of the standard transmission cell. This cell was designed for ease of operation in particular for rapid sample change. Its short length, in combination with cables with right-angle male SMA connectors, allows orientation of the TEM axis parallel to the B-axis of a wide-gap magnet. The light-blue area is the sample compartment.

Figure 5. Picture of a monolitic aluminum cell. With its perpendicular SMA connectors this 82 mm inner length construct is the longest possible cell allowing TEM ∥ B operation in our wide-gap magnet.

dielectric insulator consists of a paramagnet or diluted paramagnet.

The characteristic impedance for a loss-less coaxial line is (e.g., [13])

$$Z_0 = \sqrt{\frac{L}{C}} = \sqrt{\frac{\mu'}{\varepsilon'}}\frac{\ln(b/a)}{2\pi} = 60\sqrt{\frac{\mu_R'}{\varepsilon_R'}}\ln\frac{b}{a} \quad (ohm) \quad (12)$$

in which 2b is the inner diameter of the outer conductor and 2a is the diameter of the inner conductor (Fig. 1a). With the universal standard of $Z_0 = 50$ ohm for RF equipment we can set up a spectrometer in which the transmission cell is designed as a coaxial device under test (DUT) connected to a vector network analyzer (VNA, see below). Increased design flexibility comes from the extension that the two-conductor geometry may vary from circular coaxial to circular eccentric [14] affording

$$Z_0 \cong 60\sqrt{\frac{\mu_R'}{\varepsilon_R'}}\ln\frac{b(1-\xi^2)}{a} \quad (13)$$

with an eccentricity $0 \leq \xi < 1$ (cf Fig. 1b), and from the notion that 'the' dielectric of the line in practice will be a (diamagnetically diluted) paramagnetic sample, s, contained in a diamagnetic cylindrical holder, h, of finite wall dimension and with an electric permittivity different from that of the sample (Fig. 1c) whose individual impedances are taken to be additive according to their fractional volume V as

$$Z_0 = Z_S V_S + Z_h V_h \quad (14)$$

The attenuation of a TEM wave in a dielectric is [12,15]

$$\alpha = \frac{\pi\sqrt{2\delta_{R'}}}{\lambda_0}\sqrt{\sqrt{1+\tan^2\delta}-1} \quad (neper/m) \qquad (15)$$

in which λ_0 is the wavelength of the microwave in vacuo,

$$\delta'' = \varepsilon_R'' \mu_R' + \varepsilon_R' \mu_R'' \qquad (16)$$

$$\delta' = \varepsilon_R' \mu_R' + \varepsilon_R'' \mu_R'' \qquad (17)$$

and the loss tangent, $\tan\delta$, is

$$\tan\delta = \frac{\delta''}{\delta'} \qquad (18)$$

With the Taylor expansion for $\tan^2\delta \ll 1$

$$\sqrt{1+x} \cong 1 + \frac{x}{2} - .... \qquad (19)$$

and, converting from nepers to decibels by multiplication with $20\log(e)$, the attenuation becomes

$$\alpha \cong \frac{27.3\sqrt{\delta'}}{\lambda_0}\tan\delta \quad (dB/m) \qquad (20)$$

This formal description encompasses a variety of physical processes of relevance for our spectroscopy including dielectric losses through the dilutant and other insulators, propagation of higher-order modes, resistive losses in the inner and outer conductors, reflective losses by impedance mismatches, and resonance attenuation via permeability absorption by the paramagnetic solute.

For this latter *extra* attenuation induced at electron paramagnetic resonance, assuming the absence of paraelectric resonance in the solute (extra resonance $\varepsilon_R'' = 0$), eqn (20) becomes

$$\alpha_{paramagnetic} \cong 27.3\frac{\sqrt{\varepsilon_R'\mu_R'}}{\lambda_0}\frac{\mu_R''}{\mu_R'} \quad (dB/meter) \qquad (21)$$

The average power from a sinusoidal field that flows through a cylindrical segment of unit length of a coaxial cable along the direction of TEM propagation is a constant (e.g., [16], p. 72):

$$P = \int_a^b \int_0^{2\pi} \frac{VI}{2\pi r^2 \ln(b/a)}rdrd\phi = VI \qquad (22)$$

whence it follows that the EPR absorption for a paramagnet homogeneously distributed over the insulator volume of a coaxial transmission cell is linear in the length, l, of the cell. It is also linear in the concentration, c, of the paramagnet since μ_R'' was defined in terms of the volume magnetic susceptibility. Remarkably, the EPR absorption is independent of the diameter of the cell for a given characteristic line impedance (i.e. for a fixed ratio of inner and

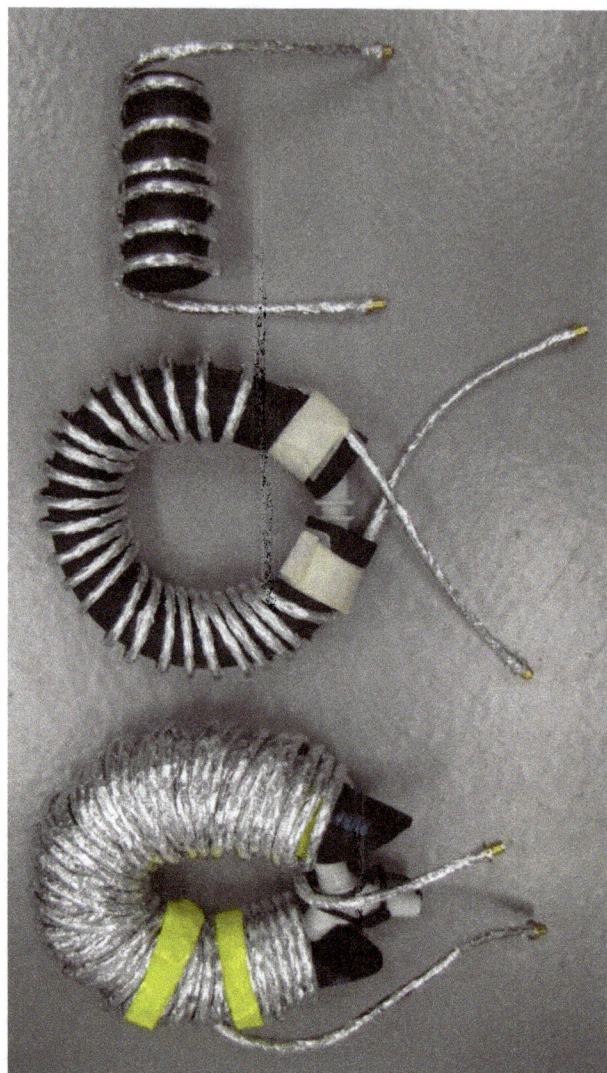

Figure 6. Long helical cells with continuously varying B₁ versus B angle. The sample-compartment lengths are 75, 300, and 1215 cm, respectively. The coax is wound around a supporting piece of elastomer into a helix or a toroidal helix.

outer conductor radii b/a defining $Z_0 = 50\ \Omega$). In an analogy with optical spectroscopy one could call this the transmission-EPR equivalent of Beer's law, which can be written in terms of a microwave-intensity ratio as

$$\frac{I_{out}}{I_{in}} = e^{-\alpha'}cl \quad (\text{dim }ensionless) \qquad (23)$$

or as an absorbance

$$A(\lambda_0) = \alpha' cl \qquad (24)$$

in which α' is the molar attenuation obtained by multiplying α with the number of paramagnetic particles per volume. For maximal EPR signal intensity this expression alone suggests us to make the paramagnet as concentrated as possible and to make the cell as long as geometric constraints (e.g., by the EPR magnet) would allow. Its independence of cell diameter also implies that

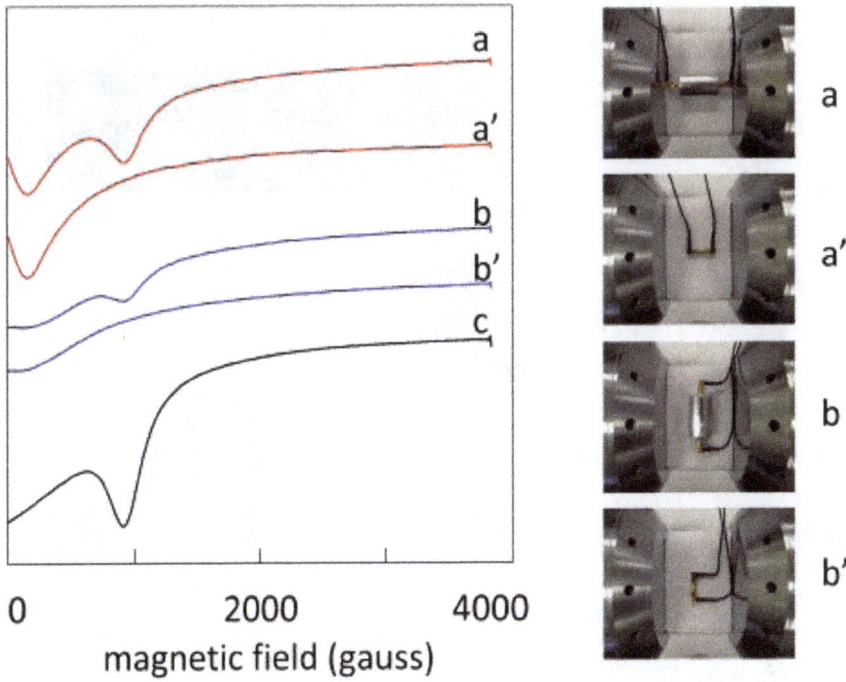

Figure 7. Transmission EPR of solid MnSO₄·H₂O in a 42 mm standard cell. Traces a and a' are for TEM ∥ B orientation, and traces b and b' are for TEM ⊥ B. The primed traces a' and b' are baselines from short-circuited leads without transmission cell. Trace c is for the 82 mm aluminum cell in TEM ∥ B orientation.

one can maximally economize on amount of sample by minimizing the diameter of the cell down to the technical limits of miniaturization. The inverse wavelength relation in Eq 21 implies that $S = 1/2$ EPR signal intensity is proportional to the microwave frequency.

This simple picture becomes a complex design problem of trade-offs when combined with the other, non-resonance related, phenomena of attenuation. Of paramount importance are the losses in the dielectric(s), which, under the assumptions that $\mu_R' \approx 1$ and $\mu_R'' \approx 0$, are usually given as (e.g., [16])

$$\alpha_{dielectric} \cong 27.3 \frac{\sqrt{\varepsilon_R'}}{\lambda_0} \tan \delta_\varepsilon \quad (dB/meter) \qquad (25)$$

with $\tan\delta_\varepsilon = \varepsilon_R''/\varepsilon_R'$. Since this α is also linear in the length of the cell (and independent of its diameter) we can only increase the length to a point where overall output power becomes too weak to be handled reliably by the vector network analyzer detection system. Thus total input power and stability of the source and the detailed electronic characteristics of the detector become important boundary conditions in the design problem. Furthermore, since $\alpha_{dielectric}$ is also linear in the microwave frequency, eqn (25) suggests that the optimal length of the transmission cell – all other conditions being equal – will be inversely proportional to the microwave frequency.

Occurrence of other attenuation mechanisms brings the line diameter back into the cell design. Higher-order modes can be sustained in coaxial lines above a limiting microwave frequency: the cutoff frequency v_c. The first mode to appear is the TE_{11} mode whose cutoff is approximatively given by (e.g., [16])

$$v_{cutoff} \approx \frac{c}{\pi(a+b)\sqrt{\mu_R'\varepsilon_R'}} \qquad (26)$$

Since partial conversion of the TEM mode to higher-order modes is detected as loss, because they are largely reflected at the connection of the EPR cell to the detector system, the sum of cell radii a+b, and therefore the total diameter is physically limited especially for high-permittivity dielectrics such as water.

The TEM wave is solely transmitted in the dielectric (and dispersively in the paramagnet); penetration into the conductors only leads to resistive losses (e.g., ref. 12)

$$\alpha_{resistive} = \sqrt{\frac{\varepsilon}{\mu} \cdot \frac{\pi v \mu_M'}{\sigma_M}} \left(\frac{1}{b} + \frac{1}{a}\right) \frac{1}{2\ln(b/a)} \quad (neper/meter) \qquad (27)$$

in which ε and μ apply to the dielectric, μ_M' is the real permeability of the conducting metal, and σ_M is the metal's conductivity in siemens/m. Note that resistive loss is proportional to the square root of the frequency, and is therefore usually outdone at increasing frequency by dielectric loss which is linear in the frequency. However, resistive loss is also inversely proportional to the line diameter, and will thus pose an excessive-loss limit (or even a limit of power breakdown by arcing) to miniaturization of the transmission cell.

Connection of two line segments with different characteristic impedance values leads to mismatch loss by reflection, which is usually defined (e.g., [13]) in terms of the voltage reflection coefficient, $0 \le |\Gamma| \le 1$,

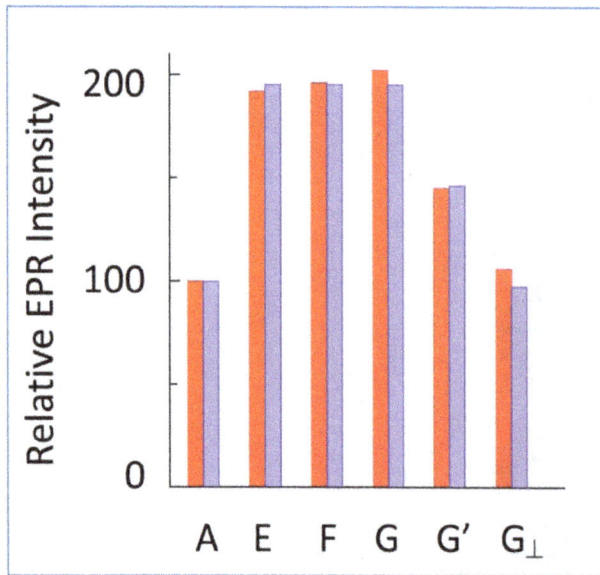

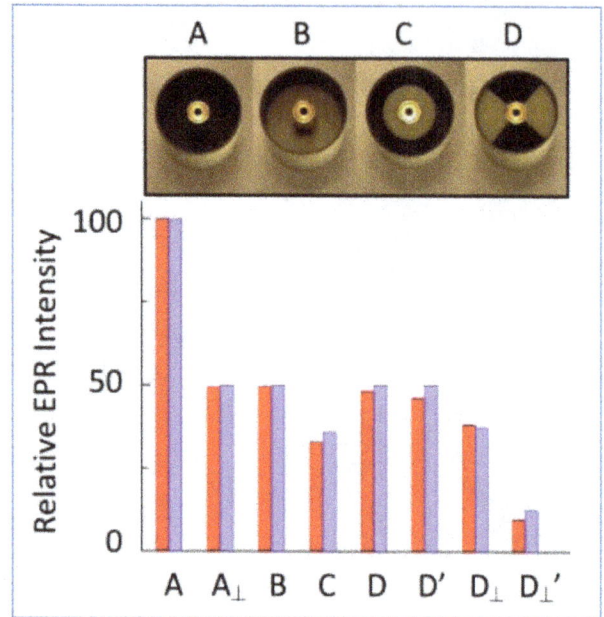

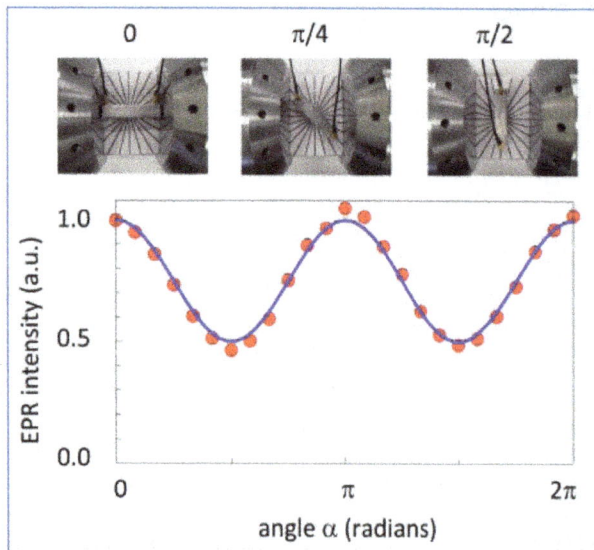

Figure 8. Signal intensity as a function of sample-compartment geometry. The 42 mm standard cell is partially filled with teflon spacers to displace paramagnetic sample. Bar A is the Mn(II) signal amplitude for a fully sample-filled cell in TEM ∥ B orientation. Red bars are measured intensities and blue bars are theory-predicted. Subscript ⊥ indicates TEM ⊥ B orientation. In B a teflon spacer occupies half of the cell; in C the spacer is a cylinder around the inner conductor over the full cell length; in D two sector spacers occupy opposite positions in 'butterfly' orientation along the full length of the cell. The lower panel shows a full 360 degree rotation experiment for the 'butterfly' setup with intensity data fitted to eqn (30).

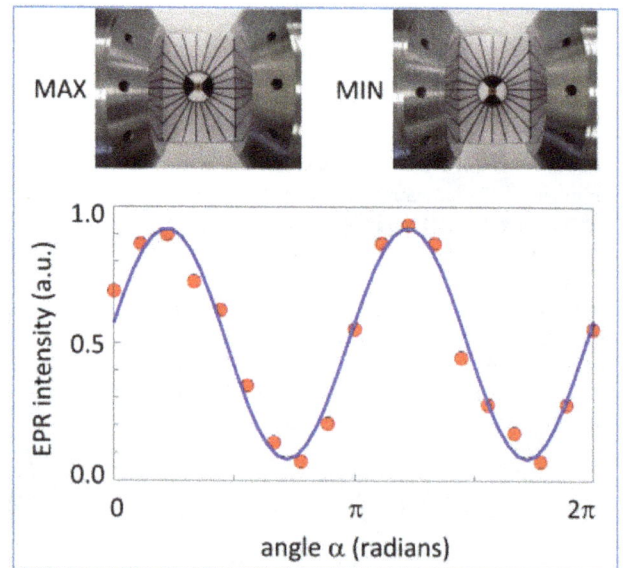

Figure 9. Signal intensity as a function of TEM versus B orientation. Bar A is the Mn(II) reference signal from the 42 mm standard cell. Red bars are measured intensities and blue bars are theory-predicted. Bars E, F, and G are from 82 mm Al cells with inner diameter of the outer conductor 8, 6, and 4 mm, respectively. Bars G' and G⊥ are for the 4 mm i.d. cell with TEM axis at 45 and 90 degrees versus B. The lower panel shows a full 360 degree rotation experiment for the 4 mm i.d. cell with intensity data fitted to eqn (30).

Since we intend to replace the dielectric of a regular coax with a variety of (dilute) paramagnetic samples as solutions, powders, or frozen solutions, in a variety of dielectric containers, there will not be a simple, generic design solution to the minimization of return losses. A related problem is in the non-ideal geometry of the coax transmission cell itself where, e.g., variation of the excentricity of the inner conductor leads to a variation of characteristic impedance (eqn (13)), and thus to distributed reflections over the length of the cell.

$$\Gamma = \frac{Z_{out} - Z_{in}}{Z_{out} + Z_{in}} \qquad (28)$$

as a positive return loss

$$\alpha_{return\ loss} = -10 \log \Gamma^2 \quad (dB) \qquad (29)$$

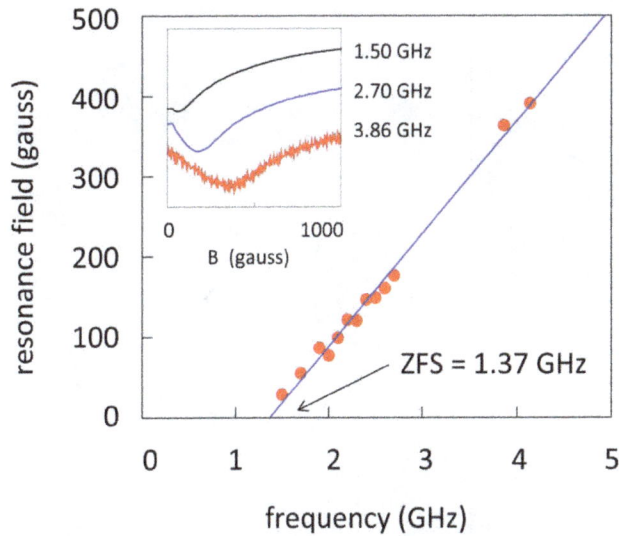

Figure 10. Frequency-dependent EPR signal assigned to S = 1 Ni(0) in male SMA connectors. The minimum of the broad signal (determined as the zero crossing of the derivative) was followed with frequency decreasing in 0.1 GHz steps until disappearance in zero field. Two points around 4 GHz were obtained with an S-band bridge as source and power-meter detection; two points at 9.1 and 10.0 GHz obtained with an X-band bridge (not shown) were also on the straight-line fit that extrapolates to a zero-field splitting of 1.37 GHz or 0.0457 cm^{-1}.

Finally, the permeability of the paramagnet, μ_R, in eqn (21) (and therefore α' in the transmission eqn (23)) is of course a function of the quantum mechanical transition probability of the spin system. In conventional EPR spectroscopy the resonator geometry is chosen such that the microwave magnetic component B' is perpendicular to the static field vector B, B' $\perp$ B, ensuring, e.g., maximal transition probability for S = 1/2 systems. Sometimes the parallel configuration B' || B is employed predominantly to make otherwise forbidden transitions allowed in integer-spin systems [17]. The relevance of this choise of geometry for the present research lies particularly in the fact that neither one of these unique orientations is easily implemented with a coax cell in an electromagnet. Although we will explore a few simple examples of short-length cells with B' $\perp$ B, for reasons of sensitivity the majority of the developed transmission cells will have helical or helical-torroidal geometries, which implies a continuous variation of the angle between B' and B over the propagation vector in the cell.

In S = 1/2 systems the transition probability, w, for B' at an angle θ with respect to B is proportional to $\sin^2\theta$ [18], and, defining the transition probability for B' $\perp$ B as unity (w $\equiv$ 1), a simple solution to the path-integrated EPR intensity is obtained as follo ws. For a straight coaxial segment whose propagation axis is colinear with the dipolar magnetic-field axis all insulator molecules experience B' $\perp$ B. When the cell axis is perpendicular to the B axis, as in Fig. 2, we must integrate the transition probilitity w over a $\pi/2$ quadrant (or any multiple) of the circular path of B' from B' $\perp$ B (unit intensity) to B' || B (zero intensity) affording an average relative intensity of 1/2

$$w(\theta) = \int_0^{2\pi} \frac{1}{2\pi} \sin^2\theta(\overline{B'},\overline{B})d\theta = 0.5 \tag{30}$$

When a long coaxial is wound in a tight helix of small pitch and placed in the magnet with its helical axis perpendicular to B the axis of the propagation vector γ of infinitesimal slices thought the cell runs over an angle α from colinear to perpendicular and the integrated EPR intensity runs from unity to half unity, which can be formalized as:

$$W = \frac{1}{2} + \frac{1}{2}w(\alpha) = 0.75 \tag{31}$$

in which

$$w(\alpha) = \int_0^{2\pi} \frac{1}{2\pi} \cos^2\alpha(\overline{\gamma},\overline{B})d\alpha \tag{32}$$

In other words for a fully filled helical cell the overall intensity is 0.75 of that of a linear cell of the same length coaxial with the external field. Furthermore, the intensity will not change when the helix is modified into a quasi-torroidal shape by bending the helical axis into a circle whose surface is perpendicualr to B. For partially filled cells with a filled sectorial fractional volume V eqns (30)–(32) generalize to

$$W(\alpha,\theta) = \frac{1}{2}V + \frac{1}{2}w(\alpha_1,\alpha_2)w(\theta_1,\theta_2) \tag{33}$$

which may be used to maximize certain B' versus B orientations in the study of integer-spin systems.

If only a colinear subsection of the cell is filled (cf Fig. 1C), such as in a cell made up of a tube holding a liquid sample, then the relative power P_s (versus the total power P of eqn (22)) through this inner section is seen to be

$$P_s(h) = P\frac{\ln(h/a)}{\ln(b/a)} \tag{34}$$

which can be generalized for constructs with multiple segments as

$$P_s(a_i,a_j) = P_{max}\frac{\ln(a_j/a_i)}{\ln(b/a_0)} \tag{35}$$

In summary, the S = 1/2 EPR intensity in a TEM transmission experiment is defined by eqns (21), (22), (24), (33), and (35) as

$$A \propto \alpha' clv W(\alpha,\theta)P_s \tag{36}$$

in which α' is the quantum-mechanical transition probablility of the S = 1/2 system, c is the spin concentration, l is the length of the cell (e.g. measured along its conductors), υ is the microwave frequency, W is the correction factor for filled sectors (W = 0.75 for completely filled cells), and P is the correction factor for filled sub-cylinders (P = 1 for samples contained only by the outer conductor).

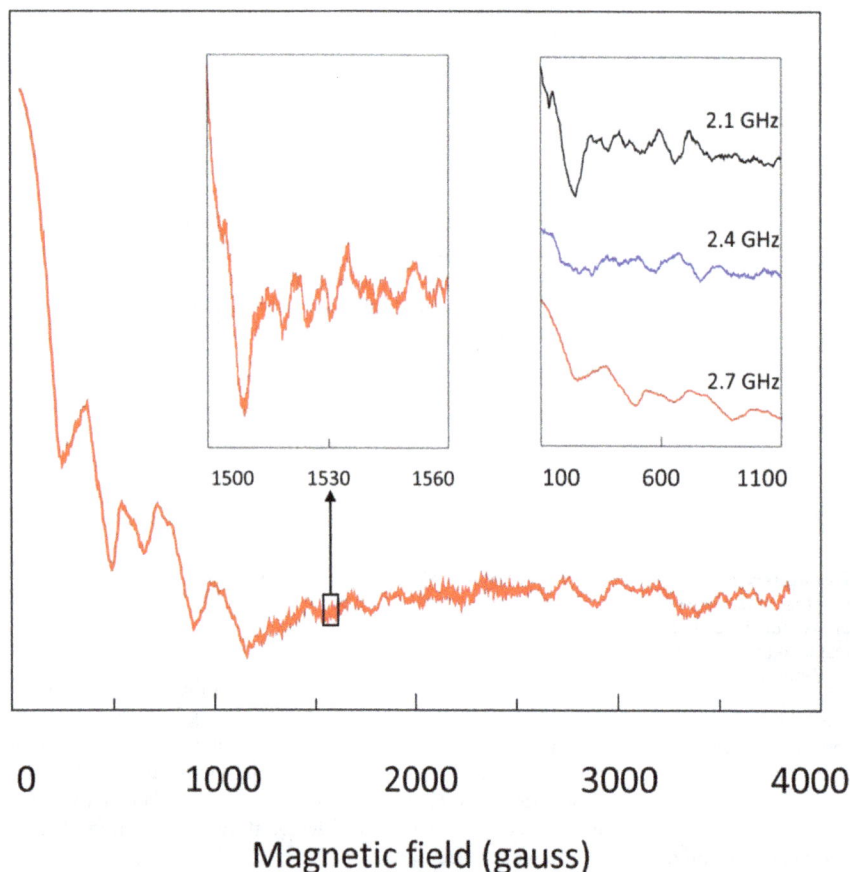

Figure 11. High-resolution transmission EPR spectra of 0.2 mol% Mn(II) in ZnSO₄·H₂O. The 220 cm cell held circa 5 g of powder of which circa 10 mg originated from MnSO₄·H₂O. Each trace is the differentiated average of circa 2500 forward scans of 100 s. The main trace was taken at 2.7 GHz; the left insert is a blow-up of a small section to illustrate detection of fine spectral details. The right insert shows very strong frequency dependence of the low-field part of the spectrum. Notethat these spectra are sums of spectra for all possible orientations of B' versus B.

Transmission EPR Spectrometer

An archetypical transmission EPR spectrometer consists of a microwave source, a coaxial transmission cell placed in the field of a magnet, and a microwave detector. In our prototype spectrometer (Fig. 3) the source is a National Instruments PXI-5641 16-bit 100 megasignals/s digital signal generator and a PXI-5610 superheterodyne upconverter affording 0.25–2700 MHz output of arbitrary shape and up to 12 dBm output power, followed by a Milmega series-2000 broadband (800–2700 MHz) RF multi-stage amplifier providing circa 50 dB power increase up to maximally 15 watt.

The transmission cells are (diluted) paramagnet-filled coax line segments whose dimensions in the TEM propagation direction are limited by the homogeneous-field volume of the dipole magnet. We used various electromagnets available from our conventional EPR spectrometers, namely (i) a Varian 9 inch magnet from our S-band spectrometer with a wide gap of 91 mm after removal of poles for short straight cells whose propagation direction can be lined up as || B, (ii) a Bruker BH 11-D magnet from one of our X-band spectrometers with a gap of 59 mm and a 15 cm diameter homogeneous-field circle to take helico-torroidally wound cells with a total propagation length up to circa 12 m, or (iii) a Varian 9 inch magnet with tapered pole caps from our Q-band spectrometer to accommodate helically wound cells of propagation length limited to circa 40 cm for rapid-scan EPR using the Q-band modulation coils that produce a rapidly time-varying

homogeneous field in a cylindrical volume of 39 mm radius and circa 3 cm height.

The detector is a National Instruments PXI-5600 down-converter and the 100 megasignals/s PXI-5142 IF digitizer for 0–2700 MHz and +20 dBm maximal unattenuated input. Together with a controling PC the source and detector units are integrated in a NI PXI-1042 chassis with 132 MB/s PCI busses, thus forming a complete digital vector network analyzer (VNA). The downconverter carries an oven-stabilized 10 MHz onboard precision reference clock also serving the digitizer through a matched coaxial line and the signal generator units through the timing and triggering chassis busses.

The scanning electromagnets are regulated using the time base units of the standard spectrometers whose start pulses are fed into the controlling program running on the embedded PC via a National Instruments USB-6221 interface. In one setting the Bruker ER 001 time base is ran at its fastest scan rate of 10 s and a second spectrum is recorded when the field returns instantaneously also in 10 s. With an intrinsic start up delay time of circa 5 s this mode of operation affords averaging of 292 single scans in one hour with only slight field oscillations at the extremes of the field scan. The high frequency stability of the VNA allows massive data acquisition over several days (e.g., over 10,000 double scans in 3 days).

In contrast to regular EPR spectroscopy no modulation of the static field is employed, which opens the possibility to rapidly scan

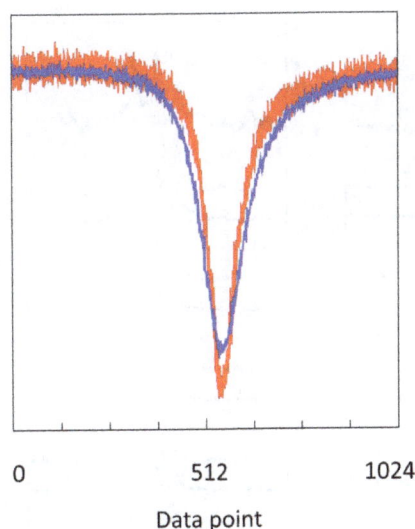

Figure 12. Rapid-scan transmission EPR to improve signal-to-noise ratio. A thick outer wall 40 cm cell was filled with solid DPPH, and a 2.7 GHz single-scan spectrum was taken in 30 s over a 959±20 gauss field range using a low-intensity microwave of −60 dBm incident power from the VNA to the amplifier to produce a spectrum with visible noise (red trace). Then a second spectrum was taken with the field at 959±0 gauss but with the field modulation unit set to 35 Hz and with a nominal peak-to-peak modulation amplitude of 40 gauss. The resulting data were mapped with a dedicated LabVIEW program to a sinusoidallly varying field and then interpolated to a 1024 point spectrum (blue trace). The increased width of the blue trace shows that the actual modulation amplitude felt short of 40 gauss, which illustrated that this experiment can be used as a convenient way to calibrate modulation coils. The rapid-scan EPR blue trace has reduced signal-to-noise (see text for details).

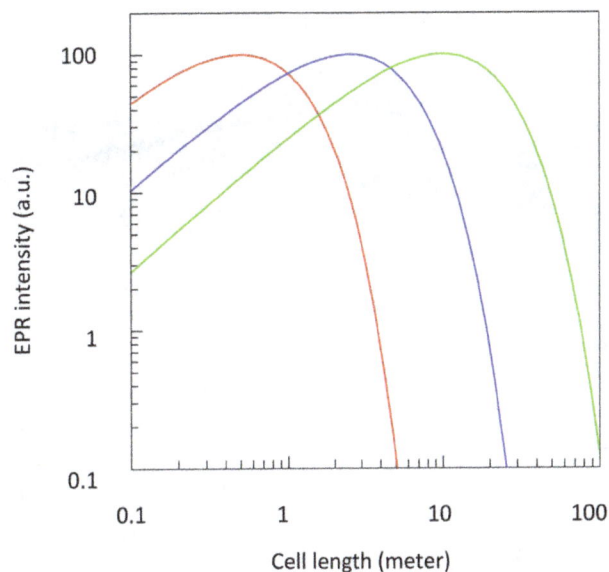

Figure 13. Transmission EPR signal from samples in lossy media as a function of cell length. These theoretical curves are based on eqn (37). Loss in dB/m is 40 (red), 8 (blue), or 2 (green). Maximal amplitudes are normalized to 100%.

the field for increased data collection rates [19–22]. For lack of dedicated equipment we emulated a rapid field scan unit using the cylindrically shaped modulation coils of our Varian Q-band spectrometer operated at 35 Hz and 40 gauss nominal peak-to-peak modulation amplitude. The unit was operated independently (no triggering) and transmission EPR data collected at maximal rate by the VNA were subsequently –phase-synchronized in a dedicated computer program.

The source unit is typically programmed to produce a monochromatic sine wave from the 0–2.7 GHz range, which is amplitude modulated to a dept of 80% at a frequency of 25 kHz with an IQ-data generation rate of 625 kHz and a bandwidth of 500 kHz. After passing through the transmission cell the signal is double side band demodulated in the detector unit at an IQ-data detection rate of 1 M signals/s and the carrier wave amplitude is recorded in Volt units.

In an alternative setup we replace the detector system with a National Instruments USB-5680 power meter for 0–6 GHz or the USB-5681 power meter for 0–18 GHz. With this broadband detector the sensitivity (signal-to-noise ratio) drops by 2–3 orders of magnitude, but we now have a simple setup to monitor insertion losses (intrinsic and/or mismatch) of transmission cells. We also have a simple and economic, be it not very sensitive, way to extend transmission measurements to the higher frequencies provided by the microwave bridges of our standard spectrometers, namely 3.8–4.1 in S-band and 9.1–10.0 GHz in X-band. The Bruker S-band bridge has an coaxial SMA exit port, and the waveguide port of X-band bridges can be fitted with a waveguide-to-coax adapter.

Cells for Transmission EPR

Conceptually the simplest possible cell is a piece of coax ending on both sides in lossless connectors and with the paramagnetic sample completely occupying the insulator space in between the inner and outer conductors. For EPR the cell should be placed in the dipolar magnet colinear with the field B in order for the microwave magnetic component B' to be perpendicular to B as required for maximal transistion probability of regular $|\Delta m_S| = 1$ EPR. This limits the length of the coaxial to less than the gap between the poles of the magnet.

A practical design approaching this idealized simple case is shown in Figure 4 for use with an electromagnet with 92 mm air gap. A polyvinylchloride cylinder closed with side lids forms an easily demountable and cleanable holder for powder samples. Each lid is centrally pierced with a 'screw on' female SMA connector whose teflon insulation is trimmed away nearly to the base plate with 1 mm remaining to hold a plastic ring to isolate the base plate from the cell's oversized (with respect to the connector) inner conductor. The extending inner lead of the connector is provided with a coaxial spring that fits to the cell's inner conductor made of a hollow silver cylinder with brass end fittings. The outer conductor is made of household aluminum foil tightly wrapped around the cell including the side lids and making electrical contact with the base plate of the connector. The ratio of the diameters b = 29 mm and a = 8 mm of the conductors was chosen for $Z_0 \approx 50\ \Omega$ with a dielectric with relative permittivity $\varepsilon_R \approx 2.3$, to accommodate ionic salts ($2 < \varepsilon_R < 5$) with reduced effective ε_R values when employed as fine-grained powders in air. No attempt was made to compensate for the small mismach from the step in conductor dimensions at the connector-cell interface. The cell is connected to the source and detector via right-angle male SMA connectors and 50 Ω coaxial cables. A range of teflon inserts allows for tests on the replacement of part of the paramagnetic powder sample from specific geometric areas of the cell. Note that we chose the cell radial dimension with a view to ease of handling

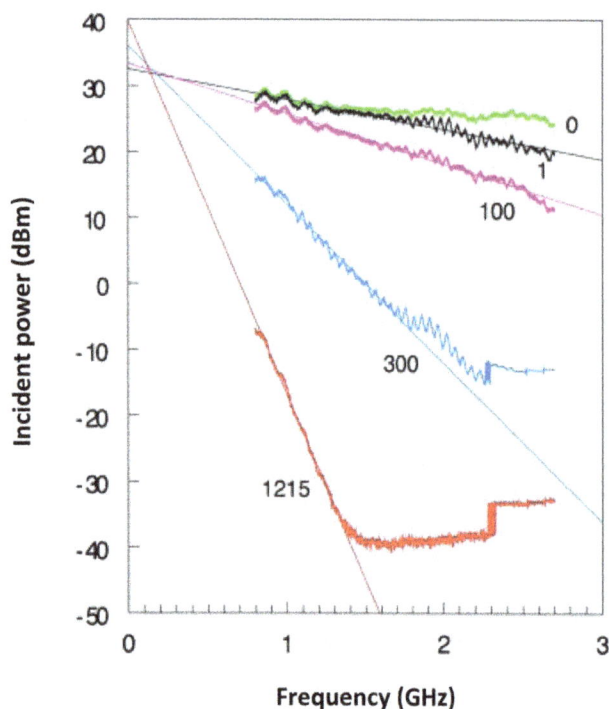

Figure 14. Power loss versus frequency in water-filled cells of different length. All cells were made of silicone tubing (4 mm o.d., 2 mm i.d.) with 1 mm diameter copper wire inner conductor (as in Fig. 6). Power from the VNA incident to the broadband amplifier was set at −30 dBm or, for the two longest cells, at −20 dBm. The three upper traces were shifted vertically over +10 dB to normalize all traces to −20 dBm incident power. The output power of the cells were measured with the broadband power meter. The green trace (0 cm) is for no cell and reflects losses in the brandless elongation cables of 1.5 m each. The black trace (1 cm) additionally shows losses due to impedance mismatch at the cell's ends. Each trace was collected in circa 15 min as 1024 9-times averaged data points in the 0.8–2.7 GHz range.

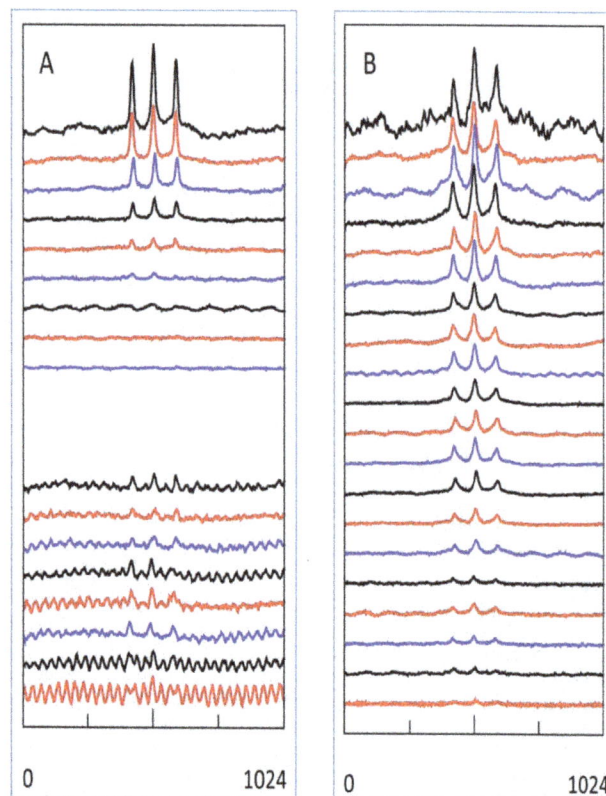

Figure 15. Dependence on frequency and on cell length of aqueous tempo tranmission EPR. Cells (cf Fig. 6) of 1215 cm (A), 300 cm (B) and 75 cm (C) were filled with 1 mM HO-tempo in 100 mM KCl. Spectra were collected at 0.1 GHz intervals from 0.8 GHz (upper trace) till 2.7 GHz (lower trace) as 500 averages of 10+10 s forward-reverse scans (i.e. circa 200 minutes per spectrum) over 200 gauss around the resonance field for g = 2 (i.e. 284±100 gauss for 0.8 GHz to 959±100 gauss for 2.7 GHz). In terms of maximal signal-to-noise the 12 m cell is seen to be optimal for frequencies approximately ≤ 1.1 GHz and the 3 m cell is optimal for circa 1.1–2.1 GHz. The 75 cm cell does not afford sufficient signal under the used conditions.

only. The cutoff frequency is predicted (cf eqn (26)) to be at the high-end limit of our frequency range for $\varepsilon_R' \approx 3.7$.

The cell can of course also be operated in other that coaxial orientation with respect to the external dipolar magnetic field. What this means, in particular, for perpendicular orientation is illustrated in Fig. 2. The orientation of the microwave B_1 field changes from perpendicular to the external dipolar field (B' ⊥ B) to parallel (B' || B) over a quarter B_1 circle. For EPR of S = 1/2 systems and half-integer spin S = n/2 systems in the weak-field limit (i.e. effective S = 1/2 systems) this means that the transition probability runs continuously from its full value to zero [3], and when integrated over all orientations the total spectral amplitude will be 0.75 of that obtained with the cell in coaxial orientation with the external field (eqn (31)).

In this geometry the cell is also suitable for the detection of EPR from non-Kramers systems with their strongly mixed ground spin manifolds producing $|\Delta m_S| = 0$ transitions when B' || B [1]. The resulting spectra will not be directly comparable to those obtained with the well-known parallel-mode rectangular X-band resonator [17] because the coaxial cell sustains the complete spectrum of all B' versus B orientations.

As a variant to the standard cell of Fig. 4 we also constructed a set of cells from aluminum as shown in Fig. 5. The solid aluminum functions as a holder and outer conductor to a machined

cylindrical space in which a copper wire inner conductor runs, which is connected to two female SMA bulkhead connectors perpendicular to the cell body so that a maximal cell length of 82 mm is achieved. With the screwed-on closing plates of 4 mm thickness the whole cell just fits into the electromagnet's 92 mm gap. To check (in)dependence of signal amplitude on coax diameter three cells were made with outer conductor inner diameters of 8, 6, and 4 mm and inner conductor diameters of 1, 0.7, and 0.7 mm, respectively.

To drastically increase the cell's sensitivity we have its length as the only practical design parameter available. For the cell to fit in between the poles of the magnet we must bend it along the TEM propagation direction, and this means that we must use flexible materials. For cells to hold aqueous solutions we chose the following parts. For the inner conductor we used polymer film insulated copper wire of typically 1 mm diameter on which we apply a transversal pull by human force to somewhat reduce bending flexibility. For the container of the liquid we used flexible silicone rubber tubing of 55 Shore A hardness and typically 2 mm inner diameter and 4 mm outer diameter. After machining one end of the copper wire into a smoothly rounded tip, we could simply push the wire through the tubing for sections up to circa 2 m. For longer sections up to 12 m we pre-filled the tubing with

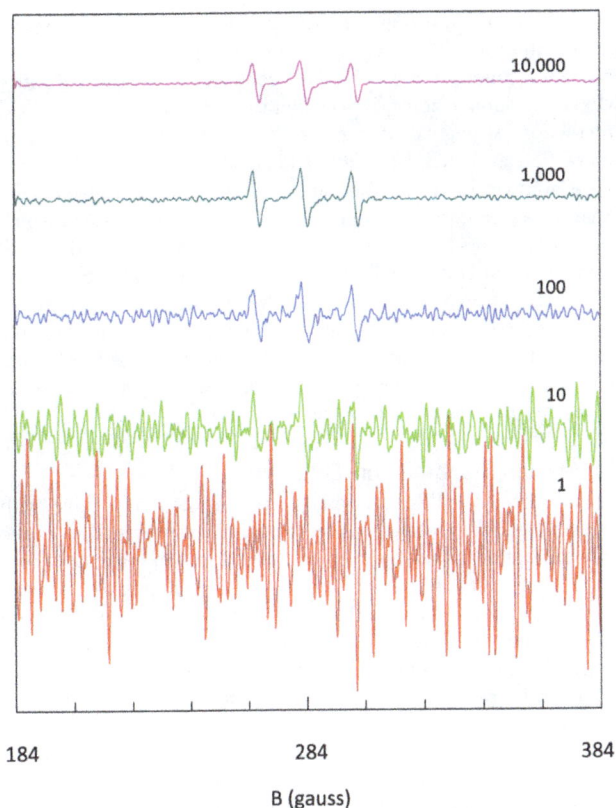

Figure 16. Estimation of tempo detection limit at 800 MHz. The 12 m cell was filled with 300 μM HO-tempo in 100 mM KCl and increasing numbers of averages were collected of 10+10 s forward-return scans. From the differentiated spectrum of 10,000 averages (67 hours data colletion) a detection limit of circa 5 μM was determined.

household dishwasher detergent to reduce mechanical resistance. After placement of the wire the detergent was washed out with water and the cell was dried with filtered compressed air.

For the outer conductor we used 0.025 mm thick household aluminum foil cut into 4×30 cm strips, which we wrapped around the container at an angle of circa 70 degrees to create multiple overlaps.

We then cautiously shaped the cell into a helix by subsequent wrapping around cylindrical templates of decreasing diameter from circa 12–3 cm diameter. To further confine the space taken up by the cell to the cylindrical box shape of the homogeneous field between the poles of the electromagnets we wrapped the helix around a piece of elastomer (copper pipe insulation of circa 35 mm o.d.) with a piece of insulated copper wire inside, and we carefully bended the helix into a quasi toroidal, or doughnut shaped, structure (Fig. 6).

To fill cells with dilute aqueous paramagnetic samples we used plastic syringes with pipette tips that we trimmed so that they would easily fit around the inner conductor wire and tightly fit into the silicone tubing. After removal of the syringe and filling the cell up to the rims with a Hamilton syringe we used 1 cm long tubing sections of i.d. = 1 mm and o.d. = 3 mm as end stoppers for the liquid container.

We removed the insulation from the copper-wire ends and trimmed them to be soldered to female SMA bulkhead connectors. The soldered connections and the stoppers were then put into a collar of transversally cut 2/4 mm silicone tubing, and the

aluminum wrap was extended to reach the connector's outer conductor. To replace a sample we had to unwrap and desolded the connectors, flush out the sample and wash the cell with water, and then re-fill and re-construct.

Intensity Dependence on Cell Geometry

The $S = 5/2$ d^5 system high-spin Mn(II) in solid $MnSO_4 \cdot 4H_2O$ has been used in early days of EPR history to record the very first S-band spectrum at 2.93 GHz using a cavity filled with 173 grams of the salt in powder form [23]. We re-visited this compound as a convenient carrier of strong paramagnetism. In the Theory section, above, equations were formulated describing the relative intensity of transmission EPR versus cell geometric properties and cell orientation in the external magnetic field. We employed the short standard cells described in the previous section in combination with the high EPR signal intensity from pure $MnSO_4$ powder to check these predictions.

The 4.2 cm standard cell of Fig. 7 filled with ca 4 gram powder and placed with its TEM propagation axis parallel to the external magnetic field (therefore B' ⊥ B) afforded the 2.7 GHz spectrum presented in Fig. 7a. The Mn(II) gives rise to a broad peak around g ≈ 2.00 at a resonance field of circa 965 gauss, i.e. hyperfine and zero-field structure is not resolved in the pure compound (cf ref. 23). A second peak of similar intensity and width is observed close to zero field, however this turns out to be a background signal because re-measurement after removal of the cell and short-circuiting the coaxial cables ending in right-angle male connectors with a female-female SMA connector, re-produces the low-field peak while the Mn(II) peak has disappeared (trace and picture a' of Fig. 7). We then turned the cell over 90° to a position such that the propagation axis was perpendicular to the external field, and this creates a distribution of orientation between B' and B from perpendicular to parallel as earlier depicted in Fig. 2. This resulted in the spectrum of trace b in Fig. 7 (and the baseline spectrum b') in which both the Mn(II) peak and the low-field peak have reduced intensity. The last spectrum (trace c) was obtained with the 8.2 cm aluminum cell (4 mm diameter; 1.8 gram powder) in TEM || B orientation. The amplitude of the Mn(II) signal is increased because the cell is longer, and the amplitude of the baseline signal is decreased because the male connectors to the cell are straight and are perpendicular to the external field (Fig. 3–2).

This qualitative monitoring of intensity was made quantitative in the following experiments. As seen in the upper panel of Fig. 8, when the standard cell was rotated over 90 degrees the signal intensity halved, and when the sample volume was halved by insertion of a teflon filler the intensity was also halved. A different filler over the full length of the cell but radially covering only the space from the inner conductor halfway towards the outer conductor afforded a relative intensity of 36% (experiment C) consistent with eqn (34).

A more complex sample-space displacement was realized with the 'butterfly' shaped filler covering two opposing 90° sectors, or quadrants, minus the space occupied by the inner conductor (experiment D), which also resulted in a volume reduction by half. With in-line orientation (TEM || B; B' ⊥ B) the Mn(II) signal was simply halved compared to the fully filled cell. However, when the propagation axis was set perpendicular to B (resulting in the B' versus B orientation distribution of Fig. 2) the EPR intensity depended on the orientation of the sample quadrants with respect to this distribution. With the sample sectors closest to the poles of the magnet the intensity was maximal (37.4% according to eqn (30) integrated twice over the interval $-\pi/4$ to $\pi/4$) while rotation over 90 degrees afforded minimal intensity (12.6%). A complete angular dependence of this effect over a full circle rotation is given in the lower panel of Fig. 8.

In the upper panel of Fig. 9 the intensity of the Mn(II) signal is seen to be linear in the length of the cell when the standard 42 mm cell is compared to three different 82 mm aluminum cells consistent with eqn (22). In contrast, the signal is independent of the diameter of the aluminum cell over the range 4–8 mm. Rotation of the aluminum cell over 45 and 90 degrees away from the in-line orientation afforded an amplitude reduced to 75 and 50%, respectively consistent with eqn (30). A complete angular dependence of this effect over a full cricle rotation is given in the lower panel of Fig. 9.

Integer-spin Transmission EPR

The low-field background signal in Fig. 7 originates in the male SMA connectors. Removal of the female-female connector and short-circuiting the males with a small wire inner conductor and an aluminum foil wrap outer conductor led to a large impedance-mismatch loss, but the signal persisted. The signal was maximal when the right-angle connectors had their connecting part parallel to B. The same signal was obtained with straight male connectors, shortened with a female-female elongation connector, with their propagation axis parallel to B. When the cheap, brandless connecting cables were replaced with high-quality cables, namely male-to-right-angle-male assemblies with 36 inch LMR240 optimally for 1–4 GHz or with 24 inch HS086 optimally for 6–18 GHz (Fairview Microwave Inc, Allen, Tx), very similar spectra were obtained from right-angle and from straight connectors.

Electrical contact pins in SMA connectors are frequently made of high conductivity beryllium copper alloy containing significant amounts (of the order of 2%) of nickel; also, other parts of SMA connectors can be electroplated with nickel. Nickel atom has electronic configuration $[Ar]3d^84s^2$, and nickel metal is ferromagnetic. Broad Ni(0) EPR signals have been observed, e.g., from nickel-plated brass [24], from nickel nanoparticles in La_2NiO_4 [25], from metallic nickel powder in KCl [25], and from annealed silica glass plates implanted with $^{58}Ni^+$ [26].

We assign the baseline signals also to Ni(0) as the only reasonable candidate for magnetic material in SMA connectors. Ni(0) EPR has to our knowledge never been interpreted in terms of a spin Hamiltonian. From the multi-frequency experiment in Fig. 10 it is clear that the spectrum is from a transition that moves into zero-field at 1.37 GHz. With reference to the electronic ground state of Ni(0) a reasonable interpretation of this data would be an $S = 1$ triplet with axial zero-field splitting $D = 1.37$ GHz and an extensive distribution in spin-Hamiltonian parameters g, D, E. In a broader perspective the experiment of Fig. 10 suggests that transmission EPR is potentially a useful novel approach to the study of integer-spin systems.

Doped-powder Spectra

The transmission EPR of solid powders in long helical cells should be not as challenging as aqueous solution experiments because dielectric losses will be strongly reduced due to their smaller ε_R'-values of circa 2–5. As a test we diluted the $MnSO_4 \cdot H_2O$ of the experiments in Figs (7–9) to 0.2% in the diamagnetic host $ZnSO_4 \cdot H_2O$ and we filled a 220 cm long cell with 5 g of the doped material (i.e. 10 mg of $MnSO_4$). The cell was wound in a helix of 14 cm heigth and 32 mm outer diameter with elongations on both ends of unfilled sections in order to keep the SMA connectors outside the magnetic field. No Z_0-optimization was attempted.

In a 4000-gauss scan at 2.7 GHz the single, broad line (Fig. 7) of pure $MnSO_4 \cdot H_2O$ now resolves into a very complex (first

derivative) spectrum (Fig. 11) with hundreds of lines and widely varying intensity. The spectrum is also strongly dependent on the frequency (inset to Fig. 11). The complexity is expected since Mn(II) in similar ionic hosts exhibits hyperfine and zero-field interaction strenghts of the order of 10% of the Zeeman interaction at X-band [27,28], and so all three interactions will be of comparable magnitude at the frequencies employed here. Analysis will require development of appropriate software in which the energy matrix for the spin multiplet is diagonalized for all orientations of B' versus B, but the results in Fig. 11 attest to the potential of transmission EPR for the collection of high-quality multi-frequency data of high-spin systems, be it – for the time being – at the expense of long data-collection times: each spectrum in Fig. 11 is the averages of circa 2500 scans of 100 s, i.e. some three days of measuring time per spectrum.

Rapid-scan Transmission EPR

Transmission EPR spectroscopy is not yet competitive with regular cw-EPR spectroscopy in terms of sensitivity. One of the possible options for improvement would be an increased data acquisition rate. Rapid-scan EPR has been proposed as a modification of regular EPR initially at 250 MHz [19,20] and subsequently at X-band [21,22]. The concept involves replacement of magnetic-field modulation at 100 kHz and phase-sensitive detection by rapid field scanning over a few gauss at a rate up to 50 kHz and direct unmodulated detection of the EPR signal. Since in our transmission EPR setup no magnetic-field modulation is employed, rapid scanning of the field is readily implementable, and the amplitude-modulated detection scheme is not affected.

For proof of principle we constructed a 40 cm thin cell filled with 2,2-diphenyl-1-picrylhydrazyl (DPPH) stable radical, tightly wound in a small helix to fit inside the cylinder carrying the coils of a Varian Q-band modulation unit. Single scans of 30 s over 40 gauss afforded very strong signals, and the source power before injection in the broadband amplifier had to be reduced to −60 dBm (i.e. 1 nanowatt) to produce a spectrum with significant noise (red trace in Fig. 12). Subsequently, the slow field scan was switched off, a ±20 gauss modulation was applied at 35 Hz using the Varian E-line low-frequency modulation unit, and the amplitude-modulated carrier wave value was collected for 30 s. Then the data were mapped with adjustable phase by visual inspection onto a 40 gauss field sinusoidal sweep to produce the blue trace in Fig. 12. The increased width of the DPPH signal is due to the fact that the actual modulation field fell short of the set value of ±20 gauss, and was measured with an AC gauss meter to be ±16.4 gauss. The 35 Hz scan produces 1050 spectra in 30 s, which would imply a signal-to-noise increase of $\sqrt{1050} \approx 32$. However, the maximal data collection rate of the present setup is 60,000 points per 30 s while for the slow scan 3000 points were collected, and so the effective signal-to-noise increase is only $\sqrt{20} \approx 4.5$. In summary, rapid field scanning is a relatively simple addition to improve signal-to-noise ratios towards regular cw-EPR values, but this will require increased data collection rates through improvements in the VNA hardware and software.

Aqueous Solution Transmission EPR

The most challenging transmission-EPR experiment by far is the detection of signals from paramagnets dissolved in aqueous solutions. Water has a high complex electric permittivity at RF frequencies and so the EPR effect is measured against a massive background of non-resonant absorption of microwaves. In regular EPR this problem is tackled by choosing a sample geometry in which overlap with the electric RF component is minimized, but in the fundamental TEM mode of coaxial structures electric and magnetic RF components are inseparable, which makes high power loss per unit length of transmission line unavoidable. Furthermore, with the high water value of $\varepsilon_R' \approx 80$ and the boundary condition $Z_0 \approx 50\ \Omega$, filling the sampe area of a coaxial completely with water leads to impractical radial dimensions according to eqn (12). For example, taking $\mu_R' \approx 1$, an outer cell diameter of 4 mm would require an inner conductor with a mechanically and electrically unrealistic diameter of 0.0023 mm.

We addressed this problem by filling up a major part of the total sample space in between the conductors with diamagnetic tubing to enclose an aqueous solution (Fig. 1C), which afforded an overall Z_0 of circa $50\ \Omega$ even with an inner conductor of 1 mm diameter. The response of the coaxial transmission cell as a function of its length is now a trade-off between increasing power absorption by EPR versus increasing power loss by dielectric dissipation. Thus, for a given cell construction (material and dimensions) operated at a fixed microwave frequency and a fixed insertion power level the cell exhibits an optimum in the signal-to-noise ratio as function of its length as illustrated in Fig. 13, in which it is assumed that EPR absorption is linear in the cell length and the dielectric power loss in dB per meter has an exponential detrimental effect on the EPR signal

$$A \propto x10^{-0.05nx} \qquad (37)$$

in which x is the length in meters and n is the attenuation in dB per meter. Note, however, the schematic nature of this illustration in which curve amplitudes have been adjusted to the same arbitrary value because their actual relation for three different cells would depend on many more paramaters than the few in eqn (37). Furthermore, actual signal-to-noise levels would be difficult to predict since noise from the signal generator is proportional to the insertion power (resulting in a constant EPR S/N ratio over a certain power range) and noise in the detector becomes dominant only below a treshold power.

Furthermore, eqn (25) predicts the loss to be proportional to the frequency. We can map this behaviour experimentally with cells of different length, filled with water, in which we insert RF at a fixed power and then measure the output power with a calibrated broadband power meter as a function of microwave frequency (Fig. 14). We found the loss in dB to be linear in the frequency with contributions from connecting cables (0 meter cell length), from impedance mismatch (0.01 meter), and from the cells themselves (1 to 12 meter).

To explore how this complex trade-off between paramagnetic and dielectric absorption works out in practical detection limits of transmission EPR we measured the spectrum from a dilute aqueous solution of the nitroxide spin label HO-tempo (4-hydroxy-2,2,6,6-tetramethylpiperidine 1-oxyl) as a function of cell length and of microwave frequency. The S = 1/2 spectrum of Tempo and its derivatives in the rapid-tumbling limit at ambient temperature consists of an isotropic line split into a triplet by ^{14}N (I = 1) hyperfine interaction and with a linewidth determined by multiple unresolved proton hyperfine interactions [3], i.e. a frequency-independent spectral shape. We constructed a long cell of 4 mm diameter and 12 m length, whose volume, when wound up in a toroidal helix, came close to the maximum that could be accomodated in the homogeneous-field space of the electromagnet with 59 mm gap and 17 cm pole diameter. The silicon-rubber tubing of the cell (1 mm wall thickness; and holding a 1 mm diameter inner conductor) was filled with 27 ml of an aqueous solution of 1 mM HO-Tempo and 100 mM KCl. Equivalent cells were constructed of length 3 and 0.75 m. Each cell was subjected to a constant RF power level such that incident power from the

broadband amplifier was $\geq$ 20 dB below amplifier saturation and power reflected to the amplifier was negligible. Multi-frequency data with 0.1 GHz intervals were collected as 500 averages of 10+10 seconds scans (Fig. 15).

A cell length of 12 m turns out to be close to optimal for frequencies near the low-end limit of 800 MHz of the broadband RF amplifier. With increasing frequency the signal-to-noise ratio rapidly deteriorates due to dielectric loss of power (cf Fig. 15 A), and above 1.2 GHz it drops below unity. With a cell length of 3 m the spectrum is readily detected over the whole available frequency range of 0.8–2.7 GHz with a broad plateau of approximately constant S/N between circa 1.1 and 2.1 GHz. With another four-fold reduction in cell length to 75 cm this plateau appears to shift to higher frequencies, but the spectra suffer from low EPR signal intensity. As a general conclusion a cell length of the order of a meter appears to be optimal for application over a broad range of microwave frequencies, where the data in Fig. 15 predict – by extrapolation – applicability in terms of practical power levels up to X-band.

To set a sensitivity standard for comparison with regular EPR and as a reference for further development of transmission EPR we signal-averaged the spectrum of 0.3 mM HO-tempo at 800 MHz in the 12 m cell up to 10,000 times. The spectra were collected as 20,000 point arrays, smoothed through a Savitzky-Golay filter (3-rd order polynomial) such that no resolution was lost after final converion to 1024 point arrays, which were discretely differentiated (second order central method) to obtain the first-derivative EPR spectra in Fig. 16.

Signal-to-noise ratios in conventional EPR are generally determined by conversion of the maximum noise amplitude to RMS noise by division by a factor of 2.5 [29]. With this common definition we find for the present setup a transmission EPR detection limit (i.e. S/N = 1) of circa 5 µM HO-Tempo in aqueous solution at 0.8 GHz. From the data in Fig. 16 it can be seen that a similar detection limit is obtainable for frequencies up to at least 2 GHz with a 3 m cell.

Experiment

Chemicals: manganese(II)sulfate monohydrate 99% and zinc sulfate heptahydrate ACS 99.0–103.0% were from Alfa Aesar, Karlsruhe, Germany. For the 0.2% Mn-doped zinc sulfate preparation 59 mg of the Mn(II) salt and 49.9 g of the Zn(II) salt were dissolved in water (Milli-Q; resistivity >18 MΩ·cm) to a total metal concentration of 1 M, and the solution was heated in a crystallizing dish to 130°C for 48 h. The residual solid was crushed to flakes with a spatula and once more heated to 130°C for 48 h, and then ground in a mortar to a fine powder with a yield of circa 31 g (i.e. the monohydrate). KCl pro analysi $\geq$ 99.5% was from Merck Chemicals via VWR International, Amsterdam, The Netherlands. HO-Tempo, or 4-hydroxy-2,2,6,6-tetramethylpiper-idinooxy 98% free radical was from Acros Organics via Fisher Scientific, Landsmeer, The Netherlands. DPPH, or 2,2-diphenyl-1-picrylhydrazyl was from Aldrich via Sigma-Aldrich Chemie B.V., Zwijndrecht, The Netherlands.

Equipment has been described in detail, above, in the sections on Transmission EPR spectrometer and on Cells for transmission EPR.

Conclusion

For nearly seven decades regular cw-EPR spectroscopy has been carried oud with single-frequency resonators usually combined with magnetic-field modulation. Comparison with broadband transmission EPR could incite a paradigm change both fundamentally (number of frequencies is unlimited) and practically (change of frequency is easy and cheap). In principle all molecular EPR spectra are frequency-dependent, and their rigorous analysis could benefit very significantly from the possibility of 2D (frequency-field) data collection with a single machine. The present work is an attempt to open up research into this field.

Compared with the few documented previous attempts to employ coaxial EPR cells the present setup appears to be a major improvement. Inspection of our 0.8 GHz three-line spectrum of 0.3 mM HO-Tempo in water (Fig. 16) suggests several orders-of-magnitude increase in signal-to-noise ratio compared to the six-line coaxial reflection spectrum of 700 mM manganese(II) in pure $MnCl_2 \cdot 4H_2O$ taken at 9.5 GHz [7]. Also, our 2.7 GHz spectrum of 0.2 mol% Mn(II) in $ZnSO_4 \cdot H_2O$ (Fig. 12) has similar signal-to-noise but better resolution than the field-modulated zero-field spectrum of 1.5 mol% Mn(II) in $Mg(NH_4)_2(SO_4)_2 \cdot 6H_2O$ taken as a frequency scan from 1 to 8 GHz [5].

Cylindrical resonators can be made frequency-tunable when constructed with moving parts, and this principle has been often applied in the past for tuning over a small bandwidth for example with P-band (circa 15 GHz) and Q-band (circa 35 GHz) cavities. A high-frequency broadband version (40–60 GHz) has been described for flat solid samples [30]. A broadband form has been recently described for tuning over a 14–40 GHz frequency range extendable down to 4 GHz by insertion of sets of dielectric plates [31]. The setup has been tested with pure Mn_{20} and V_6 molecular magnets but sensitivity has not yet been documented for doped solids or for dilute liquid samples. It would be interesting to competitively compare this approach with the one proposed by us in terms of sensitivity and practical handling while the two methodologies develop.

In the present work signal intensity was found to be independent of radial cell dimension (at constant ratio of inner and outer conductor radii), but no attempt was made yet to minimize sample volume by decreasing cell radius. Miniaturization, possibly with lithographic technology, should not only increase absolute sensitivity, but it will also shift the cutoff frequency for loss via higher-order modes towards higher values. Significant future improvements in concentration sensitivity can be reasonably expected to be attainable with increased data-collection rates based on a combination of improvements in rapid-scan hardware, VNA hardware, CPU hardware, and in coding of software and programmable hardware. These options and also variable-temperature cryogenic applications are under our active investigation.

In summary, as a method under development transmission EPR in the present study has already shown value in certain nice areas in particular for half-integer and integer-spin systems with relatively small zero-field splittings. Further development towards a generally applicable spectroscopy should in our view be focused on improvement of sensitivity through rapid detection schemes, reduction of sample size through cell miniaturization, and extension towards higher microwave frequencies where properly constructed coaxial structures are in principle applicable up to circa 50 GHz.

Acknowledgments

Martijn Smeulers from National Instruments Netherlands B.V. is acknowledged for help with signal-detection programming of the VNA. Arno van den Berg and his crew of the fine-mechanical workshop in our Department helped with the construction of transmission cells.

Author Contributions

Conceived and designed the experiments: WRH. Performed the experiments: WRH. Analyzed the data: WRH. Contributed reagents/materials/analysis tools: WRH. Wrote the paper: WRH.

References

1. Abragam A, Bleaney B (1970) Electron paramagnetic resonance of transition ions. Clarendon Press, Oxford.
2. Weil JA, Bolton JR (2007) Electron paramagnetic resonance; elementary theory and practical applications. 2nd edition, Wiley, Hoboken, NJ.
3. Hagen WR (2009) Biomolecular EPR spectroscopy, CRC press Taylor & Francis Group, Boca Raton, FL.
4. Kochelaev BI, Yablokov YV (1995) The beginning of paramagnetic resonance. World Scientific, Singapore.
5. Bramley R, Strach SJ (1983) Chem. Rev. 83: 49–82. Electron paramagnetic resonance spectroscopy at zero magnetic field.
6. Delft CD, Bramley R (1997) J. Chem. Phys. 107: 8840–8847. Zero-field electron magnetic resonance spectra of copper carboxylates.
7. Rubinson KA (1989) Rev. Sci. Instrum. 60: 392–395. Broadband (up to 10 GHz) electron-paramagnetic-resonance spectrometer: cw implementation with direct detection.
8. Rubinson KA, Koscielniak J, Berliner LJ (1995) J. Magn. Reson. 117: 91–93. Modified, short-circuited coaxial-line resonators for CW-EPR.
9. Rubinson KA, Cook JA, Mitchell JB, Murugesan R, Krishna MC, et al. (1998) J. Magn. Reson. 132: 255–259. FT-EPR with a nonresonant probe: use of a truncated coaxial line.
10. Hagen WR (1999) Coord. Chem. Rev. 190–192: 209–229. High-frequency EPR of transition ion complexes and metalloprpoteins.
11. Slichter CP (1990) Principles of magnetic resonance, 3rd enlarged and updated edition, Springer, New York. Corrected 2nd printing 1992.
12. Von Hippel AR (1954) Dielectrics and waves. Wiley, New York. New edition (facsimile) 1995, Artch House Microwave Library, Artech House, Boston. Reprint on demand 2010.
13. Pozar DM (2012) Microwave engineering, 4th edition, Wiley, Hoboken, NJ, USA.
14. Lewis IAD, Wells FH (1959) Millimicrosecond pulse techniques, 2nd Ed. Pergamon Press, London.
15. Weir WB (1974) Proc. IEEE 62: 33–36. Automatic measurements of complex dielectric constant and permeability at microwave frequencies.
16. Rizzi PA (1988) Microwave engineering; passive circuits. Prentice Hall, Englewood Cliffs, NJ, USA.
17. Hagen WR (1982) Biochim. Biophys. Acta 707: 82–98. EPR of non-Kramers doublets in biological systems; characterization of an S = 2 system in oxidized cyrochrome c oxidase.
18. Kneubühl FK, Natterer B (1961) Helv. Phys. Acta 34: 710–717. Paramagnetic resonance intensity of anisotropic substances and its influence on line shapes.
19. Stoner JW, Szymanski D, Eaton SS, Quine RW, Rinard GA, et al. (2004) J. Magn. Reson. 170: 127–135. Direct-detected rapid-scan EPR at 250 MHz.
20. Tseitlin M, Rinard GA, Quine RW, Eaton SS, Eaton GR (2011) J. Magn. Reson. 211: 156–161. Rapid frequency scan EPR.
21. Kittell AW, Camenisch TG, Ratke JJ, Sidabras JW, Hyde JS (2011) J. Magn. Reson. 211: 228–233. Detection of undistorted continuous wave (CW) electron paramagnetic resonance (EPR) spectra with non-adiabatic rapid sweep (NARS) of the magnetic field.
22. Mitchell DG, Quine RW, Tseitlin M, Eaton SS, Eaton GR (2012) J. Magn. Reson. 214: 221–226. X-band rapid-scan EPR of nitroxyl radicals.
23. Cummerow RL, Halliday D (1946) Phys. Rev. 70: 433. Paramagnetic losses in two manganous salts.
24. Griffiths JHE (1946) Nature 158: 670–671. Anomalous hihg-frequency resistance of ferromagnetic metals.
25. González-Calbet JM, Vallet-Regi M, Sayagués MJ, Sánchez RD, Causa MT (1994) J. Mater. Sci. 9: 176–179. EPR and magnetization of La_2NiO_4.
26. Isobe T, Park SY, Weeks RA, Zuhr RA (1995) J. Non-Cryst. Solids 189: 173–180. The optical and magnetic properties of Ni^+-implanted silica.
27. Misra SK, Kahrizi M (1983) Phys. Rev. B 28: 5300–5302. EPR of Mn^{2+}-doped single crystals of $ZnK_2(SO_4)_2 \cdot 6H_2O$ and $NiK_2(SO_4)_2 \cdot 6H_2O$: Mn^{2+}-Ni^{2+} exchange constant.
28. Morin G, Bonnin D (1999) J. Magn. Reson. 136: 176–199. Modeling EPR powder spectra using numerical diagonalization of the spin Hamiltonian.
29. Ernst RR (1966) Sensitivity enhancement in magnetic resonance. Advances in magnetic resonance Vol 2, 1–135. Academic Press, New York. [in Gorlaeus; S/N = 2.5].
30. Seck M, Wyder P (2000) Rev. Sci. Instr. 69: 1817–1822. A sensitive broadband high-frequency electron spin resonance/electron nuclear double resonance spectrometer operating at 5–7.5 mm wavelenght.
31. Schlegel C, Dressel M, van Slageren J (2010) Rev. Sci. Instr. 81: 093901. Broadband electron spin resonance at 4–40 GHz and magnetic fields up to 10 T.

3

Testing the Applicability of Nernst-Planck Theory in Ion Channels: Comparisons with Brownian Dynamics Simulations

Chen Song[1,2], Ben Corry[1]*

1 School of Biomedical, Biomolecular and Chemical Sciences, The University of Western Australia, Perth, Australia, **2** Department of Theoretical and Computational Biophysics, Max Planck Institute for Biophysical Chemistry, Göttingen, Germany

Abstract

The macroscopic Nernst-Planck (NP) theory has often been used for predicting ion channel currents in recent years, but the validity of this theory at the microscopic scale has not been tested. In this study we systematically tested the ability of the NP theory to accurately predict channel currents by combining and comparing the results with those of Brownian dynamics (BD) simulations. To thoroughly test the theory in a range of situations, calculations were made in a series of simplified cylindrical channels with radii ranging from 3 to 15 Å, in a more complex 'catenary' channel, and in a realistic model of the mechanosensitive channel MscS. The extensive tests indicate that the NP equation is applicable in narrow ion channels provided that accurate concentrations and potentials can be input as the currents obtained from the combination of BD and NP match well with those obtained directly from BD simulations, although some discrepancies are seen when the ion concentrations are not radially uniform. This finding opens a door to utilising the results of microscopic simulations in continuum theory, something that is likely to be useful in the investigation of a range of biophysical and nano-scale applications and should stimulate further studies in this direction.

Editor: Jörg Langowski, German Cancer Research Center, Germany

Funding: This work was supported by the National Health and Medical Research Council, Australia. The funders had no role in study design, data collection and analysis, decision to publish, or preparation of the manuscript.

Competing Interests: The authors have declared that no competing interests exist.

* E-mail: ben.corry@uwa.edu.au

Introduction

Biological ion channels are membrane bound proteins responsible for rapidly moving ions across the cell membrane. They play a major role in the transmission of electrical signals within the brain, nervous system and muscles, and their malfunction is associated with a range of diseases [1]. Therefore, understanding them at the molecular level and relating their structure to their function is essential for improving our knowledge about these fundamental components of biology and in finding treatments to ion channel related diseases. One important step in this direction is to be able to predict the ion conductance for a given structure and much research has taken place into finding efficient means of doing this.

Accompanying the rapid progress of experimental techniques, especially driven by the emergence of more and more high resolution structures of ion channels, there have been a lot of efforts to perform theoretical studies on the ion channels because such studies can provide experimentally unaccessible insights. For example, molecular dynamics (MD) simulations have been widely used to give atomic level insight into the function of channels, such as the steps involved in ion conduction [2], possible gating mechanisms [3–7] and how selective transport can arise in these pores [8–10]. MD has even been used to simulate ion conduction with an external electric field up to a microsecond timescale [11–14]. However, directly predicting the channel conductance

using MD is very computationally demanding which makes calculating statistically meaningful values of ion conductance unreachable for most investigations.

Brownian dynamics (BD) simulations provide an alternative method for predicting the conductance of a given structure [15–21]. In these, only some atoms (usually the ions) are simulated explicitly, moving in a stochastic manner under the influence of random and frictional forces in addition to electrostatic or average forces arising from other ions and the protein. By adopting approximations such as considering the protein and water as continuous dielectric media, BD can be easily used to simulate the motion of ions on the microsecond timescale. Therefore, many ion conduction events can be observed and statistically meaningful conductances can be determined. But, such approximations also have drawbacks. For example protein motions and fluctuations are usually ignored, and highly detailed atomic interactions such as that between the ions and water are mostly unaccounted for.

Continuum theories provide another computationally efficient method for calculating channel currents. In these ionic flux is generally determined from the Nernst-Planck (NP) equation (drift-diffusion) that was well established for bulk electrolytes. While the NP equation has long been applied to studying ion channels [22,23] it requires prior knowledge of the electrostatic potential and ion concentrations as well as extension to multi-ion permeation [24]. The most common way of overcoming this is to combine the NP equation directly with Poisson's equation yielding the so called Poisson-Nernst-Planck (PNP) theory [25], which has also been widely used in the last two decades [26–30]. The use of PNP theory in ion channels was motivated by macroscopic ion transport studies wherein the ions are also considered as continuous charge distributions. By using PNP, one can calculate the ion concentration, electrostatic potential, and ion flux in a single short calculation on a desktop computer. Therefore, the continuum approaches require much less simulation time than microscopic approaches such as MD. However, previous work by Corry et al. has shown that the simple implementation of PNP is flawed at the microscopic scale due to the over simplistic representation of the few ions in the channels by their mean field properties, and particularly by the overestimation of the shielding of forces on permeating ions by counter ions [31]. Although there has been some effort to improve PNP theory by introducing additional terms to the PNP equations or using explicit ions in the calculation [32–35], the results are still not satisfying and the number of open parameters make it less attractive if the aim is to determine the likely conductance of a given structure. There are several good reviews about the use of MD, BD and PNP methods for studying ion channels which are recommended for further reading [36–38].

Since the main reason for the failure of the PNP theory in ions channels is the incorrect prediction of ion concentration in narrow pores [31], it is worth investigating whether the Nernst-Planck theory can still be used if the ion concentrations can be determined in a more reliable manner. Is the Nernst-Planck theory when used alone applicable for use in narrow ion channels if the ion concentration and the potential could be correctly obtained? If so, then alternative approaches for determining channel conductances may be possible that can balance computational cost and accuracy. Indeed, there has been some pioneering work in this direction, such as the calculation of ion concentration by using Monte Carlo (MC) or density functional theory [39,40], and the use of ion concentrations obtained from MD or Monte Carlo (MC) methods directly within the NP equation [41–43]. The motivation of this kind of combination is that it is hoped that shorter simulation times are required to estimate the ion concentration and diffusion coefficient (which can then be used in the NP equation) than would be required to directly predict ion currents. For example, Allen et al. used molecular dynamics simulations and the umbrella sampling method to calculate the potential of mean force (PMF) and ion concentration in the gramicidin channel, and then used the NP equation to estimate the maximum conductance of the channel, something that took less computational effort than directly simulating the ion current [43]. However, despite its use in this context, the primary mystery of whether the NP theory is valid in the microscopic world remains unresolved. This is an essential problem that must be solved before further effort in this direction are carried out.

Therefore, we aimed to test if NP theory is applicable in narrow ion channels by combining it with BD simulations. That is, we determined the time averaged ion concentration and electrostatic potential in the channel directly from BD and used these as input to perform NP calculations from which we determined the channel current. The reason for conducting the calculation in this way is that the current can also be determined directly from the BD simulations. Thus, the NP and BD simulations will be utilising consistent concentrations and potential, but determining the conductance in two different ways. In this way we can directly check if the current obtained from the continuum calculation is the same as that found using explicit simulations of the ions. While BD simulations have been shown to be able to reliably predict channel currents in a number of cases, this is not critical to the present

study. Rather, we aim to see if the continuum approach can provide results in accord with that found when employing explicit ions. To test if the NP equation is valid in various situations, we performed our tests in a series of sequentially more complex channel models: cylindrical channels without dielectric boundaries, cylindrical channels with dielectric boundaries, cylindrical channels with dielectric boundaries and fixed charges in the channel wall, non cylindrical channels and a realistic model of the transmembrane (TM) domain of the mechanosensitive channel of small conductance (MscS) derived from a recently determined crystal structure [44]. The aim of using these different channel models is to examine if the accuracy of the NP equation is influenced by the channel radius, the channel shape, the channel occupancy, the rate of change of ion concentrations or forces in the pore, or differences in the cation and anion concentrations. In the results section we show that general agreement between the two approaches is found in all situations although discrepancies arise when the concentration of one ion is much lower than the other, before we discuss the potential applications and limitations of the proposed method of calculating channel currents.

Methods

Nernst-Planck theory

The NP electrodiffusion equation is widely used in the continuum theory of non-equilibrium processes such as ion transport, and can be written as follows:

$$\mathbf{J}_v = -D_v(\nabla n_v + \frac{z_v e n_v}{kT}\nabla\Phi),\qquad(1)$$

where $\mathbf{J}_v$ is the flux of each ion species, D_v, $z_v e$, and n_v are diffusion coefficient, charge, and number density of the ions of species v, respectively. Φ is the electrostatic potential (ESP) in this case. In our 1D case, it can be written as:

$$J_v = -D_v(\frac{dn_v}{dz} + \frac{z_v e n_v}{kT}\frac{d\Phi}{dz})\qquad(2)$$

To evaluate the ion fluxes, there are three main parameters or variables that need to be determined. The first is the diffusion coefficient of each ion species. Many previous studies keep this variable as an open parameter that can be adjusted to fit the experimentally determined conductance values, but this approach is not satisfying if the aim of the study is to determine the likely conductance of a given channel structure. In some other cases the values of the diffusion coefficients have been determined directly from MD simulations which show this to be position dependent [19,45–49]. In general, the value usually decreases by 30%~50% in the interior of the channel compared to that in bulk water. But in some studies, a value lower than 10% of the bulk value was obtained [6,50], which leaves the determination of the diffusion coefficient rather uncertain and highly system dependent. The second variable is the number density n_v (in SI units), which is related to the ion concentration c_v (in moles/liter) through $n_v = 10^3 N_A c_v$. Finally, the third variable is the ESP, Φ. In the most widely used version of PNP theory, the ion concentration and ESP are obtained by simultaneously numerically solving Poisson's equation and stationary NP equation iteratively [27]. But, as noted previously, the mean field approximation implicit in this encounters problems in narrow channels [31]. Alternative approaches have used MD [43] or MC [41,42] methods to get the ion concentration for input to the NP equation, but none of these studies had a clear way to tell if the combination of these methods is reliable.

Brownian dynamics

BD simulations have been successfully applied to determine channel currents and ion conduction events in various ion channels in recent years [18,19,51–54]. In BD, the motion of individual ions is traced explicitly, but the water and protein atoms are treated as continuous dielectric media [17,18]. In these simulations the channel is usually taken to be a rigid structure during the simulation (see [55] for an exception), and partial charges are assigned to the protein based upon the atomic positions. Ions are given starting positions in or around the channel and the motion of these ions under the influence of electric and random forces is then traced using the Langevin equation.

In the present case, most of the channel models were made from idealised shapes and a small number of partial charges were added at specific positions described in the results section. The one exception was for the studies of the MscS channel in which the pore was centred on the z-axis and a smooth water-protein boundary of the channel was defined by rolling a 1.4 Å sphere representing the water molecule along the surface. The boundary was symmetrised by taking only the minimum radius at each z-coordinate, and then the curve was rotated by 360° to obtain a three-dimensional channel structure with radial symmetry. In this case partial charges were assigned using the CHARMM27 all atom parameter set [56].

In all cases, 16 pairs of Na^+ and Cl^- were randomly distributed in 30 Å reservoirs that mimic the intra- and extra-cellular solution to bring the ion concentration to 300 mM. A time step of 100 fs was used and the trajectory was saved every 100 steps. Electrostatic forces were precalculated by assigning dielectric constants to the protein, channel interior and bulk water and solving Poisson's equation using an iterative method [57] and stored in tables to speed up the simulation [58]. While the dielectric constant in the channel is uncertain, we follow previous studies that have shown the best results in channels of this dimension are obtained assuming dielectric constants of 2 for the protein and 60 for the channel interior [54,59–62]. While the dielectric constant of the bulk water is likely to be closer to 80, for computational ease it is also set to 60 and the Born energy barrier for the ion to move between the dielectric constants of 80 and 60 is included as an additional force as in the previous studies above. We note that the exact choice is not critical for this study provided a consistent set of parameters is used in both the BD and NP calculations. The current is determined directly from the number of ions passing through the channel. In all cases described here an electric field of 20 mV/nm was applied to create a membrane potential along the z direction by incorporating the electric field into the solution of Poisson's equation, rather than simply applying forces on the ions. The boundaries between channel and water was treated as rigid walls from which ions elastically scatter, i.e., when the ions get to the channel boundary as close as 0.55 Å, the radial velocities of the ions would be multiplied by −1 while the axial velocities keep unchanged. The ions thus only move in the water environment and the ion-ion interaction can be calculated from Coulomb's law with an additional short range potential that reproduces the ion-ion radial distribution function found in all atom MD simulations [51]. All BD simulations were run for 1.6 µs. More details about the BD simulation methodology can be found in previous studies [17,18,52].

The combination of Brownian dynamics with Nernst-Planck theory

In order to test the validity of the NP equation, we incorporated the ion concentrations and potential found from BD simulations into the NP equation (BD-NP) to determine the channel current for comparison with those found directly from BD. Thus, a method of combining the results of BD with the NP equation needed to be determined for this study. As mentioned above, three quantities are needed for NP calculations: the diffusion coefficient, ion concentration and potential. In our tests we derive each of these directly from the corresponding BD simulations.

Since we only need to make sure that the same diffusion coefficients are used for both the BD and BD-NP methods, we can choose any arbitrary value for this as it should not affect our final comparison. To make things simpler, we adopted the diffusion coefficients of ions in bulk water for both BD and BD-NP calculations, which are 1.33×10^{-9} m^2/s for Na^+ and 2.03×10^{-9} m^2/s for Cl^- respectively.

The ion concentrations are calculated from the BD trajectories. For each channel model, a 1.6-µs BD simulation was performed. The first 0.2 µs was assigned as equilibration and not considered for data analysis. The latter 1.4-µs BD simulation trajectory was utilised to calculate the one dimensional (1D) ion concentration with a grid spacing of 0.5 Å, which was then implemented to NP equation for further calculation. Please refer to the supporting information to find more details about this (Text S1 section S1.1, and figure S1).

To make sure that the electrostatic potential determined from the BD simulations is consistent with the ion concentration, we proceeded in two steps. First we fixed the value at the end points of the calculation region to be that found from solving Poisson's equation (as done for calculating the force in BD). Next we determined the values in between by solving the stationary NP equations which enforces that the flux though the channel is the same at all points along its length:

$$\nabla \mathbf{J}_v = 0. \qquad (3)$$

Further details of the implementation of this strategy can be found in Text S1, section S1.2. The ESP could be determined in other ways, for example by solving Poisson's equation at each snapshot of the BD trajectory and averaging, but the approach described above is less sensitive to slight fluctuations in the average potential which are amplified when calculating the flux (please cf Text S1 section S3, figure S3 and figure S4).

The diffusion coefficients of ions, the ion concentrations and the potentials determined from the BD simulations were put into the NP equation 2 to calculate the currents as described in the Text S1, section S1. In all these calculations, we adopted a grid spacing of 0.5 Å which gives the most stable prediction of ion currents (please cf table S1).

Results and Discussion

BD vs BD-NP for passive cylindrical channels

We started our test with the simplest model — a cylindrical channel as shown in figure 1 with no dielectric boundaries (in this case a dielectric constant of 60.0 is used throughout). We term this a 'passive' channel to reflect the fact that there are no induced forces on ions from the channel walls and ions simply elastically scatter from the water/channel interface. The channel has a cylindrical shape spanning from $z = -20$ Å to $z = 20$ Å with the central axis of the channel aligned on the z axis. A series of such models were built with the radii of the channels ranging from 3 Å

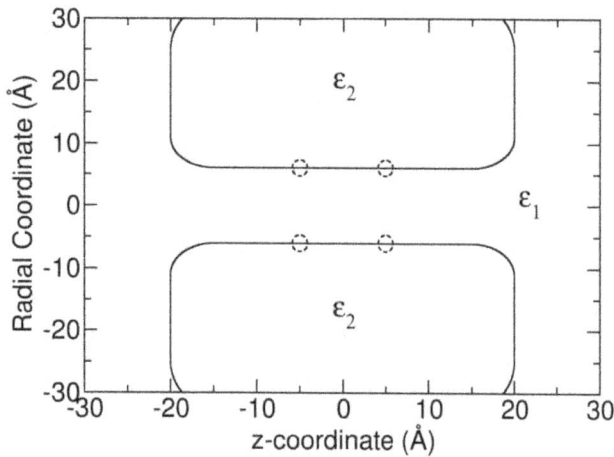

Figure 1. The cylindrical channel model. A 6-Å-radius model is shown here. The dashed circles show the positions of the charged rings in the charged cylindrical channel.

to 15 Å. In the NP calculations, values of the ion concentration and potential in the segment between $-15\,\text{Å} \le z \le 15\,\text{Å}$ was considered, although the choice of the segment was found not to influence the results provided we avoided including the reservoirs where the channel shape changes rapidly.

An example of ESP in a passive cylindrical channel with radius of 6.0 Å is shown in figure 2a with the dotted line. Our method of calculating the ESP accounts for not only the external applied

electric field, but also the dielectric boundary and fixed charges in the system. But, in this case, since there is neither a dielectric boundary nor charge for the passive channel, the potential changes linearly through the pore. Meanwhile, [Na$^+$] and [Cl$^-$] are also shown in figure 2a, as calculated from the last 1.4 µs BD trajectory. The concentrations are fairly flat in the channel, however, [Na$^+$] shows a slight decrease and [Cl$^-$] a slight increase along the direction of the electric field caused by the build up of concentration on the membrane surface around the ends of the channel. The current carried by Na$^+$ and Cl$^-$ found using each method is shown in figure 3a. As can be seen, the BD-NP results match pretty well with the BD results at all the channel radii studied. Even in the narrow channels the current is reproduced with a high degree of accuracy indicating that the concept of combining BD and NP in this way to determine the channel current is reasonable. The agreement in this case is not surprising given that the PNP theory also predicts accurate currents at all radii in these passive channels [31].

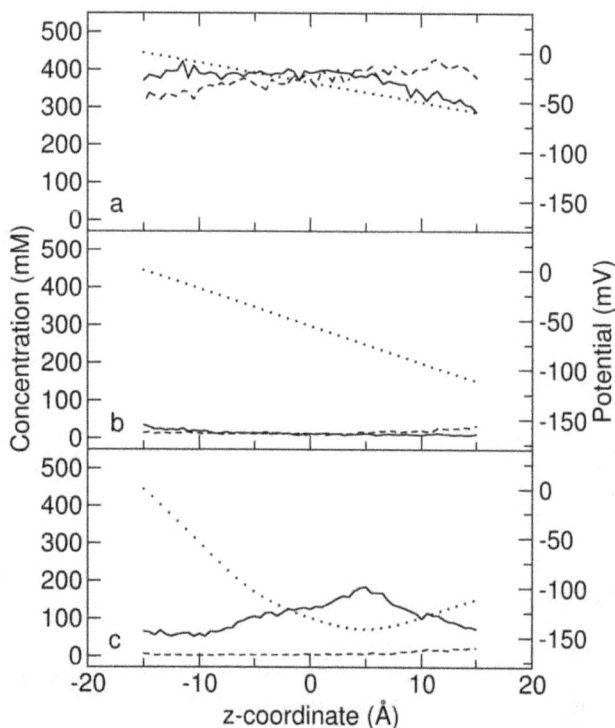

Figure 2. The ion concentration and ESP of the (a) passive, (b) real and (c) charged cylindrical channels with a radius of 6 Å. The concentration of Na$^+$ and Cl$^-$ are shown with solid and dashed lines respectively, and the ESP is shown with the dotted line. The ESP shown is that obtained in the absense of mobile charges.

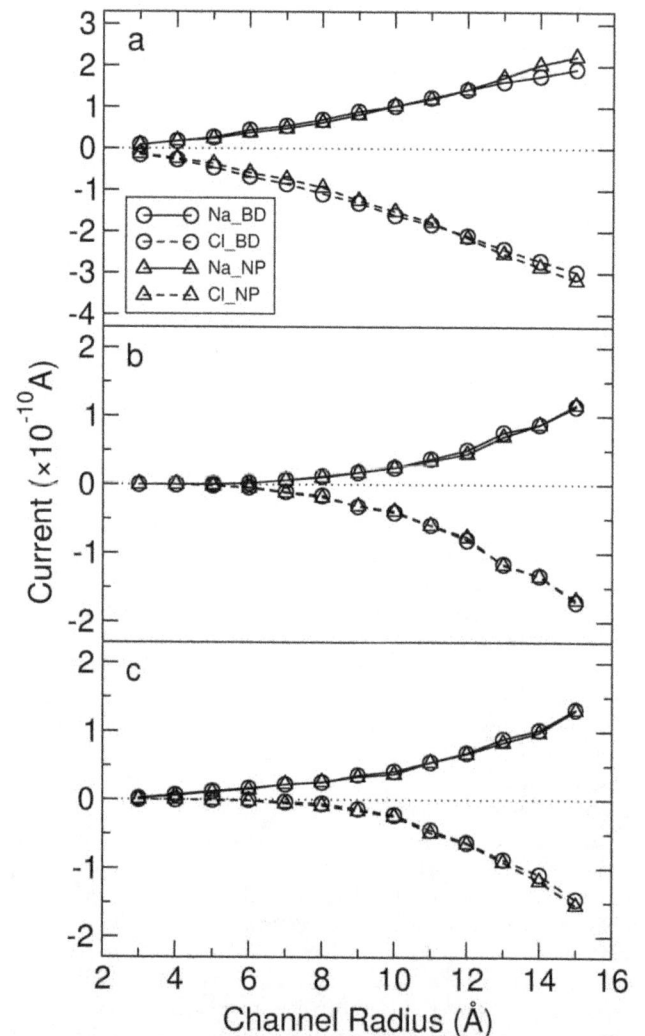

Figure 3. The currents of Na$^+$ and Cl$^-$ through the (a) passive, (b) real and (c) charged cylindrical channels of differing radius under 20 mV/nm electric field found using BD simulations and the BD-NP method. The error bars are smaller than the size of the symbol and therefore not shown here.

BD vs BD-NP for real cylindrical channels

To make further tests in more realistic channels, we utilized 'real' cylindrical channels for which the dielectric constant of water ε_1 was set to 60.0, while the dielectric constant of the channel ε_2 was set to 2.0 as shown in figure 1. This means that the channel body is now more distinct from the water and there will be induced surface charges on the channel boundary in the presence of ions. As there are no permanent charges in this case (as would arise from partial charges on the protein atoms) we are able to study the effect of the dielectric boundary in isolation. All the other parameters were the same as those for the passive cylindrical channels.

The ESP and ion concentration from the BD trajectory for a real cylindrical channel with radius 6 Å are shown in figure 2b. The ESP also decreases linearly like in the passive cylindrical channels because there are no point charges on the channel, but in this case there is a larger potential drop due to the existence of the dielectric boundary. The ion concentrations show very low values in the channel interior, exhibiting a distinct difference from those in passive channels. This is expected as in this case, the low value of the dielectric constant in the protein leads to induced surface charges on the dielectric boundary that have the same sign as the conducting ions and repel the ions from the channel wall, effectively creating a dehydration barrier for ions to enter the pore.

The ion currents calculated from BD and BD-NP methods are shown in figure 3b. Although the currents are lower than in the corresponding passive channels, the results from the two different methods still match well. This is a significant finding, especially when recalling that the PNP theory completely fails in the narrow channels used here [31] his reinforces the fact that the failure of PNP in narrow channels originates from the incorrect prediction of ion concentration calculated by the combination of Poisson's equation and the NP equation. If the ion concentration can be obtained from more accurate method, such as BD simulations here, then the NP theory is able to accurately predict the current for these channels.

BD vs BD-NP for charged cylindrical channels

So far we have considered fairly simple channel models in which the ion concentration and potential vary smoothly throughout the pore and in which the channel is either passive, wide, or narrow but containing very few ions. In most realistic cases none of these conditions will hold and it is important to check if more rapid fluctuations in ion concentration, ESP or multiple occupancy influence the accuracy of the NP results. For example, in the classical model, the atoms in proteins carry partial charges, and often the presence of charged rings or functional groups at specific positions near the pore is used to control ion permeation and select between different ion types. The presence of such charges can create more rapid changes in the ion concentration, ESP as well as multiple occupancy.

To mimic this effect and study how the BD-NP method behaves under this more complex situation, we built 'charged' cylindrical channels. All the parameters for these charged channels are the same as those for the real cylindrical channels, except that there are two charged rings in the channel. As shown in figure 1, the dashed circles at $z = -5.0$ and $z = 5.0$ show the positions where 16 point charges were manually fixed at the channel boundary. At each position, 8 point charges each with a charge of -0.09 e were uniformly distributed at the channel boundary. Therefore, each of the two rings has a net charge of -0.72 e, which is expected to make it easier for cations to enter the channel than anions [63]. These point charges were treated statically to mimic charged

atoms, as often seen in ion channels, rather than intending to represent the dielectric polarization.

The ESP from electrostatic calculations for a 6-Å-radius charged cylindrical channel is shown in figure 2c. It is obvious that there is a potential well located at around $z = 5.0$ due to the combined effect of the two charged rings and the membrane potential. Correspondingly, $[Na^+]$ has a maxima at this position due to the electrostatic interactions with the charged rings. In contrast, $[Cl^-]$ remains at very low values throughout the channel. The charged rings do act to form a selectivity filter by attracting more cations into the channel and repelling anions.

The ion currents for all the charged cylindrical channels are shown in figure 3c. Again, the BD-NP results generally match well with those from BD simulations. Furthermore, the negatively charged channels do have cation selectivity, which is especially obvious when the channel radius is small. This is very encouraging which means that the BD-NP method is applicable to all the cylindrical channels, even if the channels are narrow, charged and selective or if there are non-monotonic ion concentrations and electrostatic potentials. One thing to mention here is that when the channel is very narrow (radius ≤ 7 Å) and negatively charged, the current of the Cl^- is less accurate. This is not obvious in figure 3c because those values are $1 \sim 2$ orders less than those of Na^+. We will discuss the importance of this later in the paper.

We also tested whether the exact value of the dielectric constant influences the reliability of the BD-NP method. To this end we have repeated all the tests for the passive, real and charged cylindrical channels with water dielectric constant set to be 80.0, and the results are found to be as good as those described above (shown in figure S2). Therefore, we believe that the BD-NP method is capable of predicting ion fluxes and currents as well as BD simulations themselves in cylindrical channels, irrespective of the channel radius and the choice of dielectric constant.

BD vs BD-NP for more complex 'catenary' channels

It is possible that the success of the 1D BD-NP approach lies in part due to the simple cylindrical shapes being employed and any deviation from such simple shapes is more likely to stress the 1D calculation. To examine if BD-NP works for channels with more complex shapes, we did further tests on a 'catenary' channel model. The channel structure is shown in figure 4. The middle part of the channel ($-5 \leq z \leq 5$) is a cylinder which has a radius of 6 Å, while the outer parts of the channel ($-25 \leq z \leq -5$ and $5 \leq z \leq 25$) has a catenary shape with the radius changing from 6 to 12 Å. Similar to the study on the cylindrical channels, tests were made on a passive, real and charged catenary channel. For the passive catenary channel, the dielectric constants for water and channel were both set to 60.00; for the real catenary channel, the dielectric constants were set to 60.00 and 2.00 respectively; for the charged catenary channel, the dielectric constants were the same as for the real channels, plus two negatively charged rings were put on the channel boundary as shown with dashed circles in figure 4. Each ring has 8 uniformly distributed point charges with the value -0.045 e, resulting a total charge of -0.36 e per ring.

The potential and ion concentration for each catenary channel, calculated from the BD simulation, are shown in figure 5. We can see that these profiles share similar features to those for cylindrical channels except that the ion concentration at the outer parts of the channel is higher because ions can build up on the narrowing faces of the pore entrances. Also, the ion concentration in the real catenary channel is much lower than in the passive channel due to the induced surface charges at the boundary. In the negatively charged catenary channel, the ion concentration of Na^+ is much

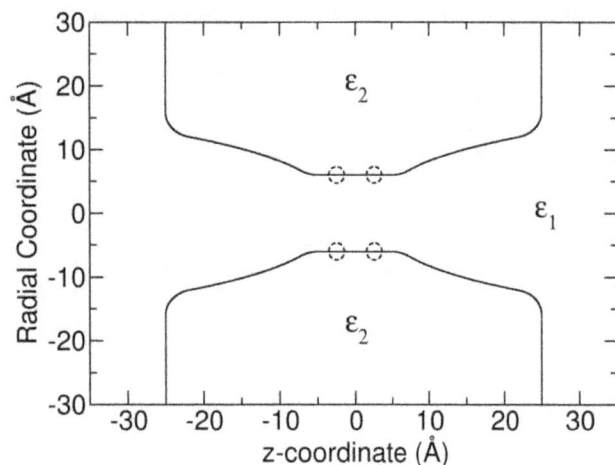

Figure 4. The catenary channel model. The dashed circles show the positions of the charged rings.

higher than Cl^- due to the electrostatic interaction, especially in the middle narrow part of the pore.

The currents determined with the BD and BD-NP methods are shown in table 1. We can see that the BD-NP method still works well in general. The biggest difference arises for Cl^- in the charged catenary channel where the BD-NP calculation gives a value about 70% higher than the BD result which is discussed in more detail below. Apart from this, the BD-NP method seems not affected by the shape and the change of the radius of the channel, which means NP could be valid in more generic channels with

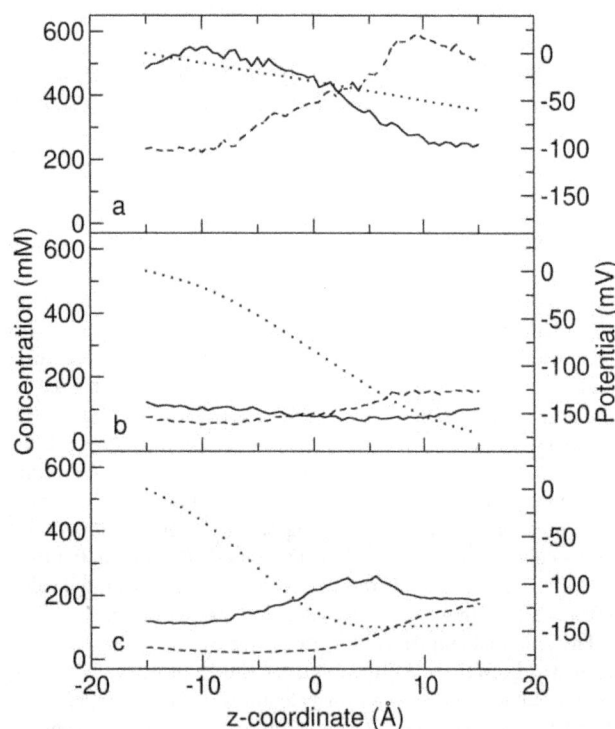

Figure 5. The ion concentration and ESP of the (a) passive, (b) real and (c) charged catenary channels. The concentration of Na^+ and Cl^- are shown with solid and dashed lines respectively, and the ESP is shown with the dotted line.

complex shapes. Additional tests with wider radius and different charges at the boundary were also performed, which showed that the BD-NP method works better in wider charged catenary channels and the amount of the charges can affect the accuracy of the results. All the BD-NP results presented above used the central segment of the channel $-15 \leq z \leq 15$ in the calculations and thus included the region where the pore radius is changing. The choice of calculation region was not found to be important to our results as those found using the regions $-20 \leq z \leq 20$ or $-10 \leq z \leq 10$, are almost identical to those presented above. The only time that the results differed was when we included the ends of the channel and the sharp radius increase at the start of the reservoirs that occurs at $z = \pm 25$. The fact that the BD-NP method works well in the situation where the channel radius is not constant is very encouraging considering the fact that we are doing 1D BD-NP calculations. The additional tests results are shown in table S2.

BD vs BD-NP for the transmembrane domain of MscS

Finally we tested the BD-NP method for a more realistic channel model — the TM domain of MscS — as a first step to practical applications. MscS is one kind of mechanosensitive channel that opens in response to mechanical forces in the lipid bilayer. In this work, we only took the TM domain of the protein (PDB entry 2vv5 [44]) as illustrated in figure 6a and performed 1.6 μs BD simulation on it. The radius of the channel is shown in figure 6b and is complex in shape and the channel is highly charged with a total charge of 35 e, which provides an ideal model to test under a very complex realistic situation including large concentrations and thus multiple ion occupancy.

The ESP and ion concentration from BD simulation are shown in figure 7. There is a large potential difference across the chosen segment ($-20 \leq z \leq 20$), about 350 mV. The concentration of Cl^- is much higher than Na^+ and even much higher than the bulk concentration 300 mM in some particular locations of the channel interior ($5 \leq z \leq 20$) due to the high positive charges on the protein. The ion currents from BD simulations are -4.11×10^{-12} and 2.54×10^{-10} for Na^+ and Cl^- respectively, showing an anion selectivity of the TM domain. The ion current from BD-NP calculation are -6.69×10^{-12} and 2.93×10^{-10} for Na^+ and Cl^- respectively. Therefore, the BD-NP method overestimates the current about 63% for Na^+ and 15% for Cl^- when comparing to the BD simulation results.

The effects of shape and charge

From the above results, we can see a trend: the BD-NP method becomes less accurate when increasing the complexibility of the channel. Two factors might be responsible for this: the shape of the channel and the charge distribution on the channel. Exploring to what extent the two factors affect the accuracy may direct us to the way to improve the method.

To see how the shape of the channel affects the accuracy of the BD-NP method, we can first compare the 'passive' channels without dielectric boundaries or charge distributions. For all the passive cylindrical channels, the results of BD-NP match well with BD as shown in figure 3a, which means the radius of the channel is not a key factor that influence the accuracy of the results. When changing the shape to the 'catenary' channel, the results from BD-NP and BD alone still match well as shown in table 1. To further verify this point, we ran an additional BD simulation on a 'passive' TM domain of MscS, i.e., the shape of TM domain of MscS (as shown in figure 6b) was utilized to generate a channel without any dielectric boundary or charge. In this simulation, the Na^+ currents calculated from BD and BD-NP are -7.17×10^{-11} and -7.74×10^{-11} A, and the Cl^- currents calculated from BD

Table 1. Currents through the catenary channels (A).

	passive		real		charged	
	Na$^+$	Cl$^-$	Na$^+$	Cl$^-$	Na$^+$	Cl$^-$
BD	7.43×10^{-11}	-1.01×10^{-10}	2.39×10^{-11}	-3.89×10^{-11}	4.94×10^{-11}	-1.26×10^{-11}
BD-NP	6.38×10^{-11}	-9.45×10^{-11}	2.56×10^{-11}	-4.55×10^{-11}	4.83×10^{-11}	-2.13×10^{-11}

and BD-NP are 1.05×10^{-10} and 1.04×10^{-10} A respectively. The result indicates that even in a very complex shape like in a real ion channel, the results from the two methods are very close. Therefore, we believe that the shape of the channel does not have a major influence of the accuracy of the BD-NP method.

As mentioned above, when the cylindrical channel is narrow and charged, the current predicted by the NP equation for the ion of lower concentration is less accurate. This is a sign that the charges on the channel might be affecting the accuracy of the NP calculation. To further study this effect, we can examine the results of the catenary and MscS channel. For the catenary channel, when changing the channel from passive to charged, the accuracy clearly decreased especially for the Cl$^-$ which is the minority ion type as shown in table 1. For the MscS TM domain, we can see similar trend in table 2. It seems that the charge distribution on the channel does have a clear influence on the accuracy of the BD-NP

method. To further understand this, we examined how the current passing through the 6-Å-radius cylindrical channel changed as we slowly increase the charge on the pore wall. As shown in table 3, we can see that as the charge on the channel increases from -0.36 e per ring to -2.88 e per ring, the deviation in the Na$^+$ current predicted from NP compared to BD increases from -3.35% to 35.36%. Interestingly, the current of the minority ion type Cl$^-$ are very different from the BD results, however, the absolute magnitude of the Cl$^-$ is also $1 \sim 2$ orders smaller than for Na$^+$. When the charges on the channel is -5.76 e per ring, the NP equation does not have a solution for the Cl$^-$ current any more, though the deviation for the Na$^+$ current is -25.85% and still in a reasonable range.

The above analysis shows that the charge distribution on the channel has a much greater affect on the accuracy of the NP results than the shape of channel, but the reasons for this deviation are yet to be established. The most likely reason for the inaccuracy is that the presence of permanent charges creates a non-uniform ion distribution in the channel. We adopted a 1D approximation in the NP calculations, and it can be expected that a smooth, uniform ion distribution would give the best results. But, if the channel has a negatively charged ring, for example, then there would be a high cation distribution and low anion concentration near the channel boundary. The 1D NP calculation does not capture this and only uses the average concentration at any position along the channel. We believe that this difference is the key factor that causes the deviation between the BD and NP currents.

Although there are obvious discrepancies between the BD and NP results, we still believe the NP equation is applicable for estimating currents in the majority of cases. Firstly, although the

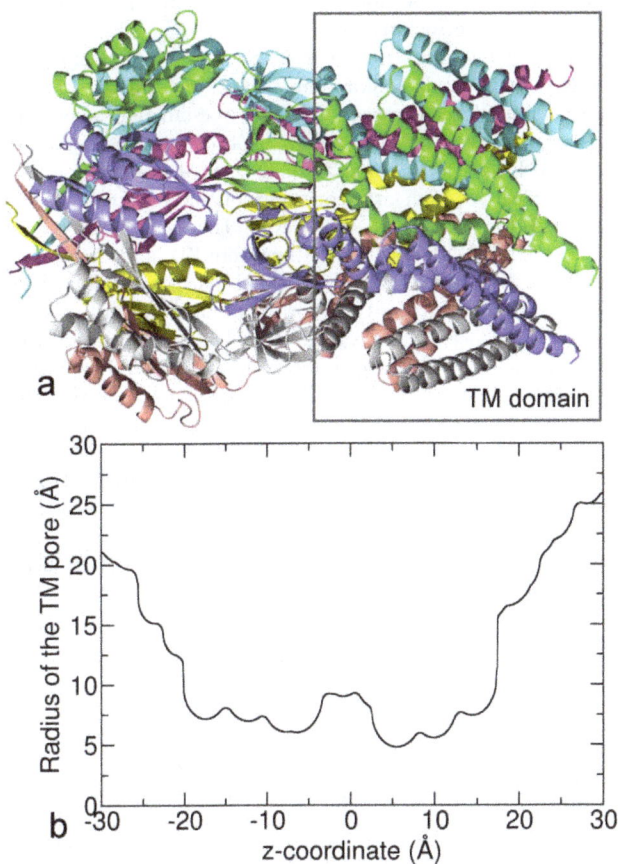

Figure 6. The model of the mechanosensitive channel MscS. (a) The structure of MscS with the TM domain marked with the rectangular box. (b) The radius of the TM domain of MscS.

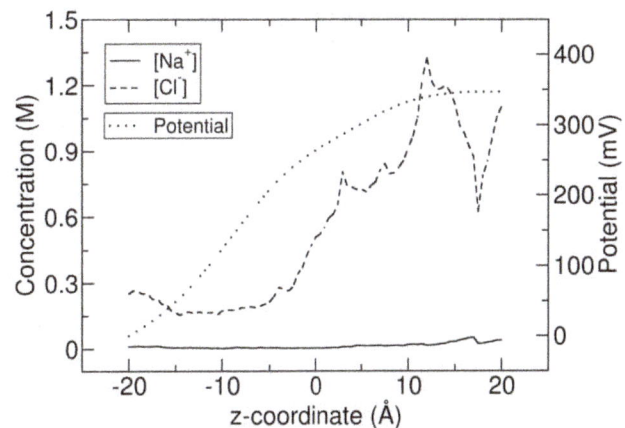

Figure 7. The ion concentration and ESP in the TM domain of MscS. The concentration of Na$^+$ and Cl$^-$ are shown with solid and dashed lines respectively, and the ESP is shown with the dotted line.

Table 2. Currents through the TM domain of MscS (A).

	passive		real		charged	
	Na$^+$	**Cl$^-$**	**Na$^+$**	**Cl$^-$**	**Na$^+$**	**Cl$^-$**
BD	-7.17×10^{-11}	1.05×10^{-10}	-4.23×10^{-11}	6.51×10^{-11}	-4.11×10^{-12}	2.54×10^{-10}
BD-NP	-7.74×10^{-11}	1.04×10^{-10}	-5.90×10^{-11}	8.33×10^{-11}	-9.48×10^{-12}	4.63×10^{-10}

percentage difference is greatest for the minority ion types, the total current is mostly dictated by the majority ion whose conductance is predicted more accurately. Secondly, these results are much more accurate than equivalent PNP results. Finally, the ability to predict the current to within 30% in the worst case scenarios still allows for the qualities such as the conductance state of the channel to be determined. Although PNP can also be used to estimate the conductance of a wide channel, it becomes less reliable for narrow ones [31]. The BD-NP method appears to have a greater range of validity, being able to estimate the magnitude of currents in both narrow and wide pores. If one extends the NP calculation to 3D, the accuracy could probably be enhanced, but this is beyond the scope of the present study.

Success, limitation and perspective

We have examined a range of different ion channel models to test the validity of the BD-NP approach including passive, real and charged cylindrical channels with various radii and more complex channel shapes to explore under what conditions the NP theory is still applicable. Although there are some deviations, the BD-NP and BD results show overall good agreement. Results are especially good when the model channel conducts ions with currents larger than 10 pA suggesting that the NP theory can be used to obtain estimates of channel currents provided that the ion concentration can be precisely obtained beforehand.

It is important to consider the reasons for both the similarities of the currents found from BD and those found using BD-NP as well as the differences. In BD, the forces acting on each ion are determined at each timestep in the simulation based upon the positions of all the ions in the system at that time. In contrast, the NP equation was derived from macroscopic Smoluchowski equation in which the average force is calculated in a mean-field manner. In the case of BD, the current is calculated using the instantaneous forces on explicit ions, while in NP current is determined from the time averaged concentration and potential. Thus, although the mean properties found in the BD simulations are consistent (indeed the same) as those employed in the NP calculation, one need not expect identical results. Furthermore,

additional differences can be expected to arise since the NP calculations in this work were performed under one dimensional approximation, while the BD simulations were three dimensional. It is not surprising, therefore, that there are some deviations between the results from the two methods. On the contrary, it is quite surprising to see such good general agreement suggesting that the mean field approach is capturing the important physics in most cases. The two cases in which the worst results were obtained using BD-NP suggest some possible limitations in the mean field approach. In both the charged catenary channel and the MscS TM domain, where the concentration of one ion type is extremely low while the other is large, the current of the minority ion type is overestimated by BD-NP, most likely as a result of non-uniform distribution of this type of ions in the channel. We also want to point out that this deviation is not due to insufficient sampling of the ion concentration, as extending the BD simulation three times longer for the charged catenary channel gave no improvement. Therefore, the ion distribution can give an indication of cases where potential errors may arise.

Having noted the conditions when the worst results were obtained, it is worth pointing out that in the majority of cases studied BD-NP can usually give good estimation about ion conductance, with an error below 30% comparing to the BD results (below 10% in most cases). Even in the worst cases, the current of the ion with large concentration and conductance were estimated to within this same level. The general agreement between the BD-NP results and those from BD alone implies the validity of NP in microscopic scale, and that it is possible to use mean-field approximations to study ion channel currents provided that the ion concentration is accurately obtained. However, when the channel is highly charged, the accuracy of the 1D approach decreases. The results from the BD-NP method are clearly better than analogous PNP results which overestimate the currents in channels with dielectric boundaries [31]. Although PNP performed best in charged channels, even in these cases the BD-NP approach appears more accurate. More recent PNP studies have attempted to include dielectric self energy to improve the predicted currents [32–35], but this improvement primarily occurs by better

Table 3. Currents through the charged 6-Å-radius cylindrical channels (A).

Charge per ring (e)	−0.36		−0.72		−1.44		−2.88		−5.76	
Ion type	Na$^+$	Cl$^-$	Na$^+$	Cl$^-$	Na$^+$	Cl$^-$	Na$^+$	Cl$^-$	Na$^+$	Cl$^-$
BD	6.17×10^{-12}	-8.00×10^{-13}	1.70×10^{-11}	-1.03×10^{-12}	2.51×10^{-11}	-8.00×10^{-13}	3.30×10^{-11}	-4.57×10^{-13}	3.69×10^{-11}	-3.43×10^{-13}
BD-NP	5.96×10^{-12}	-2.51×10^{-12}	1.48×10^{-11}	-2.00×10^{-11}	2.84×10^{-11}	-5.56×10^{-12}	4.47×10^{-11}	-6.49×10^{-12}	2.74×10^{-11}	NA
Dev	−3.35%	213.63%	−12.76%	94.47%	13.19%	595.50%	35.36%	1320.57%	−25.85%	NA

'NA' means there is no solution for the stationary NP equation for this case.

predicting ion concentrations. By utilising accurately determined concentrations directly in the calculation, the BD-NP method can be expected to outperform even the modified PNP methods.

One should also keep it in mind that our NP calculations have all been conducted in 1D - that is only average concentrations and potentials in the axial direction of the channel were considered (see SI section S1 for more details). This makes the success of the method even more surprising and suggests that it could be further improved by extension to 3 dimensions. This also explains why it is that when the ion distribution is more uniform and the channel shape is more smooth, the agreement of the two methods is better. One of the main reasons we did not do the 3D calculation here is that accurate results require the concentration to be well sampled at all points in the channel. In 3D this generally requires longer simulations. It should also be noted that the NP calculation is extremely efficient. Once the ion concentration, the potential difference and the diffusion coefficient are known, it takes less than a minute to get the ion current on a single PC.

There are also some limitations in our proposed method for combining BD with NP. The most significant one is that the potentials calculated from [Na$^+$] and [Cl$^-$] do not match exactly, which is a compromise to make sure we can get constant ion current values through the channel. More results and discussion about this can be found in Text S1, section S3. Another limitation is that the segment chosen to do the NP calculation must be part of the channel (not including the bulk) in order to avoid sudden changes of radius, which is a shortcoming resulting from the 1D approximation. But fortunately, the specific choice of the segment does not matter as long as it is part of the channel proper.

The combination of NP and BD itself is not very exciting as the BD simulations themselves are already able to yield ion currents. Indeed, by the time the concentration is well determined from the BD simulation we already have a statistically reliable conductance value. Thus, there is no need to resort to the more approximate BD-NP method at this stage. Our purpose in doing these calculations was not to propose BD-NP itself as a useful approach to calculating channel currents, but rather to test if NP theory is still valid in narrow ion channels. By making our comparison of the BD-NP results to those from BD alone we can compare currents determined from exactly the same underlying concentration and potential data, that is we know what current we should expect to get from the NP calculation and we can directly test the mean field approximation.

Our encouraging finding is that the NP theory appears to be applicable at the microscopic scale, and our study presents a good example about how microscopic simulations can be related to continuum theory calculations. This method can be easily extended to 3D version as long as ion concentration could be obtained in a more efficient way. We believe that this direction could be further pursued to find a more useful way of getting reliable channel conductances from detailed microscopic simulations. For example, one might want to try combining other methods for determining the ion concentrations in the channel (but that cannot themselves reliably predict channel currents) with NP. The first natural consideration is MD. Predicting channel currents is difficult in MD due to restrictions on the timestep and the computational power required. However, MD can explicitly account for the interactions between water molecules, and can more easily account for protein flexibility than BD, both of which may be important considerations in determining ion permeation. In principle, the ion concentration, diffusion coefficient and electrostatic potential could all be obtained from MD simulations, which could be taken as inputs for further NP calculations to predict the ion conductance. This would be useful if it could be done using shorter simulations than those needed to directly simulate multiple conduction events in the MD simulations themselves. The trickiest problem to overcome, however, is to work out how to reliably calculate the ion concentration as it can be hard to get sufficient sampling of ions in the channel using MD. To do this, some advanced simulation techniques such as umbrella sampling might be needed to get more statistically meaningful values, ideally to produce a potential of mean force (PMF) that can be employed in the NP calculation, an approach that has been tried previously by Allen et al. in gramicidin [43,64]. By combining these PMFs with the NP equation, our results show that it is possible that the ion current could be estimated. Furthermore, it may be possible to use a single PMF to predict the current values under different voltages which could further save computational costs. One thing to keep in mind is that the ion occupancy probability is related to the free energy by an exponential relation. Thus, any uncertainty in the PMF (which are usually > 1 kcal/mol) will be amplified when determining the ion concentration to use in the NP equation, which could in turn result in a poor estimate of the ion current. Furthermore, there could be additional problems if multiple ions are resident in the channel which would be likely to require longer simulations to accurately sample all positions or to get multi-ion PMFs. Alternatively, Monte-Carlo approaches rather than dynamic ones may allow for more efficient sampling of the ion concentration as they can cover configurational space more efficiently [39,40]. We suggest that further tests need to be carried out to see if a worthwhile means of combining MD or MC with NP to calculate channel currents can be devised. The PNP method might also be further improved, and our results here show that this ultimately requires the method to be able to determine the ion concentration accurately.

The systematic simulations and tests of the BD-NP method conducted here show that NP equation can be used to estimate ion currents provided they incorporate accurate ion concentrations and potentials. The accuracy could be probably further enhanced if a 3D NP equation is adopted. After verifying the validity of the NP theory in this way, the door is open to find efficient ways of combining microscopic and continuum approaches to predict ion channel currents. In this context, we believe that our study has provided a solid cornerstone for further effort in this direction.

Supporting Information

Figure S1   The sketch of 1D NP calculation.

Figure S2   The currents through the cylindrical channels when setting the dielectric constant of water to be 80.

Figure S3   The Na$^+$ currents for each sampling point calculated using the potential from Poisson's equation. This example is from the real cylindrical channel of 6-Å radius.

Figure S4   Potential profiles calculated from Poisson's equation (BD) and our strategy (NP_Na$^+$ and NP_Cl$^-$), for a real cylindrical channel of 6-Å radius.

Table S1   The influence of grid spacing on the NP current.

Table S2   The influence of segment selection on NP current.

Acknowledgments

We thank Prof. Bert de Groot for helpful discussions.

Author Contributions

Conceived and designed the experiments: CS BC. Performed the experiments: CS. Analyzed the data: CS BC. Wrote the paper: CS BC.

References

1. Ashcroft FM (2006) From molecule to malady. Nature 440: 440–447.
2. Bernèche S, Roux B (2001) Energetics of ion conduction through the K+ channel. Nature 414: 73–77.
3. Cheng X, Lu B, Grant B, Law RJ, McCammon JA (2006) Channel Opening Motion of α7 Nicotinic Acetylcholine Receptor as Suggested by Normal Mode Analysis. J Mol Biol 355: 310–324.
4. Beckstein O, Sansom MSP (2004) The influence of geometry, surface character, and flexibility on the permeation of ions through biological pores. Phys Biol 1: 42–52.
5. Beckstein O, Biggin PC, Bond P, Bright JN, Domene C, et al. (2003) Ion channel gating: insights via molecular simulations. FEBS Lett 555: 85–90.
6. Zhu F, Hummer G (2010) Pore opening and closing of a pentameric ligand-gated ion channel. Proc Natl Acad Sci USA 107: 19814–9.
7. Nury H, Poitevin F, Van Renterghem C, Changeux JP, Corringer PJ, et al. (2010) One-microsecond molecular dynamics simulation of channel gating in a nicotinic receptor homologue. Proc Natl Acad Sci USA 107: 6275–6280.
8. Noskov SY, Bernèche S, Roux B (2004) Control of ion selectivity in potassium channels by electrostatic and dynamic properties of carbonyl ligands. Nature 431: 830–4.
9. Thomas M, Jayatilaka D, Corry B (2007) The predominant role of coordination number in potassium channel selectivity. Biophys J 93: 2635–43.
10. Wang HL, Cheng X, Taylor P, McCammon JA, Sine SM (2008) Control of Cation Permeation through the Nicotinic Receptor Channel. PLoS Comput Biol 4: e41.
11. Crozier PS, Henderson D, Rowley RL, Busath DD (2001) Model Channel Ion Currents in NaCl-Extended Simple Point Charge Water Solution with Applied-Field Molecular Dynamics. Biophys J 81: 3077–3089.
12. Crozier P, Rowley R, Holladay N, Henderson D, Busath D (2001) Molecular Dynamics Simulation of Continuous Current Flow through a Model Biological Membrane Channel. Phys Rev Lett 86: 2467–2470.
13. Yang Y, Henderson D, Busath D (2003) Applied-field molecular dynamics study of a model calcium channel selectivity filter. J Chem Phys 118: 4213.
14. Jensen MO, Borhani DW, Lindorff-Larsen K, Maragakis P, Jogini V, et al. (2010) Principles of conduction and hydrophobic gating in K+ channels. Proc Natl Acad Sci USA 107: 5833–8.
15. Cooper K, Jakobsson E, Wolynes P (1985) The theory of ion transport through membrane channels. Prog Biophys Mol Biol 46: 51–96.
16. Bek S, Jakobsson E (1994) Brownian dynamics study of a multiply-occupied cation channel: application to understanding permeation in potassium channels. Biophys J 66: 1028–1038.
17. Li SC, Hoyles M, Kuyucak S, Chung SH (1998) Brownian dynamics study of ion transport in the vestibule of membrane channels. Biophys J 74: 37–47.
18. Chung SH, Allen TW, Kuyucak S (2002) Conducting-state properties of the KcsA potassium channel from molecular and Brownian dynamics simulations. Biophys J 82: 628–645.
19. Im W, Roux B (2002) Ion permeation and selectivity of OmpF porin: a theoretical study based on molecular dynamics, Brownian dynamics, and continuum electrodiffusion theory. J Mol Biol 322: 851–869.
20. Bernèche S, Roux B (2003) A microscopic view of ion conduction through the K+ channel. Proc Natl Acad Sci USA 100: 8644–8648.
21. Cheng MH, Coalson RD, Tang P (2010) Molecular Dynamics and Brownian Dynamics Investigation of Ion Permeation and Anesthetic Halothane Effects on a Proton-Gated Ion Channel. J Am Chem Soc 132: 16442–16449.
22. Levitt DG (1986) Interpretation of biological ion channel flux data–reaction-rate versus continuum theory. Annu Rev Biophys Biophys Chem 15: 29–57.
23. Andersen OS (1989) Kinetics of ion movement mediated by carriers and channels. Methods Enzymol 171: 62–112.
24. Levitt DG (1987) Exact continuum solution for a channel that can be occupied by two ions. Biophys J 52: 455–466.
25. Levitt DG (1991) General continuum theory for multiIon channel. Biophys J 59: 271–277.
26. Eisenberg R (1996) Computing the field in proteins and channels. J Membr Biol 150: 1–25.
27. Kurnikova MG, Coalson RD, Graf P, Nitzan A (1999) A Lattice Relaxation Algorithm for Three-Dimensional Poisson-Nernst-Planck Theory with Application to Ion Transport through the Gramicidin A Channel. Biophys J 76: 642–656.
28. Hollerbach U, Chen DP, Busath DD, Eisenberg B (2000) Predicting Function from Structure Using the PoissonNernstPlanck Equations: Sodium Current in the Gramicidin A Channel. Langmuir 16: 5509–5514.
29. Hwang H, Schatz GC, Ratner Ma (2006) Ion current calculations based on three dimensional Poisson-Nernst-Planck theory for a cyclic peptide nanotube. J Phys Chem B 110: 6999–7008.
30. Bolintineanu DS, Sayyed-Ahmad A, Davis HT, Kaznessis YN (2009) Poisson-Nernst-Planck models of nonequilibrium ion electrodiffusion through a protegrin transmembrane pore. PLoS Comput Biol 5: e1000277.
31. Corry B, Kuyucak S, Chung SH (2000) Tests of Continuum Theories as Models of Ion Channels. II. PoissonNernstPlanck Theory versus Brownian Dynamics. Biophys J 78: 2364–2381.
32. Luchinsky DG, Tindjong R, Kaufman I, McClintock PVE, Eisenberg R (2009) Self-consistent analytic solution for the current and the access resistance in open ion channels. Phys Rev E 80: 021925.
33. Jung YW, Lu B, Mascagni M (2009) A computational study of ion conductance in the KcsA K(+) channel using a Nernst-Planck model with explicit resident ions. J Chem Phys 131: 215101.
34. Corry B (2003) Dielectric Self-Energy in Poisson-Boltzmann and Poisson-Nernst-Planck Models of Ion Channels. Biophys J 84: 3594–3606.
35. Koumanov A, Zachariae U, Engelhardt H, Karshikoff A (2003) Improved 3D continuum calculations of ion flux through membrane channels. Eur Biophys J 32: 689–702.
36. Kuyucak S, Andersen OS, Chung SH (2001) Models of permeation in ion channels. Rep Prog Phys 64: 1427–1472.
37. Chung SH, Corry B (2005) Three computational methods for studying permeation, selectivity and dynamics in biological ion channels. Soft Matter 1: 417–427.
38. Roux B, Allen T, Bernèche S, Im W (2004) Theoretical and computational models of biological ion channels. Q Rev Biophys 37: 15–103.
39. Boda D, Henderson D, Busath DD (2001) Monte Carlo Study of the Effect of Ion and Channel Size on the Selectivity of a Model Calcium Channel. J Phys Chem B 105: 11574–11577.
40. Boda D, Fawcett WR, Henderson D, Sokolowski S (2002) Monte Carlo, density functional theory, and Poisson-Boltzmann theory study of the structure of an electrolyte near an electrode. J Chem Phys 116: 7170.
41. Gillespie D, Boda D (2008) The anomalous mole fraction effect in calcium channels: a measure of preferential selectivity. Biophys J 95: 2658–2672.
42. Gillespie D, Boda D, He Y, Apel P, Siwy ZS (2008) Synthetic nanopores as a test case for ion channel theories: the anomalous mole fraction effect without single filing. Biophys J 95: 609–619.
43. Allen TW, Andersen OS, Roux B (2004) Energetics of ion conduction through the gramicidin channel. Proc Natl Acad Sci USA 101: 117–122.
44. Wang W, Black SS, Edwards MD, Miller S, Morrison EL, et al. (2008) The structure of an open form of an E. coli mechanosensitive channel at 3.45 A resolution. Science 321: 1179–83.
45. Hummer G (2005) Position-dependent diffusion coefficients and free energies from Bayesian analysis of equilibrium and replica molecular dynamics simulations. New J Phys 7: 34.
46. Wang Z, Fried JR (2007) A hierarchical approach for predicting the transport properties of the gramicidin A channel. Soft Matter 3: 1041–1052.
47. O'Keeffe J, Cozmuta I, Bose D, Stolc V (2007) A predictive MD-NernstPlanck model for transport in alpha-hemolysin: Modeling anisotropic ion currents. Chem Phys 342: 25–32.
48. Noskov SY, Im W, Roux B (2004) Ion permeation through the alpha-hemolysin channel: theoretical studies based on Brownian dynamics and Poisson-Nernst-Plank electrodiffusion theory. Biophys J 87: 2299–2309.
49. Cozmuta I, O'Keeffe JT, Bose D, Stolc V (2005) Hybrid MD-Nernst Planck model of α-hemolysin conductance properties. Mol Simulat 31: 79–93.
50. Mamonov AB, Kurnikova MG, Coalson RD (2006) Diffusion constant of K+ inside Gramicidin A: a comparative study of four computational methods. Biophys Chem 124: 268–78.
51. Corry B, Allen TW, Kuyucak S, Chung SH (2001) Mechanisms of permeation and selectivity in calcium channels. Biophys J 80: 195–214.
52. Corry B (2006) An energy-efficient gating mechanism in the acetylcholine receptor channel suggested by molecular and brownian dynamics. Biophys J 90: 799–810.
53. Song C, Corry B (2009) Role of acetylcholine receptor domains in ion selectivity. BBABiomembranes 1788: 1466–1473.
54. Song C, Corry B (2010) Ion conduction in ligand-gated ion channels: Brownian dynamics studies of four recent crystal structures. Biophys J 98: 404–411.
55. Chung SH, Corry B (2007) Conduction properties of KcsA measured using Brownian dynamics with flexible carbonyl groups in the selectivity filter. Biophys J 93: 44–53.
56. MacKerell AD, Jr., Bashford D, Bellott M, Dunbrack RL, Jr., Evanseck JD, et al. (1998) All-atom empirical potential for molecular modelling and dynamics studies of proteins. J Phys Chem B 102: 3586–3616.
57. Hoyles M, Kuyucak S, Chung SH (1998) Solutions of Poisson's equation in channel-like geometries. Computer Phys Commun 115: 45–68.

58. Hoyles M, Kuyucak S, Chung SH (1998) Computer simulation of ion conductance in membrane channels. Phys Rev E 58: 3654–3661.
59. Chung S, Allen T, Hoyles M, Kuyucak S (1999) Permeation of ions across the potassium channel: Brownian dynamics studies. Biophys J 77: 2517–2533.
60. Corry B, O'Mara M, Chung SH (2004) Conduction Mechanisms of Chloride Ions in ClC-Type Channels. Biophys J 86: 846–860.
61. O'Mara M, Cromer B, Parker M, Chung SH (2005) Homology Model of the GABA$_A$ Receptor Examined Using Brownian Dynamics. Biophys J 88: 3286–3299.
62. Ng JA, Vora T, Krishnamurthy V, Chung SH (2008) Estimating the dielectric constant of the channel protein and pore. Eur Biophys J 37: 213–222.
63. Corry B, Chung S (2006) Mechanisms of valence selectivity in biological ion channels. Cell Mol Life Sci 63: 301–315.
64. Allen TW, Andersen OS, Roux B (2006) Molecular dynamics - potential of mean force calculations as a tool for understanding ion permeation and selectivity in narrow channels. Biophys Chem 124: 251–67.

4

Self-Rotation of Cells in an Irrotational AC E-Field in an Opto-Electrokinetics Chip

Long-Ho Chau[1], Wenfeng Liang[2], Florence Wing Ki Cheung[3], Wing Keung Liu[3]*, Wen Jung Li[1,2,4]*, Shih-Chi Chen[1], Gwo-Bin Lee[5]

1 Centre for Micro and Nano Systems, The Chinese University of Hong Kong, Hong Kong, **2** State Key Laboratory of Robotics, Shenyang Institute of Automation, Chinese Academy of Sciences, Shenyang, China, **3** School of Biomedical Sciences, Faculty of Medicine, The Chinese University of Hong Kong, Hong Kong, **4** Department of Mechanical and Biomedical Engineering, City University of Hong Kong, Hong Kong, **5** Department of Power Mechanical Engineering, National Tsing Hua University, Hsinchu, Taiwan

Abstract

The use of optical dielectrophoresis (ODEP) to manipulate microparticles and biological cells has become increasingly popular due to its tremendous flexibility in providing reconfigurable electrode patterns and flow channels. ODEP enables the parallel and free manipulation of small particles on a photoconductive surface on which light is projected, thus eliminating the need for complex electrode design and fabrication processes. In this paper, we demonstrate that mouse cells comprising melan-a cells, RAW 267.4 macrophage cells, peripheral white blood cells and lymphocytes, can be manipulated in an opto-electrokinetics (OEK) device with appropriate DEP parameters. Our OEK device generates a non-rotating electric field and exerts a localized DEP force on optical electrodes. Hitherto, we are the first group to report that among all the cells investigated, melan-a cells, lymphocytes and white blood cells were found to undergo self-rotation in the device in the presence of a DEP force. The rotational speed of the cells depended on the voltage and frequency applied and the cells' distance from the optical center. We discuss a possible mechanism for explaining this new observation of induced self-rotation based on the physical properties of cells. We believe that this rotation phenomenon can be used to identify cell type and to elucidate the dielectric and physical properties of cells.

Editor: Aristides Docoslis, Queen's University at Kingston, Canada

Funding: This project was funded by a start-up fund from the City University of Hong Kong (project no. 9610216). The project was also partially supported by The Chinese University of Hong Kong's graduate assistantship to Mr. Long-Ho Chau. The funders had no role in study design, data collection and analysis, decision to publish, or preparation of the manuscript.

Competing Interests: The authors have declared that no competing interests exist.

* E-mail: ken-liu@cuhk.edu.hk (WKL); wenjli@cityu.edu.hk (WJL)

Introduction

Single-cell manipulation plays an important role in the biomedical research fields of culturing, drug development, physiology and replication. Existing methods for manipulating a single biological cell are applied in microfluidic devices. The successful application of mechanical based techniques such as hydrodynamic flow [1–2], laminar flow control [3–4] and micromechanical filters [5] have been successfully demonstrated in cell transportation and separation by considering the size and morphology of cells. The channel fabricated while applying these techniques is exclusively designed for particles and cells of a specific size and provides them with a desired path to move along. However, the device may not perform properly if larger or smaller particles are injected into the system for transportation or separation. In many such cases, engineers must redesign the channel and micro-fabricate the device again. Electrokinetic based methods including (AC) electro-osmosis (EO) [6], electrophoresis (EP) [7], dielectrophoresis (DEP) [8–10] and isoelectrophoresis (IEP) [11] can precisely transport bulky cells to a desired position according to their size and dielectric properties. In all such methods, chips with patterned electrodes are required to guide the cells in the specified direction. These methods, however, cannot be employed to manipulate a single cell without an appropriately

designed channel. They also require complex fabrication processes such as metal deposition, lithography and channel fabrication. Moreover, although they guide cells along a specific track, they cannot manipulate them into other positions.

An enhanced DEP-based technology called optical dielectrophoresis (ODEP) has recently been developed to manipulate cells and particles using focused light and an AC E-field [12–14]. Under ODEP, an opto-electrokinetics (OEK) device is used to perform cell/particle manipulation. The device consists of a smooth photoconductive layer and a conductive chip. A localized DEP force is generated across the OEK device when an optical image emitted from a light source, such as a laser or projector, is projected onto the photoconductive layer. Because the position of the DEP force varies with the location of the image, the optical image acts as a cell manipulation controller. Hence, cells can move in any direction within a void volume with sufficient DEP force.

ODEP has been used to successfully manipulate nano-/micro-particles/biological cells. The literature has shown that the general motion of particles in an OEK device is purely translational [13–15], with the E-field generated being irrotational. However, according to YL Liang et al. (2011), yeast cells undergo translation and rotation at the same time from the dark-field region to the image spot [16]. The same study also demonstrated Ramos cells generating self-rotation in an OEK device with the aid of a well. Both rotation observations were performed in a rotating E-field. In this paper, we present our recent findings on cell rotation among some of the cells examined in an irrotational E-field using an OEK device. White blood cells (WBCs) and melan-a cells were observed to undergo self-rotation near the image spot at an appropriate applied frequency and voltage. Neither physical barriers nor other mechanical barriers, such as a well, were applied to assist the rotation of the cells.

Theory

An ODEP system consists of optics and DEP components. Optics, i.e. projected light or images, control the position of the DEP force in an OEK device. When there is no optical image (light), no AC E-field is generated across the device. When light is projected onto a particular area of the photoconductive layer, a localized DEP field is generated across the illuminated area and the conductive layer. The average time it takes for a DEP force to act on a round cell in an OEK device is expressed in Equation (1) with the assumption that the cell suspended in the medium is perfectly spherical [17]:

$$F_{DEP} = \pi \varepsilon_m R^3 \, \mathrm{Re}[K(\omega)] \nabla |E^2| \qquad (1)$$

where ε_m is the dielectric permittivity of the medium, R is the radius of the particle, $K(\omega)$ is the Clausius-Mossotti (CM) factor, ω is the applied angular frequency across the medium and E is the electric field. An animal cell consists of a nucleus, cytoplasm and other organelles and is surrounded by a membrane. It can be considered a single-shell model rather than a homogeneous sphere, as shown in Figure 1. The CM factor is therefore expressed as [18]:

$$K(\omega) = - \left[\frac{\omega^2(\tau_s \tau_m + \tau_p \tau'_m) + j\omega(\tau'_m - \tau_m - \tau_s) - 1}{\omega^2(2\tau_s \tau_m + \tau_p \tau'_m) - j\omega(\tau'_m + \tau_m + 2\tau_s) - 2} \right] \qquad (2)$$

where $\tau'_m = c_s R / \sigma_m$, $\tau_m = \varepsilon_m / \sigma_m$, $\tau_p = \varepsilon_p / \sigma_p$ and $\tau_s = \varepsilon_s / \sigma_s$. $\mathrm{Re}[K(\omega)]$ ranges from -0.5 to 1 and its sign determines the direction of the DEP force. If $\mathrm{Re}[K(\omega)] > 0$, a positive DEP force is generated in such a way that the particles/cells tend to move toward a strong E-field region. If $\mathrm{Re}[K(\omega)] < 0$, a negative DEP is generated, and the particles are thus pushed toward a weak E-field region.

Several studies in the literature have demonstrated that particular cells can rotate in a two-phase-shifted DEP E-field and in a travelling wave DEP due to electrorotation. The real part of Equation (2) determines the direction of the DEP force while the imaginary part determines the electrorotation of the particles in the rotating E-field. DEP theory states that the action of an externally applied E-field on a polarizable particle results in the formation of an induced dipole moment. When a dipole sits in a uniform E-field, each charge on the dipole is parallel to the field; hence, it experiences a torque. If the direction of the field vector changes, the induced dipole moment vector must realign itself with the E-field vector, which causes particle rotation to occur. The torque of the particle in a rotating E-field [17] is expressed as

$$\Gamma_{ROT} = -4\pi \varepsilon_m R^3 \, \mathrm{Im}[K(\omega)] |E^2| \qquad (3)$$

Electrorotation occurs if the E-field has a spatially dependent phase, otherwise $\mathrm{Im}[K(\omega)] = 0$. Dielectrophoresis and electrorotation are tools commonly applied to measure the dielectric

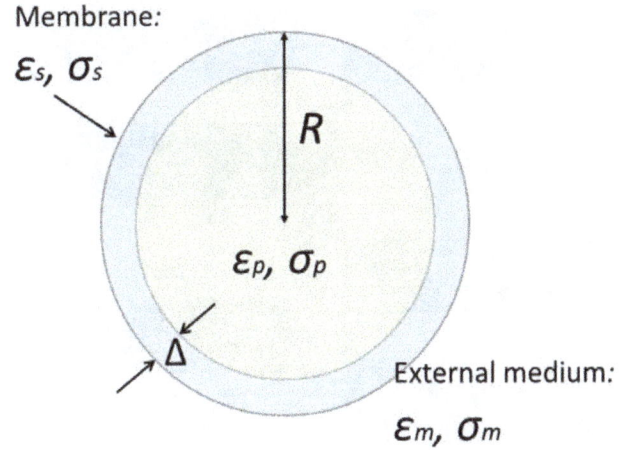

Figure 1. The single-shell model. A spherical particle with radius R, permittivity ε_p and conductivity σ_p, which is covered by a uniform layer of thickness $\Delta << R$, permittivity ε_s and conductivity $\sigma_s < \sigma_p$, and surrounded by a solution of permittivity ε_m and conductivity σ_m.

properties of biological cells [19]. The data obtained can be used to identify different cell types or to form a cell library. Scientists and engineers often use these electrical parameters to manipulate cells, such as in cell separation and transportation using DEP or ODEP.

In this study, our research group observed the self-induced rotation of melan-a, lymphocytes and WBCs in an ODEP system. The cells could rotate during stationary and translational motion in a non-uniform E-field. In a notable finding, the AC E-field generated in the OEK device did not rotate to cause electrorotation; hence, no torque was generated according to Equation (3).

Materials and Methods

Instrumentation

The ODEP system employed in this study is shown in Figure 2 and Figure S1. The OEK chip was placed on a 2D stage integrated with an optical microscope (Nikon ECLIPSE TE2000-U). Light was projected from a commercial projector (DELL 1510X) and passed through the condensing lens (Leica X/0.15), indium tin oxide (ITO) glass and a liquid medium before being focused on the hydrogenated amorphous silicon (a-Si:H) surface. Image patterns of light were controlled by a computer. Cell motion was recorded using a high-speed camera (PCO 1200S) with a 40× objective lens. The top and bottom ITO glass chips were connected to an AC signal generator and a CRO. In each experiment, a drop of cell solution was injected into the OEK device and there was no net fluid flow in the micro-chamber. The frequencies applied for cell manipulation ranged from 10 kHz to 3 MHz and the voltages applied were from 0 V to 20 V, peak to peak. All experiments were performed at room temperature.

Chemical and cell preparation

Immortalized mouse melanocyte, melan-a pigment cells (a gift from Prof. D.C. Bennett at St. George's Hospital Medical School in London, UK) [20] and non-pigment cells, including the mouse RAW 264.7 macrophage cell line (ATCC TIB-71), mouse peripheral white blood cells and lymphocytes were suspended in 0.2 M sucrose in deionized water. The measured conductivity of the sucrose was 0.37 mSm^{-1}. The experiments were performed immediately after the fresh cells were prepared.

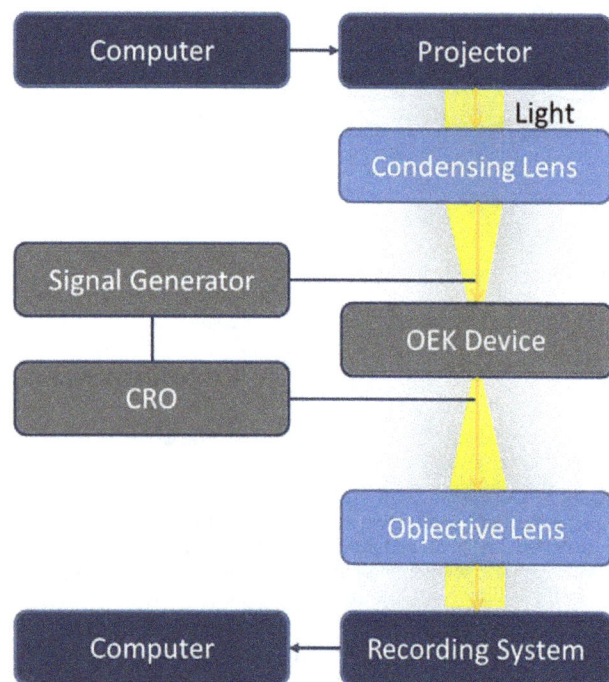

Figure 2. Illustration of an ODEP system. Experimental setup for manipulating cells with opto-electrokinetic device.

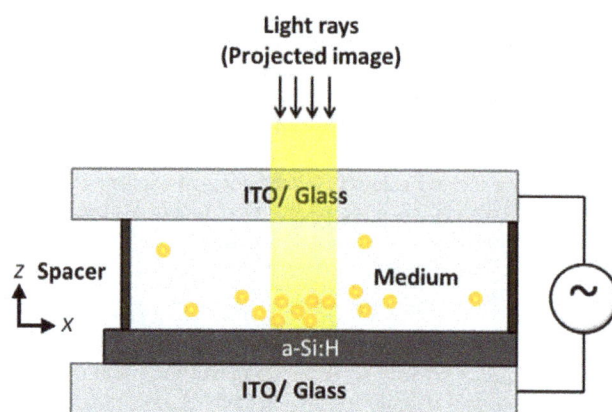

Figure 3. The opto-electrokinetics device (OEK). An illustration of the OEK used to manipulate biological cells. The patterned optical image is focused by a condenser lens and projected onto the hydrogenated amorphous silicon surface.

Fabrication of opto-electrokinetics device

a) Fabrication of OEK photoconductive layer. The fabrication of the glass-ITO-a-Si:H structure employed in this study was described by Yen-Heng Lin et al. (2010). [14]A 0.3 µm a-Si:H thin film was coated on the ITO layer. It was further processed by etching part of the a-Si:H for electrical connection. A 5 mm×8 mm area of a-Si:H was patterned through standard photolithography and dry-etching using the Oxford Plasma Lab 80 Plasma Etching System with 2% oxygen, 12.5% CF_4 gas, a 30 mTorr etching chamber and 6-minute plasma exposure. The chip was then rinsed and cleaned with acetone and DI water before being dried by nitrogen gas.

b) Fabrication of OEK device. The OEK device was assembled by combining an amorphous silicon a-Si-coated glass and an ITO-coated glass with the SU-8 intermediate layer as shown in Figure 3. The SU-8 layer (SU-8 2035, MicroChem, Newton, MA, USA) served as a micro-fluidic channel for cell-medium transportation and cell separation. The ITO glass was prepared by sputtering 600 Å ITO on clean glass for 10 minutes. A few through-holes were then created in the ITO glass to serve as channel ports. The fabrication process for the SU-8 micro-fluidic channel consisted of several steps. First, an uncross-linked SU-8 negative photoresist was spun on an a-Si glass at 500 rpm for

10 seconds, 1000 rpm for 60 minutes and 5000 rpm for 3 seconds to obtain a thickness of 50 µm, which was followed by a two-step soft baking process on a hotplate at 65°C for 3 minutes, then at 95°C for 6 minutes. The soft-baked SU-8 was exposed to UV light using a Karl Suss MA4 mask aligner for 70 seconds. Subsequently, the exposed SU-8 layer was hard-baked following a two-step approach: it was initially baked at 65°C for 1 minute, then at 95°C for 5 minutes. The resulting SU-8 layer was cooled to room temperature and kept for 4 hours before being subjected to a developing process to strengthen the adhesion of the SU-8 and a-Si layers. In the final step, the cross-linked SU-8 was developed, rinsed with isopropanol (IPA) and dried by nitrogen gas.

The a-Si glass/SU-8 structure, thus prepared, was sealed to the ITO glass by employing the SU-8 imprinting method shown in Figure S2. First, a cleaned transparency was spin-coated with a 10 µm SU-8 2010 layer at 500 rpm for 10 seconds and 3000 rpm for 30 seconds. The patterned SU-8 structure was then imprinted on the uncross-linked SU-8 coated transparency to allow the bulge surface to make good contact with the SU-8. The transparency was then detached from the SU-8 structure. The structured SU-8/a-Si glass and the ITO glass were then aligned and brought into contact. They were then clamped tightly and baked in a 150°C oven for 30 minutes.

Results and Discussion

Investigation of cell properties and translation behavior under ODEP

a) WBCs. Peripheral white blood cells (leukocytes) collected from male adult C57 mice consist of different types of cells including lymphocytes, macrophages and polymorphonucelar cells, which can be categorized into two main types: granulocytes

Table 1. The response of white blood cell samples in the DEP field.

Applied Frequency (Hz) in 20 Vpp		<50 kHz	100 kHz	200 kHz <
DEP Force	WBCs	Negative (100%)	Negative (50%) & Positive (50%)	Positive (90%)
	h-WBCs	Negative (100%)	Negative (50%) & Positive (50%)	Positive (90%)
	s-WBCs	Negative (100%)	Negative (70%) & Positive (30%)	Negative (50%) & Positive (50%)

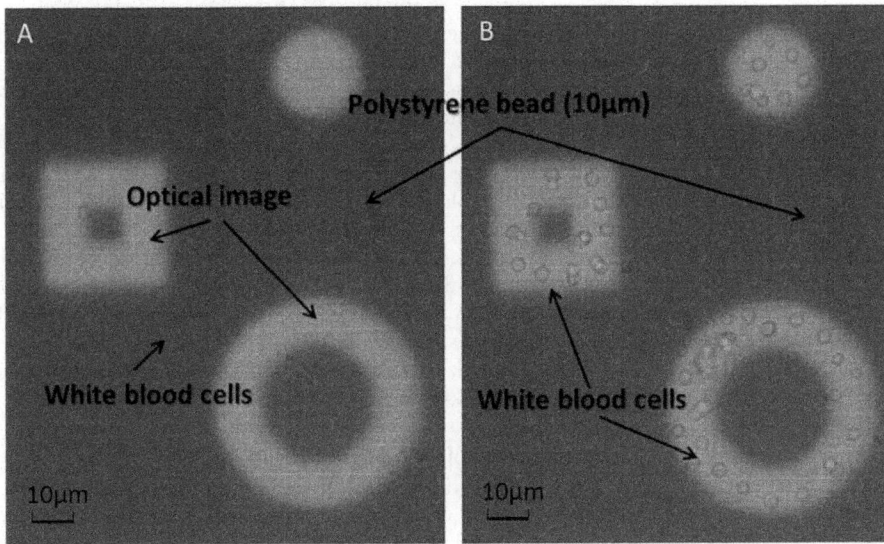

Figure 4. DEP force acting on the peripheral WBCs. Three optical images were projected onto an a-Si:H surface. A 10 μm diameter micro polystyrene bead acted as a control. (A) Peripheral WBCs experienced a negative DEP force at the applied frequency of 50 kHz and a voltage of 10 Vpp. They lay in the dark-field region and between the square and the ring image. (B) When the frequency was 200 kHz, 10 Vpp, a positive DEP force caused the cells to shift into the image. The polystyrene beads stayed in the same position in both cases because they only experienced a negative DEP force in 0.2 M sucrose solution.

and agranulocytes. Granulocytes are white blood cells that contain granules in their cytoplasm; neutrophils, basophils and eosinophils belong to this group. Agranulocytes are white blood cells that do not contain granules, such as monocytes and lymphocytes. Here, we discuss the rotation and translation behavior of three kinds of cells under the ODEP system: white blood cells, lymphocytes and macrophages.

Three primary mouse blood samples were collected from the abdominal aorta in saline solution, with or without heparin, and from the spleens of C57 mice. White blood cells were obtained after the lysis of red blood cells in ammonium chloride solution. WBC in saline without heparin (WBCs), with heparin (h-WBCs) and from the spleen (s-WBCs) were analyzed in an OEK device. Frequencies ranging from 5 kHz to 300 kHz at 20 Vpp were applied to the OEK chip and the results are shown in Movie S2 and summarized in Table 1. All samples experienced a negative DEP force when the frequency was less than 50 kHz. The cells moved away from the virtual image when a negative DEP force was applied, as shown in Figure 4. When the frequency was

increased to 100 kHz, 50% of WBCs, 50% of h-WBCs and 30% of s-WBCs experienced a positive DEP force. As such, they were attracted from the dark-field region to the bright-field area (virtual image). As the cells translated from the dark-field region toward the image spot, they also generated rotational motion. The rotation axis was perpendicular to the E-field and the cells rotated toward the image. When the applied frequency reached 200 kHz, most of the WBCs and h-WBCs invaded the image spot by means of a positive DEP force. However, only half of the s-WBCs experienced a positive DEP force at 200 kHz and above, which indicated that half of the s-WBCs did not have a positive DEP frequency spectrum in the studied medium.

b) Macrophages. Mouse macrophages from the RAW 264.7 cell line have previously been investigated by our group [21]. Their diameter is approximately 10 μm and they can only be manipulated by a negative DEP force in the 50 kHz to 150 kHz frequency range. When 10 Vpp and 150 kHz was applied to an OEK device, the macrophage left the image at a speed of 2 μm/s.

Table 2. The response of melan-a cells and macrophages to the DEP field.

Applied Frequency (Hz) in 20 Vpp		<25 k	25 k–50 k	50 k–100 k	100 k–300 k	>300 k
DEP Force	Melan-a	Positive DEP	Positive DEP	Positive DEP	Positive DEP	No Response
		All cells were strongly attracted to the image and then burst.	80% of the cells were attracted to the image. Other cells stayed stationary near the image.	Some cells were attracted to the image and some cells stayed stationary near the image.	70% of the cells moved towards the image and then stayed stationary near the image.	No movement of the cells was observed.
	Macrophage	No Response	No Response	Weak negative DEP	Negative DEP	No Response
		No movement of the cells was observed.	No movement of the cells was observed.	Cells started to repel from the image slowly.	Cells moved away from the image at the speed of 2 μm/s at 150 kHz.	No movement of the cells was observed.

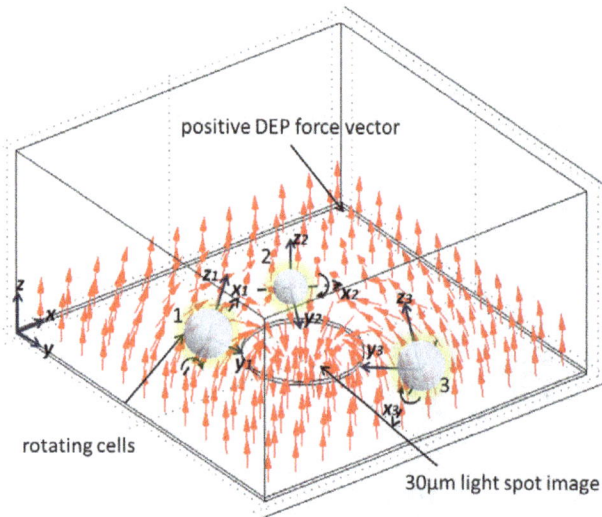

Figure 5. Illustration of cell rotation in an OEK. The cells in the dark-field region rotated toward the 30 µm spot image. Their z-axes were normal to the positive DEP force vectors. The axis of rotation was at the x-axis perpendicular to the E-field.

We found that the macrophages did not self-rotate at any applied frequency or voltage.

c) Lymphocytes. Lymphocytes have a large nucleus that usually occupies 80% of the whole cell [22]. They have a spherical shape and range in size from 6 µm to 7 µm. We tested frequencies ranging from 25 kHz to 3 MHz. The lymphocytes experienced a negative DEP force below 40 kHz, causing them to repel from the optical image and rotate in the dark-field region. When the frequency was increased to between 50 kHz and 800 kHz, the cells were subjected to a positive DEP force and tended to move toward the optical image. However, they eventually stopped some distance from the image and underwent self-rotation in that position. For example, the cells shifted from 150 µm at the image spot to 65 µm away from the image spot if the frequency changed from 40 kHz to 60 kHz. We believe that there was a 'repulsive force' around the image boundary which is possibly generated by counter-rotating electrothermal flow structures around the image. This 'repulsive force' is apparently stronger than the attractive DEP force acting on the cells when the cells are close to the image, which causes the cells from further translation in to the image. At 140 kHz, the cells experienced the strongest DEP force; some cells moved inside the spot, whereas others stayed 30 µm from the spot. For frequencies above 800 kHz, the cells did not translate and/or rotate in any direction because there was insufficient DEP force acting on them.

d) Melan-a. Melan-a is an immortalized mouse melanocyte line. Melanocytes are melanin-producing cells located in the

bottom layer (the stratum basale) of the skin's epidermis. Melanin is commonly found in most organisms. It determines the color of the skin, hair and iris. Melan-a cells vary in size from 8 µm to 10 µm and appear to have a sphere-like shape when suspended in a medium, although their membrane is not smooth. Our group has previously examined their translational motion under ODEP [21]. The responses of melan-a cells were found to be very similar to those of lymphocytes in that some melan-a cells did not shift into the image spot when a positive DEP force was applied to them, as shown in Movie S1. These cells stayed at a particular position around the image spot and underwent self-rotation; for instance, at 20 Vpp and 100 kHz, they rotated at 90 rpm and remained at a distance of 10 µm to 15 µm from the image spot. Table 2 summarizes the responses of the melan-a and macrophage cells in the OEK device.

Investigation of cellular self-rotation behavior

As previously described, lymphocytes and melan-a cells in the dark-field region were found to rotate near the image spot under both positive and negative DEP forces. The rotation axes for both cells were perpendicular to the applied E-field and the direction of rotation was toward the image spot, as shown in Figure 5. In this section, we analyze the relationship of cell rotational speed to applied voltage and frequency.

a) Rotation Measurement. Cell motion was captured using a CCD camera with a fixed frame rate of 64.5/s. The images captured were processed by resizing the image and focusing on the cells as shown in Figure 6. The modified cell images were then imported into rotation tracking software developed by our group. The algorithm used for the tracking process first identifies the feature points of each imported cell image before calculating the length differences among the images in comparison with the first imported image and generating a mapping similarity index for the cell ranging from 0 to 1 (with 1 meaning identical and 0 meaning not correlated). For example, the first imported image was assigned an index of 1 as a reference, the second image was calculated a value of 0.9973 because its orientation differs from that of the first image, the third image was calculated a value of 0.9964 and for a complete cell rotation loop, its index was 0.9963, compared with the first image. We can eventually identify the rotation period by looking at the similarity index between sequences and calculating the length of the frames.

b) Rotational Speed vs. Voltage. Figure 7 shows the rotational speed of melan-a cells at applied voltages from 0 Vpp to 20 Vpp at 40 kHz. The melan-a cells located 65 µm from the light spot were recorded. The cells were originally at rest from 0 Vpp to 1 Vpp. They started self-rotating at 15 rpm at 2 Vpp, and increased their spinning speed non-linearly as the applied voltage rose. At 20 Vpp, the melan-a cells rotated at 400 rpm. In the same plot, the lymphocytes started to rotate at 2 Vpp at a speed of 7.4 rpm. The rotational speed was non-linearly proportional to the applied voltage. At 20 Vpp, the rotational speed of the cells was as fast as 363 rpm.

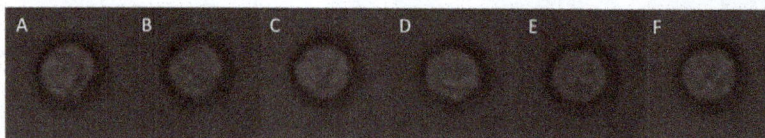

Figure 6. Rotation of melan-a. Time lapse for a melan-a cell completing one revolution with the applied frequency of 40 kHz at 6 Vpp as recorded by a CCD camera after grey-scale treatment. The similarity coefficients were 1 (A), 0.9973 (B), 0.9964 (C), 0.99656 (D), 0.9951(E) and 0.9963(F). Images were taken 0.155 seconds apart. The rotational speed was recorded as 65 rpm.

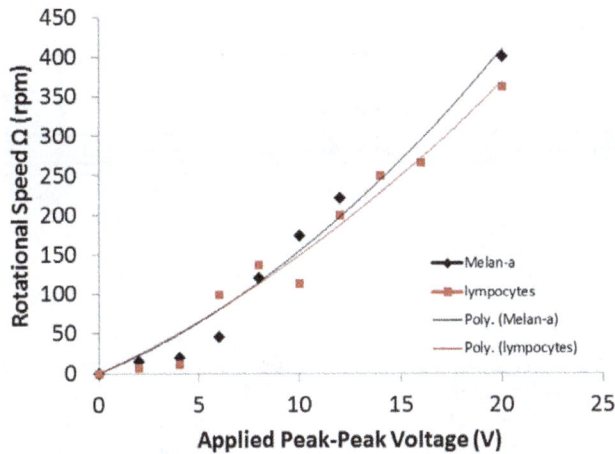

Figure 7. Rotational speed of the melan-a and lymphocytes in 0.2 M sucrose solution versus applied voltage from 0 Vpp to 20 Vpp at 40 kHz.

These results show that the rotational speed of specific cells depended on the AC voltage applied. The plots indicate a quadratic relationship between rotational speed and applied voltage, with the initial speed being zero when no voltage is applied. The relationship was found to be similar to that reported in traditional DEP setups [19,23–24].

c) **Rotational Speed vs. Frequency.** Figure 8 shows the rotational speed of melan-a cells versus applied frequency at 20 Vpp. The cells rotated at applied frequencies between 25 kHz and 800 kHz. At frequencies under 25 kHz, the melan-a cells moved swiftly into the image spot and eventually burst. If the applied frequency was too high, such as over 800 kHz, the cells neither moved toward nor were repelled by the spot, and they did not rotate in either the bright-field or the dark-field regions. The

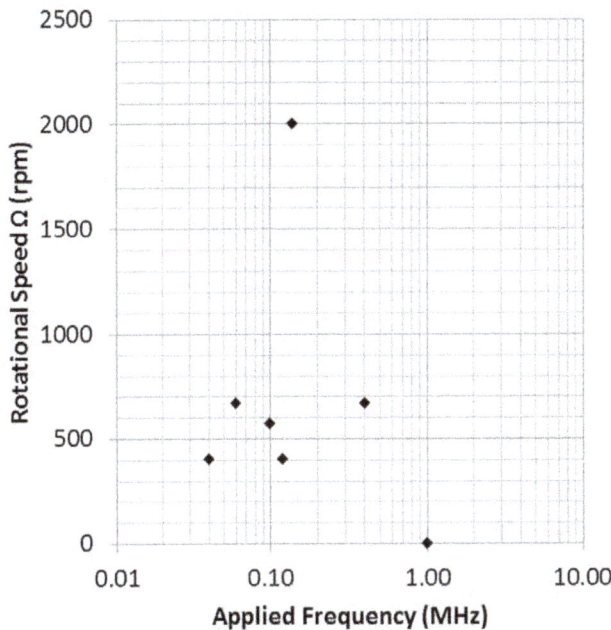

Figure 8. Rotational speed of the melan-a in 0.2 M sucrose solution versus applied frequency from 40 kHz to 1 MHz at 20 Vpp.

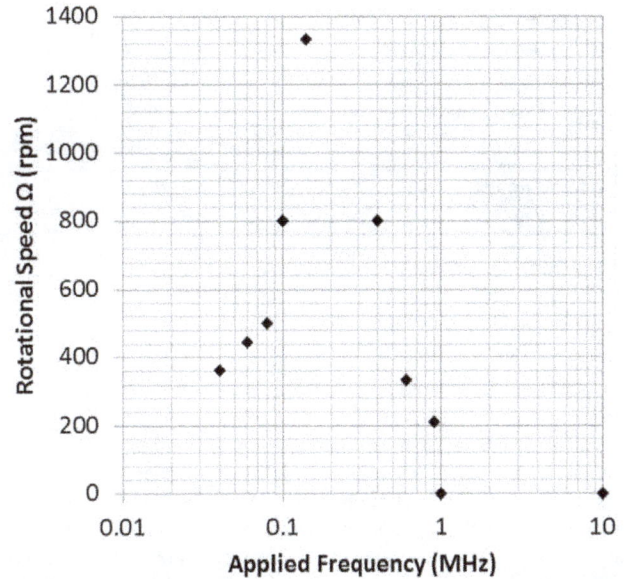

Figure 9. Rotational speed of the lymphocytes in 0.2 M sucrose solution versus applied frequency from 40 kHz to 1 MHz at 20 Vpp.

plot in Figure 8 shows that the fastest rotational speed of 2000 rpm was achieved when the applied frequency was 120 kHz.

The lymphocytes also yielded results similar to those of the melan-a cells, as Figure 9 reveals. Frequencies over 1 MHz did not drive the cells to undergo translation and rotation. The cells started to rotate at a frequency of 800 kHz and reached the maximum speed of 1350 rpm at 120 kHz. The rotational speed of the cells decreased at frequencies from 120 kHz to 40 kHz. Frequencies below 20 kHz, which cause the heating effect, generated bubbles on the a-Si:H surface.

Figures 8 and 9 show that rotational speed is also associated with frequency. Both types of cells attained their highest rotational speed at a specific frequency and had no response in high frequency ranges. It is noticeable that the rotational speed of cells versus applied frequency is found to be very similar to the typical electrorotation spectrum [8,25]. Hence, it is believed that the spinning effect is associated with DEP, but in an irrotational AC E-field.

d) **Cell rotation inside the bright-field region.** When melan-a cells entered the image spot, they did not rotate as they had in our previous research [21]. However, we observed the rotation of white blood cells and lymphocytes in the bright-field region. Table 3 shows the rotation analysis of WBCs, h-WBCs and s-WBCs. In the negative DEP region, all of the sample cells were in the dark-field region, with 95% of WBCs and h-WBCs undergoing rotation and 80% of s-WBCs rotating. The difference in the percentage of cells rotating indicates that the composition of the WBCs extracted from the blood vessels and spleens of mice were not the same. As the frequency increased from 50 kHz to 200 kHz, all of the WBCs and h-WBCs and 50% of the s-WBCs moved into the bright-field region under the influence of a positive DEP force. We observed that some cells in the image spot rotated. However, the proportion of each type of cell in the image spot differed, with 90% of WBCs, 90% of h-WBCs and 30% of S-WBCs rotating in the light image.

Lymphocytes (b-/t- and NK-) also rotated in the bright-field region. Three lymphocytes were initially trapped in the image spot

Table 3. Rotation behavior of white blood cell samples in the DEP field.

Applied Frequency (Hz) in 20 Vpp		<50 kHz		200 kHz<	
		Dark-field Region	**Bright-field Region**	**Dark-field Region**	**Bright-field Region**
Induced rotation	WBCs	95% rotate	N/A	95% rotate	90% rotate
	h-WBCs	95% rotate	N/A	95% rotate	90% rotate
	s-WBCs	80% rotate	N/A	90% rotate	30% rotate

and one of them rotated at a constant speed. The rotating lymphocyte was isolated from the other cells and was moved to the cell-free area. A light spot with the same diameter and applied frequency and voltage was projected onto the rotating cell, but it did not rotate at all, even when the frequency was altered. The experiment was repeated and similar results were observed, indicating that the rotation of cells in the bright-field region was affected by the neighboring cells. Holzapfel et al. (1982) reported similar findings, specifically that two adjacent protoplast cells in the DEP field caused each other to rotate [24]. The author stated that cells can rotate only through cell-cell interaction or the local disturbance of the E-field. However, because cell-cell interaction can lead to a local disturbance in the E-field, the dipole moment of one cell interacts with the dipole moment of another cell to cause rotation.

Despite the above findings, a single cell can rotate alone in the dark-field region. The physical mechanism driving cell rotation in ODEP is still unclear. We hypothesize that rotation is due to the uneven distribution of mass within cells, such as the large nucleus incline in a lymphocyte or the uneven distribution of melanin in melan-a cells. When a dipole moment is created on the cell, it aligns the cell parallel to the E-field. If the center of gravity (C.G.) of the cell is not at its geometric center, the dipole moment may drive the cell to rotate continuously with sufficient DEP force. Hence, the physical properties of lymphocytes and melan-a cells may cause them to undergo self-rotation.

These explicit results show the induced rotation of the cells influenced by optically-induced DEP. With further experiments to explore how cells' rotational characteristics could be affected by their volume, morphology, organelles, and electrical properties, this observed self-rotation phenomenon could potentially be used to identify and separate different cell types. For example, we foresee that an automated system could be developed for the selection of cells in different cell-cycle phases. There are four phases in a cell cycle– G_1 gap phase, synthesis (S) phase, G_2 gap phase, and the mitosis (M) phase, and they are correlated to different dielectrics, such as changes in relative electrical permittivity [26], and physical properties, such as volume and size [8]. Therefore, based on our current results, it is entirely possible to segregate cells from the four cell-cycle phases by ODEP, and the segregation process could be automated by implementing computer-vision algorithms for real-time cell-rotation motion analysis and differentiation.

Conclusion

In this paper, we describe the kinetics of WBCs, lymphocytes, macrophages and melan-a cells in an OEK device subjected to different AC voltages and frequencies.

The E-field generated in the OEK device was non-linear and irrotational. However, most of the WBCs and melan-a cells

rotated. We hypothesize that the self-rotation of cells is due to the uneven distribution of their mass, such that once a dipole moment is created on a cell, it aligns itself to move parallel to the E-field. However, because a cell's center of mass is not located precisely at its center, the cell will continue to undergo rotation. This phenomenon is similar to that of an AC motor.

Much experimental work remains to be performed to derive a complete theoretical model for cell rotation in ODEP. Nevertheless, the cell rotation phenomenon observed here has great potential for use in constructing biomarkers for different cells. In addition to its application to cell identification, it may also be used to characterize cells in terms of their dielectric and mechanical properties, given that our results show rotational speed and translation motion to be inextricably linked to applied voltage and frequency.

Supporting Information

Figure S1   The ODEP system. The actual ODEP system setup used to manipulate cells in our experiments.

Figure S2   Fabrication procedure for the OEK device.

Movie S1   Induced rotation of melan-a and lymphocytes near the image spot. A video recording showing the rotation of melan-a cells and lymphocytes around the 30 μm light spot in an OEK with the applied voltage of 20 Vpp and the applied frequency of 40 kHz.

Movie S2   Induced rotation of white blood cells. A video recording showing the rotation and manipulation of white blood cells in an OEK with the applied voltage of 20 Vpp and the applied frequency from 50 kHz to 200 kHz. Cells at 50 kHz were under negative DEP force and at 200 kHz were trapped in the optical image by positive DEP force. Cells were observed rotation in both bright and dark regions.

Acknowledgments

The authors would like to thank Mr. Tak-kit LAU and Mr. Kai-wun LIN of the Networked Sensors and Robotics Laboratory, The Chinese University of Hong Kong (CUHK) for their help with the visual system.

Author Contributions

Conceived and designed the experiments: WJL L-HC. Performed the experiments: L-HC. Analyzed the data: L-HC WL. Contributed reagents/materials/analysis tools: FWKC WKL G-BL S-CC. Wrote the paper: L-HC.

References

1. Bang H, Chung C, Kim J, Kim S, Chung S, et al. (2006) Microfabricated fluorescence-activated cell sorter through hydrodynamic flow manipulation. Microsystem Technologies 12: 746–753.

2. Davis JA, Inglis DW, Morton KJ, Lawrence DA, Huang LR, et al. (2006) Deterministic hydrodynamics: Taking blood apart. PNAS 103: 14779–14784.

3. Roman G, Chen Y, Viberg P, Culbertson A, Culbertson CT (2007) Single-cell manipulation and analysis using microfluidic devices. Analytical and Bioanalytical Chemistry 387: 9–12.

4. Wu Z, Hjort K, Wicher G, Fex Svenningsen A (2008) Microfluidic high viability neural cell separation using viscoelastically tuned hydrodynamic spreading. Biomed Microdevices 10: 631–638.

5. Wu HW, Lin XZ, Hwang SM, Lee GB (2009) A microfluidic device for separation of amniotic fluid mesenchymal stem cells utilizing louver-array structures. Biomedical Microdevices 11: 1297–1307.

6. Wu J, Ben Y, Battigelli D, Chang HC (2005) Long-range AC electroosmotic trapping and detection of bioparticles. Ind Eng Chem Res 44: 2815–2822.

7. Minerick AR, Ostafin AE, Chang HC (2002) Electrokinetic transport of red blood cells in microcapillaries. Electrophoresis 23: 2165–2173.

8. Kim U, Shu CW, Dane KY, Daugherty PS, Wang JYJ, et al. (2007) Selection of mammalian cells based on their cell-cycle phase using dielectrophoresis. PNAS 104: 20708–20712.

9. Park K, Suk HJ, Akin D, Bashir R (2009) Dielectrophoresis-based cell manipulation using electrodes on a reusable printed circuit board. Lab on a Chip 9: 2224–2229.

10. Chrimes A, Khoshmanesh K, Stoddart P, Kayani A, Mitchell A, et al. (2012) Active control of silver nanoparticles spacing using dielectrophoresis for surface-enhanced Raman scattering', Anal Chem 84: 4029–4035.

11. Vahey MD, and Voldman J (2009) High-throughput cell and particle characterization using iso-dielectric separation. Anal Chem 81: 2446–2455.

12. Chiou PY, Wong W, Liao JC, Wu MC (2004) Cell addressing and trapping using novel optoelectronic tweezers. 17th IEEE International Conference on Micro Electro Mechanical Systems: 21–24.

13. Liang W, Wang S, Qu Y, Dong Z, Lee GB, et al. (2011) An equivalent electrical model for numerical analyses of ODEP manipulation. IEEE International Conference on Nano/Micro Engineered and Molecular Systems (NEMS): 825–830.

14. Lin YH, Lee GB (2010) An integrated cell counting and continuous cell lysis device using optically induced electric field. Sensors and Actuators B: Chemical 145: 854–860.

15. Lin YH, Lin WY, Lee GB (2009) Image-driven cell manipulation. IEEE Nanotechnology Magazine 3: 6–11.

16. Liang YL, Huang YP, Lu YS, Hou MT, Yeh JA (2010) Cell rotation using optoelectronic tweezers. Biomicrofluidics 4: 043003.

17. Morgan H, Green NG (2002) AC Electrokinetics: Colloids and nanoparticles. 1st ed. Research Studies Pr.

18. Li M, Qu Y, Dong Z, Wang Y, Li WJ (2008) Limitations of Au particle nanoassembly using dielectrophoretic force: A parametric experimental and theoretical study. IEEE Transactions on Nanotechnology: 477–479.

19. Gimsa J, Marszalek P, Loewe U, Tsong TY (1991) Dielectrophoresis and electrorotation of neurospora slime and murine myeloma cells. Biophysical Journal 60: 749–760.

20. Bennett DC, Cooper PJ, Hart IR (1987) A line of non-tumorigenic mouse melanocytes, syngeneic with the B16 melanoma and requiring a tumour promoter for growth. Int J Cancer. 39(3): 414–418.

21. Chau LH, Ouyang M, Liang W, Lee GB, Li WJ, et al. (2012) Inducing self-rotation of Melan-a cells by ODEP. IEEE International Conference on Nano/Micro Engineered and Molecular Systems: 195–199.

22. Loiko VA, Ruban GI, Gritsai OA, Berdnik VV, Goncharova NV (2007) Mononuclear cells morphology for cells discrimination by the angular structure of scattered light. 10th Conference on Electromagnetic and Light Scattering by Non-spherical Particles: 105–108.

23. Ouyang M, Zhang G, Li WJ, Liu WK (2011) Self-induced rotation of pigmented cells by dielectrophoretic force field. IEEE International Conference on Robotics and Biomimetics (ROBIO): 1397–1402.

24. Holzapfel C, Vienken J, Zimmermann U (1982) Rotation of cells in an alternating electric field: Theory and experimental proof. Journal of Membrane Biology 67: 13–26.

25. Huang Y, Holzel R, Pethig R Wang XB (1992) Differences in the AC electrodynamics of viable and non-viable yeast cells determined through combined dielectrophoresis and electrorotation studies. Phys Med Biol 37: 1499.

26. Asami K, Takahashi K, Shirahige K (2000) Progression of cell cycle monitored by dielectric spectroscopy and flow-cytometric analysis of DNA content. Yeast 16: 1359–1363.

5

Long-Distance Axial Trapping with Focused Annular Laser Beams

Ming Lei[9], Ze Li[9], Shaohui Yan, Baoli Yao*, Dan Dan, Yujiao Qi, Jia Qian, Yanlong Yang, Peng Gao, Tong Ye

State Key Laboratory of Transient Optics and Photonics, Xi'an Institute of Optics and Precision Mechanics, Chinese Academy of Sciences, Xi'an, China

Abstract

Focusing an annular laser beam can improve the axial trapping efficiency due to the reduction of the scattering force, which enables the use of a lower numerical aperture (NA) objective lens with a long working distance to trap particles in deeper aqueous medium. In this paper, we present an axicon-to-axicon scheme for producing parallel annular beams with the advantages of higher efficiency compared with the obstructed beam approach. The validity of the scheme is verified by the observation of a stable trapping of silica microspheres with relatively low NA microscope objective lenses (NA = 0.6 and 0.45), and the axial trapping depth of 5 mm is demonstrated in experiment.

Editor: Christof Markus Aegerter, University of Zurich, Switzerland

Funding: This research is supported by the National Basic Research Program (973 Program) of China under Grant No. 2012CB921900, and the Natural Science Foundation of China (NSFC) under Grant Nos. 61275193, 61205123. The funders had no role in study design, data collection and analysis, decision to publish, or preparation of the manuscript.

Competing Interests: The authors have declared that no competing interests exist.

* E-mail: yaobl@opt.ac.cn

9 These authors contributed equally to this work.

Introduction

Since its first demonstration in 1986 by Ashkin *et al.* [1], optical tweezers has been serving as a powerful tool for microscopic trapping and manipulating, providing a stimulus to many research fields, such as in physics [2], biology [3] and colloid [4].

In an optical tweezers, stable trapping requires that the gradient force overcomes the scattering force, the former of which depends on the gradient of the intensity of the focused fields and the latter increases with increasing energy flow. Generally, stably axial trapping is more difficult than lateral trapping because of the intensively axial scattering force exerted by the axial energy flow in the focal region. As a result, the reduction of the axial scattering force is a key factor in stably axial trapping. Different approaches to improving axial trapping efficiency have been demonstrated to date. For larger particles, the use of higher-order Laguerre-Gaussian (LG) beam modes can improve axial trapping efficiency [5,6]. But it should be noted that, although the LG beam is a hollow beam in intensity shape, it has a spiral phase distribution, which makes it have very different focusing property rather than other hollow beams with homogeneous phase distribution. Raktim Dasgupta et al. [7] utilized LG_{01} mode to trap silica microspheres at 200 μm axial distance. Rodrigo et al. [8,9] demonstrated 3D trapping with long working distance via counter propagating light fields. Recently, radially polarized beam is shown to produce a vanishing axial Poynting vector component on the optical axis in the focal region, leading to a higher axial trapping efficiency in comparison with the linearly polarized beam or circularly polarized beam [10–14]. Double-ring radially polarized beams can further make the enhancement of the axial trapping efficiency [15]. Bowman [16] and Thalhammer et al. [17]. demonstrated a

"macro-tweezers" approach, and the optical mirror trap was created after reflection of two holographically shaped collinear beams on a mirror. Although all these approaches are promising, their implementation demand some special optical elements such as spiral phase plate or spatially varying retarders, which limits their applicability. Ashkin [18] predicated that the use of an obstructed beam could increase the axial trapping efficiency of a dielectric particle since the annular intensity distribution enhanced the contribution of rays with a large angle of convergence, that would decrease the axial scattering force. Gu and Morrish [19] proved that Mie metallic particles were axially trapped with a centrally obstructed Gaussian (TEM_{00}-mode) beam focused by a high NA objective lens. Commonly, an opaque disk is used in the obstructed beam approach. This allows only those rays converging at large angles to be focused. The maximal axial trapping efficiency increases with the size of the center obstruction, but most of the incident light will be lost in this geometry. In addition, most of the axial trapping techniques developed so far utilize objective lenses with high numerical aperture (NA>1), which enable higher spatial resolution and axial trapping efficiency. However, high NA objective lenses are generally designed with a very short working distance (typically less than 0.2 mm), that means in many cases that the available space is too small to move the sample axially. High NA objectives also suffer from spherical aberrations when used for imaging in aqueous solutions. The spherical aberration introduced by the refractive difference between glass and water will produce a degradation of the imaging performance and inevitably limit the trapping depth in the aqueous medium [20,21].

In this paper, we investigate theoretically the optical trapping efficiency on dielectric particles and make comparison of the axial

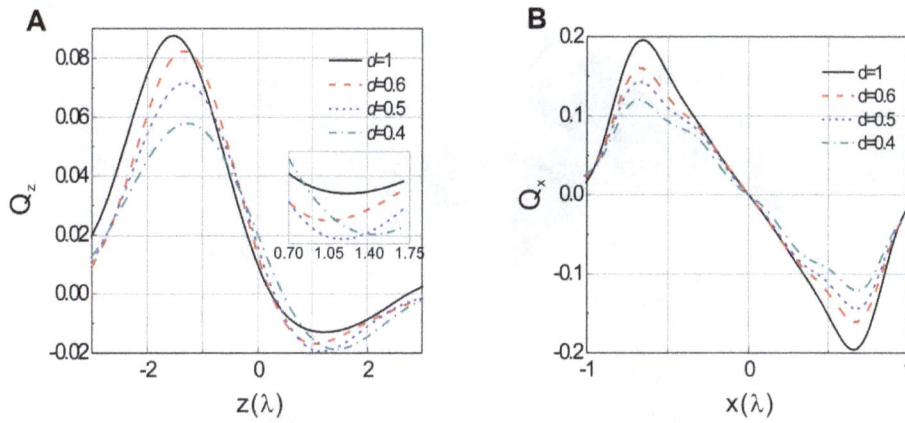

Figure 1. Calculated trapping efficiencies of a spherical particle by annular beams with different widths. A. Axial trapping efficiency, **B.** lateral trapping efficiency. The trapping wavelength is 491 nm, the sphere radius is 2 μm, and the numerical aperture of the objective is 0.6.

and lateral trapping efficiencies for different widths of the annular beams by using vectorial diffraction theory. Long working distance objectives from Nikon Inc. and Edmund Optics Inc. (working distance are 8 mm and 13 mm, respectively) are used to focus the parallel annular laser beam to near diffraction-limited focal spot and realize the three-dimensional optical trapping. The parallel annular beam is generated by a telescopic pair of axicon lenses with transmittance of nearly 100%. The axial trapping depth of 5 mm is realized in the experiment. Theoretical calculations and experimental results are in good agreement.

Theoretical Calculation

In this section, we present the numerical results to show how an annular beam improve the axial trapping efficiency. The optical trapping force F on the particle can be calculated by vectorial diffraction theory based on the electromagnetic scattering theory [11,22]. Instead of using the force F, we generally use a dimensionless quantity, trapping efficiency Q, to characterize how effective the optical trapping is. The trapping efficiency Q is defined, in component form, as

$$Q_i = \frac{\langle F_i \rangle}{(n_1 P/c)}, i = x, y, z, \qquad (1)$$

where P is the power of the incident laser beam, c is the speed of light in vacuum, n_1 is the refractive index of surrounding medium, and $\langle F_i \rangle$ is the time-averaged trapping force. Throughout the

paper, the wavelength is set to be $\lambda = 491$ nm, the refractive index of the particle is $n_2 = 1.5$ and that of the surrounding medium is $n_1 = 1.33$, the radius of the spherical particle is $a = 2$ μm. For the convenience of discussion, a parameter d, which can be regarded as the normalized width of the annular beams is introduced and defined as $d = (r_2 - r_1)/r_2$, where r_1 and r_2 are the inner radius and outer radius of the annular beam, respectively. It can be seen that with the increasing of the value of d, the width of the annulus increases accordingly.

The axial trapping efficiency Q_z is calculated based on the vectorial diffraction theory [23,24]. Figure 1 (a) shows the calculated curves of the axial trapping efficiencies of the annular beams with different d focused by a NA = 0.6 objective. For small values of d, the maximal backward trapping efficiency Q_{max} increases with increasing of d. This corresponds to the cases of $d = 0.4$ and 0.5. With further increasing of d, Q_{max}, however, decreases, as can be seen for $d = 0.6$ and 1.0 curves. This implies that there is a critical value of d, say d_0, for which Q_{0max} assumes its maximum. The occurrence of d_0 is a result of the competition between the scattering force and the gradient force in the focal region. A annular with small values of d can offer a higher portion of rays with large converging angle of focusing, leading to a reduction of the scattering force. However, a smaller value of d also results in the extent of the axial focal depth, which will reduce the axial gradient force. Therefore, it is important to find an optimal value of d in practice. From Figure 1(a) we can see that the maximal backward axial trapping efficiency at the optimal annulus

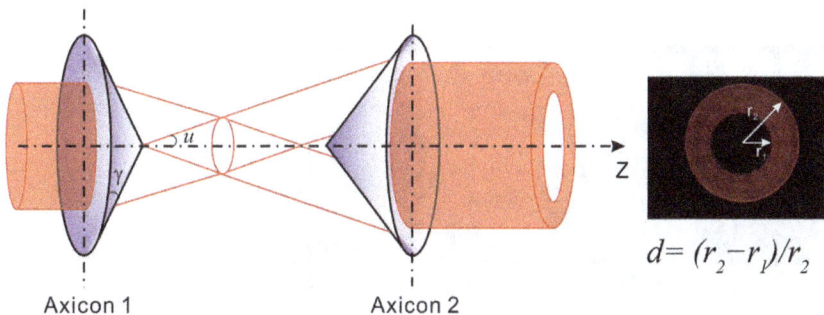

Figure 2. Axicon-pair for generation of parallel annular beam.

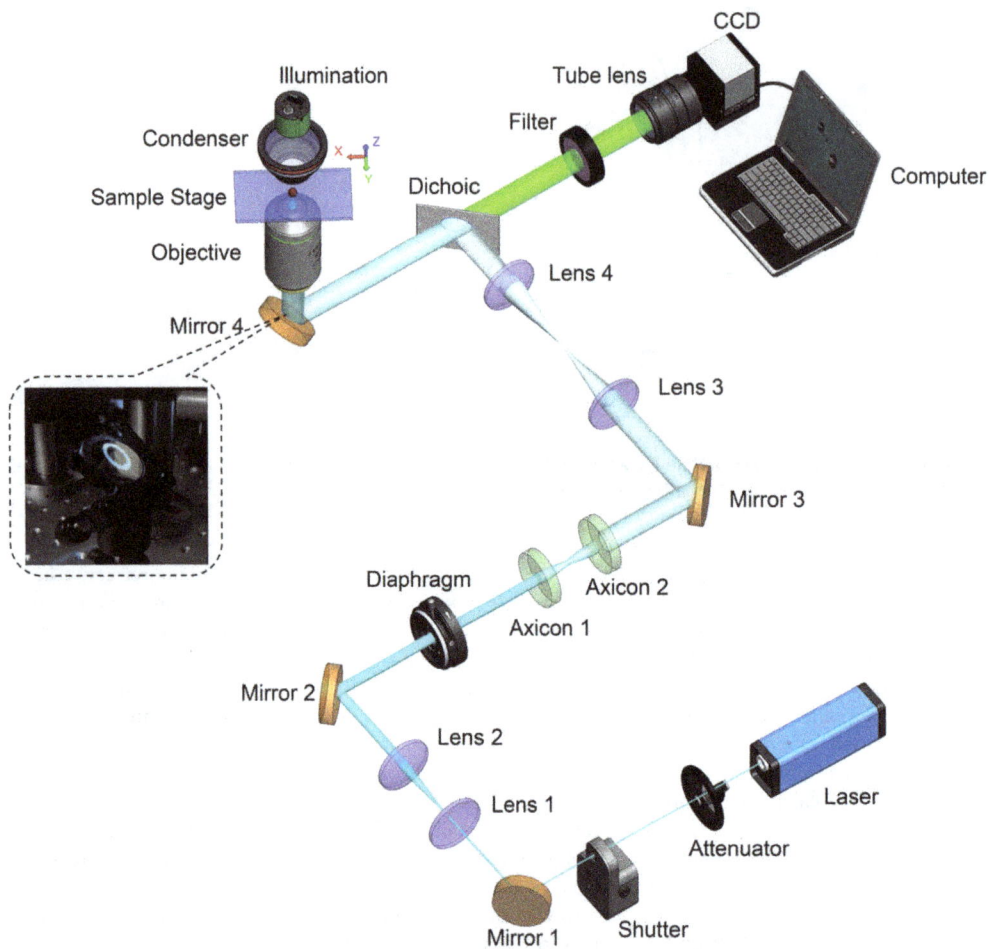

Figure 3. Experimental layout of the trapping system. The inset gives the annular beam shape captured on Mirror 4.

($d = 0.5$) has an improvement up to 49.2% in comparison with the TEM$_{00}$ beam ($d = 1$).

Figure 1(b) gives the lateral trapping efficiency for different values of d. All the while, the maximal lateral trapping efficiency falls with decreasing of the values of d. This is because that the gradient force is dominant due to the multiple refraction on the surface of the trapped dielectric particle [25], i.e., there will be much loss of lateral trapping force when d is smaller. However, the reduction of lateral trapping efficiency has little influence on stably lateral trapping in practice.

Methods

Axicon (conical lens) is a well known wavefront division optical element to transform a Gaussian beam into a Bessel beam, and has already found applications in optical trapping [26] and multifunctional darkfield microscopy [27]. Shao et al. [28] utilized a 3-axicon approach to generate a three-dimensional ring-shaped trap. We adopt here a telescopic pair of axicons to transform a Gaussian beam into a parallel annular beam. As shown in Figure 2, a collimated Gaussian beam incident on the base of the first axicon is deviated cylindrically toward the optical axis due to refraction. All deviated rays independently propagate along different direc-

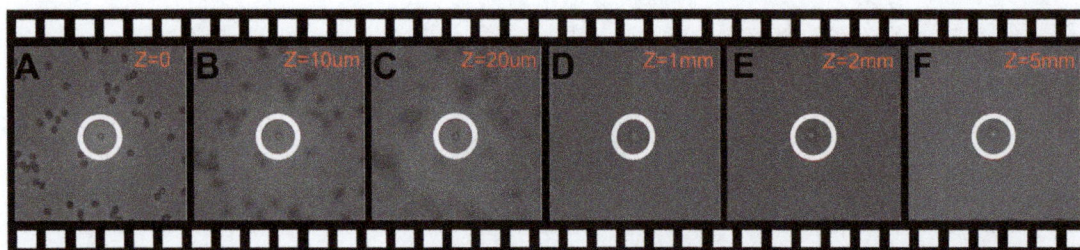

Figure 4. Trapping and moving a silica microsphere in 5 mm depth axially with the focused annular beam. A–F show the trapped microsphere at different axial planes. Using a long working distance microscope objective (20X/NA0.6) (Video S1).

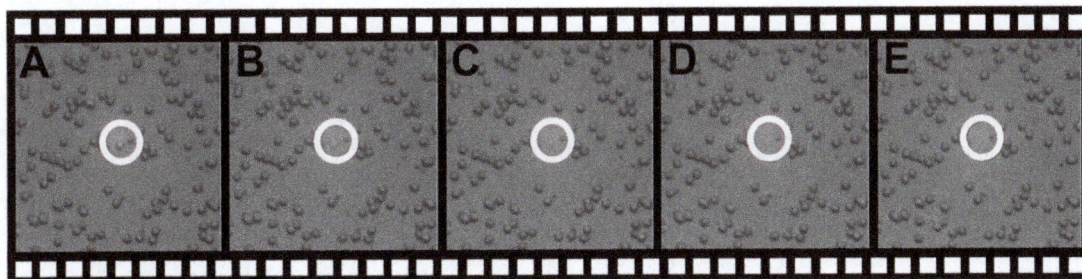

Figure 5. Pushing a silica microsphere out of the focus with the focused TEM$_{00}$ beam. Using a long working distance microscope objective (20X/NA0.6). (Video S2).

tions and form a hollow cone beam, and then incident on the conical surface of the second axicon with a ring-shape of intensity distribution. Two axicons with the same open angle are arranged tip to tip to ensure the output beam behind the second axicon parallel to the optical axis. Adjusting the relative distance between the two axicons corresponds to changing the diameter of the output parallel annular beam [29]. By adjusting the diameter of the output parallel annular beam to overfill the back aperture of the objective len, a near diffraction-limited focus can be obtained. Considering the light transmittance, the open angle γ of the axicon is usually designed very small (less than 10 degree), giving a good approximate calculation of the divergence angle $u \approx (n-1)\gamma$, where n is the refractive index of the axicon. We used two axicons from Thorlabs Inc. with the open angle $\gamma = 5°$ and the refractive index $n = 1.5$. One of the advantages of using the axicon-pair geometry is the very high transmittance, which can be nearly 100% in theory. Considering the transmittance of the antireflection coating of the two axicons, the total conversion efficiency can still reach to 90% in practice.

Figure 3 shows the experimental layout of our trapping system. A diode-pumped solid-state laser (Calypso 491, Cobolt AB Inc., Sweden) working at wavelength of 491 nm is expanded by a telescope formed by Lens 1 and Lens 2. After beam expanding, the collimated beam passes through the telescopic pair of axicon to be transformed into a parallel annular beam. The parallel annular beam then passes through another telescope comprised of Lens 3 and Lens 4, and then is reflected by a dichroic mirror (short-pass 475 nm) into the back aperture of the long working distance 20X objective (EO M Plan HR, NA0.6, 13 mm working distance, Edmund Optics Inc., USA) or a 20X objective (Plan Fluor, NA0.45, 8 mm working distance, Nikon Inc., Japan). Sliding the Lens 4 will slightly change the divergence of the annular shaped beam and ensure that it can be focused on the sample exactly. The sample is mounted on a motorized XYZ stage (MP-285, Sutter Instrument Inc., Canada) that can be driven either manually or by

a programmable software interface. A USB CCD camera (DMK 41BU02, 1280×960 pixels, 4.65 µm×4.65 µm pixel size, The Imaging Source Europe GmbH, Germany) is employed to record the trapping process, with an lens (50 mm, F/1.8, Nikon Inc, Japan) serving as a tube lens. The CCD camera allows imaging speed as high as 30 fps at a resolution of 1280×960 pixels. A long-pass filter (long-pass 505,) is inserted in front of the CCD camera to block the trapping laser beam. The maximal laser power delivered on the sample is about 10 mW. Silica microspheres (4 µm in diameter, Polysciences Inc., USA) immersed in water in a sample chamber are used as test samples.

Results and Discussion

The axial trapping force F_z on a trapped silica microsphere suspended in water was measured by observing the maximal translation speed of the motorized stage at which the particle fell out of the trap. The force is calculated by the Stokes law $F = 6\pi\eta a v$, where a is the radius of a trapped particle, v is the maximal translation speed, and η is the viscosity of the surrounding medium. The maximal axial trapping efficiency $Q_{z,max}$ is then calculated by Eq.(1). In the trapping experiment, the upward motion of the motorized stage is chosen, because the trap is weakest in that direction in the inverted microscope due to the opposing action between the scattering force and the gradient force. The experiment is repeated several times to reduce the error. The measured maximal axial trapping efficiency is 0.0104. Figure 4 (Video S1) demonstrates actually an axial movement of a trapped silica bead in 5 mm depth with a 20X/NA0.6 objective. The axial movement of the particle is controlled by a motorized stage with resolution of 200 nm, and the moving speed and position are controllable with the software. Figure 4A–F show the trapped bead at different axial planes. According to the working distance of the microscope objective, the maximal axial trapping depth can theoretically reach up to 13 mm.

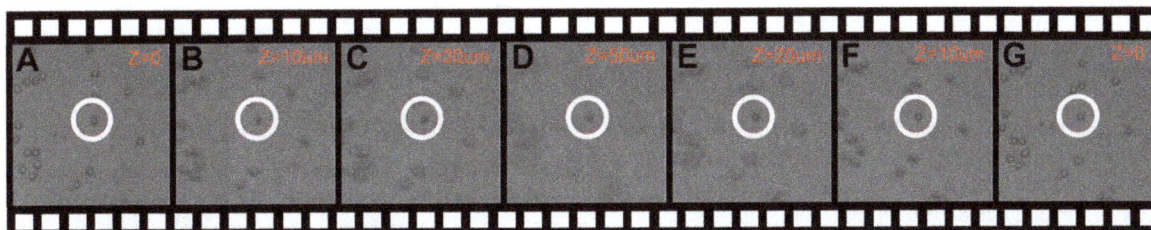

Figure 6. Trapping and moving a silica microsphere axially by an objective lens with much lower NA(0.45). A–D moving upward, **E–G** moving backward. (Video S3).

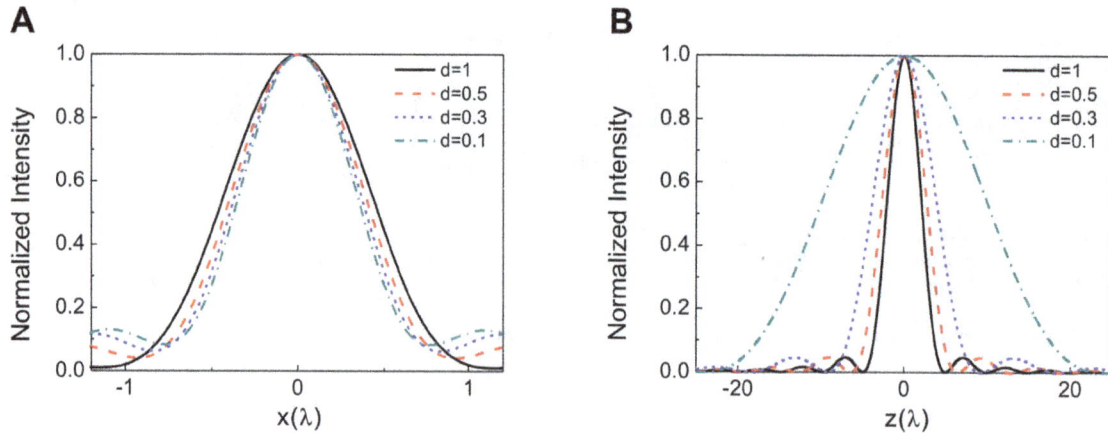

Figure 7. Simulated intensity distributions of the focused annular beam with different widths. A. Lateral intensity distribution, **B.** axial intensity distribution. The numerical aperture of the objective is 0.6.

As expected, the trapping force caused by the TEM_{00} Gaussian beam focused by the same objective failed to trap the same particle, and always pushed the particle out of the focus. This process is demonstrated in Figure 5 (Video S2).

We also demonstrate that it is even possible to trap the particles by an objective with much lower NA(0.45), as shown in Figure 6 (Video S3).

Gu et al. [19] proposed a method for three-dimensional optical trapping of metallic Mie particles using an obstructed laser beam based on the geometrical optics model. The maximal axial trapping efficiency increases with increasing of the size of the center obstruction in their calculation. However, the geometrical optics model ignores the light intensity distribution in the focal region, and can only be employed as a good approximation to compute the force on particles which are much larger than the wavelength of the incident light. We tried decreasing the width of the annular beam to observe the impact on the axial trapping efficiency, but contrary to the calculation based on geometrical optics model, the axial trapping efficiency will decrease as the annular width keeps decreasing after the optimal value $d = 0.5$. We think this is because that the width of the annular beam has an impact on the intensity distribution near the focal region. Figure 7 gives the simulated lateral and axial intensity distributions for the annular beam with different widths focused by a $NA = 0.6$ objective. It can be seen that the lateral intensity distribution keeps almost unaltered with different widths. While for axial direction, the thin annular light field will lead to the focus to be stretched a lot along the axial direction, thereby reducing the axial trapping efficiency. This phenomenon was also mentioned by Kitamura et al. [30] The theoretical calculation of optical trapping force in our simulation is based on the vectorial diffraction theory, which provides a more rigorous solution of the scattering field and is not limited by the size and shape of particles in contrast with the geometrical optics model.

Trapping stiffness is a common parameter to quantify the stability of an optical tweezers system. It can be measured and calculated by using the probability distribution method. Generally, there are two main routes to measure the motion of trapped particles, i.e., quadrant photodiode (QPD) [31] and video-based particle tracking [32]. Due to the fluctuation of trapped particles is normally in the order of nanometers, such measurement typically needs oil-immersion objective with high magnification. In our presented system, a 20X objective is used to trap 4 μm-diameter particles. The resolution of our imaging system is 230 nm/pixel approximately, which is barely enough for the position measurement algorithm. From our trapping videos, we just estimate that the lateral stiffness of our tweezers system is about 3×10^{-6} N/m @ 20 mW. For integrity, we also applied a trapping force calibration technique as described above. The measured maximal axial trapping efficiency is consistent with the theoretical calculation.

Optical trapping of using large NA objectives has some limitations such as extremely short working distance, narrow field of view, and tight focusing with high power density, which might cause heating and optical damage to the biological specimen. In some applications, a large field of view and long working distance are highly desirable. Long working distance objectives are particularly useful for industrial inspection such as wafer probing and flaw detection. The utilization of such kind of objective in optical trapping may open the possibility of exploring large volume with optical tweezers. Due to the long distance imaging capability, the axicon-pair-based optical tweezers could also be a useful tool in various biological researches, for example, examining specimens in vitro through thick glass walls, where the objective lens must be protected against environmental hazards such as heat, vapors, and volatile chemicals by a thick coverslip.

Conclusion

We have proposed a long distance axial trapping approach with focused annular laser beam based on a telescopic pair of axicon. The optical trapping efficiencies on dielectric particles for different widths of the annular beams have been calculated by using vectorial diffraction theory. By trapping silica microspheres suspended in water, we have demonstrated that the system has the advantages of higher efficiency and longer trapping range over the conventional axial trapping geometry.

Supporting Information

Video S1 Trapping and moving a silica microsphere in 5 mm depth axially with the focused annular beam.

Video S2 Pushing a silica microsphere out of the focus with the focused TEM_{00} beam.

Video S3 Trapping and moving a silica microsphere axially by an NA 0.45 objective lens.

Author Contributions

Conceived and designed the experiments: ML BY. Performed the experiments: ML ZL. Analyzed the data: ZL SY DD. Contributed reagents/materials/analysis tools: YQ JQ YY PG TY. Wrote the paper: ML ZL SY.

References

1. Ashkin A, Dziedzic JM, Bjorkholm JE, Chu S (1986) Observation of a single-beam gradient force optical trap for dielectric particles. Opt. Lett 11: 288–290.
2. Higuchi T, Pham QD, Hasegawa S, Hayasaki Y (2011) Three-dimensional positioning of optically trapped nanoparticles. Appl. Opt 50: H183–H188.
3. Carmon G, Feingold M (2011) Rotation of single bacterial cells relative to the optical axis using optical tweezers. Opt. lett 36: 40–42.
4. Gutsche C, Elmahdy MM, Kegler K, Semenov I, Stangner T, et al. (2011) Micro-rheology on (polymer-grafted) colloids using optical tweezers. J. Phys.: Condens. Matter 23: 184114.
5. Simpson NB, McGloin D, Dholakia K, Allen L, Padgett MJ (1998) Optical tweezers with increased axial trapping efficiency. Journal of Modern Optics 45: 1943–1949.
6. O'Neil AT, Padgett MJ (2001) Axial and lateral trapping efficiency of Laguerre-Gaussian modes in inverted optical tweezers. Opt. Commun 193: 45–50.
7. Dasgupta R, Verma RS, Ahlawat S, Chaturvedi D, Gupta PK (2011) Long-distance axial trapping with Laguerre–Gaussian beams. Appl. Opt 50: 1469–1476.
8. Rodrigo PJ, Daria VR, Glückstad J (2004) Real-time three-dimensional optical micromanipulation of multiple particles and living cells. Opt. Lett 29: 2270–2272.
9. Rodrigo PJ, Daria VR, Glückstad J (2005) Four-dimensional optical manipulation of colloidal particles. Appl. Phys. Lett 86: 074103.
10. Zhan QW (2004) Trapping metallic Rayleigh particles with radial polarization. Opt. Express 15: 3377–3382.
11. Yan SH, Yao BL (2007) Radiation forces of highly focused radially polarized beam on spherical particles. Phys. Rev A 76: 053836.
12. Michihata M, Hayashi T, Takaya Y (2009) Measurement of axial and transverse trapping stiffness of optical tweezers in air using a radially polarized beam. Appl. Opt 48: 6143–6151.
13. Iglesias I, Saenz JJ (2012) Light spin forces in optical traps: comment on "Trapping metallic Rayleigh particles with radial polarization". Opt. Express 20: 2832–2834.
14. Donato MG, Vasi S, Sayed R, Jones PH, Bonaccorso F, et al. (2012) Optical trapping of nanotubes with cylindrical vector beams. Opt. Lett 37: 3381–3383.
15. Yao BL, Yan SH, Ye T, Zhao W (2010) Optical trapping of double-ring radially polarized beam with improved axial trapping efficiency. Chin. Phys. Lett 27: 108701.
16. Bowman R, Jesacher A, Thalhammer G, Gibson G, Ritsch-Marte M, et al. (2011) Position clamping in a holographic counterpropagating optical trap. Opt. Express 19: 9908–9914.
17. Thalhammer G, Steiger R, Bernet S, Ritsch-Marte M (2011) Optical macro-tweezers: trapping of highly motile micro-organisms. J Opt. 13: 044024.
18. Ashkin A (1992) Forces of a single-beam gradient laser trap on a dielectric sphere in the ray optics regime. Biophy J 61: 569–582.
19. Gu M, Morrish D (2002) Three-dimensional trapping of Mie metallic particles by the use of obstructed laser beams. J Appl. Phys. 91: 1606–1612.
20. Reihani SNS, Charsooghi MA, Khalesifard HR, Golestanian R (2006) Efficient in-depth trapping with an oil-immersion objective lens. Opt. Lett 31: 766–768.
21. Vermeulen KC, Wuite GJL, Stienen GJM, Schmidt CF (2006) Optical trap stiffness in the presence and absence of spherical aberrations. Appl. Opt 45: 1812–1819.
22. Yan SH, Yao BL (2007) Transverse trapping forces of focused Gaussian beam on ellipsoidal particles. J Opt. Soc. Am B 24: 1596–1602.
23. Ganic D, Gan XS, Gu M (2004) Exact radiation trapping force calculation based on vectorial diffraction theory. Opt. Express 12: 2670–2675.
24. Ganic D, Gan XS, Gu M (2005) Optical trapping force with annular and doughnut laser beams based on vectorial diffraction. Opt. Express 13: 1260–1265.
25. Gu M, Morrish D, Ke PC (2000) Enhancement of transverse trapping efficiency for a metallic particle using an obstructed laser beam. Appl. Phys. Lett 77: 34–36.
26. Garces-Chavez V, McGloin D, Melville H, Sibbett W, Dholakia K (2002) Simultaneous micromanipulation in multiple planes using a self-reconstructing light beam. Nature 419: 145–147.
27. Lei M, Yao BL (2008) Multifunctional darkfield microscopy using an axicon. J Biomed. Opt 13: 044024.
28. Shao B, Esener SC, Nascimento JM, Berns MW, Botvinick EL, et al. (2006) Size tunable three-dimensional annular laser trap based on axicons. Opt. Lett 31: 3375–3377.
29. Golub I, Tremblay R (1990) Light focusing and guiding by an axicon pair generated tubular light beam. J Opt.Soc. Am B 7: 1264–1267.
30. Kitamura K, Sakai K, Noda S (2010) Sub-wavelength focal spot with long depth of focus generated by radially polarized, narrow-width annular beam. Opt. Express 18: 4518–4525.
31. Sørensen KB, Flyvbjerg H (2004) Power spectrum analysis for optical tweezers Rev. Sci. Instrum. 75: 594–612.
32. Otto O, Czerwinski F, Gornall JL, Stober G, Oddershede LB, et al. (2010) Real-time particle tracking at 10,000 fps using optical fiber illumination. Opt. Express 18: 22722–22733.

6

Proton-Binding Sites of Acid-Sensing Ion Channel 1

Hiroshi Ishikita[1,2]*

1 Career-Path Promotion Unit for Young Life Scientists, Graduate School of Medicine, Kyoto University, Kyoto, Japan, 2 Precursory Research for Embryonic Science and Technology, Japan Science and Technology Agency, Saitama, Japan

Abstract

Acid-sensing ion channels (ASICs) are proton-gated cation channels that exist throughout the mammalian central and peripheral nervous systems. ASIC1 is the most abundant of all the ASICs and is likely to modulate synaptic transmission. Identifying the proton-binding sites of ASCI1 is required to elucidate its pH-sensing mechanism. By using the crystal structure of ASIC1, the protonation states of each titratable site of ASIC1 were calculated by solving the Poisson-Boltzmann equation under conditions wherein the protonation states of all these sites are simultaneously in equilibrium. Four acidic-acidic residue pairs—Asp238-Asp350, Glu220-Asp408, Glu239-Asp346, and Glu80-Glu417—were found to be highly protonated. In particular, the Glu80-Glu417 pair in the inner pore was completely protonated and possessed 2 H$^+$, implying its possible importance as a proton-binding site. The pK_a of Glu239, which forms a pair with a possible pH-sensing site Asp346, differs among each homo-trimer subunit due to the different H-bond pattern of Thr237 in the different protein conformations of the subunits. His74 possessed a pK_a of $\approx$6–7. Conservation of His74 in the proton-sensitive ASIC3 that lacks a residue corresponding to Asp346 may suggest its possible pH-sensing role in proton-sensitive ASICs.

Editor: Meni Wanunu, University of Pennsylvania, United States of America

Funding: This work was supported by the JST PRESTO program, Grant-in-Aid for Science Research from the Ministry of Education, Science, Sport and Culture of Japan (21770163), Special Coordination Fund for Promoting Science and Technology of MEXT, and Takeda Science Foundation. The funders had no role in study design, data collection and analysis, decision to publish, or preparation of the manuscript.

Competing Interests: The author has declared that no competing interests exist.

* E-mail: hiro@cp.kyoto-u.ac.jp

Introduction

Acid-sensing ion channels (ASICs) are proton-gated cation channels that exist throughout the mammalian central and peripheral nervous systems. ASIC1a, ASIC1b, ASIC2a, ASIC2b, ASIC3, and ASIC4 have been already identified [1]. With the exception of acid-insensitive ASIC2b and ASIC4 when expressed as a homo-trimer [2], most ASICs are activated by a decrease in the extracellular pH, i.e. increase of proton concentration. Activation of these channels by protons plays an important role in physiological and pathological processes such as nociception, mechanosensation, synaptic plasticity, and acidosis-mediated neuronal injury [3]. Thus, ASICs are important pharmacological targets in neurological diseases.

In the central nervous system, ASIC1 is the most abundant of all the ASICs and is likely to modulate synaptic transmission, memory, and fear conditioning [4]. A recently identified crystal structure of ASIC1 from chicken at 1.9 Å resolution [5] reveals the geometry of a transmembrane domain, which comprises 2 transmembrane helices, and another 5 domains, namely, finger, thumb, knuckle, palm, and β-ball (Figure 1). The finger and thumb domains, located considerably away from the transmembrane domain, are of particular interest because their domain interface was proposed to be a possible pH-sensing site of ASIC1, based on the observations of the crystal structure and mutational studies by Jasti et al. [5] (Figure 1). The following pH-sensing mechanism wherein Asp346 plays a key role has been proposed: protonation/deprotonation of Asp346 of the thumb domain could alter the interaction with its pair partner Glu239 on the finger domain, resulting in the translocation of the thumb domain. Since the thumb domain is linked with the transmembrane domain, movement of the thumb domain causes reorientation of the transmembrane domain and modifies channel

gating (see supplementary information: http://www.nature.com/nature/journal/v449/n7160/extref/nature06163-s2.htm in Ref. [5]). On the other hand, other residues have also been proposed to play an important role in pH sensitivity, mainly by mutational studies. In ASIC2a, mutation of a residue that corresponds to His74 of ASIC1 resulted in pH-sensing deficiency [6,7]. The residue pair Asp79-Glu80 is conserved in all pH-sensitive ASICs. Mutations of the corresponding residue pair in ASIC3 enhance the rate of channel inactivation [8] and those in ASIC2a render the channels proton insensitive [7]. To identify the pK_a values or the protonation states of these residues is, therefore, a starting point in an effort to understand the pH-sensing mechanism of proton-sensitive ASICs.

In the present study, by using the ASIC1 crystal structure [5] and solving the linear Poisson-Boltzmann (LPB) equation, the protonation probabilities of all the titratable sites identified in the crystal structure (51 Arg, 75 Asp, 101 Glu, 73 Lys, 72 Tyr, 15 His, and 3 pairs of N/C-terminal residues) were calculated. In this system, the protonation states of each titratable site are affected by the protonation states of all the titratable sites, accomplishing a completely equilibrated system in terms of the protein protonation pattern. The protonation states of the residues and their pK_a values were obtained and compared among the homo-trimer subunits A, B, and C of ASIC1. In the present study, the computational conditions and procedures employed in previous studies on other ion [9,10] and proton [11,12] channels were consistently used.

Results and Discussion

Acidic-acidic residue pairs in ASIC1

In the ASIC1 crystal structure, there exist 4 pairs of unusually close acidic-acidic residues: (a) Asp238-Asp350, (b) Glu220-

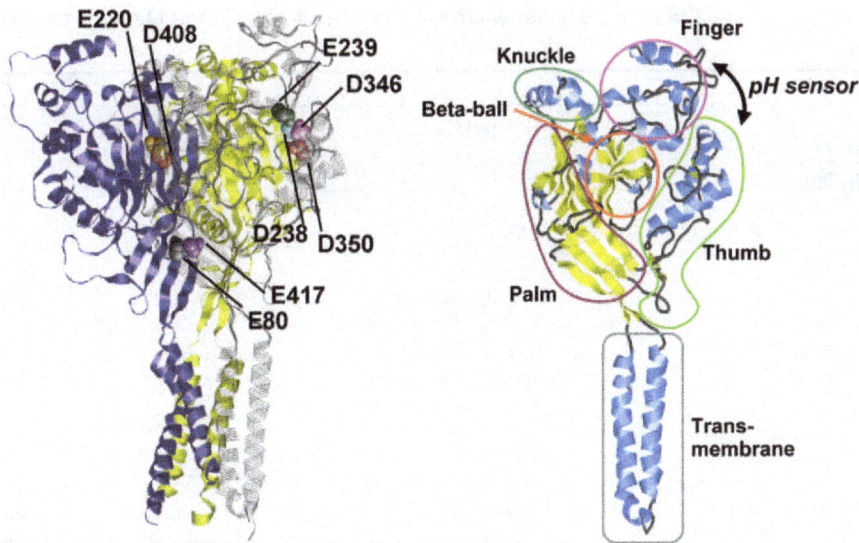

Figure 1. Overview of ASIC1 [5]. Acidic-acidic residue pairs (only in subunit A) are depicted as spheres. Subunit A is depicted as gray strands, and subunits B and C are depicted as blue and yellow ribbons, respectively.

Asp408, (c) Glu239-Asp346, and (d) Glu80-Glu417 (Figure 1). Since the carboxyl O-O atom distances in these pairs range between 2.8 and 3.0 Å, it was proposed that at least 1 of the acidic residues in each pair should be protonated, and that they form the primary sites for pH sensing in ASIC1 [5]. Indeed, the calculated titration curves of these pairs of acidic residues showed unusually high protonation probability (compare, for instance, with the titration curves in Ref. [12]), indicating that each pair has at least 1 protonated acidic residue.

(a) **Asp238-Asp350 and Glu220-Asp408 pairs.** In the Asp238-Asp350 pair, Asp238 was mostly protonated and Asp350 was completely deprotonated at all the pH investigated (pH 5–9) (Figure 2). In the Glu220-Asp408 pair, although Glu220 was considerably protonated, Asp408 was more protonated than Glu220. It appears from Figure 2 that this pair of acidic residues binds a total of 1–1.5 H^+ at pH 5–9.

(b) **Glu239-Asp346 pair.** In the Glu239-Asp346 pair, Glu239 was permanently protonated at pH 5–9 (Figure 2). The Glu239-

Asp346 pair is located at the interface between the finger and the thumb domains of the ASIC1 protein, considerably away from the conducting pore of the transmembrane domain (Figure 1). Nevertheless, the mutation of Asp346 to Asn resulted in a significant shift in the pH_{50} value (i.e., pH of half-maximal activation) on the pH-dose-response curve; thus, Asp346 was proposed to play a key role in the pH-sensing mechanism of ASIC1 [5].

The pH_{50} value on the pH-dose-response curve of the wild-type ASIC1 protein has been experimentally determined to be 6.7 [5]. Interestingly, in the present study, among all the acidic-acidic residue pairs, Asp346 was the only residue with an apparent pK_a obtained from the pH-dependent titration curve (Figure 2) of $\approx$6–6.5. Furthermore, the pK_a of Asp346 obtained from protonation energy at pH 7 (see the method section for definition) was also 6.7 (subunit A in Table 1), and this value was in agreement with the experimentally measured pH_{50} value of 6.7 [5]. Thus, assuming that the pH-dependent domain movement plays a key role in pH sensing, Asp346 can be most likely the pH-sensing site of ASIC1 in terms of its remarkable pK_a value.

(c) **Glu80-Glu417 pair.** Notably, both Glu80 and Glu417 were completely protonated at pH 5–9, implying that this pair possesses $\approx$2 H^+ (Figure 2). Three pairs of Glu80-Glu417 were located on the inner pore of the ASIC1 trimer, and they formed a ring comprising 6 acidic residues (Figure 3). Although all crystal waters were absent during the computations (see Materials and Methods for discussion), this completely protonated state of the Glu80-Glu417 pair can be anticipated based on the existence of a water molecule in the ASIC1 crystal structure at an H-bonding distance from the carboxyl oxygen atom of Glu80 (O_{Glu80}-O_{water} distance = 2.8 Å). Without this water molecule, the positioning of these acidic residues at very small distances (Figure 3) would be unusually energetically unstable to exist in the crystal structure. The following 2 factors may be considered responsible for this unusually high protonation state of the acidic-acidic ring: (i) the specific arrangement of the acidic-acidic residues interacting ring that comprises 6 acidic residues, and (ii) the unfeasibility of solvation in the inner pore where these acidic residues are located.

The 6 acidic residues that form a ring in the pore may be considered to significantly favor protonation of each acidic residue

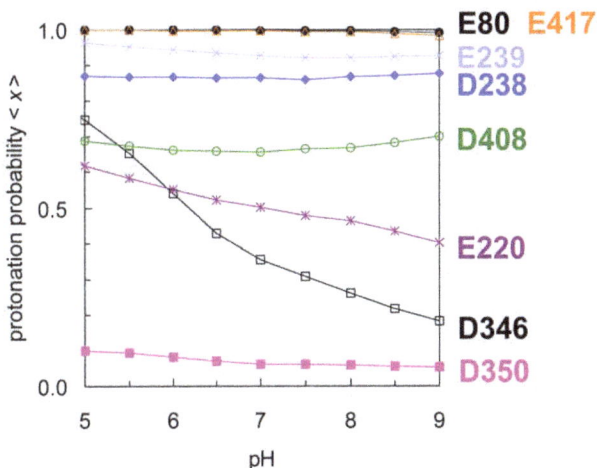

Figure 2. Protonation probabilities $<x>$ [21] of the acidic-acidic residue pairs in ASIC1.

Table 1. The pK_a of 4 acidic-acidic residue pairs obtained from protonation energy at pH 7 (see Materials and Methods for the definition).

Subunit	residue	pK_a			residue	pK_a		
		A	B	C		A	B	C
	Asp238	7.8	7.6	7.4	Asp350	5.8	6.2	6.5
	Glu220	7.0	**8.9**	7.2	Asp408	7.3	**11.5**	7.1
	Glu239	8.1	8.1	**9.7**	Asp346	6.7	7.0	7.3
	Glu80	9.7	9.8	9.6	Glu417	9.5	9.2	9.6

Asp238-Asp350, Glu239-Asp346, Glu220-Asp408, and Glu80-Glu417 in ASIC1 (pK_a). These pK_a values were obtained by calculating the protonation energy required to protonate the residue by 0.5 H$^+$ at pH 7. Therefore, the values may differ from the apparent pK_a values obtained from the titration curves (see the method section for the definition) in Figure 1.

because each residue in the ring has 2 acidic residues on either side (Figure 3, right). To investigate whether this arrangement is a primary factor for the unusually high protonation state of the Glu80-Glu417 pair, only the 6-residue ring (3 Glu80-Glu417 pairs) was isolated from the ASIC1 protein, solvated the acidic ring into an aqueous solution (represented by the dielectric constant $\varepsilon_w = 80$), and titrated it at pH 5–9. From the titration curves (File S1), the apparent pK_a values of Glu80 and Glu417 were determined to be ≈ 5.3 and ≈ 6–6.3, respectively. Although these pK_a values are slightly higher than the standard pK_a value for Glu in an aqueous solution (4.4), the present study revealed that the isolated Glu80-Glu417 ring itself is not highly protonated in an aqueous solution at pH 7. It is noteworthy that the protonation states of the isolated Glu80-Glu417 ring in an aqueous solution remained unchanged when their H-bonding partners were added, namely, Gln277 and Gln279 (data not shown).

Further, the protonation states of the 6-residue ring in the uncharged ASIC1 protein were also investigated where all the atomic charges, except those of the Glu80-Glu417 pairs, were set to 0. In this uncharged protein environment, the 6 acidic residues were found to be completely protonated at all pH investigated (pH 5–9; data not shown). In the uncharged ASIC1, although the influence of the atomic charges of the proteins is absent, the protein volume still prevents solvation of the 6 titratable acidic residues.

From this result, it can be concluded that the loss of solvation in the ASIC1 protein environment is a factor required to explain the unusually high protonation state of the acidic-acidic ring in ASIC1. In general, in the inner pore of channel proteins, the loss of solvation (rather than repulsive interactions) is the major contributor toward destabilizing the charged groups [9,13]. While the ASIC1 pore is formed by a number of charged and polar residues, it appears that bulk water access is limited there, similar to other ion-channel proteins. It can also be concluded that the unfeasibility of solvation is a primary factor for the unionized, highly protonated states of the Glu80-Glu417 pairs.

Glu80 is conserved in all the pH-sensitive ASICs. Mutations of the corresponding residue in ASIC3 enhance the rate of channel inactivation [8], while those of the residues in ASIC2a render the channels proton insensitive [7]. Thus, the highly protonated Glu80 demonstrated in the present study on ASIC1 may imply its possible role as a proton-binding site in ASICs.

Asymmetry of the subunits A, B, and C in the ASIC1 trimer form

Although ASIC1 comprises 3 homo-trimer subunits A, B, and C, the ASIC1 crystal structure shows marked differences among

their subunit conformations. In particular, different conformations of the transmembrane domains among the 3 subunits have been reported (see Figure 2b in Ref. [5]). It is debatable whether the structural asymmetry is inherent to the trimer [5]. In the present study, the generation of atomic coordinates of H atoms led to different H-bonding networks among these residues. Consequentially, different protonation states of the Glu220-Asp408 and Glu239-Asp346 pairs among the 3 subunits were observed (Table 1).

(a) **Glu220-Asp408 pair.** The pK_a values obtained from protonation energy at pH 7 of both the residues in the Glu220-Asp408 pair of subunit B are higher by ≈ 2–4 than those of subunits A and C (Table 1). This is due to the proximity of the side chain O atom of Gln271 to the carboxyl O atom of Asp408 in subunit B (O-O distance = 3.0 Å) in the ASIC1 crystal structure (Figure 4, right). In contrast, the side chain N atom of Gln271 forms an H bond with the carboxyl O atom of Asp408 in subunits A and C, stabilizing the ionized state of Asp408 (Figure 4, left) and thus lowering its pK_a value. Although the different conformation of the Gln271 side chain is a potentially interesting finding, currently, it cannot be conclusively determined whether this is functionally relevant because of the limited resolution of 1.9 Å [5] in the ASIC1 crystal structure at which N and O atoms are unlikely to be distinctly distinguished.

(b) **Glu239-Asp346 pair.** In the Glu239-Asp346 pair, the pK_a value obtained from protonation energy at pH 7 of Glu239 of subunit C was higher by 2 than the corresponding values of subunits A and B (Table 1). Therefore, a significantly different H-bond pattern involving mainly Glu239, Thr237, and Thr240 was found. The hydroxyl group of Thr237 formed a weak H bond with 1 of the carboxyl O atoms of Glu239 in subunits A and B (Figures 5A and 5B, respectively), while the corresponding H bond was absent in subunit C (O$_{Thr237}$-O$_{Glu239}$ distance = 3.7–3.8 Å; Figure 5C). Thus, the ionized state of Glu239 in subunits A and B can be stabilized, leading to the downshift in the pK_a value of Glu239 as compared to that in subunit C.

Furthermore, there was another significant difference in protein conformations among the subunits: not only the H-bond pattern but also the orientation of Thr240 differed significantly between subunit A and subunits B and C. In subunit A, the Thr240 side chain was oriented toward the hydroxyl O atom of Thr237, forming an H bond (O$_{Thr240}$-O$_{Thr237}$ distance = 3.5 Å; Figure 5A). However, the H bond between the Thr240 and Thr237 side chains observed in subunit A was absent in subunit B, because the hydroxyl O atom of Thr240 is considerably closer to the carbonyl O atoms of the protein backbone at Thr237 and Thr240 (O$_{Thr240(OH)}$-O$_{Thr237(CO)}$ distance = 3.1 Å, O$_{Thr240(OH)}$-O$_{Thr240(CO)}$ distance = 3.1 Å) than to

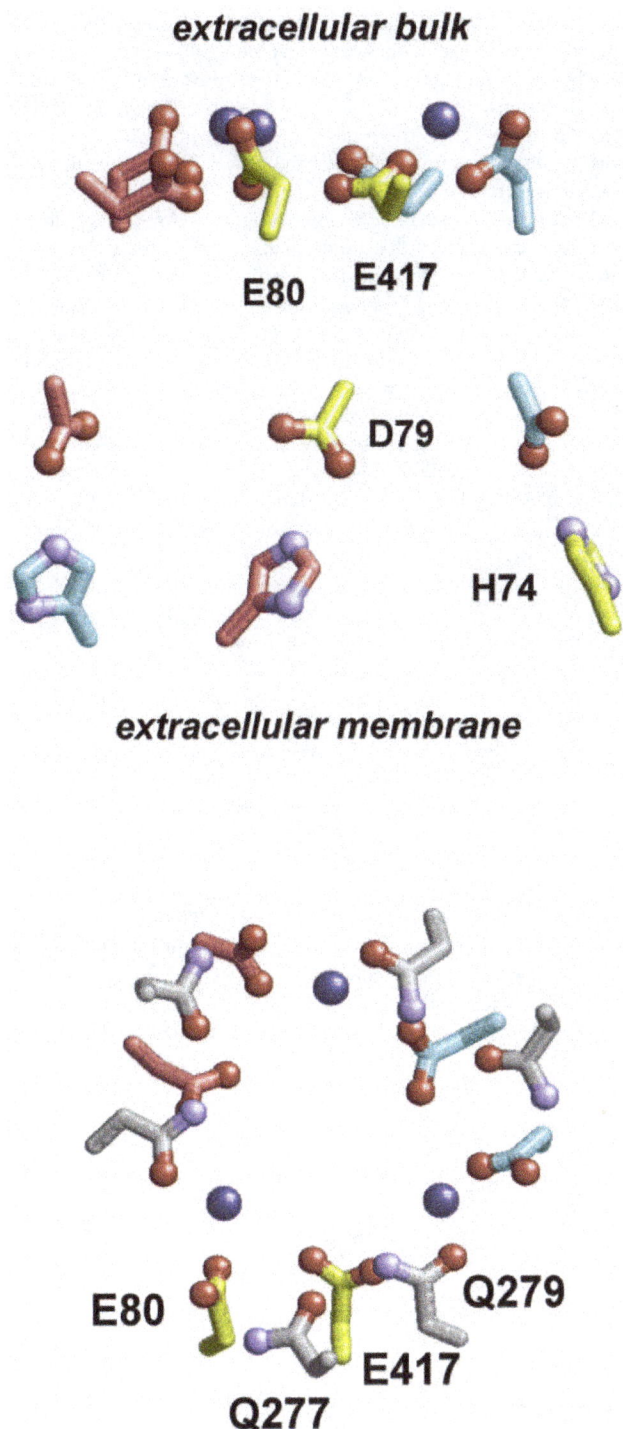

extracellular bulk

E80   E417

D79

H74

extracellular membrane

E80   Q279

E417

Q277

Figure 3. An acidic-residue ring that comprises Glu80 and Glu417. (Left) side view and (right) top view from the extracellular bulk. Carbon atoms of the residues that belong to the same subunit are depicted in the same color (yellow, pink, and cyan), while those of Gln277 and Gln279 are in white in all the subunits. Gln277 and Gln279 are depicted as sticks. The water oxygen atoms at Glu80 are depicted as blue spheres.

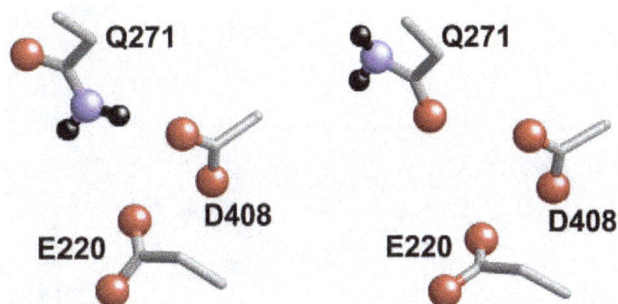

Q271   Q271

D408   D408

E220   E220

Figure 4. Orientation of the Gln271 side chain with respect to the Glu220-Asp408 residue pair. H atoms are depicted as black spheres.

and Thr240 (Figure 5B). In subunit C, Thr237 did not form an H bond with Glu239 but formed an H bond with Thr240, because the donor-acceptor distance of the H bond is lesser in the latter (O_{Thr240}–O_{Thr237} distance = 3.3 Å) than in the former (O_{Thr237}–O_{Glu239} distance = 3.8 Å). Therefore, the ionized state of Glu237 is less stable in subunit C than in subunits A and B because of the lack of the H bond from Thr237 to Glu237 (Figure 5C), leading to the upshifting of its pK_a by 1.6 (Table 1). In contrast to the significant difference in the pK_a of Glu239 (which was affected by the H-bond pattern of Thr237), the pK_a of the pair-partner residue Asp346 did not vary significantly (Table 1). This was due to the identical protonation state of its pair-partner residue Glu239 (i.e., Asp346 was essentially protonated in each subunit and its protonation state did not alter among the subunits).

Of the residues that participate in the H-bonding network of the Glu239-Asp346 pair (i.e., Glu239, Asp346, Thr237, and Thr240), Glu239 and Thr240 are highly conserved residues in ASICs. On the other hand, Asp346 and Thr237 are highly conserved in all ASICs but ASIC3 (see supplementary information in Ref. [5]). In ASIC3, Asp346 and Thr237 are replaced with Ser and Met (rat and mouse) or His and Asn (human), respectively. The fact that Asp346 and Thr237 are simultaneously replaced in ASIC3 implies that these 2 residues probably function cooperatively in ASICs, as indicated in the highly associated H-bonding network shown in Figure 5. Assuming that Asp346 plays a key role in the proposed pH-sensing mechanism of ASIC1 [5], Thr237 may also cooperate with Asp346 in ASIC1 pH sensing. In addition, the absence of the residues corresponding to Asp346 and Thr237 of ASIC1 in ASIC3 suggests that the proposed pH-sensing role of Asp346 in ASIC1 [5] does not hold true for the pH-sensing mechanism of ASIC3 (the pH-sensing mechanism has been further described later in the Discussion).

The only ASIC1 crystal structure currently available was obtained at a low pH and has been proposed to represent a thermodynamically favorable desensitized state [5]. The desensitized state can transform into the open state and the closed state [14]. The different H-bonding pattern and protein conformations of residues Glu239, Asp346, Thr237, and Thr240 of each subunit revealed in the present study may imply possible variations of the ASIC1 conformations in other channel states (i.e., including the open and closed states).

Residues that possess pK_a near $pH_{50} = 6.7$

Since the pH_{50} value on the pH-dose-response curve of the wild-type ASIC1 protein is 6.7 [5], it is worthwhile to reveal residues that possess pK_a value of ≈ 6.7 for understanding the pH-

the hydroxyl O atom of Thr237 ($O_{Thr240(OH)}$–$O_{Thr237(CO)}$ distance = 3.7 Å). Thus, the Thr240 side chain, in turn, formed an H bond with the carbonyl O atoms of the protein backbone at Thr237

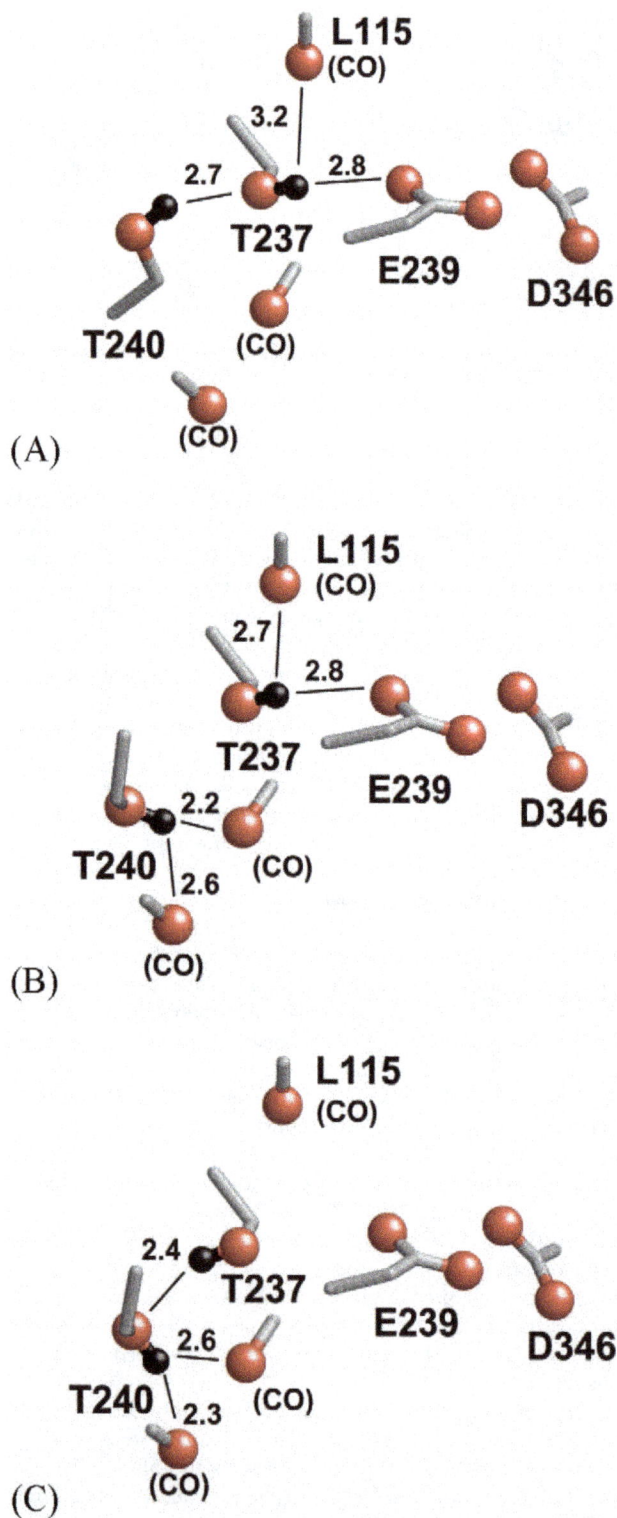

Figure 5. Discrepancy of the H-bonding network that involves the Glu239-Asp346 pair. (A) subunit A, (B) subunit B, and (C) subunit C. The H atoms are depicted as black spheres. Solid lines indicate the orientation of the H bond, and the numbers indicate the length between the H atom and the H-bond donor O atom in Å.

sensing mechanism of ASIC1. Among the acidic residues that are not involved in the acidic-acidic residue pairs, only Glu299 and Glu343 (apparent pK_a of ≈ 7 and 6 obtained from the titration curves, respectively) showed their pK_a values at similar levels (File S1). However, since these residues are located on the protein surface, they are probably not likely to be involved in the pH-sensing mechanism of ASIC1.

Among the acidic-acidic residue pairs, only Asp346 was found to possess pK_a within this range in the present study (Table 1). Although Asp346 is highly conserved among ASICs, it is replaced with a nontitratable residue (Asn) in ASIC1 from lamprey. Interestingly, ASIC1 from lamprey, which lacks Asp346, is proton insensitive, which is in contrast to the proton-sensitive ASIC1 from chicken [15]; this fact may support the proposed role of Asp346 in the pH sensing mechanism of ASIC1 from chicken [5]. However, it should be noted that Asp346 is not conserved even in the proton-sensitive ASIC3. Therefore, it might be speculated that either (i) ASIC3 has a pH-sensing mechanism different from that of ASIC1 or (ii) both ASIC1 and ASIC3 have another common residue(s) that also plays a pH-sensing role.

Regarding non-acidic residues, His74 (Figure 6, subunit C) and His111 (File S1) were found to possess an apparent pK_a of $\approx 6-7$ in the present study, although the latter was located on the protein surface. It is noteworthy that no other His residues have their pK_a values at this range of pH, irrespective of the fact that the pK_a of an isolated His residue is generally ≈ 7 (File S1). Mutation of a His residue corresponding to His74 to Ala in ASIC2a resulted in the deficiency of pH sensitivity [6,7]. This His residue can also be observed in the proton-sensitive ASIC2a but not in the proton-insensitive ASIC2b. His74, therefore, has been proposed to be involved in the activation of ASIC2a by protons [7]. The ASIC1 crystal structure revealed that His74 is located on the inner cavity surface on the 3-fold axis (Figure 3) [5]. Thus, His74 is probably the residue that can directly tune the gating path in terms of the protonation states of the conducting pore near pH_{50}. Interestingly, His74 is also conserved in the proton-sensitive ASIC3 that lacks Asp346. Since both ASIC1 and ASIC3 exhibit the highest proton affinity among the ASICs, the presence

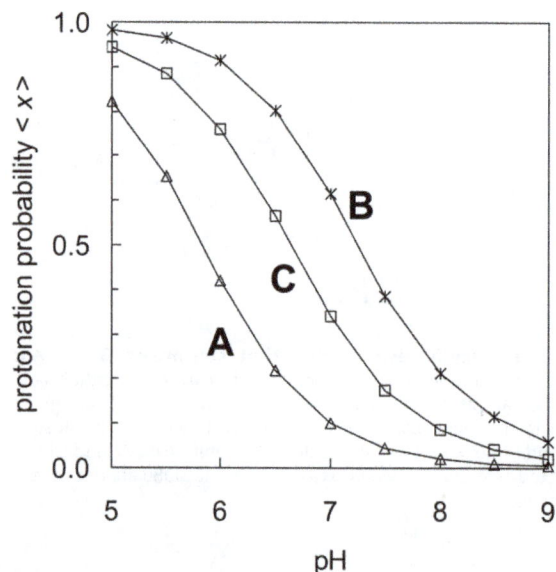

Figure 6. Protonation probabilities $<x>$ of His74 in subunits A, B, and C.

of a residue corresponding to His74 in ASIC3 may explain why ASIC3 is capable of sensing pH without possessing an acidic residue corresponding to Asp346.

Materials and Methods

Atomic coordinates and charges

For performing computations of the ASIC1 trimer form, the crystal structure of ASIC1, comprising subunits A, B, and C, from chicken at 1.9 Å resolution was used (protein data bank [PDB] code: 2QTS) [5]. The atomic coordinates were obtained using the same procedures used in previous studies on channel proteins [9,10,11,12]. The positions of H atoms were energetically optimized with CHARMM [16] by using the CHARMM22 force field [17]. While carrying out this procedure, the positions of all non-H atoms were fixed, and the standard charge states of all the titratable groups were maintained, i.e., the basic and acidic groups were considered protonated and deprotonated, respectively. All the other atoms whose coordinates were available in the crystal structure were not geometrically optimized. Atomic partial charges of the amino acids were adopted from the all-atom CHARMM22 parameter set [16].

Protonation pattern

The present computation is based on the electrostatic continuum model created by solving the LPB equation with the MEAD program [18]. To facilitate a direct comparison with previous computational results, identical computational conditions and parameters such as atomic partial charges and dielectric constants were used (e.g., Refs. [9,10,11,12,19]). The ensemble of the protonation patterns was sampled using the Monte Carlo (MC) method with the Karlsberg program (Rabenstein, B. *Karlsberg online manual*, http://agknapp.chemie.fu-berlin.de/karlsberg/). The dielectric constant was set to $\varepsilon_p = 4$ inside the protein and to $\varepsilon_w = 80$ for solvent and protein cavities corresponding to water. All computations were performed at 300 K, pH 5–9, and an ionic strength of 100 mM. The LPB equation was solved using a 3-step grid-focusing procedure with a starting grid resolution of 2.5 Å, an intermediate grid resolution of 1.0 Å, and a final grid resolution of 0.3 Å. MC sampling yields the probabilities $[A^-]$ and $[AH]$ of the deprotonated and protonated states of the titratable residue A, respectively.

Dielectric volume

As a general and uniform strategy, all crystal waters were removed during the computations (for instance, Refs. [9,10,11,12,19]) due to the lack of experimental information on hydrogen atom positions. Cavities resulting from the removal of crystal waters were uniformly filled with a solvent dielectric medium of $\varepsilon = 80$. Thus, effectively, the effect of the removed water molecules was compensated for implicitly by the high value of the dielectric constant in these cavities. A discussion on the appropriate value of the dielectric constant in proteins for electrostatic energy computations can be found in Ref. [20].

Definition of pK$_a$

a) Apparent pK$_a$ obtained from the titration curve. Protonation probability $\langle x \rangle$ is defined as $\langle x \rangle = [AH]/([AH]+[A^-])$ [21]. Once the residue titration curve ($\langle x \rangle$ versus pH) is obtained by changing the pH of bulk aqueous solution and then plotting $\langle x \rangle$ of the residue, the pK$_a$ value can be obtained as the pH at the point where $\langle x \rangle = 0.5$ on the titration curve. In this manuscript, this pK$_a$ is called "*apparent pK$_a$ obtained from the titration curve.*"

b) pK$_a$ obtained from protonation energy at pH 7. However, titration of acidic-acidic pairs often yields a titration curve that never decreased to $\langle x \rangle = 0.5$ at the investigated pH. Even if the titration curve reaches $\langle x \rangle = 0.5$, the slope at $\langle x \rangle = 0.5$ may be too gentle to determine a unique *apparent pK$_a$* value appropriately. In such a case, one could use an alternative value of pK$_a$ that can be obtained by calculating the energy required to yield 0.5 H$^+$ protonation (i.e., $\langle x \rangle = 0.5$) of the residue at pH 7 (*pK$_a$ obtained from protonation energy at pH 7*). When determining this pK$_a$ value, the focusing residue possesses $\langle x \rangle = 0.5$, while all the other titratable sites possess protonation states equilibrated at pH 7. This pK$_a$ value can always be obtained uniquely at a specific pH of bulk water.

Since only a single residue of the protein cannot be experimentally titrated, in general, the apparent pK$_a$ obtained from the titration curve rather than the pK$_a$ obtained from protonation energy is considered to correspond to an experimentally determined pK$_a$. The discrepancy between the pK$_a$ values defined above cannot be ignored, particularly when the focusing site is subjected to an unusually large influence of the surrounding charged residues (e.g., close positioning of the acidic-acidic residue pair). Nevertheless, in cases wherein such a strong charge influence from the surrounding residues can be ignored, the pK$_a$ obtained from protonation energy at pH 7 is often essentially identical to the apparent pK$_a$ obtained from the titration curve and provides relevant information of the protonation state (see, for instance, Ref. [12]).

In the present study, using the Henderson-Hasselbalch equation, pK$_a$ obtained from protonation energy at pH 7 was calculated as the formal pH at which the concentrations of $[A^-]$ and $[AH]$ are equal. The procedures to obtain pK$_a$ of the titratable residues are identical to those used to determine the redox potential for redox-active groups; the Nernst equation is applied in the latter case [21]. Therefore, the accuracy of the present pK$_a$ computations is directly comparable to that of former computations of redox-active cofactors (e.g., [22,23]). From this analogy, the numerical error of pK$_a$ computation can be estimated to be ≈ 0.2 pH units. Systematic errors, which typically relate to specific conformations that may differ from the given crystal structures, can be considerably larger sometimes. Since the calculations in present study were performed under the same conditions as our previous pK$_a$ computation for other channel proteins, further details on error estimates and comparisons with the previous results can be obtained in Refs. [9,10,11,12].

Supporting Information

File S1 Protonation probabilities of titratable residues.

Author Contributions

Conceived and designed the experiments: HI. Performed the experiments: HI. Analyzed the data: HI. Contributed reagents/materials/analysis tools: HI. Wrote the paper: HI.

References

1. Wemmie JA, Price MP, Welsh MJ (2006) Acid-sensing ion channels: advances, questions and therapeutic opportunities. Trends Neurosci 29: 578–586.
2. Lingueglia E, de Weille JR, Bassilana F, Heurteaux C, Sakai H, et al. (1997) A modulatory subunit of acid sensing ion channels in brain and dorsal root ganglion cells. J Biol Chem 272: 29778–29783.
3. Xiong ZG, Pignataro G, Li M, Chang SY, Simon RP (2008) Acid-sensing ion channels (ASICs) as pharmacological targets for neurodegenerative diseases. Curr Opin Pharmacol 8: 25–32.
4. Wemmie JA, Chen J, Askwith CC, Hruska-Hageman AM, Price MP, et al. (2002) The acid-activated ion channel ASIC contributes to synaptic plasticity, learning, and memory. Neuron 34: 463–477.
5. Jasti J, Furukawa H, Gonzales EB, Gouaux E (2007) Structure of acid-sensing ion channel 1 at 1.9 A resolution and low pH. Nature 449: 316–323.
6. Baron A, Schaefer L, Lingueglia E, Champigny G, Lazdunski M (2001) Zn^{2+} and H^+ are coactivators of acid-sensing ion channels. J Biol Chem 276: 35361–35367.
7. Smith ES, Zhang X, Cadiou H, McNaughton PA (2007) Proton binding sites involved in the activation of acid-sensing ion channel ASIC2a. Neurosci Lett 426: 12–17.
8. Cushman KA, Marsh-Haffner J, Adelman JP, McCleskey EW (2007) A conformation change in the extracellular domain that accompanies desensitization of acid-sensing ion channel (ASIC) 3. J Gen Physiol 129: 345–350.
9. Ishikita H, Knapp E-W (2007) Protonation states of ammonia/ammonium in the hydrophobic pore of ammonia transporter protein AmtB. J Am Chem Soc 129: 1210–1215.
10. Ishikita H (2007) Modulation of the protein environment in the hydrophilic pore of the ammonia transporter protein AmtB upon the GlnK protein binding. FEBS Lett 581: 4293–4297.
11. Ishikita H, Knapp E-W (2005) Induced conformational change upon Cd^{2+} binding at photosynthetic reaction centers. Proc Natl Acad Sci USA 102: 16215–16220.
12. Ishikita H, Knapp E-W (2005) Oxidation of the non-heme iron complex in photosystem II. Biochemistry 44: 14772–14783.
13. Kato M, Pisliakov AV, Warshel A (2006) Barrier for proton transport in aquaporins as a challenge for electrostatic models: the role of protein relaxation in mutational calculations. Proteins 64: 829–844.
14. Chen X, Kalbacher H, Grunder S (2006) Interaction of acid-sensing ion channel (ASIC) 1 with the tarantula toxin psalmotoxin 1 is state dependent. J Gen Physiol 127: 267–276.
15. Coric T, Zheng D, Gerstein M, Canessa CM (2005) Proton sensitivity of ASIC1 appeared with the rise of fishes by changes of residues in the region that follows TM1 in the ectodomain of the channel. J Physiol 568: 725–735.
16. Brooks BR, Bruccoleri RE, Olafson BD, States DJ, Swaminathan S, et al. (1983) CHARMM: a program for macromolecular energy minimization and dynamics calculations. J Comput Chem 4: 187–217.
17. MacKerell AD, Jr., Bashford D, Bellott RL, Dunbrack RL, Jr., Evanseck JD, et al. (1998) All-atom empirical potential for molecular modeling and dynamics studies of proteins. J Phys Chem B 102: 3586–3616.
18. Bashford D, Karplus M (1990) pK_a's of ionizable groups in proteins: atomic detail from a continuum electrostatic model. Biochemistry 29: 10219–10225.
19. Ishikita H (2010) Origin of the pK_a shift of the catalytic lysine in acetoacetate decarboxylase. FEBS Lett 584: 3464–3468.
20. Schutz CN, Warshel A (2001) What are the dielectric constants of proteins and how to validate electrostatic models? Proteins 44: 400–417.
21. Ullmann GM, Knapp E-W (1999) Electrostatic models for computing protonation and redox equilibria in proteins. Eur Bophys J 28: 533–551.
22. Ishikita H, Knapp E-W (2005) Redox potential of cytochrome c550 in the cyanobacterium *Thermosynechococcus elongatus*. FEBS Lett 579: 3190–3194.
23. Ishikita H (2007) Contributions of protein environment to redox potentials of quinones in flavodoxins from *Clostridium beijerinckii*. J Biol Chem 282: 25240–25246.

7

Embryonic Carcinoma Cells Show Specific Dielectric Resistance Profiles during Induced Differentiation

Simin Öz[1], Christian Maercker[2,3]*, Achim Breiling[1]*

1 Division of Epigenetics, DKFZ-ZMBH Alliance, German Cancer Research Center, Heidelberg, Germany, 2 Mannheim University of Applied Sciences, Mannheim, Germany, 3 Genomics and Proteomics Core Facilities, German Cancer Research Center, Heidelberg, Germany

Abstract

Induction of differentiation in cancer stem cells by drug treatment represents an important approach for cancer therapy. The understanding of the mechanisms that regulate such a forced exit from malignant pluripotency is fundamental to enhance our knowledge of tumour stability. Certain nucleoside analogues, such as 2′-deoxy-5-azacytidine and 1β-arabinofuranosylcytosine, can induce the differentiation of the embryonic cancer stem cell line NTERA 2 D1 (NT2). Such induced differentiation is associated with drug-dependent DNA-damage, cellular stress and the proteolytic depletion of stem cell factors. In order to further elucidate the mode of action of these nucleoside drugs, we monitored differentiation-specific changes of the dielectric properties of growing NT2 cultures using electric cell-substrate impedance sensing (ECIS). We measured resistance values of untreated and retinoic acid treated NT2 cells in real-time and compared their impedance profiles to those of cell populations triggered to differentiate with several established substances, including nucleoside drugs. Here we show that treatment with retinoic acid and differentiation-inducing drugs can trigger specific, concentration-dependent changes in dielectric resistance of NT2 cultures, which can be observed as early as 24 hours after treatment. Further, low concentrations of nucleoside drugs induce differentiation-dependent impedance values comparable to those obtained after retinoic acid treatment, whereas higher concentrations induce proliferation defects. Finally, we show that impedance profiles of substance-induced NT2 cells and those triggered to differentiate by depletion of the stem cell factor OCT4 are very similar, suggesting that reduction of OCT4 levels has a dominant function for differentiation induced by nucleoside drugs and retinoic acid. The data presented show that NT2 cells have specific dielectric properties, which allow the early identification of differentiating cultures and real-time label-free monitoring of differentiation processes. This work might provide a basis for further analyses of drug candidates for differentiation therapy of cancers.

Editor: Austin John Cooney, Baylor College of Medicine, United States of America

Funding: This work was supported by the Ministerium für Wissenschaft, Forschung und Kunst, Baden Württemberg (Kooperatives Promotionskolleg Krankheitsmodelle und Wirkstoffe) (http://mwk.baden-wuerttemberg.de/). The funders had no role in study design, data collection and analysis, decision to publish, or preparation of the manuscript.

Competing Interests: The authors have declared that no competing interests exist.

* E-mail: a.breiling@dkfz.de (AB); c.maercker@hs-mannheim.de (CM)

Introduction

The induction of differentiation by treatment with natural ligands and synthetic drugs represents an important approach for cancer therapy [1,2]. Tumours are thought to originate from cells with stem cell characteristics that have acquired aberrant gene expression patterns, mostly due to genetic and/or epigenetic mutations, which destabilise the homeostasis of cellular proliferation and differentiation [1,3]. Cancer is thus characterised by a block in differentiation and by the induction of uncontrolled proliferation [3]. The identification and characterisation of substances that induce differentiation in human cancer cells therefore represents an important aspect in the development of novel cancer therapies.

A prominent example for a differentiation inducing drug is 2′-deoxy-5-azacytidine (decitabine, DAC), that has been suggested to induce differentiation by DNA demethylation [4]. A compound closely related to decitabine, 1β-arabinofuranosylcytosine (cytarabine, araC), induces differentiation without inhibiting DNA methylation [5]. DAC, araC and the structurally related drug 5-

azacytidine (AZA), are used for the treatment of myeloid leukaemias, a group of diseases that is characterised by a differentiation block of precursor cells [6,7]. While the precise molecular modes of action of these drugs are still not well understood, nucleoside analogues can be incorporated into DNA and thereby trigger DNA damage or other stress response pathways [8]. Indeed, we have recently shown that both DAC and araC induce neuronal differentiation in the embryonal carcinoma (EC) cell line NTERA2 D1 (NT2) by triggering degradation of OCT4 and other stem cell proteins via DNA damage pathways [9].

NT2 EC cells express high levels of stem cell specific transcription factors (especially OCT4 and NANOG), Polycomb Group (PcG) proteins and DNA methyltransferases. The cells also show significant levels of non-CpG methylation, a DNA mark restricted to pluripotent cells that is strongly reduced upon differentiation induction with all-trans-retinoic acid (RA), a conserved intercellular signaling molecule found in most vertebrates [10]. NT2 cells have not only been shown to differentiate along the neuronal lineage, but also show mesodermal and ectodermal lineage potential and thus represent a valuable human cancer stem

A

B

C

D

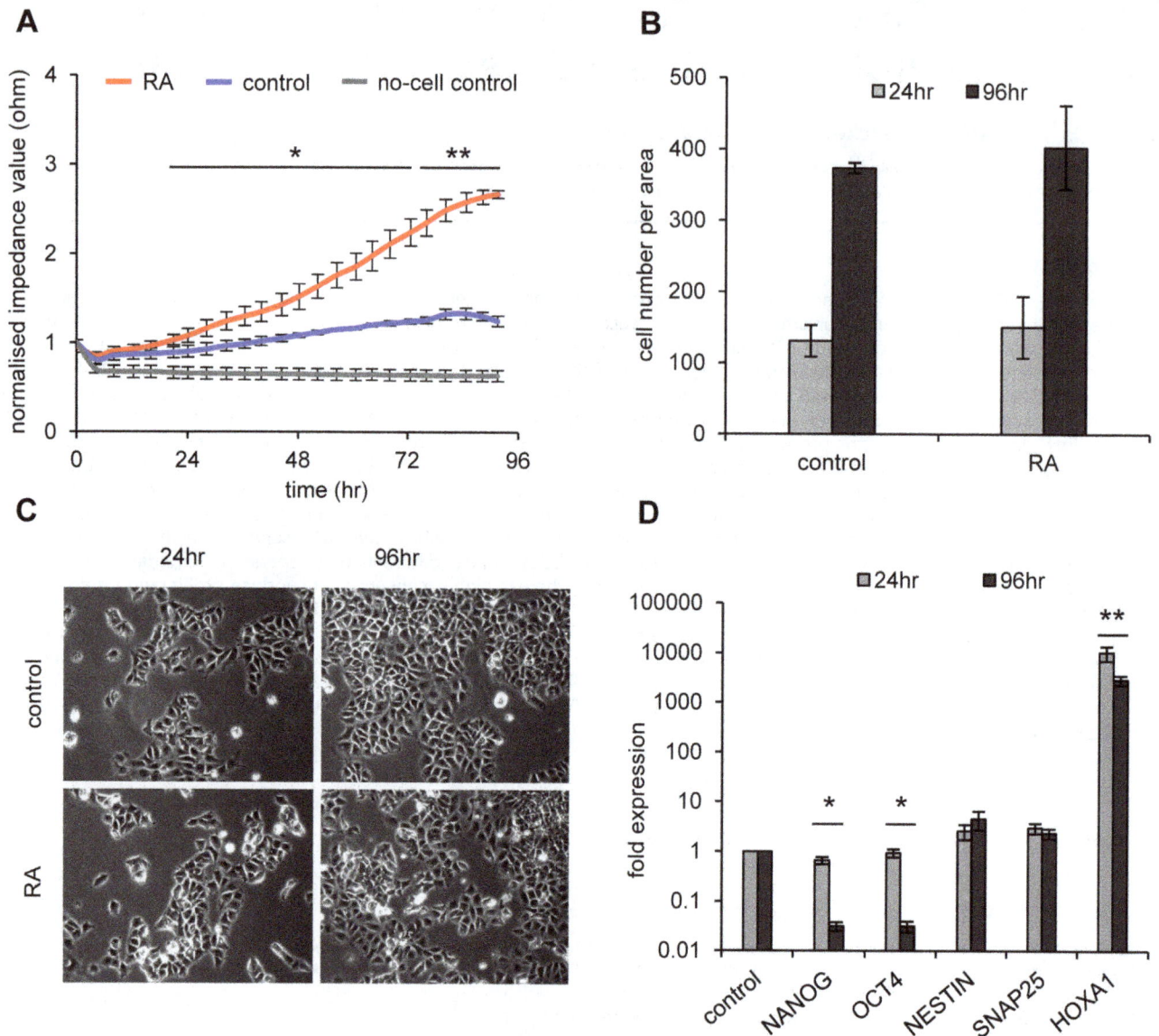

Figure 1. Retinoic acid induced neuronal differentiation of NT2 EC cells. (A) Impedance profiles comparing RA-induced (10 µM - red) and untreated NT2 cells (blue) during a 4 day period. The mean of three independent experiments is shown. Standard deviations are indicated by error bars every four hours. Measurements were executed at 45 kHz in 5-minute intervals for 96 hours. Normalised resistance values were compared by two-tailed Student's t-test. After 20 hours of RA treatment differences in impedance values start to become statistically significant (*p$<$0.05, **p$<$0.005). Black lines show regions with significant differences in respect to the untreated cell control. **(B)** Average cell numbers of three replicates of untreated and RA-treated NT2 cells after 24 and 96 hours do not differ significantly. Standard deviations are indicated by error bars. **(C)** Microscopic images (10× magnification) of NT2 control cells and NT2 cells treated with RA for 24 and 96 hours. No clear differentiation phenotype becomes apparent for the RA treatment. **(D)** qRT-PCR expression analysis of stem cell factors *NANOG*, *OCT4* and the differentiation markers *NESTIN*, *SNAP25* and *HOXA1* in RA- treated and control cells after 24 and 96 hours of treatment. Data is shown in logarithmic scale. Only *HOXA1* is prominently induced by retinoic acid at both time points. The stemness genes are only found reduced after 96 hours of RA treatment. All qRT-PCR measurements were repeated at least three times and internally normalised to the corresponding *β-actin* values. Standard deviations are indicated by error bars. Two-tailed student's t-test showed significant differences when comparing expression levels of *OCT4*, *NANOG* and *HOXA1* at 24 hours with the expression levels at 96 hours. (*p$<$0.05, **p$<$0.005).

cell model system [11,12]. Cultures exposed to differentiation-inducing substances are usually rather heterogeneous and show a mixture of neuronal, ectodermal and mesodermal features [11–14]. Induction of differentiation with the natural ligand retinoic acid results in visible morphological changes only after prolonged treatment of at least three days [9,15]. Changes in marker gene expression are even more delayed. Efficient reduction of stem cell factors or induced expression of neuronal markers becomes

apparent only after several days of RA treatment [9,13,15]. In order to screen drug libraries for differentiation-inducing substances a fast method for early-identification of cellular differentiation is thus desirable.

Electrical cell-substrate impedance sensing (ECIS) is a label-free, non-invasive monitoring technique to study the formation of cell-matrix as well as cell-cell contacts during cell proliferation, cell migration, metastasis, wound healing, cellular differentiation and cancer development [16–18]. The method is based on the phenomenon that living cells behave as dielectric particles and thus alter the electrode impedance after attachment to a micro-electrode surface. Impedance measurements at the electrode-cell interface are influenced by increasing cell number, increased adhesion, morphological changes and cell spreading [19]. We have previously used this non-invasive assay to measure imped-ance profiles of differentiating mesenchymal stem cells [20]. Mesenchymal stem cells (MSCs) induced for adipogenesis or osteogenesis *in vitro*, showed characteristic changes in dielectric properties, that were already visible within 24 hours. ECIS is thus a reliable tool for real-time monitoring of stem cell differentiation [20,21].

To study immediate effects on impedance values, we analysed the onset of drug-induced differentiation in NT2 cells by ECIS. Already after 20 hours of retinoic acid induction we found a significant increase of impedance values. The slope/time ratios of the dielectric resistance profiles positively correlated with the employed concentration of RA. Further experiments determined the concentrations of nucleoside drugs that induced impedance changes with slope/time ratios comparable to those obtained with retinoic acid. These differentiation-specific effects could be separated from cytotoxicity. Finally, we show that differentiation induction by nucleoside drugs and retinoic acid is mainly caused by the reduction of the levels of stemness factors, in particular OCT4. Taken together, our work provides a basis for further real-time studies in living cells evaluating drug candidates as differentiation inducing agents for cancer therapy.

Materials and Methods

Cell Culture and Drug Treatment

The human cell line NT2 D1 [22,23] was a kind gift from Peter W. Andrews (University of Sheffield). Cell line authentication was provided by LGC Standards (Teddington, report tracking no. 71008933). Cells were maintained in Dulbecco's Modified Eagle Medium (DMEM) supplemented with 10% FCS (Invitro-gen), 200 U/ml penicillin (Gibco) and 200 µg/ml streptomycin (Gibco). NT2 cells were induced to differentiate (if not mentioned otherwise) with 10 µM all-trans retinoic acid (Sigma), 1 µM deoxycytidine (Sigma), 1 µM azacytidine (Sigma), 1 µM decita-bine (Sigma), 1 µM cytarabine (Sigma), 5 mM hexamethylene bisacetamide (Sigma) and 50 µM fibroblast growth factor 2 (Novitec) in 5% CO_2 at 37°C.

RNA Isolation and qRT PCR

Total RNA was isolated from NT2 D1 cells using the Trizol reagent (Invitrogen) or the RNeasy kit (Qiagen), following the manufacturer's recommendations. Total RNA (500 ng) was re-verse transcribed using Superscript III (Invitrogen). Quantitative RT-PCR was performed utilising the LightCycler 480 System (Roche). 1 µl of cDNA was used for 10 µl PCR reaction using Absolute QPCR SYBR Green Mix (Thermo Scientific) under following conditions: 1 cycle at 95°C for 15 min followed by 50 cycles at 95°C for 15 s, at 60°C for 40 s. All samples were measured in triplicates. Cycle threshold numbers for each amplification were measured with the LightCycler 480 software, and relative expression values were calculated and normalised using *β-actin* as an internal standard. For RT-primer sequences see Table S5.

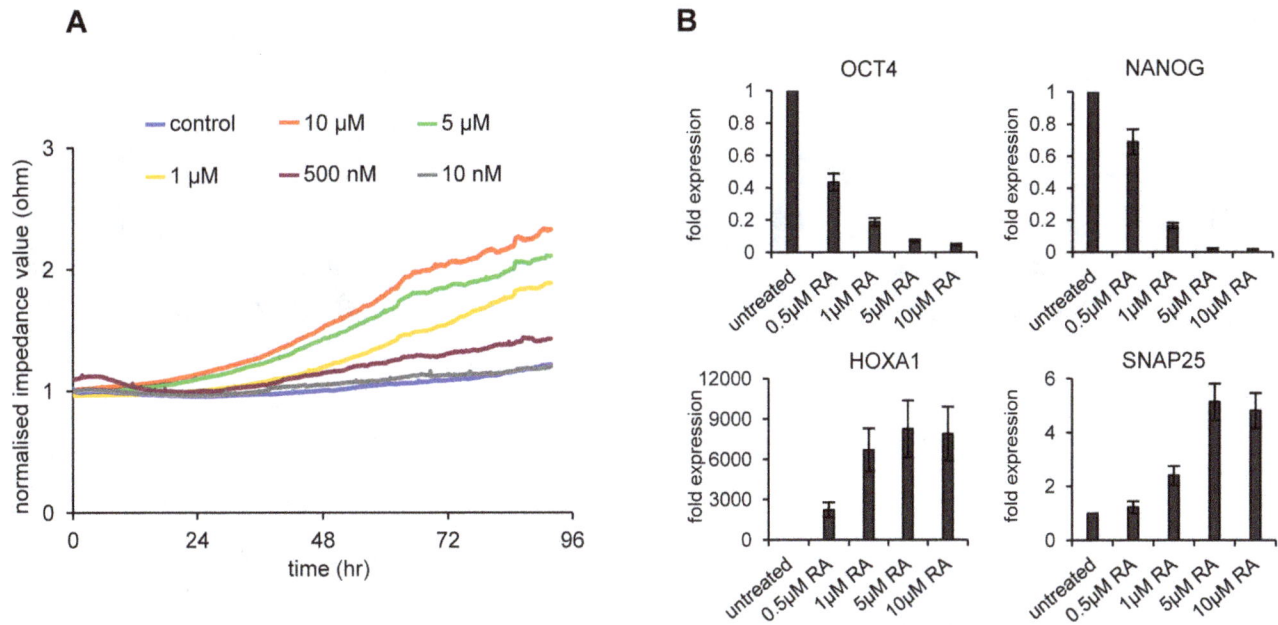

Figure 2. Induced concentration-dependent differentiation of NT2 cells by RA. (A) Impedance profiles comparing induction profiles of different RA concentrations during a 4 day period. Measurements were executed at 45 kHz in 5-minute intervals for 96 hours. The mean of three independent experiments is shown. Standard deviations are not shown to avoid crowding of the diagram. For single diagrams including standard deviations and statistical tests for these data sets see Fig. S1. **(B)** qRT-PCR expression analysis of stem cell factors *NANOG*, *OCT4* and the differentiation markers *HOXA1* and *SNAP25* and in RA- treated and control NT2 cells after 96 hours of treatment. The concentration of RA employed correlates negatively with the expression of stem cell factors, but positively with the expression of differentiation markers. All qRT-PCR measurements were repeated at least three times and internally normalised to the corresponding *β-actin* values. Standard deviations are indicated by error bars.

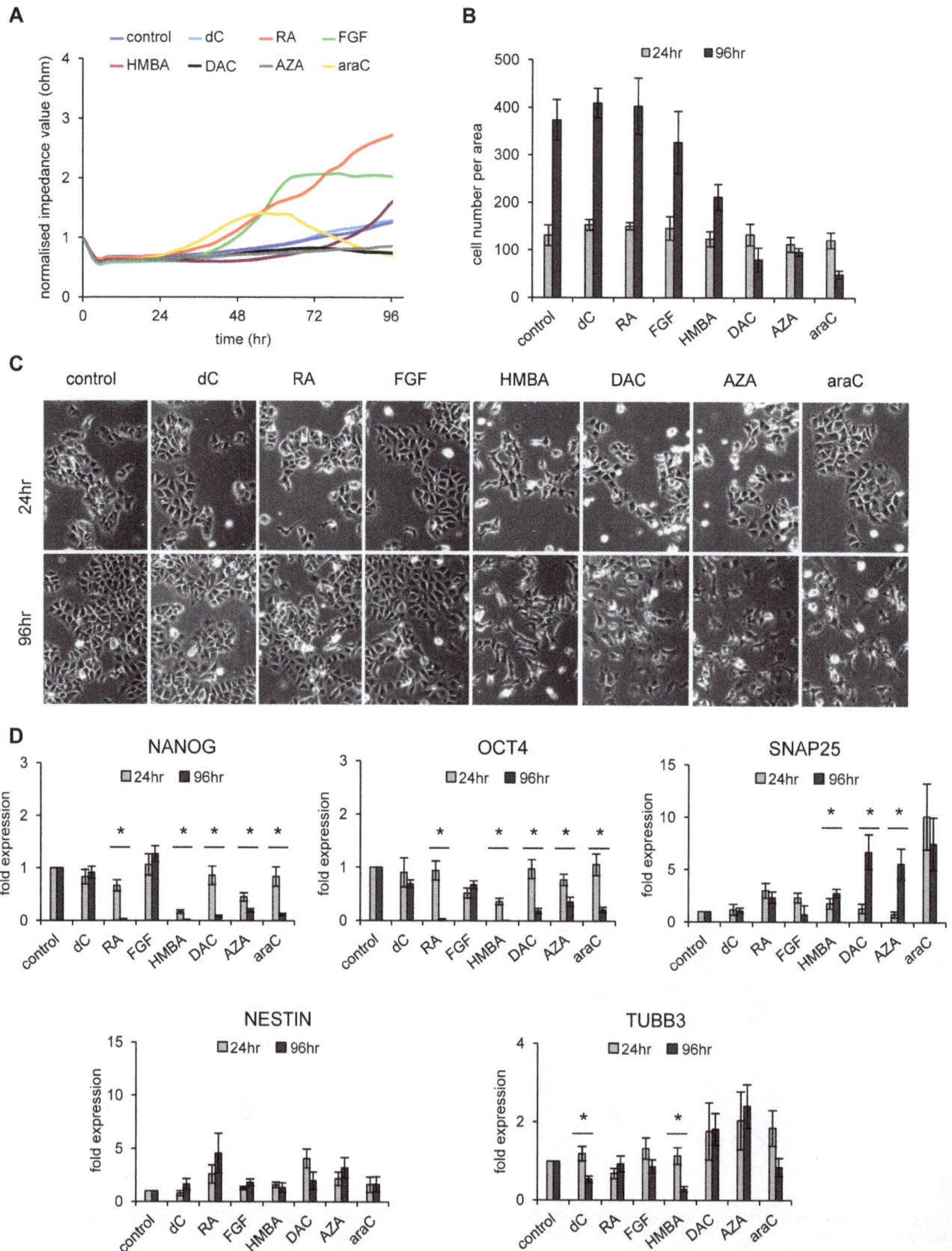

Figure 3. Induced differentiation by a defined panel of drugs. (**A**) Impedance profiles comparing NT2 cells treated with retinoic acid (RA, 10 μM), hexamethylene bisacetamide (HMBA, 5 mM), 5-azacytidine (AZA, 1 μM), deoxycytidine (dC, 1 μM), fibroblast growth factor 2 (FGF, 50 μM), 2′-deoxy-5-azacytidine (DAC, 1 μM) and 1β-arabinofuranosylcytosine (araC, 1 μM) during a 4 day period. Measurements were executed at 45 kHz in

5-minute intervals for 96 hours. The mean of three independent experiments is shown. Standard deviations are not shown to avoid crowding of the diagram. For single diagrams including standard deviations and statistical tests for these data sets see Fig. S2. (**B**) Average cell numbers of three replicates of untreated and treated NT2 cells after 24 and 96 hours. Nucleoside drugs are cytotoxic at the concentrations used and show significant growth inhibition. Standard deviations are indicated by error bars. (**C**) Microscopic images (10× magnification) of NT2 control cells and NT2 cells treated with the various substances mentioned in (A) after 24 and 96 hours of treatment. (**D**) qRT-PCR expression analysis of stem cell factors *NANOG*, *OCT4* and the differentiation markers *NESTIN*, *SNAP25* and *TUBB3* in treated and NT2 control cells after 24 and 96 hours of treatment. All qRT-PCR measurements were repeated at least three times and internally normalised to the corresponding *β-actin* values. Standard deviations are indicated by error bars. Treatments showing significant differences comparing expression levels at 24 hours with those at 96 hours are marked with an asterisk (two-tailed student's t-test; $p < 0.05$).

Impedance Measurements

NT2 D1 cells (2×10^4 per well) were seeded in triplicates or quadruplicates and grown in 400 µl DMEM supplemented with 10% FCS (Invitrogen), 200 U/ml penicillin (Gibco) and 200 µg/ml streptomycin (Gibco) on 8W10E+ ECIS Cultureware arrays (Applied Biophysics) that contain 40 250-µm gold electrodes per well. Cells were then treated as indicated. Measurements were carried out in cell culture medium (without medium changes) and the arrays were kept in the incubator with 5% CO_2 at 37°C. The arrays were measured on an ECIS^{TM} Model 1600 (Applied Biophysics) at 45 kHz in 5-minute intervals for 96 hours. Data were normalised to their starting values. Analysis was done on ECIS software based on the model developed by Giaever and Keese [19].

Staining and Microscopy

At ECIS endpoints, cells attached to the arrays were washed with 1X Phosphate Buffered Saline (PBS) and fixed with 4% paraformaldehyde. Cells were permeabilised with 0.1% TritonX-100, stained with Phalloidin Tritc (Sigma) and DAPI (Invitrogen) in PBS for 1 hour and washed. After staining, 8-well chamber tops were removed from the base slides and mounting media and cover slips were added. Fluorescence images were taken on Zeiss Axioskop 2 Plus. Phase contrast images of living cells were taken on a Leica DM-IRBE inverse microscope.

Protein Depletion via siRNAs

For the depletion of OCT4, specific ON-TARGETplus SMARTpool siRNAs (Dharmacon) were used as described [9]. In brief, NT2 cells were seeded in quadruplicate into 8W10E+ ECIS Cultureware arrays (Applied Biophysics) at a density of 2×10^4 per well, with 400 µl of medium (see above). Cells were transfected with siRNAs using the DharmaFECT1 transfection reagent (Dharmacon) according to the manufacturer's instructions. The final siRNA concentration was 50 nM. Scrambled siRNAs for negative control experiments were also obtained from Dharmacon. Impedance measurements were started immediately after transfection, as described above and carried out during a 6 day period. Medium was changed once after 3 days. Impedance peaks caused by the medium change were equalised during data analysis with the ECIS software.

Statistical Analysis and Bioinformatics

Two-tailed student's t-test was used for statistical analysis of ECIS and qRT-PCR data. To determine the slope maxima of the differentiation-induced impedance data, we applied a cubic smoothing spline with 10 degrees of freedom to the resistance values and performed a generalised cross validation to the data [24]. The steepest rise in impedance was determined as the time point with the maximum slope.

Results

Monitoring RA-induced Differentiation using Electric Impedance Sensing

RA-induced neuronal differentiation of NT2 cells [15,25] is a comparably slow process that usually requires several weeks of treatment before morphological changes become visible, even if gene expression patterns change more rapidly [15]. In order to use a non-invasive method and to directly monitor differentiation induction *in vitro*, we seeded NT2 cells (2×10^4 per well) into eight-well ECIS-arrays (8WE10+) and measured resistance changes at 45 kHz every 5-minutes in the absence or presence of RA over a four day period (96 hours). As shown in Figure 1A, untreated NT2 cells showed only a weak increase of frequency dependent impedance values over four days, which was most likely mainly caused by the increasing cell number. In striking contrast, treatment with 10 µM RA led to a significant increase of impedance values starting with 20 hours of treatment (Fig. 1A). Total cell numbers did not differ significantly between untreated and RA-treated cells at 24 or 96 hours of growth (Fig. 1B), indicating that impedance differences were not due to differing growth rates. Also, overall the morphology of both cell populations was very similar (Fig. 1C). Early onset of differentiation is usually monitored by marker gene expression using quantitative reverse transcription PCR (qRT-PCR) on total RNA isolated from growing cells. As shown in Fig. 1D, transcription of the stem cell factors *NANOG* and *OCT4* was significantly down-regulated in RA-treated cells, but only after 96 hours of RA treatment, whereas the specific differentiation markers *NESTIN*, *SNAP25* and *HOXA1* were induced. As expected, *HOXA1*, a very early and prominent marker of differentiation, showed a strong increase of expression within 24 hours (see also ref. 9). However, expression differences of the other genes investigated were barely visible by qRT-PCR within the first day after start of treatment (Fig. 1D, light grey bars). Thus, impedance measurement is a highly sensitive and robust method to follow the early onset of RA-induced differentiation.

In order to monitor the effect of RA concentration on differentiation induction, we treated NT2 cells with different concentrations of retinoic acid and registered the dielectric resistance profiles (Fig. 2A, Fig. S1). Increasing RA concentrations lead to increased resistance (as indicated by the time point of first statistical significant difference in impedance values of control experiments versus RA treatment, see Fig. S1) and also a steeper slope of the impedance profiles. This correlated with the state of differentiation, as confirmed by the measurement of marker gene expression after 96 hours of treatment (Fig. 2B). We then chose two parameters that characterise RA-induced resistance changes: the slope of the curve obtained when joining the single points of resistance measurements and the time point when the maximum slope is reached. The higher the slope and the earlier the slope maximum, the faster and stronger are the resistance changes that reflect ongoing differentiation. We therefore analysed the dataset shown in Fig. 2A by applying a cubic smoothing spline and

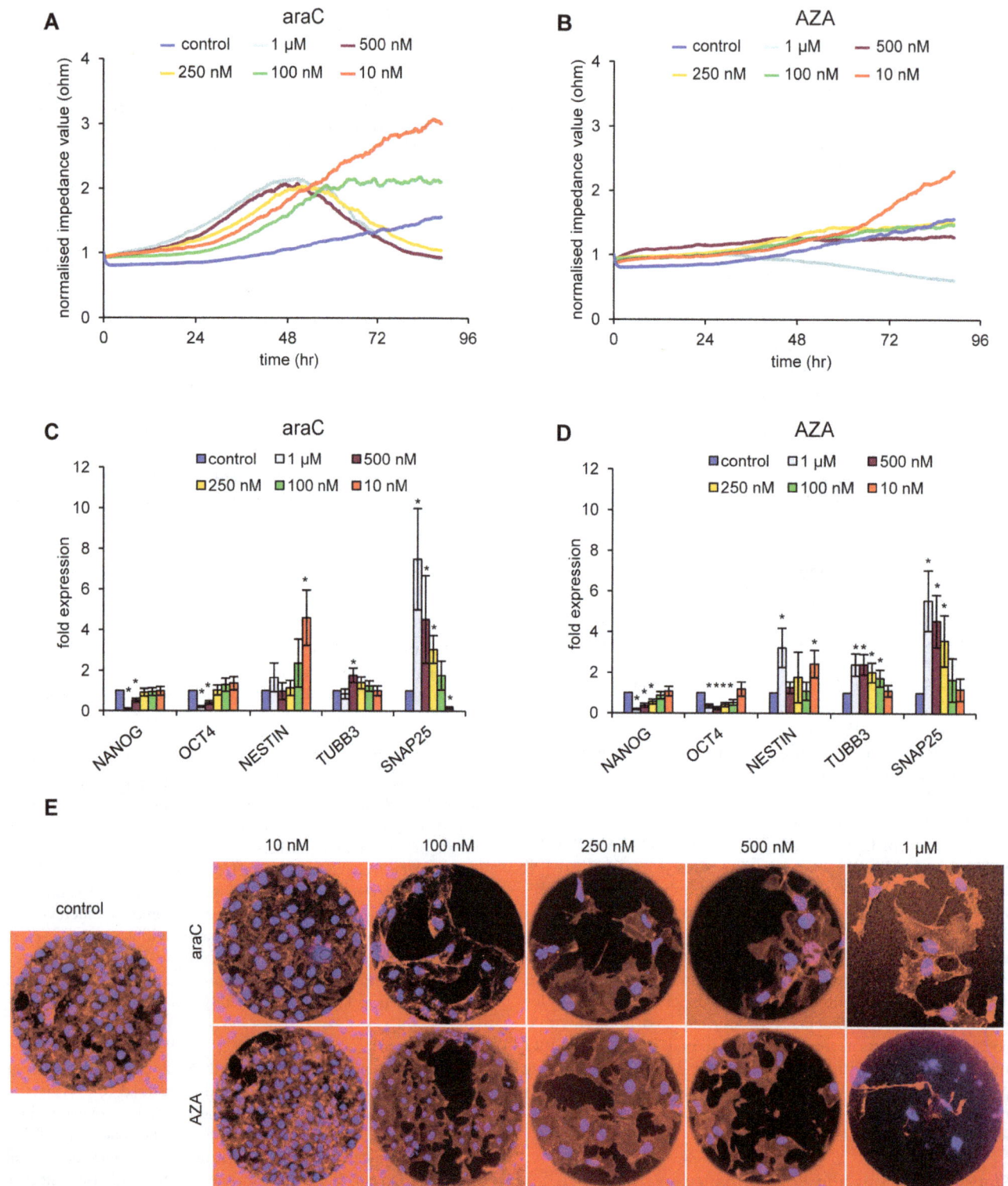

Figure 4. Induced concentration-dependent differentiation by araC and AZA. (A) Impedance profiles of NT2 cells treated with different concentrations of 1β-arabinofuranosylcytosine (araC) during a 4 day period. Concentrations above 100 nM are severely cytotoxic, which leads to a drastic drop in impedance values after 48 hours. Measurements were executed at 45 kHz in 5-minute intervals for 96 hours. One representative experiment is shown. For single diagrams showing the mean of at least three experiments including standard deviations and statistical tests see Fig. S3A. **(B)** Impedance profiles of NT2 cells treated with different concentrations of 5-azacytidine (AZA) during a 4 day period. Concentrations above 100 nM strongly induce proliferative defects, which prevents the increase of impedance values. Measurements were executed at 45 kHz in 5-minute intervals for 96 hours. One representative experiment is shown. For single diagrams showing the mean of at least three experiments including standard deviations and statistical tests see Fig. S3B. **(C)** qRT-PCR expression analysis of stem cell factors *NANOG*, *OCT4* and the neuronal differentiation markers *NESTIN*, *SNAP25* and *TUBB3* in NT2 cells treated with different concentrations of araC and control cells after 96 hours of treatment. All qRT-PCR measurements were repeated at least three times and internally normalised to the corresponding *β-actin* values. Standard

deviations are indicated by error bars. Expression levels of the respective genes showing significant differences compared with the untreated control are marked an asterisk (two-tailed student's t-test; p<0.05). (**D**) qRT-PCR expression analysis of stem cell factors *NANOG*, *OCT4* and the neuronal differentiation markers *NESTIN*, *SNAP25* and *TUBB3* in NT2 cells treated with different concentrations of AZA and control cells after 96 hours of treatment. All qRT-PCR measurements were repeated at least three times and internally normalised to the corresponding *β-actin* values. Standard deviations are indicated by error bars. Expression levels of the respective genes showing significant differences compared with the untreated control are marked an asterisk (two-tailed student's t-test; p<0.05). (**E**) Phalloidin staining of growing cultures. Flourescence images (10× magnification) of NT2 control cells and NT2 cells treated with the indicated concentrations of AZA and araC. The circular dark region is the electrode measuring area covered by the cells. Cells were stained with Phalloidin TRITC (red) and DAPI (blue).

determined for each treatment the maximum slope and the time point the maximum was reached. Then, by calculating the slope/time ratio (in order to compensate for low, but early slope maxima), a clear positive correlation between RA concentration, marker gene expression and maximum slope and time was found (Table S1). Thus the slope/time ratio can be used as an early marker for differentiation.

Electric Impedance Sensing of NT2 Cells Treated with a Panel of Differentiation Inducing Drugs

In order to expand our analyses to other differentiation-inducing substances and to monitor NT2 cells induced to differentiate into other lineages, we treated the cells with the nucleoside analogues araC, DAC and AZA and measured the impedance profiles (Fig. 3A, Fig. S2). In parallel experiments, differentiation was induced by Fibroblast Growth Factor 2 (FGF2; bFGF) and hexamethylene bisacetamide (HMBA). Expression of bFGF increases during retinoic acid induced differentiation of NT2 cells and bFGF treatment of floating spheres of NT2 cells has been shown to trigger terminal differentiation into neurons [14,25–27]. In addition, mesodermal features in aggregated NT2 cells after prolonged treatment with bFGF have also been reported [12]. HMBA treatment of NT2 cells has been shown to result in the expression of marker genes usually associated with epithelial

structures, suggesting that HMBA triggers differentiation into epidermal ectoderm [11,13].

As shown in Figure 3A (see also Fig. S2), araC treatment led to an early increase of resistance, indicating onset of differentiation. This effect was followed by a significant drop below the values of the deoxycytidine (dC) control (Fig. 3A, Fig. S2F). All three nucleoside drugs induced strong proliferation defects after 24 hours of growth (Fig. 3B), most likely by triggering DNA-damage dependent apoptotic pathways [9], which led to low cell numbers and reduced impedance values.

bFGF induced an increase in resistance after 48 hours of treatment (Fig. 3A, Fig. S2B). Cells treated with HMBA, however, showed values in the range of the untreated control. HMBA treatment also reduced the cell number in the culture, but in contrast to araC, an increase in resistance was observed after 72 hours (Fig. 3A, Fig. S2C).

The analysis of the slope/time ratios reveals similar values for RA and HMBA, reflecting a similar potential to induce differentiation (Table S2). Nevertheless, after HMBA treatment the maximum slope was reached much later, indicating a distinct differentiation pathway. bFGF and araC show higher slope/time ratios, due to an earlier onset of increasing resistance for araC and a steeper slope for bFGF, again reflecting different modes of differentiation induction. Taken together, RA, HMBA, bFGF and araC showed significant induction potentials, resulting in specific

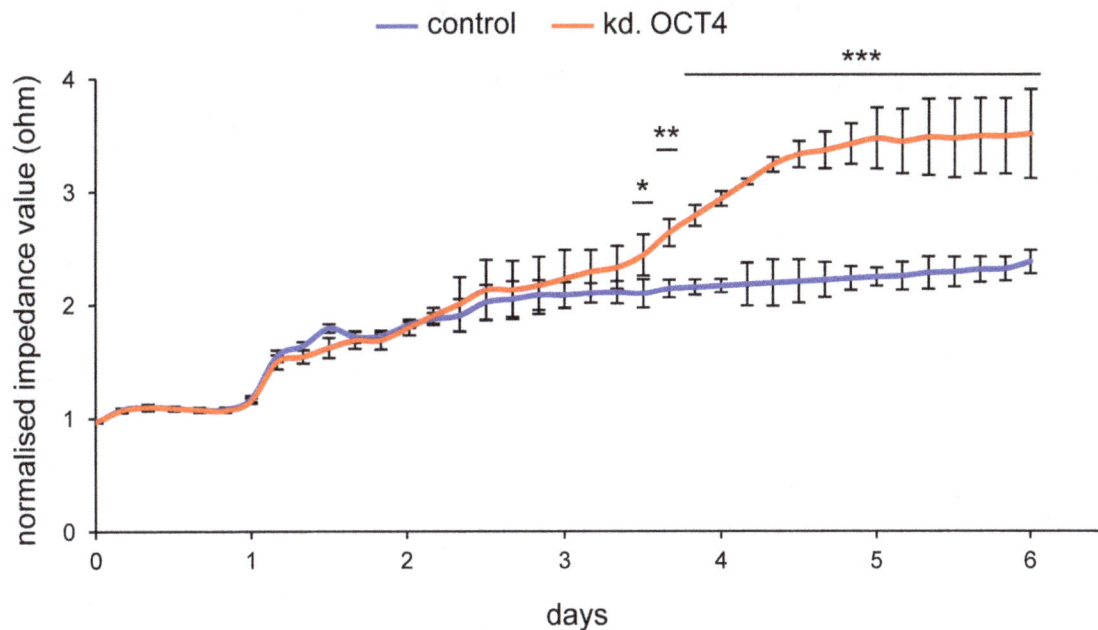

Figure 5. Induced differentiation by RNAi-mediated depletion of OCT4. Impedance profiles of control NT2 cells (blue - scrambled knock down) and NT2 cells depleted for OCT4 (red) during a 6 day period. Measurements were executed at 45 kHz in 5-minute intervals for 6 days. The mean of three independent experiments is shown. Standard deviations are indicated by error bars every four hours. Student's t-test was used for statistical analysis. Differences between control and knock down experiments in the indicated regions have been found to be statistically significant (*p<0.05, **p<0.005, ***p<0.001).

dielectric resistance profiles and slope/time ratios, whereas AZA and DAC treated cells had a similar profile as the controls after 24 hours of treatment. Nevertheless, for araC, DAC and AZA the strong reduction of surviving cells lead to declining impedance values at later time points (Fig. 3A, Fig. 3B, Fig. S2D, S2E and S2F).

As already observed for retinoic acid (Fig. 1C), the majority of differentiation-inducing factors did not induce any significant morphological differences after 24 hours of treatment (Fig. 3C). However, HMBA treated cells, which had shown a delayed increase in dielectric resistance, appeared morphologically different already after 24 hours of treatment (Fig. 3C). After 96 hours of incubation, onset of differentiation was visible for all differentiation-inducing factors (Fig. 3C). In addition, araC, DAC and AZA treated cells showed clear reduction of cell numbers due to the cytotoxicity of these compounds.

As shown by qRT-PCR in Figure 3D, stem cell factors (*OCT4* and *NANOG*) were less expressed after 96 hours of treatment with all substances (except bFGF), indicating ongoing differentiation and loss of pluripotency. Consistent with the morphological alterations, expression of both genes was only weakly reduced after 24 hours. Furthermore, differentiation markers (*SNAP25, NESTIN, TUBB3*) were only moderately increased after 96 hours of treatment (Fig. 3D). At 24 hours, PCR-based expression analysis of differentiation marker genes as well as phase contrast microscopy failed to clearly indicate onset of differentiation (Fig. 3C, 3D, light grey bars). However, the drug-specific impedance values suggest that differentiation already starts within the first day of treatment, especially with RA, bFGF and araC (Fig. 3A, Fig. S2). Thus treatment-induced early differentiation steps obviously trigger changes in cell-extracellular matrix contacts, leading to increased resistance (Fig. 1A, Fig. 2A, Fig. 3A, Table S2). This finding underscores the value of ECIS analysis, especially to analyse early differentiation states, and is also in accordance with recent *in vitro* differentiation data of MSCs, which revealed changes in impedance profiles already within the first hours of adipogenic or osteogenic differentiation [20,21]. Our data further suggest that lineage specific morphological changes influence impedance values in different ways, leading to characteristic resistance profiles and slope/time ratios.

Concentration Dependence of Drug Induced Impedance Curves

As shown in Figures 3A and 3B, the nucleoside drugs were cytotoxic at the concentrations used initially (1 µM), leading to reduced cell numbers after prolonged treatment, therefore impeding the monitoring of differentiation-dependent impedance changes. We addressed the possibility to separate their influence on differentiation induction from cytotoxic side effects by lowering the concentration of these compounds. As shown in Figs. 4 and S3, 10 nM araC induced impedance values that were similar to RA-treated cells, without triggering cell death, with a slope/time ratio comparable to the one obtained with RA at 10 µM (Table S2 and Table S3). Up to 48 hours, also araC concentrations higher than 10 nM induced increased dielectric resistance, leading to positively correlated slope maxima (Table S3). However, at later time points cytotoxicity lead to significantly reduced growth and a drop in impedance (Fig. 4A, Fig. S3). Treatment with AZA at 10 nM lead to a similar, but retarded increase in resistance, again comparable to RA at 10 µM, which is also reflected by the slope/time ratio (Fig. 4B; Fig. S3, Table S3). Thus, as with araC, 10 nM AZA triggered differentiation of NT2 cells without inducing proliferation defects. Higher concentrations of AZA were mostly toxic for the cells.

End point qPCR of stem cell and differentiation markers showed only moderate changes for low concentrations (100 nM and 10 nM) of araC and AZA (Fig. 4C and 4D). Only *NESTIN* expression was significantly increased in araC-treated cells. It should also be noted that the moderate differentiation phenotypes obtained with 10 nM araC or AZA were morphologically similar to those of RA-induced cultures (see Fig. 1C). Stronger differentiation phenotypes and gene expression changes were only obtained with higher, cytotoxic drug concentrations (Fig. 4C–E). These results further underscore the sensitivity of the ECIS method in detecting early onset of differentiation.

Electric Impedance Sensing of OCT4-depleted NT2 Cells

We have recently shown that siRNA-mediated depletion of the stem cell specific protein OCT4 induces neuronal differentiation in NT2 cells [9]. In order to analyse if reduction of OCT4 levels alone will lead to an increase of impedance levels in a similar way as retinoic acid or drug treatment, we seeded NT2 cells into ECIS-arrays and depleted OCT4 by siRNA transfection, using conditions that cause more than 90% reduction of *OCT4* mRNA levels [9]. As shown in Figure 5 increased resistance became apparent for the OCT4 depleted population after 2–3 days, reaching levels comparable to RA treatment around day 4. The delay of 2–3 days is caused by the knock down procedure, as efficient turnover of OCT4 protein is only achieved after 3 days [9], which also explains the observed increase in impedance over this period, as the cells continued to grow (Fig. 5). After substantial depletion of OCT4 was achieved at day 3, the cells started to differentiate, leading to a subsequent increase in resistance values (Fig. 5). Interestingly, when calculating the slope/time ratio (Table S4), we observed very similar values to the ones found in RA treated cells (compare Tables S1, S2 and S4). These findings show that differentiation induction of NT2 cells is triggered by the reduction of OCT4 and provide important confirmation for the argument that the observed changes in resistance were indeed caused by the onset of cellular differentiation.

Discussion

In this study we demonstrate that treatment with well characterised differentiation triggering substances induces distinct dielectric changes in differentiating NT2 cell populations. We were able to generate impedance profiles during the first days of differentiation that allow to monitor the onset of differentiation very early (after 20 hours), when other phenotypic changes or differentiation specific marker gene expression patterns are not yet apparent. Further, by calculating slope/time ratios of each data set, we obtained a measure for the degree of induced differentiation. Impedance analysis also seems to allow the correlation of lineage choices with specific resistance profiles that could enable the prediction of differentiation pathways induced by specific drugs (RA, bFGF - early max. slope) from epidermal differentiation (HMBA - late max. slope).

MSCs show specific impedance profiles during adipogenic or osteogenic differentiation, caused by the modulation of cellular contacts with the gold electrodes or the extracellular matrix [20,21]. As shown in this work, RA-induced neuronal differentiation of NT2 cells is also accompanied by specific interactions with the extracellular matrix (ECM) that change during later stages. Untreated NT2 cells express *alpha5beta1 integrin*, which specifically interacts with fibronectin, whereas NT2 neuron-like cells express *alpha3beta1 integrin* with LAMIN-5 as a ligand in the extracellular matrix [28]. During epithelial transition induced by HMBA, NT2 cells also show specific morphological characteristics. The nuclear to cytoplasm ratio is increased and the actin cytoskeleton is reorganised [29]. Impedance sensing thus seems to discriminate between these different modes of interaction with the extracellular matrix, which are specific for each differentiation pathway. Extracellular matrix molecules can also directly influence the efficiency of differentiation *in vitro* [20]. Besides cell-ECM interactions, also cell-cell contacts are of outmost importance during differentiation. Changes in cell membrane capacitance also can be described by an ECIS scan with different frequencies [21]. For example, connexins, expressed during neuronal differentiation [28] or desmosomes, expressed during epithelial differentiation [29] might come into focus in this respect.

Further we show that the previously described drug-induced differentiation using nucleoside analogues [9] induces similar impedance profiles and slope/time ratios as the natural ligand retinoic acid. The activation of differentiation and the maintenance of differentiation-specific gene expression patterns require substantial epigenetic modulation, especially changes in PcG presence, histone modification patterns and also DNA methylation [30–33]. We have previously used the DNMT inhibitor 2′-deoxy-5-azacytidine to induce hypomethylation and differentiation in NT2 cells [9]. Nevertheless, we observed even stronger differentiation induction by cytarabine, a drug that has no epigenetic modulatory potential and does not trigger any changes in DNA methylation [9]. This suggested a mechanism of drug-dependent differentiation that does not interfere directly with the epigenetic maintenance system.

RA- and nucleoside-drug induced NT2 cells showed very similar early impedance profiles. At higher concentrations of the drugs, cytotoxicity became predominant, resulting in reduced impedance. The ECIS assay allowed to determine the concentration thresholds of the tested drugs that were sufficient for the induction of differentiation without triggering cytotoxic side effects. At concentrations as low as 10 nM, araC and AZA induced differentiation-specific impedance profiles very similar to RA treatment that were stably increasing over more than three days. At concentrations above 100 nM, cytotoxicity became prominent, although surviving cells showed strong differentiation phenotypes. For both DAC and AZA, dose dependent dual mechanisms have been described, with cytotoxic and anti-proliferative effects at high doses and DNA hypomethylation at low doses [34]. Our impedance data further support this concept and may indicate that low doses of these drugs, including araC, can specifically induce differentiation in cancer stem cell populations.

Depletion of OCT4 by RNA interference induced similar resistance profiles and slope/time ratios as RA and low concentrated nucleoside drugs. This confirms the hypothesis that induced differentiation is caused predominantly by the reduction of OCT4 levels, either by proteolytic degradation (as for the nucleoside drugs) or by the transcriptional down-regulation of the gene (as for the natural ligand retinoic acid) [9]. Since cell membrane capacitance obviously is an early marker of differentiation, lower (1 kHz to 8 kHz) and higher frequencies (62.5–64 kHz) to measure the multifrequency complex impedance (Z*) might improve the characterisation of the drug induced cell inherent dielectric properties [21]. We so far only monitored impedance at 45 kHz. Measurements at other frequencies possibly will help to describe cell-cell vs. cell-matrix interactions in more detail. Nevertheless, our work shows that impedance sensing is a robust and sensitive method to describe the effects of differentiation inducing drugs by measuring the dielectric properties of cells in real-time. Using this method, very early differentiation processes could be followed within the first day

after drug treatment in non-invasive conditions. Therefore, our work can serve as a basis for a more detailed analysis of molecular effects of signalling pathways involved in cellular differentiation and for the screening for drugs that modulate cellular phenotypes.

Supporting Information

Figure S1 Induced concentration-dependent differentiation by RA. Impedance profiles comparing untreated NT2 with cells treated with 10 nM RA (**A**), 500 nM RA (**B**), 1 μM RA (**C**), 5 μM RA (**D**) and 10 μM RA (**E**) are shown. Measurements were executed at 45 kHz in 5-minute intervals for 96 hours. Each experiment was repeated at least three times. Standard deviations are indicated by error bars every four hours. Student's t-test was used for statistical analysis (*p<0.05. **p<0.005). Black lines show regions with significant differences in respect to the untreated control.

Figure S2 Induced differentiation by a panel of drugs. Impedance profiles comparing untreated NT2 with cells treated with 1 μM dC (**A**), 50 μM bFGF (**B**), 5 mM HMBA (**C**), 1 μM DAC (**D**), 1 μM AZA (**E**) and 1 μM araC (**F**) are shown. Measurements were executed at 45 kHz in 5-minute intervals for 96 hours. Each experiment was repeated at least three times. Standard deviations are indicated by error bars every four hours. Student's t-test was used for statistical analysis (*p<0.05. **p<0.005). Black lines show regions with significant differences in respect to the dC control.

Figure S3 Induced concentration-dependent differentiation by araC and AZA. (**A**) Impedance profiles comparing untreated NT2 cells (dark blue) and cells treated with 1 μM (light blue), 500 nM (purple), 250 nM (yellow), 100 nM (green) and 10 nM (red) araC. (**B**) Impedance profiles comparing untreated NT2 cells (dark blue) and cells treated with 1 μM (light blue), 500 nM (purple), 250 nM (yellow), 100 nM (green) and 10 nM (red) AZA. Measurements were executed at 45 kHz in 5-minute intervals for 96 hours. Each experiment was repeated at least three times. Standard deviations are indicated by error bars every four hours. Student's t-test was used for statistical analysis (*p<0.05. **p<0.005). Black lines show regions with significant differences in respect to the control.

Table S1 Slope maxima of RA-treated NT2 cells.

Table S2 Slope maxima of drug-treated NT2 cells.

Table S3 Slope maxima of araC- and AZA-treated NT2 cells.

Table S4 Slope maxima of OCT4-depleted NT2 cells.

Table S5 RT-Primer pairs used in this study.

Acknowledgments

We thank Francesca Tuorto for help with microscopy, Fabian Graf for help with impedance measurements and cell staining and Sebastian Bender for bioinformatic assistance. S.Ö. is PhD student of the HBGIS graduate school, Heidelberg, and member of the joint PhD program "disease models and drugs" between Heidelberg University and Mannheim University of Applied Sciences.

Author Contributions

Conceived and designed the experiments: CM AB SÖ. Performed the experiments: AB SÖ. Analyzed the data: CM AB SÖ. Contributed reagents/materials/analysis tools: CM AB. Wrote the paper: CM AB SÖ.

References

1. Sell S (2004) Stem cell origin of cancer and differentiation therapy. Crit Rev Oncol Hematol 51: 1–28.
2. Degos L (1999) Differentiating agents in the treatment of leukemia. Leuk Res 14: 717–719.
3. von Wangenheim KH, Peterson HP (2008) The role of cell differentiation in controlling cell multiplication and cancer. J Cancer Res Clin Oncol 134: 725–741.
4. Jones PA, Taylor SM (1980) Cellular differentiation, cytidine analogs and DNA methylation. Cell 20: 85–93.
5. Hatse S, De Clercq E, Balzarini J (1999) Role of antimetabolites of purine and pyrimidine nucleotide metabolism in tumor cell differentiation. Biochem Pharmacol 58: 539–555.
6. Burnett A, Wetzler M, Löwenberg B (2011) Therapeutic Advances in Acute Myeloid Leukemia. J Clin Oncol 29: 487–494.
7. Robak T, Wierzbowska A (2009) Current and emerging therapies for acute myeloid leukemia. Clin Ther 31: 2349–2370.
8. Ewald B, Sampath D, Plunkett W (2008) Nucleoside analogs: molecular mechanisms signalling cell death. Oncogene 27: 6522–6537.
9. Musch T, Oz Y, Lyko F, Breiling A (2010) Nucleoside drugs induce cellular differentiation by caspase-dependent degradation of stem cell factors. PLoS One 5: e10726.
10. Bocker MT, Tuorto F, Raddatz G, Musch T, Yang FC, et al. (2012) Hydroxylation of 5-methylcytosine by TET2 maintains the active state of the mammalian HOXA cluster. Nat Commun 3: 818.
11. Andrews PW (2002) From teratocarcinomas to embryonic stem cells. Philos Trans R Soc Lond B Biol Sci 357: 405–417.
12. Pal R, Ravindran G (2006) Assessment of pluripotency and multilineage differentiation potential of NTERA-2 cells as a model for studying human embryonic stem cells. Cell Prolif 39: 585–598.
13. Przyborski SA, Christie VB, Hayman MW, Stewart R, Horrocks GM (2004) Human embryonal carcinoma stem cells: models of embryonic development in humans. Stem Cells Dev 13: 400–408.
14. Marchal-Victorion S, Deleyrolle L, De Weille J, Saunier M, Dromard C, et al. (2003) The human NTERA2 neural cell line generates neurons on growth under neural stem cell conditions and exhibits characteristics of radial glial cells. Mol Cell Neurosci 24: 198–213.
15. Pleasure SJ, Page C, Lee VM (1992) Pure, postmitotic, polarized human neurons derived from NTera 2 cells provide a system for expressing exogenous proteins in terminally differentiated neurons. J Neurosci 12: 1802–1815.
16. Wegener J, Keese CR, Giaever I (2000) Electric cell-substrate impedance sensing (ECIS) as a noninvasive means to monitor the kinetics of cell spreading to artificial surfaces. Exp Cell Res 259: 158–166.
17. Keese CR, Bhawe K, Wegener J, Giaever I (2002) Real-time impedance assay to follow the invasive activities of metastatic cells in culture. Biotechniques 33: 842–850.
18. Hong J, Kandasamy K, Marimuthu M, Choi CS, Kim S (2011) Electrical cell-substrate impedance sensing as a non-invasive tool for cancer cell study. Analyst 136: 237–245.
19. Giaever I, Keese CR (1991) A morphological biosensor for mammalian cells. Nature 366: 591–592.
20. Angstmann M, Brinkmann I, Bieback K, Breitkreutz D, Maercker C (2011) Monitoring human mesenchymal stromal cell differentiation by electrochemical impedance sensing. Cytotherapy 13: 1074–1089.
21. Bagnaninchi PO, Drummond N (2011) Real-time label-free monitoring of adipose-derived stem cell differentiation with electric cell-substrate impedance sensing. Proc Natl Acad Sci USA 108: 6462–6467.
22. Andrews PW, Damjanov I, Simon D, Banting GS, Carlin C, et al. (1984) Pluripotent embryonal carcinoma clones derived from the human teratocarcinoma cell line Tera-2. Differentiation in vivo and in vitro. Lab Invest 50: 147–162.
23. Andrews PW (1984) Retinoic acid induces neuronal differentiation of a cloned human embryonal carcinoma cell line in vitro. Dev Biol 103: 285–293.
24. Green PJ, Silverman BW (1994) Nonparametric Regression and Generalized Linear Models: A Roughness Penalty Approach. London: Chapman & Hall. 182 p.
25. Schuldiner M, Yanuka O, Itskovitz-Eldor J, Melton DA, Benvenisty N (2000) Effects of eight growth factors on the differentiation of cells derived from human embryonic stem cells. Proc Natl Acad Sci USA 97: 11307–11312.
26. Zhang RL, Zhang L, Zhang ZG, Morris D, Jiang Q, et al. (2003) Migration and differentiation of adult rat subventricular zone progenitor cells transplanted into the adult rat striatum. Neuroscience 116: 373–382.
27. Coyle DE, Li J, Baccei M (2011) Regional differentiation of retinoic acid-induced human pluripotent embryonic carcinoma stem cell neurons. PLoS One 6: e16174.
28. Meland MN, Herndon ME, Stipp CS (2010) Expression of alpha5 integrin rescues fibronectin responsiveness in NT2N CNS neuronal cells. J Neurosci Res 88: 222–232.
29. Simoes PD, Ramos T (2007) Human pluripotent embryonal carcinoma NTERA2 cl.D1 cells maintain their typical morphology in an angiomyogenic medium. J Negat Results Biomed 6: 5.
30. Fouse SD, Shen Y, Pellegrini M, Cole S, Meissner A, et al. (2008) Promoter CpG methylation contributes to ES cell gene regulation in parallel with oct4/Nanog, PcG complex, and histone H3 K4/K27 trimethylation. Cell Stem Cell 2: 160–169.
31. Mohn F, Weber M, Rebhan M, Roloff TC, Richter J, et al. (2008) Lineage-specific polycomb targets and de novo DNA methylation define restriction and potential of neuronal progenitors. Mol Cell 30: 755–766.
32. Bracken AP, Dietrich N, Pasini D, Hansen KH, Helin K (2006) Genome-wide mapping of Polycomb target genes unravels their roles in cell fate transitions. Genes Dev 20: 1123–1136.
33. Sessa L, Breiling A, Lavorgna G, Silvestri L, Casari G, et al. (2007) Noncoding RNA synthesis and loss of Polycomb group repression accompanies the colinear activation of the human HOXA cluster. RNA 13: 223–239.
34. Stresemann C, Lyko F (2008) Modes of action of the DNA methyltransferase inhibitors azacytidine and decitabine. Int J Cancer 123: 8–13.

8

Controlled Synthesis and Microwave Absorption Property of Chain-Like Co Flower

Chao Wang[1,2]*, Surong Hu[1], Xijiang Han[2], Wen Huang[1], Lunfu Tian[1]

1 Institute of Mechanical Manufacturing Technology, China Academy of Engineering Physics, Mianyang, People's Republic of China, **2** Department of Chemistry, Harbin Institute of Technology, Harbin, People's Republic of China

Abstract

Chain-like Co flower is synthesized by simply modulating the reaction conditions via a facile liquid-phase reduction method. The morphology evolution process and transformation mechanism from particle to flower and finally to chain-like flower have been systematically investigated. [001] is the preferred growth orientation due to the existence of easy magnetic axis. The microwave loss mechanism can be attributed to the synergistic effect of magnetic loss and dielectric loss, while magnetic loss is the main loss mechanism. In addition, the special microstructure of chain-like Co flower may further enhance microwave attenuation. The architectural design of functional material morphology is critical for improving its property toward future application.

Editor: Bing Xu, Brandeis University, United States of America

Funding: This work was funded by the Natural Science Foundation of China no. 51202228 (www.nsfc.gov.cn). The funders had no role in study design, data collection and analysis, decision to publish, or preparation of the manuscript.

Competing Interests: The authors have declared that no competing interests exist.

* E-mail: wangchaohit@126.com

Introduction

The microwave absorbing material has received extensive attention due to their prospective application in electromagnetic shielding coatings, microwave darkrooms, and self-concealing weapons in industry, commerce, and military affairs [1–3]. The conventional electromagnetic loss mechanisms are magnetic loss and dielectric loss, so most of microwave absorbers are composed of magnetic and dielectric materials. However, as a kind of representative magnetic material, only a few literatures have discussed the microwave absorption property of Co due to its poor microwave absorption ability. Kato et al. has validated cobalt particles could only exhibit strong microwave absorption during very narrow frequency band [4], while Cao et al. found hollow chain-like cobalt displayed absorption ability no stronger than −11 dB [5]. Therefore, it is valuable to deeply explore the microwave absorption mechanism of Co and find the route to improve its absorption.

In our previous research [6], we reported the synthesis of differently shaped Ni and found the microwave absorbing ability could be apparently influenced by material microstructure. In addition, controllable preparation of hierarchical building components with specific structures and novel properties is now attracting much attention of scientists not just for its role in deeply comprehending the self-assembly mechanism but also for its prospective applications as functional materials [7–9]. As a sequence, here we exhibit the facile synthesis of chain-like Co flower via modulating experiment conditions and investigate the relationship between morphology and electromagnetic property.

Experiment

Materials

$CoCl_2 \cdot 6H_2O$, PVP K30, $N_2H_4 \cdot H_2O$, and ethylene glycol (EG) were bought from Guangfu Chemical Co. Ltd. (Tianjin, China) and NaOH was bought from Dalu Chemical Co. Ltd. (Tianjin, China), which were analytical grade and used as received.

Synthesis

Differently shaped Co was synthesized via modulating the content of $CoCl_2 \cdot 6H_2O$ and NaOH. In a typical experiment, 0.01–0.02 g $CoCl_2 \cdot 6H_2O$ and 0.5 g PVP were added into 22.5 mL EG at 85°C in a three-necked flask. After vigorous stirring for 10 min, 0.2–0.8 g NaOH was added and 10 min later, 8 mL $N_2H_4 \cdot H_2O$ was injected, then the reaction was maintained until the solution became clear and some black particles emerged to make sure the reaction was completed. The product was collected by centrifugation at 3000 rpm for 10 min and washed with deionized water six times and with ethanol three times to eliminate residual PVP. Finally the product was dried in a vacuum oven at 60°C for 12 h.

Characterization

The morphologies of the samples were characterized by scanning electron microscopy (SEM, FEI SIRION). The crystallite structure were determined using an XRD-6000 X-ray diffractometer (Shimadzu) with a Cu Kα radiation source ($\lambda = 1.5405$ Å, 40.0 kV, 30.0 mA). The magnetic properties were measured by a vibrating sample magnetometer (VSM, Lake Shore 7307). A HP-5783E vector network analyzer was applied to characterize relative complex permeability and permittivity during 2–18 GHz for the calculation of reflection loss (microwave absorption). A

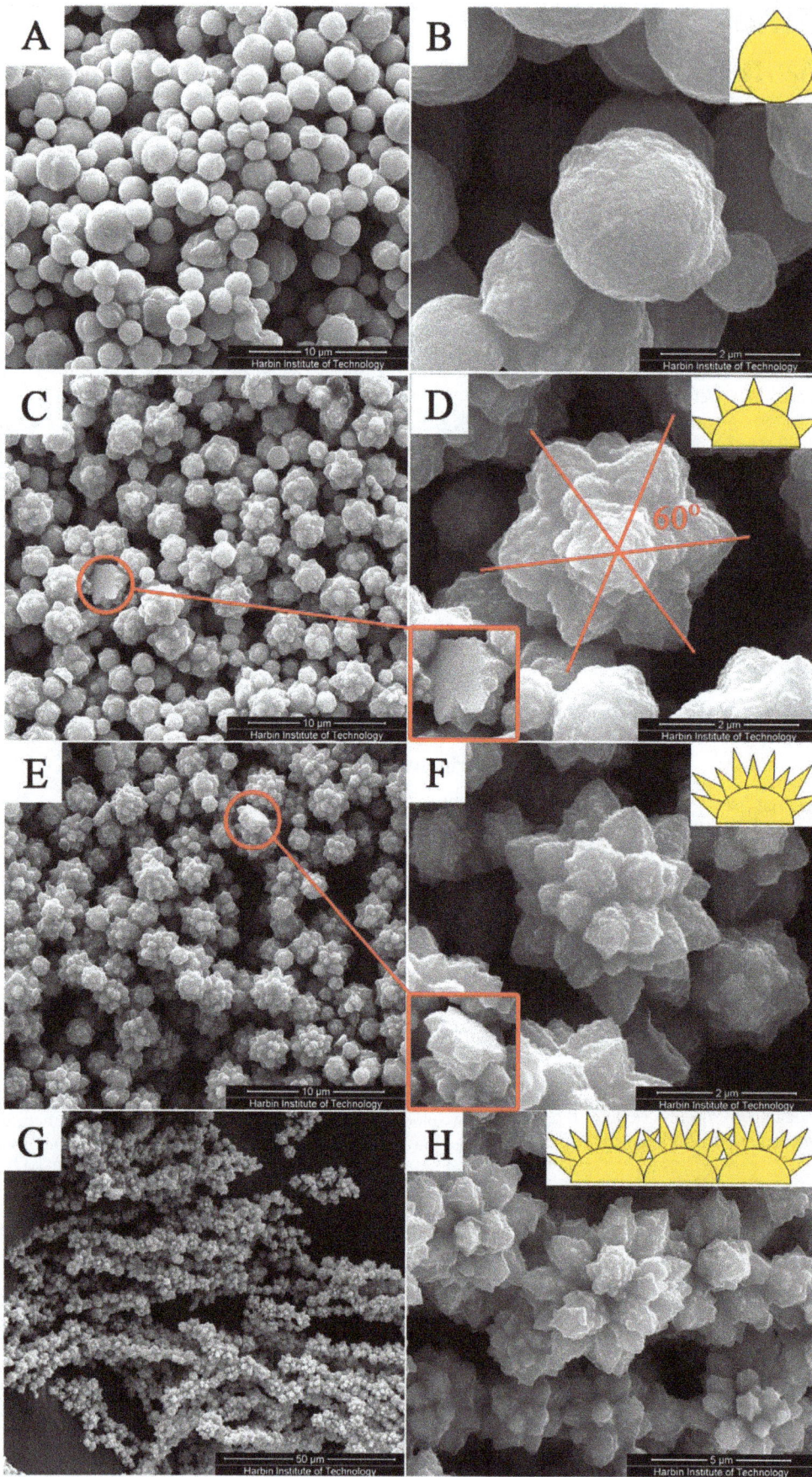

Figure 1. SEM images of the Co prepared at varying experiment conditions: (A, B: Co particles) 0.2 g, (C, D: rudiment of Co flower) 0.4 g, (E, F: Co flower) 0.8 g NaOH is added at the CoCl₂·H₂O content of 0.01 g; (G, H: chain-like cobalt flower) 0.8 g NaOH is added while the CoCl₂·H₂O content is increased to 0.02 g.

sample containing 60 wt% of Co was pressed into a ring with an outer diameter of 7 mm, an inner diameter of 3 mm, and a thickness of 2 mm for electromagnetic measurement, in which wax was used as binder.

Results and Discussion

Typical SEM images of obtained product with different magnifications are presented in Fig. 1. When only 0.2 g NaOH is added, spheric Co particles with a few short stings growing on the surface are obtained as shown in Figs. 1A and 1B. If 0.4 g NaOH is added, stings' number and length increase, resulting in the formation of Co flower rudiment as seen by Figs. 1C and 1D. In addition, some stings present a symmetrical structure as hexagon with adjacent ones displaying an angle of 60°. Once the NaOH is increased to 0.8 g, well-crystallized Co flower with the maximum and longest stings is obtained as shown in Figs. 1E and 1F. Surprisingly and interestingly, Figs. 1D and 1F demonstrate that the bottoms are flat regardless the morphology of flower. If CoCl₂·H₂O is increased from 0.01 g to 0.02 g and keep NaOH still at 0.8 g, chain-like Co flower is obtained as exhibited in Figs. 1G and 1H.

PVP is a conventional surfactant that can be used to activate certain crystal plane, which may lead to newly reduced Co atoms adsorbing on the plane of existed Co particles and finally the stings are formed. In addition, $N_2H_4·H_2O$ only exhibits strong reduction ability under alkaline condition, that's why the growth of stings is slow under low NaOH content. While increasing NaOH addition may raise the number of reduced Co atoms and promote the rise of stings' number and length and finally induce the Co flower rudiment growth. Once NaOH reaches 0.8 g, newly reduced Co atoms will be much more and enough to accelerate stings' growth and the well organized Co flower is shaped.

Under the preparation conditions of well grown Co flower, chain-like Co flower can be obtained without any external assistant magnetic field if CoCl₂·6H₂O addition is increased from

0.01 g to 0.02 g. This may be due to each Co flower generates its own magnetostatic field and the increase of $CoCl_2·6H_2O$ will result in large number of Co flowers, which enhance the magnetostatic field in the reaction system. Hence, these flowers are arranged linearly along magnetic force line to form chain-like Co flower. The following discussions are based on the measure data of these chain-like Co flower.

Fig. 2 shows an XRD pattern of obtained chain-like Co flower (upper) and the pattern from data bank entry (lower). The characteristic peaks arise at $2\theta = 41.61°$, 44.42°, 47.29°, 75.89° can be indexed to (100), (002), (101), (110) planes, respectively, of hexagonal-phase cobalt (JCPDS: 05-0727). No characteristic peaks arising from impurities of cobalt oxides or hydroxides are detected, which may be due to the floating reduction product N_2 and NH_3 can exclude air to inhibit oxidation of newly reduced Co. It is found that the peaks relative intensities corresponding to (002)/ (100) and (002)/(101) planes are significantly higher than that of standard values, indicating the preferred growth orientation is along [001] direction, which is very likely to the reported dendritic cobalt [10]. In fact, [001] direction is the easy magnetic axis of Co [11], while magnetostatic energy will be the lowest and reaction system can maintain steady if Co atoms are stacked along easy magnetic axis. Therefore, the growth rate of [001] direction is faster than others.

As shown in Fig. 3, the saturation magnetization (M_s), remanent magnetization (M_r), coercivity (H_c) of chain-like Co flower are 181.38 emu/g, 2.56 emu/g, 33.74 Oe, respectively. It should be noted that the M_s of chain-like Co flower is a little higher than that of bulk Co [12], which may arise from its special chain-like microstructure or tiny experimental error. In addition, the high M_s can induce high initial permeability which is beneficial for microwave absorption [6].

The real parts of relative complex permittivity and permeability symbolize storage ability to electromagnetic energy, while the imaginary parts represent loss ability. The low permittivity imaginary part (ε'') as shown in Fig. 4 demonstrates dielectric loss is weak. It is worthy to be noted that ε'' is a little below zero under some bands while there is not a convincing interpretation for the negative value, Deng et al. explain the phenomenon as the magnetic energy being radiated out [2], Chiu et al. point out it is meaningless and might be due to noise [13], while Chen et al. point out it does not arise from test error and needs further study [14]. The real part of relative complex permeability (μ') decreases as the increase of frequency displays slight frequency-scattering effect. The imaginary part of permeability (μ'') weak peak around 6 GHz exhibits the existence of natural resonance just as reported dendritic Co [3].

Microwave absorption ability can be preliminarily judged by dielectric loss factor ($\tan\delta_e = \varepsilon''/\varepsilon'$) and magnetic loss factor ($\tan\delta_m = \mu''/\mu'$), which are shown in Fig. 5. The larger $\tan\delta_m$ than $\tan\delta_e$ demonstrates that main loss mechanism is magnetic loss rather than dielectric loss. There are three dielectric loss factor peaks around 4, 10, and 14.5 GHz which arise from relative high ε'' and low ε' at above frequencies. The $\tan\delta_m$ rapidly increases in 2–6 GHz and undulately increases during 6–16.4 GHz, then dramatically decreases. Due to magnetic loss is main loss mechanism and $\tan\delta_m$ is large around 16.4 GHz, the microwave absorption peak may emerge at 16.4 GHz.

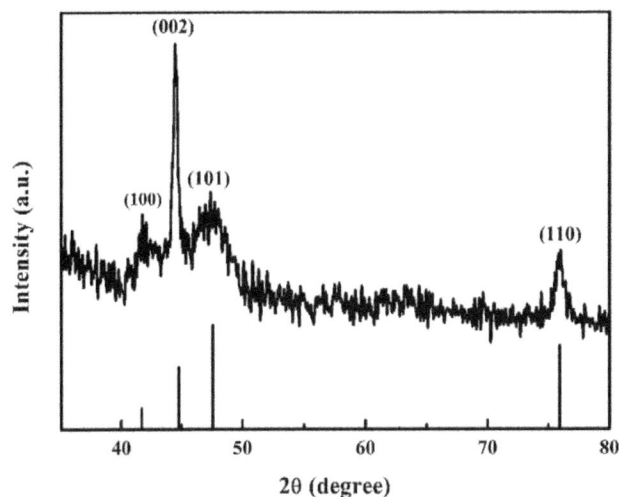

Figure 2. XRD pattern of chain-like Co flower (upper) and standard diffraction pattern of Co (JCPDS Card No. 05-0727) (lower).

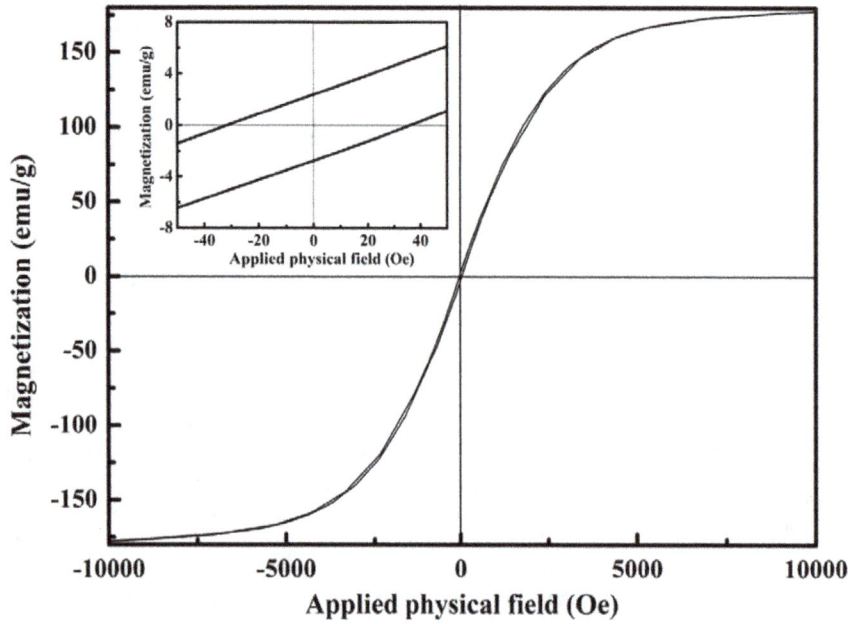

Figure 3. Magnetic hysteresis loop at room temperature for chain-like Co flower.

According to the measured data of permittivity and permeability, reflection loss (RL) usually can be calculated by following equation according to transmission line theory and assuming the absorber is attached on high reflective metal plates [15–17]:

$$\mathrm{RL} = 20 \log |(Z_{in} - Z_0)/(Z_{in} + Z_0)|$$

where Z_0 is the impedance of free space, and Z_{in} is the input characteristic impedance, which can be expressed as:

$$Z_{in} = Z_0 \sqrt{\mu_r/\varepsilon_r} \tanh\left\{j(2\pi f d/c)\sqrt{\mu_r \varepsilon_r}\right\}$$

where c is the velocity of light and d is the thickness of an absorber. The reflection loss (microwave absorption) ability of chain-like Co flower is shown in Fig. 6.

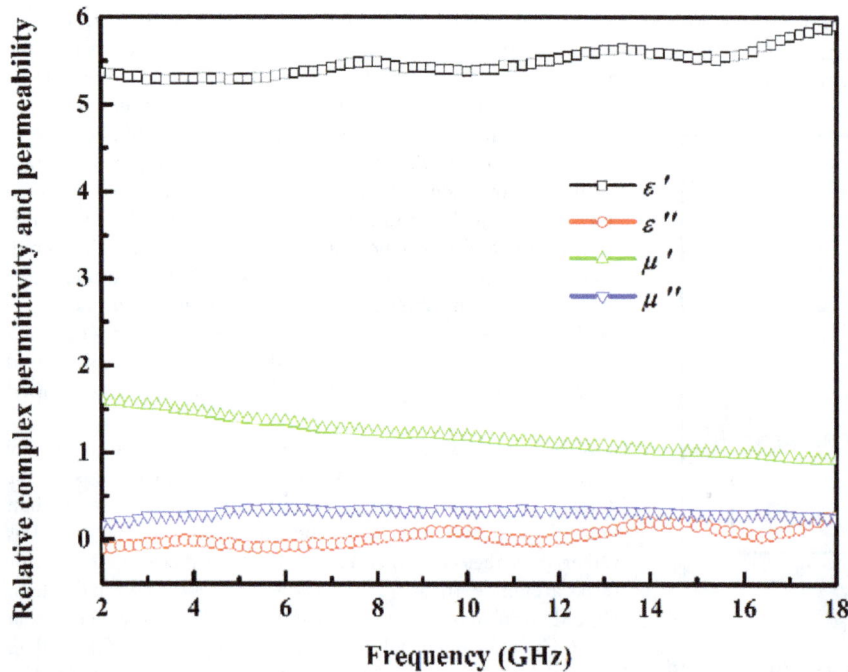

Figure 4. The relative complex permittivity (ε' and ε'') and permeability (μ' and μ'') of chain-like Co flower.

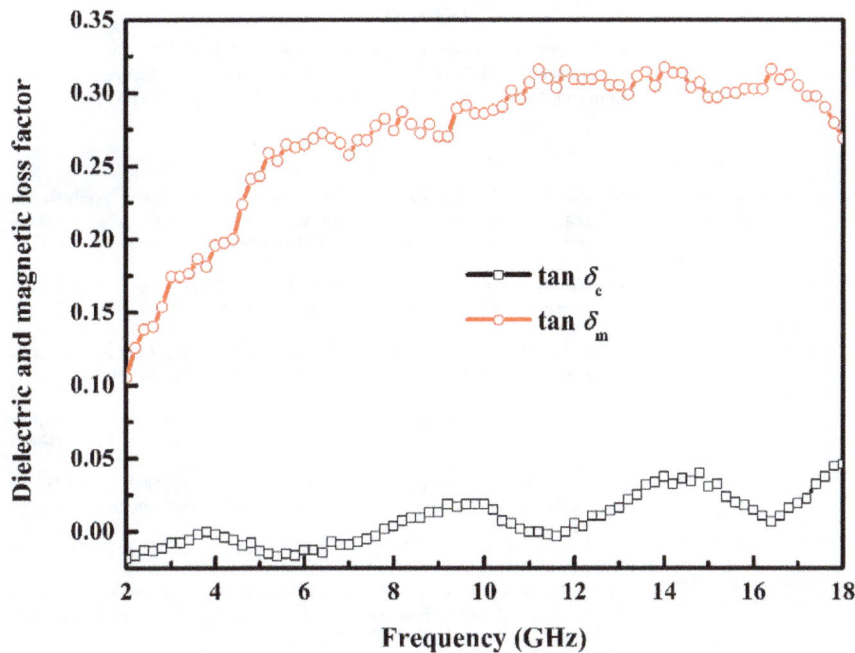

Figure 5. The dielectric loss factor (tanδ_e) and magnetic loss factor (tanδ_m) of chain-like Co flower.

Conventional measure frequency band of microwave absorbing materials is 2–18 GHz. It is reported that Co nanochains can reach reflection loss peak about −11 dB at 17.2 GHz at the thickness of 2.5 mm, and Co nanoparticles can only reach the maximum reflection loss −5 dB in 2–5 GHz at the thickness of 3.16 mm [4,5]. While chain-like Co flower at a thinner thickness of 2 mm as shown in Fig. 6 exhibits stronger reflection loss (−11.5 dB at 14.8 GHz) than the above reported Co materials, which may be illustrated from its special microstructure. There are many sharp petals that can be tuned with the incident microwave for the point discharge effect just as the action mechanism of lightning rod to lightning, and electromagnetic energy will be induced into dissipative current. Then, the chain-like structure is

beneficial for transmission of dissipative current and finally leads to energy attenuation by transforming to heat.

It should be noted that chain-like Co flower reaches the loss peak at 14.8 GHz rather than the maximum tanδ_m frequency of 16.4 GHz. Loss peak frequency is determined by loss mechanism or absorber thickness that can be characterized as times of $\lambda_m/4$ [6]. Although Co is a kind of magnetic loss microwave absorbing material, free electronic will be polarized due to its electronic conductive essentiality and dielectric loss can be induced under external alternated electromagnetic field. Therefore, dielectric loss is also one of the loss mechanisms even though it is not strong. The tanδ_e at 16.4 GHz in Fig. 5 demonstrates dielectric loss is weak, so microwave absorption can not be the strongest even though tanδ_m is the largest at this frequency. However, there is a tanδ_e peak around 14.8 GHz and tanδ_m is also very large at that frequency, so chain-like Co flower reaches the strongest microwave absorption. After 14.8 GHz, the variation tendencies of tanδ_e and tanδ_m are just opposite and result in microwave absorption becoming weaker and weaker in total. The above analysis adequately demonstrates the loss nature is coming from synergistic effect of magnetic loss and dielectric loss, while magnetic loss is the main loss mechanism.

Conclusions

In summary, we successfully realize the evolution progress from Co particle to chain-like Co flower via modulating reaction conditions. The increasing number of newly reduced Co atoms can accelerate petals' growth, and the flowers will be arranged linearly under magnetostatic force to form chain-like structure. Both magnetic loss and dielectric loss are beneficial for microwave absorption though the former is the main loss mechanism. In addition, the special sharp petal microstructure can be tuned with incident microwave and induces electromagnetic energy into dissipative current, while chain-like microstructure is beneficial for dissipative current transmission and finally leads to energy attenuation, which enhances the microwave absorption. Our research demonstrates that the architectural microstructure design

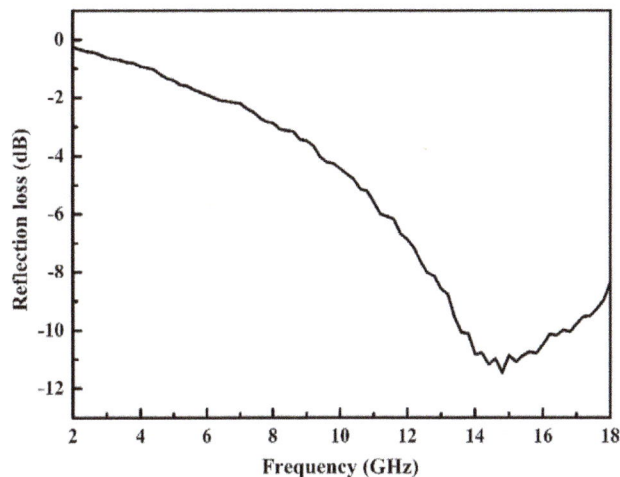

Figure 6. Absorbing ability of chain-like Co flower at 2 mm thickness.

of functional material may be a promising route to improve property towards to future application.

Author Contributions

Conceived and designed the experiments: CW XJH. Performed the experiments: CW. Analyzed the data: CW. Contributed reagents/materials/analysis tools: CW XJH. Wrote the paper: SRH WH LFT.

References

1. An ZG, Pan SL, Zhang JJ (2009) Facile Preparation and Electromagnetic Properties of Core-Shell Composite Spheres Composed of Aloe-like Nickel Flowers Assembled on Hollow Glass Spheres. J. Phys. Chem. C 113: 2715–2721.

2. Deng LJ, Han MG (2007) Microwave absorbing performances of multiwalled carbon nanotube composites with negative permeability. Appl. Phys. Lett. 91: 023119.

3. Wang C, Han XJ, Zhang XL, Hu SR, Zhang T, et al. (2010) Controlled synthesis and morphology-dependent electromagnetic properties of hierarchical cobalt assemblies. J. Phys. Chem. C 114: 14826–14830.

4. Kato Y, Sugimoto S, Shinohara K, Tezuka N, Kagotani T, et al. (2002) Magnetic properties and microwave absorption properties of polymer-protected cobalt nanoparticles. Mater. Trans. 43: 406–409.

5. Shi XL, Cao MS, Yuan J, Fang XY (2009) Dual nonlinear dielectric resonance and nesting microwave absorption peaks of hollow cobalt nanochains composites with negative permeability. Appl. Phys. Lett. 95: 163108.

6. Wang C, Han XJ, Xu P, Wang JY, Du YC, et al. (2010) Controlled synthesis of hierarchical nickel and morphology-dependent electromagnetic properties. J. Phys. Chem. C 114: 3196–3203.

7. Hong J, Stefanescu E, Liang P, Joshi N, Xue S, et al. (2012) Carbon nanotubes based 3-D matrix for enabling three-dimensional nano-magneto-electronics. PloS One 7: e40554.

8. Che RC, Zhi CY, Liang CY, Zhou XG (2006) Fabrication and microwave absorption of carbon nanotubes/$CoFe_2O_4$ spinel nanocomposite. Appl. Phys. Lett. 88: 033105.

9. Guo L, Liang F, Wen XG, Yang SH, He L, et al. (2007) Uniform magnetic chains of hollow cobalt mesospheres from one-pot synthesis and their assembly in solution. Adv. Funct. Mater. 17: 425–430.

10. Zhu LP, Xiao HM, Zhang WD, Yang Y, Fu SY (2008) Synthesis and characterization of novel three-dimensional metallic Co dendritic superstructures by a simple hydrothermal reduction route. Cryst. Growth Des. 8: 1113–1118.

11. Niu HL, Chen QW, Zhu HF, Lin YS, Zhang X (2003) Magnetic field-induced growth and self-assembly of cobalt nanocrystallites. J. Mater. Chem. 13: 1803–1805.

12. Li YL, Zhao JZ, Su XD, Zhu YC, Wang Y, et al. (2009) A facile aqueous phase synthesis of cobalt microspheres at room temperature. Colloids Surf. A 336: 41–45.

13. Chiu SC, Yu HC, Li YY (2010) High electromagnetic wave absorption performance of silicon carbide nanowires in the gigahertz range. J. Phys. Chem. C 114: 1947–1952.

14. Chen YJ, Gao P, Wang RX, Zhu CL, Wang LJ, et al. (2009) Porous Fe_3O_4/SnO_2 core/shell nanorods: Synthesis and electromagnetic properties. J. Phys. Chem. C 113: 10061–10064.

15. Yu SH, MacGillivray LR, Janiak C (2012) Nanocrystals. CrystEngComm 14: 7531–7534.

16. Ni SB, Sun XL, Wang XH, Zhou G, Yang F, et al. (2010) Low temperature synthesis of Fe_3O_4 micro-spheres and its microwaveabsorption properties. Mater. Chem. Phys. 124: 353–358.

17. Wang C, Han XJ, Xu P, Zhang XL, Du YC, et al. (2011) The electromagnetic property of chemically reduced graphene oxide and its application as microwave absorbing material. Appl. Phys. Lett. 98: 072906.

Vehicle Exhaust Gas Clearance by Low Temperature Plasma-Driven Nano-Titanium Dioxide Film Prepared by Radiofrequency Magnetron Sputtering

Shuang Yu[1], Yongdong Liang[2], Shujun Sun[1], Kai Zhang[1], Jue Zhang[1,2]*, Jing Fang[1,2]

1 Academy for Advanced Interdisciplinary Studies, Peking University, Beijing, China, **2** College of Engineering, Peking University, Beijing, China

Abstract

A novel plasma-driven catalysis (PDC) reactor with special structure was proposed to remove vehicle exhaust gas. The PDC reactor which consisted of three quartz tubes and two copper electrodes was a coaxial dielectric barrier discharge (DBD) reactor. The inner and outer electrodes firmly surrounded the outer surface of the corresponding dielectric barrier layer in a spiral way, respectively. Nano-titanium dioxide (TiO_2) film prepared by radiofrequency (RF) magnetron sputtering was coated on the outer wall of the middle quartz tube, separating the catalyst from the high voltage electrode. The spiral electrodes were designed to avoid overheating of microdischarges inside the PDC reactor. Continuous operation tests indicated that stable performance without deterioration of catalytic activity could last for more than 25 h. To verify the effectiveness of the PDC reactor, a non-thermal plasma(NTP) reactor was employed, which has the same structure as the PDC reactor but without the catalyst. The real vehicle exhaust gas was introduced into the PDC reactor and NTP reactor, respectively. After the treatment, compared with the result from NTP, the concentration of HC in the vehicle exhaust gas treated by PDC reactor reduced far more obviously while that of NO decreased only a little. Moreover, this result was explained through optical emission spectrum. The O emission lines can be observed between 870 nm and 960 nm for wavelength in PDC reactor. Together with previous studies, it could be hypothesized that O derived from catalytically O_3 destruction by catalyst might make a significant contribution to the much higher HC removal efficiency by PDC reactor. A series of complex chemical reactions caused by the multi-components mixture in real vehicle exhaust reduced NO removal efficiency. A controllable system with a real-time feedback module for the PDC reactor was proposed to further improve the ability of removing real vehicle exhaust gas.

Editor: Andrew C. Marr, Queen's University Belfast, United Kingdom

Funding: This study is supported by the Bioelectrics Inc., United States of America. The funders had no role in study design, data collection and analysis, decision to publish, or preparation of the manuscript.

Competing Interests: This study is supported by the Bioelectrics Inc., United States of America. There are no patents, products in development or marketed products to declare.

* E-mail: zhangjue@pku.edu.cn

Introduction

Vehicle exhaust gas contains hydrocarbons, nitrogen oxides, carbon dioxide, carbon monoxide, sulphur dioxide, carbon particles, fine particulate matter and small amounts of aromatic hydrocarbons (benzene) and dioxins. Among these pollutants, hydrocarbons are a major contributor to smog, especially in urban areas. Prolonged exposure to hydrocarbons can cause asthma, liver disease, lung disease and cancer. Carbon monoxide reduces the ability of blood to carry oxygen, and overexposure to carbon monoxide poisoning can be fatal. Nitrogen oxides(NO_x), which is a mixture of NO, N_2O, and NO_2, is generated when nitrogen in the air reacts with oxygen at the high temperature and pressure inside the engine. NO_x is a precursor to smog and acid rain.

In the past few decades, non-thermal plasma (NTP) has been widely used to remove volatile organic compounds (VOC) and NO_x [1–2]. However, the application of NTP has been greatly restricted by its low energy efficiency and poor CO_2 selectivity. Besides, undesirable byproducts (such as ozone), need to be further treated. Recently, these problems were solved to some extent by a

combination of non-thermal plasma with catalyst, so called plasma-driven catalysis. As a promising technology, this technique integrates the advantages of high selectivity from catalysis and fast ignition from plasma, which maintained high energy efficiency and mineralization rate with low by-product formation [3–4].

Plasma-driven catalysis (PDC) is a physical and chemical reaction. The reactive species from plasma, such as ions, electrons, excited atomics, molecules and radicals, generate considerable micro-discharge on the surface of the dielectric. These reactive species, especially high energy electrons, contain a large amount of energy, which will activate nearby catalyst and lower the activation energy of the reaction. Internal transition of high-energy particles will generate ultraviolet radiation [5–6]. If absorbed energy is greater than the band gap, the electron inside the semiconductor will be excited with a transition from the valence band to the conduction band, which will form electron-hole pairs and induce a series of further redox reactions. Photo-excited holes have a strong ability to obtain electrons, resulting in reacting with the hydroxide ions together with water adsorbed on the catalyst surface and then generating hydroxyl radicals, which leads to a further oxidation of pollutants. Compared with common catalysts, plasma-driven catalysts have many unique advantages, such as high distribution of reactive species, decreased energy consumption, enhanced catalytic activity and selectivity as well as the reduction of the sensitivity to poison [7–8].

Compared with non-thermal plasma, the addition of catalysts could significantly enhance the VOC removal efficiency with increased CO_2 selectivity and carbon balance, while the byproducts, such as O_3 and organic compounds were dramatically reduced, which was mainly due to increased amount of O formed from O_3 destruction [9]. Besides, plasma-driven catalysis has also been used in NO_x removal. It was reported that (plasma generated) ozone, hydroxyl radicals and atomic oxygen played important roles in the oxidation of NO to NO_2 [10]. Many efforts have been made to purify VOC or NO_x gas by using plasma-driven catalysis [11–13]. However, those studies focused on only one specific polluted gas or some simulated gases, which were quite different from real vehicle exhaust gas from a launched car. In fact, complex components in the exhaust interacted with each other. For example, according to the molecular dynamics theory, NO removal efficiency depends heavily on the content of HC and O_2 [14]. Meanwhile, partially oxidized hydrocarbons and peroxy radicals (RO_2) will in return react with NO and strongly influence NO_2 formation rates [10]. Thus the application of the technique for examining the vehicle exhaust removal rate has a practical significance.

In previous studies, catalyst material can be introduced into the reactor in several ways, such as coating on the reactor wall or electrodes, as a packed-bed (granulates, coated fibers, pellets) or as a layer of catalyst material (powder, pellet, granulates, coated fiber) [8]. What researches worried about was the deactivation of catalyst [15,16]. The catalyst contacted with the high electrode directly in all the studies above. What's more, most catalysts were prepared with liquid phase method, which contained too many complex chemical steps, even toxic or organic gas evaporating into the air [16,17,18,19].

In this study, a novel and special structure plasma-driven catalysis device was proposed. The PDC reactor was a coaxial DBD reactor with three dielectric barrier layers and two copper electrodes. The middle dielectric barrier was designed to separate the catalyst from the high voltage electrode. The catalyst TiO_2, prepared by RF magnetron sputtering, was coated on the outer surface of the middle quartz tube. The electrodes surrounded the outer surfaces of corresponding dielectric barrier in a spiral way to prevent the damage to the catalyst for too much heat from microdischarges in PDC reactor. Then the PDC reactor was

employed to treat the real vehicle exhaust. In order to explain the result, a simple and intuitive method–optical emission spectroscopy was conducted, which was different from chemical kinetics analysis in previous studies. This study also analyzed the practical problems for the application of PDC technique in real vehicle exhaust and proposed some solutions.

Materials and Methods

The PDC Reactor

As shown in Fig. 1, the proposed PDC reactor was a coaxial DBD reactor with three quartz tubes as dielectric barrier layers and two copper electrodes. The inner electrode attached to the inner quartz tube with a spiral rotation, of which the helix width and the pitch were 6 mm and 2 mm, respectively, while the outer electrode attached to the outer quartz tube in the same way as the helix width of 9 mm and the pitch of 4 mm. The nano-titanium dioxide film prepared by radio frequency magnetron sputtering (JGP450 High vacuum magnetron sputtering, China) was coated on the outer surface of middle quartz tube. During the process of coating TiO_2 film, the Ti target was set at the bottom of the vacuum chamber to avoid a large amount of impurities groups into TiO_2 film, and the inner quartz rotated by a constant angular velocity of 0.15 r/s, guaranteeing a well-distributed film on the same circle of the reactor. The proportion of oxygen and argon was about 1:1 with the power of 150W at 1 Pa pressure. The XRD detection given in Fig. 2 indicates that the titanium dioxide has an approximate composition of 85% anatase and 15% rutile forms of TiO_2. The discharge of the whole PDC reactor was shown in Fig. 3, with the excited voltage of 7.78 kV and the power of 2.75W. In order to know the surface features of the TiO_2, AFM detection was conducted on three points on the middle quartz tube with different distances of 20 mm, 50 mm, 90 mm from the same side, respectively. It was worth mentioning that the tube rotated at a certain speed of 0.15 r/s, which would help to form the uniform film in the same circle. Fig. 4 shows the AFM photograph of the uniform surface of the TiO_2. The length of the whole PDC reactor was 180 mm, and the outside diameter of the outer quartz tube was 18 mm, which constituted a compact reactor.

Clearance System

In this study, the experiment included exhaust gas source collection, treatment and components detection. The real exhaust gas was acquired from a jeep (Charokee-type Beijing Jeep 2500, made in China in 1999), in which the engine with inline four-cylinder was fed by #93 gasonline, in the condition of

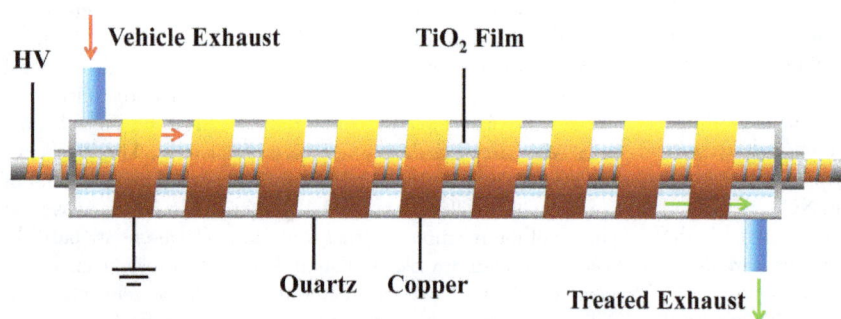

Figure 1. A schematic diagram of plasma-driven catalysis reactor. The PDC reactor was a coaxial DBD reactor with three quartz tubes as dielectric barrier layers and two copper electrodes. The outer and inner electrodes attached to the surfaces of the corresponding quartz tubes in a spiral rotation, respectively. The length of the whole reactor was 180 mm. The outside diameters of the quartz tubes were 18 mm, 10 mm and 6 mm, respectively.

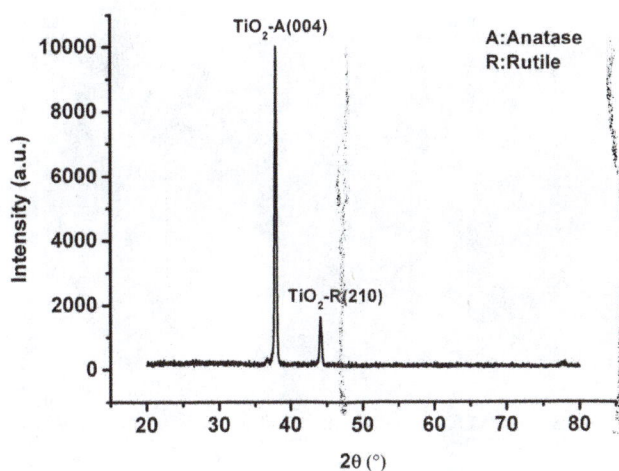

Figure 2. XRD pattern of prepared TiO₂ film. The titanium dioxide has an approximate composition of 85% anatase and 15% rutile forms of TiO₂.

temperature/humidity 90°C/60% (monitored by MINGLE Hygrometer TH101B, China). The contents of components in the vehicle exhaust before and after treatment with the PDC reactor were detected by an exhaust gas analyzer (CV-5Q, Tianjin Shengwei Inc. Tianjin, China), which serves officially as a standard exhaust analysis in Beijing, China.

Clearance Process

At the beginning of each experiment, the jeep was started to let the vehicle exhaust gas steadily emit while the exhaust gas analyzer was turned on for detection. Then the exhaust was induced into the PDC reactor, which was excited for low temperature plasma later at the voltage of 8–10 kV. As shown by the detected spectrum in Fig. 5, many significant peaks can be observed in spectrum scope of working plasma inside the PDC reactor ranged from 250 nm to 520 nm, which satisfies the required wavelengths of TiO₂ catalysis.

We then conducted experimental comparisons among three different groups, including a control group without any plasma treatment, a PDC group with TiO₂, and a NTP group which had an identical structure to the PDC reactor but without the catalyst TiO₂. Five tests were conducted for each group. In each test, after engine reaching steady state, the concentrations of different components of the vehicle exhaust were recorded sequentially at an interval of 20 s within 3 minutes for three groups. Then plasma was excited in both PDC and NTP reactors which the vehicle exhaust gas was lead into. The concentrations of different components were recorded at an interval of 20 s within another 9 minutes for these two groups.

Electrical Measurement

The PDC reactors were ignited by an AC high voltage power supply equipped with a transformer, which controlled the input power of the plasma generator (CTP-2000K). AC high frequency high voltage exported from the generator was applied to the PDC reactor, providing excitation power for the reactor. The excitation power was measured through the output voltage and current detection in the generator. Applied high voltage (V) was measured with a 1000:1 high voltage probe (TEKtronix, P6015A). V–Q Lissajous method was used to determine the discharge power in the PDC reactor. The charge Q was determined by measuring the voltage across the capacitor of 0.47 uF connected in series to the ground line of the PDC reactor. The voltage across this capacitor is proportional to the charge. The signals of applied voltage and charge were recorded with a digitizing oscilloscope (Tektronix, MSO2024) by averaging 62.5 k scans. The discharge power (P_{dis}) was evaluated from the area of V–Q parallelogram by multiplying the frequency. Specific input energy (SIE), which is defined as the energy input per unit gas-flow rate, can be obtained as follows:

$$SIE(J/L) = \frac{p_{dis}(watt)}{gas\,flow\,rate(L/\min)} \times 60 \qquad (1)$$

In this study, both the minimum excited state and steady state of working PDC reactor were measured according to the method above.

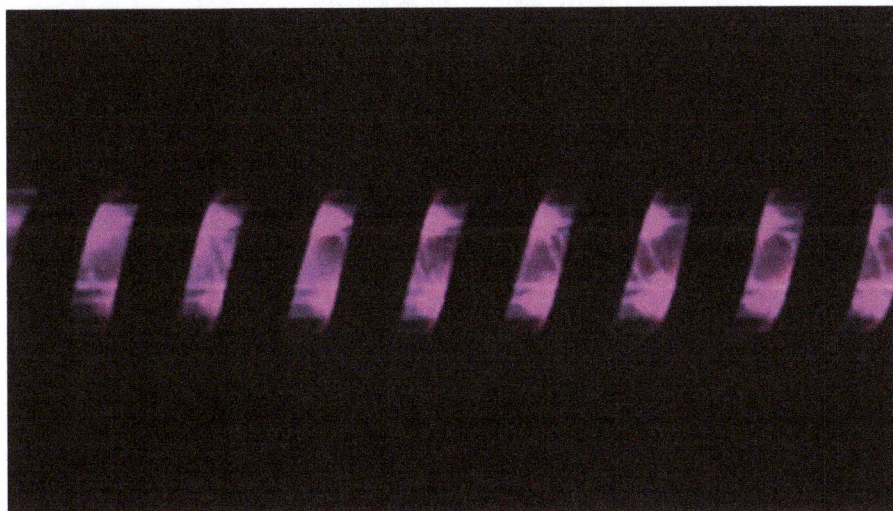

Figure 3. Discharge of plasma-driven catalysis reactor. The excited voltage and power were about 7.78 kV and 2.75W, respectively.

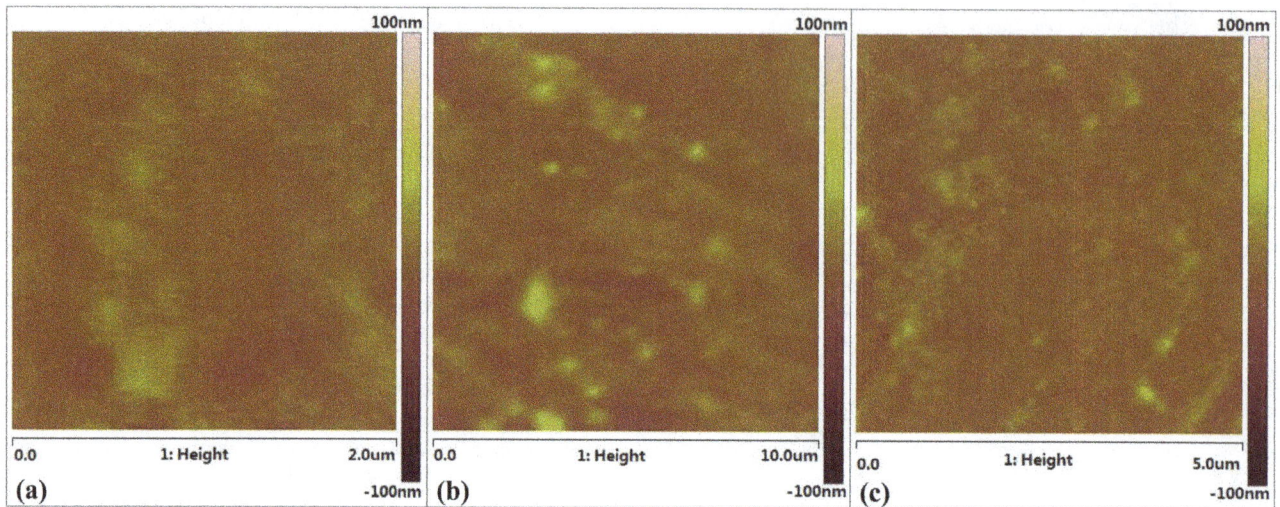

Figure 4. AFM photograph of prepared TiO₂ film. AFM detections on three points on the middle quartz tube with different distances of (a) 20 mm, (b) 50 mm and (c) 90 mm from the same side.

Optical Emission Spectroscopy

Optical emission spectroscopy (OES) is one of the most widely used diagnostic methods for low-temperature plasmas [20]. Optical emission spectrometry involves applying electrical energy in the form of spark generated between an electrode and a metal sample, whereby the vaporized atoms are brought to a high energy state within a so-called "discharge plasma". These excited atoms and ions in the discharge plasma create a unique emission spectrum specific to each element. Thus, a single element generates numerous characteristic emission spectral lines. Therefore, the light generated by the discharge can be said to be a collection of the spectral lines generated by the elements in the sample. This light is split by a diffraction grating to extract the emission spectrum for the target elements. The intensity of each emission spectrum depends on the concentration of the element in the sample. Detectors (photomultiplier tubes) measure the presence or absence of the spectrum extracted for each element and the intensity of the spectrum to perform qualitative and quantitative analysis of the elements [21,22]. The vehicle exhaust gas was induced into the PDC and NTP reactors, respectively. And the plasmas in the two reactors were excited with the same

voltage and power. The probe was put at the same point of each reactor. Then the spectrums were obtained.

Results

Vehicle Exhaust Clearance

Figure 6(a) and (b) show that both of the two reactors have a positive effect on the reduction of HC and NO. Moreover, the removal efficiency of PDC reactor was significantly higher than that of NTP reactor. The vehicle exhaust removal efficiency η was estimated as follow:

$$\eta(\%) = \frac{c_{in} - c_{out}}{c_{in}} \times 100\% \qquad (2)$$

where c_{in}, c_{out} are inlet and outlet concentration of the certain component, respectively.

According to Eq.1, the removal efficiency in the PDC reactor for HC and NO are as follows:

$\eta_{NO} = 64.5\% \pm 1.8\%$, and $\eta_{HC} = 32.1\% \pm 1.3\%$.

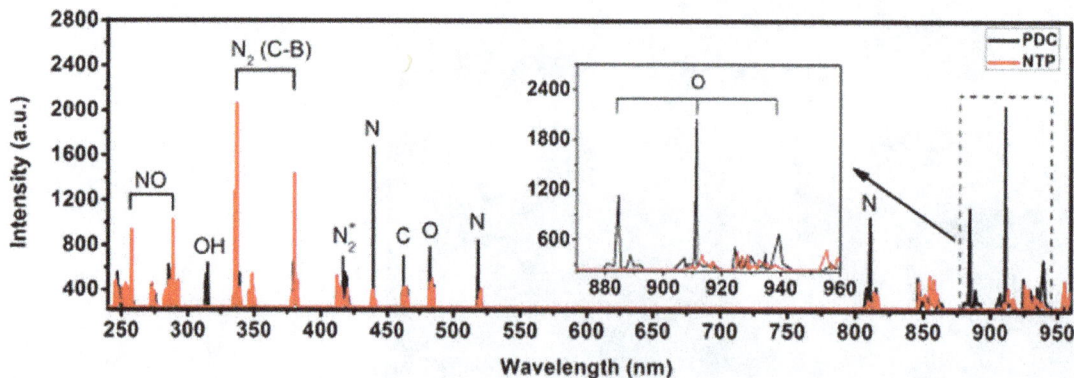

Figure 5. Spectrum of the discharge by PDC and NTP reactor. The appearance of O spectrum can be observed between 870 nm and 960 nm for wavelength in PDC reactor.

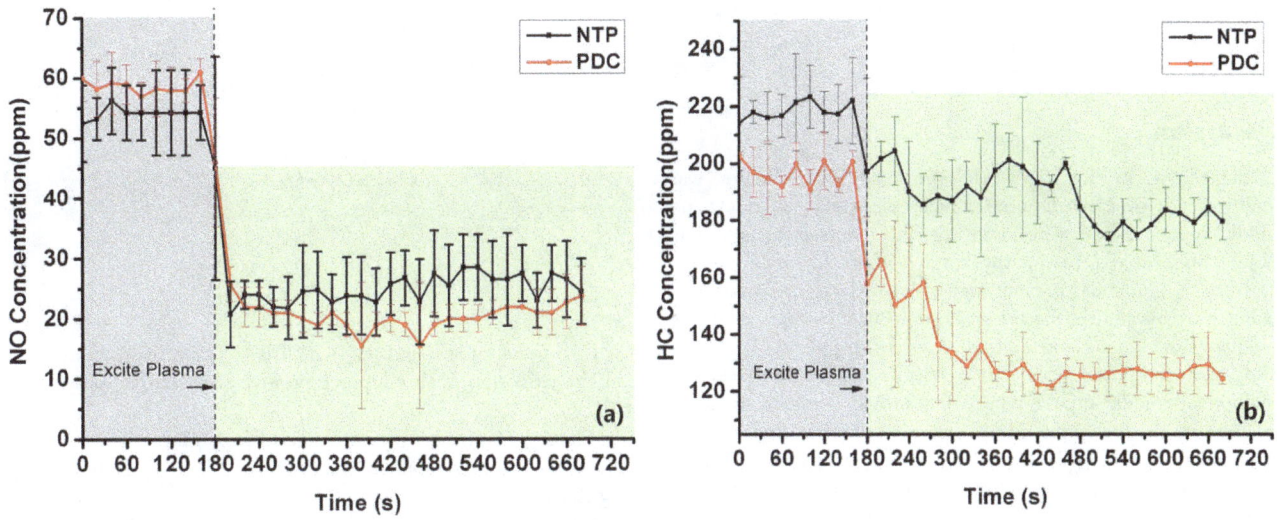

Figure 6. Real-time removal results of (a) NO removal and (b) HC removal. The removal efficiency for HC in the PDC reactor was $\eta_{NO} = 64.5\% \pm 1.8\%$, and the removal efficiency for NO in the PDC reactor was $\eta_{HC} = 32.1\% \pm 1.3\%$

Electrical Measurement

Experimental data indicated that our vehicle exhaust flow rate was about 590 sccm. The voltage and current of the PDC reactor were measured through capacitor sampling and resistance sampling, respectively. The V-Q Lissajous method was used to determine the discharge power. Both of the minimum excited state and steady state of PDC reactor were measured. All the electrical measurement results were shown in Fig. 7. At the minimum excited state, the effective voltage and current were 7.78 kV and 0.35 mA. Specific input energy was about 279.66 J/L. At the

steady state, the effective voltage and current were 8.06 kV and 0.35 mA. Specific input energy was about 289.83 J/L.

Optical Emission Spectrum

The optical emission spectrum result was shown in Fig. 5. The OH, NO and N emission lines are visible. The appearance of O emission lines can be observed between 870 nm and 960 nm for wavelength in PDC reactor [23,24,25,26]. Considering the facts that the removal efficiency of HC treated by PDC was much higher than that of NTP, while the removal efficiency of NO by

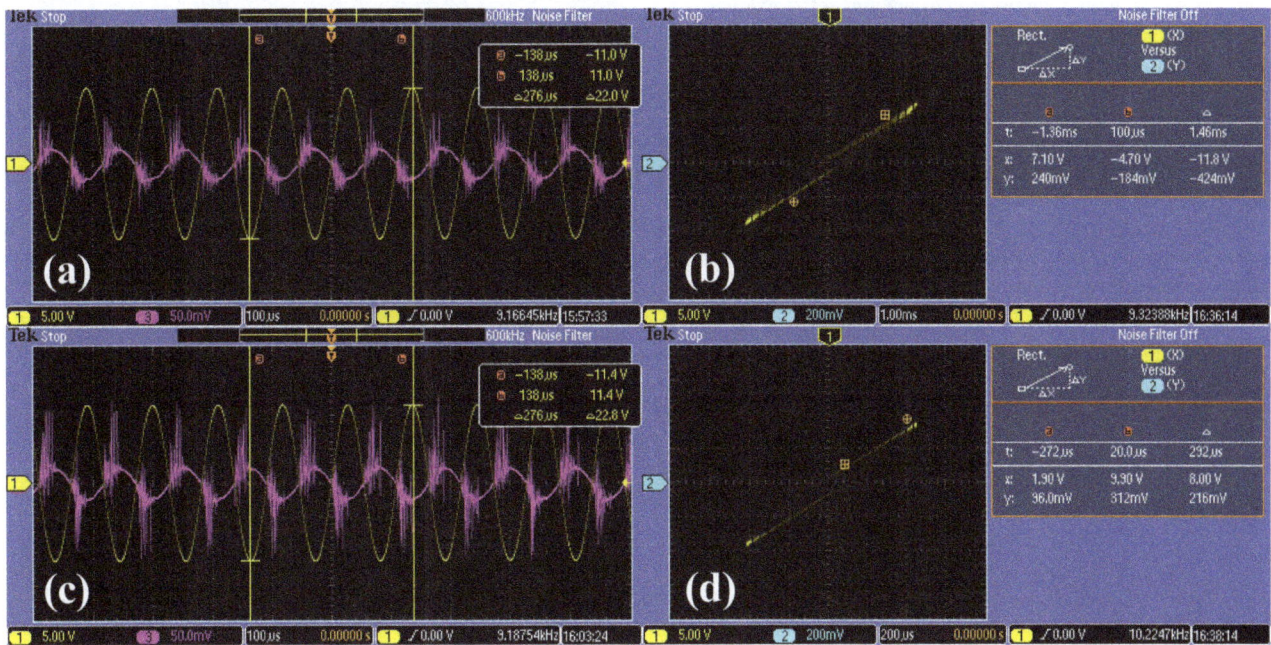

Figure 7. Electrical measurement of the PDC reactor. (a) The voltage and current of the PDC reactor at the minimum excited state. **(b)** Lissajous figure of the PDC reactor at the minimum excited state. **(c)** The voltage and current of the PDC reactor at the stable working state. **(d)** Lissajous figure of the PDC reactor at the stable working state.

PDC was only a little higher than that of NTP, we hypothesized that O might have made a significant contribution to the much higher HC and NO removal efficiencies.

Discussion

First of all, in this study, the proposed PDC reactor was designed with three quartz tubes as dielectric layers. The outer and middle quartz tubes were both dielectric barrier layers. The middle one could separate the catalyst from the high voltage electrode, which could prevent high voltage electrode from being oxidized by oxygen during the reactions. On the other hand, the middle quartz tube could increase the area for depositing more TiO_2 film compared with a bare electrode. Furthermore, the addition of middle quartz tube could provide more chances to generate microdischarges [27], which would increase catalyst surface temperature [28], enhance the dispersion of active catalytic components [29,30] and influence the stability with catalytic activity of the exposed catalyst material [31]. All above would promote catalytic VOC removal efficiency. However, hot spots can be formed in PDC reactors as a result of localizing heating by intense microdischarges, which might lead to the damage to the high voltage electrode and catalyst [32]. In order to avoid too many hot spots, each electrode surrounded the outer surfaces of corresponding dielectric barrier in a spiral way. The method of preparing TiO_2 film was RF magnetron sputtering without any toxic or organic gas evaporating into the air. Continuous operation tests indicated that stable performance without deterioration of catalytic activity could last for more than 25 h.

Secondly, the removal efficiency result was further explained through the optical emission spectrum approach, a simple and intuitive method different from chemical kinetics analysis in previous studies. The optical emission spectrum result showed that the appearance of O emission lines can be observed between 870 nm and 960 nm for wavelength in PDC reactor. It is our suggestion that the enhanced performance of hydrocarbon destruction was mainly due to a great amounts of atomic oxygen (O) formed, primarily from catalytically O_3 decomposition. Compared with NTP reactor, TiO_2 film acting as a semiconductor oxide catalyst provided a large number of free electron-hole pairs and consequently promoted the oxidation-reduction reaction.

The emission of O_3 from the NTP reactor was harmful to both human health and global environment. The addition of catalyst could significantly enhance the HC destruction with an increased O formation while the byproducts O_3 from the plasma were dramatically reduced [9]. Basically, O_3 formation in the NTP reactor proceeded via a two-step process [33]: formation of atomic oxygen and recombination of atomic oxygen with oxygen molecule (Eqs. (3)-(5)):

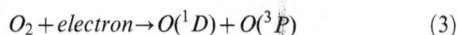

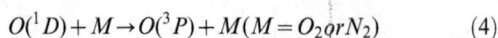

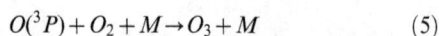

$$O_2 + electron \rightarrow O(^1D) + O(^3P) \qquad (3)$$

$$O(^1D) + M \rightarrow O(^3P) + M (M = O_2 or N_2) \qquad (4)$$

$$O(^3P) + O_2 + M \rightarrow O_3 + M \qquad (5)$$

where $O(^1D)$ and $O(^3P)$ represent the excited and ground state oxygen atom, respectively.

It has been also reported that O_3 can be decomposed by catalysts into molecular oxygen via atomic oxygen and peroxides (Eqs. (6)-(8)) [34], where * denotes an active site on the catalyst surface:

$$O_3 + * \rightarrow O* + O_2 \qquad (6)$$

$$O* + O_3 \rightarrow O_2 + O_2* \qquad (7)$$

$$O_2* \rightarrow O_2 + * \qquad (8)$$

In general, atomic oxygen, which is highly active and involved in HC oxidation, is also imposed positive effect on NO destruction [9].

Thirdly, in the presence of catalyst TiO_2 film, when there is a faster rate of oxidation of hydrocarbons (Fig. 6b) there is not a significant increasing reduction of NO (Fig. 6a) correspondingly. Since the gas employed in this study was real vehicle exhaust gas containing many different kinds of compositions, there would be some reactions that did not occur in only one specific gas or some simulated gases. Actually, those complex components in the vehicle exhaust interacted with each other during the PDC process, which would remarkably influence the removal efficiency of NO and HC. It is worth mentioning that there exists a dynamic equilibrium between NO and HC, that HC decomposition will lead to the formation of NO [14]. Besides, it is known that both •OH and O_3 play an important role in NO removal. However, since O_3 is an •OH scavenger, partial O_3 and •OH will react with each other as follow [35]:

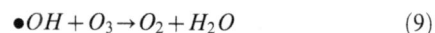

$$•OH + O_3 \rightarrow O_2 + H_2O \qquad (9)$$

Thus both contents of O_3 and •OH will decrease with an increased O_2 content. It has been reported in a previous literature that high O level will definitely lead to a conversion back to NO, and decrease the the removal efficiency of NO [36]. Although several studies have indicated that plasma-driven catalysis technique was quite effective in removing NO or hydrocarbon [37,38,39,40,41], according to our results and analysis above, the clearance rate of NO would be reduced if mixed with HC as well as other gas components.

The complex chemical reactions among gas compositions in the vehicle exhaust gas are briefly illustrated in Fig. 8. For example, higher NO removal efficiency is under the condition of lower content of HC or decreased O_2 content [14], meanwhile, NO can be removed by the formation of NO_2 through the reaction with partially oxidized hydrocarbons and peroxyl radicals (RO_2) [10]. Thus, the removal efficiency of HC could not be as high as that of NO in our experiment. Besides, many fundamental components from working plasma, including ozone, hydroxyl radicals and atomic oxygen also play an important role in the oxidation of NO to NO_2 [10]. When hydrocarbon was treated by the plasma discharge, partially oxidized hydrocarbons ($C_xH_yO_z$) and peroxy radicals (RO_2) reacting with NO will be generated and strongly influenced NO_2 formation rate. Meanwhile, NO_2 reacted with the catalyst TiO_2 film, while partially oxidized hydrocarbons were consumed during selective catalytic reduction, producing CO_2, N_2, and H_2O, which are environment-friendly products [42]. Finally, •OH radicals can convert the formed NO_2 into HNO_3 [10] with the existence of H_2O [9]:

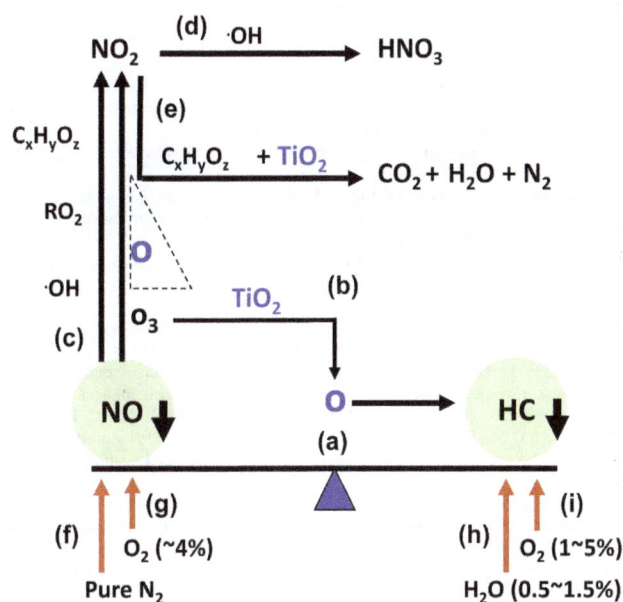

Figure 8. Chemical reactions gas compositions in the vehicle exhaust gas. (a) HC decomposition will lead to the formation of NO [14]. (b) The addition of catalysts could generate more single atomic oxygen from O_3 destruction, which contributed to the HC decomposition [9]. (c) single atomic oxygen, ozone, OH, oxidized hydrocarbons and peroxyl radicals played an important role in the oxidation of NO to NO_2 [10], and high O level will definitely lead to a conversion back to NO [36]. (d) •OH radicals can convert the formed NO_2 into HNO_3 [10] with the existence of H_2O [9]. (e) NO_2 reacted with the catalyst TiO_2 film, while partially oxidized hydrocarbons is consumed during selective catalytic reduction, producing CO_2, N_2, and H_2O. (f) Nitrogen in air can help keep the high removal efficiency of NO [45]. (g) NOx decomposition by plasma was known to be possible only if the oxygen content was less than about 4% [45,46]. (h) The highest carbon balance and CO_2 selectivity for HC destruction were obtained with water vapor content between 0.5 and 1.5% [24]. (i) The optimal oxygen ranges between 1% and 5% for VOC with NTP [44].

$$e + H_2O \rightarrow e + \bullet OH + H \qquad (10)$$

$$O(^1D) + H_2O \rightarrow e + 2\bullet OH. \qquad (11)$$

Based on the complex chemical reactions mentioned above, the removal efficiency of NO and HC should have been much higher if the PDC reactor was used to remove only NO or HC at the same amount of electricity consumption.

Besides, there were some other factors affecting the removal efficiency of NO and HC by using the proposed PDC reactor. On one hand, water vapor existing in the vehicle exhaust would reduce the vehicle exhaust removal efficiency. Although the drying modules had been assembled in our experiments, water vapor could not be removed completely and it would generate a considerable number of •OH, leading to enhancement of NO conversion but decrease of HC removal efficiency with an increased incompletely oxidizing byproducts [10]. Meanwhile, water vapor can make catalyst deactivate through poisoning its active sites, annihilating high energetic electrons and depress the HC destruction through competing to be absorbed by the catalyst

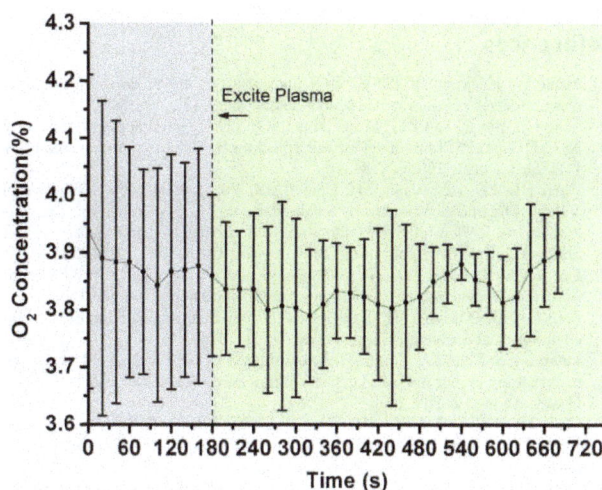

Figure 9. The O_2 content in the PDC reactor during the whole process. The oxygen content in the exhaust ranged roughly from 3.5% to 4.2% all the time.

[43]. It has been reported that the highest carbon balance and CO_2 selectivity were obtained with water vapor content between 0.5 and 1.5% [17]. In further study, water vapor could be well controlled to further improve the exhaust removal efficiency.

On the other hand, similar to the presence of water vapor, the oxygen content in the vehicle exhaust that affects significantly the discharge performance plays a key role in the occurring chemical reactions. It has been mentioned that the optimal oxygen ranges between 1% and 5% for VOC with NTP [44], while NO_x decomposition by plasma was known to be possible only if the oxygen content was less than about 4% [45,46]. In this study, the oxygen content in the exhaust ranged roughly from 3.5% to 4.2% all the time as shown in Fig. 9, which could benefit the PDC reactor for HC removal. However, the oxygen content above could have also benefited the PDC reactor for NO removal efficiency, but it was unreal for the exhaust gas containing VOC [14]. The effect of oxygen content on the NO and HC removal efficiency in this study was obviously reflected in the removal result as shown in Fig. 6.

In order to enhance the ability of the application of PDC reactor in vehicle exhaust gas, a reliable automatic control system will be expected. Based on the measured proportion of HC and NO in real-time, vehicle exhaust can be further removed effectively by the feedback control system that adaptively induces the air outside into PDC reactor. Particularly, the oxygen in air can be used to modulate the appropriate proportions among HC, NO and O_2 while the nitrogen in air can help keep the high removal efficiency of NO [45]. In this way, the real vehicle exhaust gas from cars will be well controlled with high efficiency in real-time.

Acknowledgments

The support of Professor Zhang Guling and Ms. Wang Ke in Minzu University of China for preparing TiO_2 film is gratefully acknowledged.

Author Contributions

Organized and helped with the design of the research: JZ JF. Conceived and designed the experiments: JZ. Performed the experiments: SY YL SS KZ. Analyzed the data: SY YL. Contributed reagents/materials/analysis tools: SY JF JZ KZ. Wrote the paper: SY JZ.

References

1. Mista W, Kacprzyk R (2008) Decomposition of toluene using non-thermal plasma reactor at room temperature. Catal. Today 137: 345–349.
2. Song CL, Bin F, Tao ZM, Li FC, Huang QF (2009) Simultaneous removals of NO_x, HC and PM from diesel exhaust emissions by dielectric barrier discharges. J. Hazard. Mater. 166: 523–349.
3. Chen HL, Lee HM, Chen SH, Chang MB, Yu SJ, et al. (2009) Removal of Volatile Organic Compounds by Single-Stage and Two-Stage Plasma Catalysis Systems: A Review of the Performance Enhancement Mechanisms, Current Status, and Suitable Applications. Environ. Sci. Technol.43: 2216–2227.
4. Fan X, Zhu T, Wang M, Li X (2009) Removal of low-concentration BTX in air using a combined plasma catalysis system. Chemosphere 75: 1301–1306.
5. Francke KP, Miessner H, Rudolph R (2000) Plasmacatalytic processes for environmental problems. Catal. Today 59: 411–416.
6. Futamura S, Zhang A, Einaga H, Kabashima H (2002) Involvement of catalyst materials in nonthermal plasma chemical processing of hazardous air pollutants. Cataly. Today 72: 259–265.
7. Chen HL, Lee HM, Chen SH, Chao Y, Chang MB (2008) Review of plasma catalysis on hydrocarbon reforming for hydrogen production–Interaction, integration, and prospects. Appl. Phys. B: Environmental 85: 1–9.
8. Durme JV, Dewulf J, Leys C, Langenhove HV (2008) Combining non-thermal plasma with heterogeneous catalysis in waste gas treatment: A review. Appl. Phys. B: Environmental 78: 324–333.
9. Huang HB, Ye DQ, Dennis YCL (2011) Abatement of Toluene in the Plasma-Driven Catalysis: Mechanism and Reaction Kinetics. IEEE Trans. Plasma Sci. 39: 877–882.
10. Maciuca A, Dupeyrat CB, Tatibouët JM (2008) Synergetic effect by coupling photocatalysis with plasma for low VOCs concentration removal from air. Appl. Catal. B: Environ. 125: 432–438.
11. Rappe KG, Hoard W, Aardahl CL, Park PW, Peden CHF, et al. (2004) Combination of low and high temperature catalytic materials to obtain broad temperature coverage for plasma-facilitated NO_x reduction. Catal. Today 89: 143–150.
12. Oh SM, Kim HH, Einaga H, Ogata A, Futamura S, et al. (2006) Zeolite-combined plasma reactor for decomposition of toluene. Thin Solid Films 506: 418–422.
13. Holzer F, Roland U, Kopinke FD (2002) Combination of non-thermal plasma and heterogeneous catalysis for oxidation of volatile organic compounds: Part 1. Accessibility of the intra-particle volume. Appl. Catal. B: Environ. 32: 163–181.
14. Kima HH, Ogat A (2011) Nonthermal plasma activates catalyst: from current understanding and future prospects. Phys. J. Appl. Phys. 55, 13806.
15. Harling AM, Kim HH, Futamura S, Whitehead J C (2007) Temperature Dependence of Plasma-Catalysis Using a Nonthermal, Atmospheric Pressure Packed Bed; the Destruction of Benzene and Toluene. J. Phys. Chem. C 2007, 111, 5090–5095.
16. Kim HH, Oh SM, Ogata A, Futamura S (2004) Decomposition of benzene using Ag/TiO2 packed plasma-driven catalyst reactor: influence of electrode configuration and Ag-loading amount. Catalysis Letters Vol. 96, Nos. 3–4, July.
17. Huang HB, Ye DQ, Dennis YCL (2011) Plasma-Driven Catalysis Process for Toluene Abatement: Effect of Water Vapor. IEEE Trans. Plasma Sci. 39: 576–580.
18. Ding HX, Zhu AM, Lu FG, Xu Y, Zhang J, et al. (2006) Low-temperature plasma-catalytic oxidation of formaldehyde in atmospheric pressure gas streams. J. Phys. D: Appl. Phys. 39: 3603–3608.
19. Harling AM, Demiduyk V, Fischer SJ, Whitehead JC (2008) Plasma-catalysis destruction of aromatics for environmental clean-up: Effect of temperature and configuration. Applied Catalysis B: Environmental 82: 180–189.
20. Zhu X, Pu Y (2010) Optical emission spectroscopy in low-temperature plasmas containing argon and nitrogen: determination of the electron temperature and density by the line-ratio method. J. Phys. D: Appl. Phys. 43: 403001.
21. Fantz U (2006) Basics of plasma spectroscopy. Plasma Sources Sci. Technol. 15: S137–S147.
22. Staack D, Farouk B, Gutsol AF, Fridman AA (2006) Spectroscopic studies and rotational and vibrational temperature measurements of atmospheric pressure normal glow plasma discharges in air. Plasma Sources Sci. Technol. 15 (2006) 818–827.
23. Walsh JL, Kong MG (2008) Contrasting characteristics of linear-field and cross-field atmospheric plasma jets. Appl. Phys. Lett. 93, 111501.

24. Lee YH, Yi CH, Chung MJ, Yeom GY (2001) Characteristics of He/O_2 atmospheric pressure glow discharge and its dry etching properties of organic materials. Surf. Coat. Technol. 146–147: 474–479.
25. Lofthus A, Krupenie PH (1977) The spectrum of molecular nitrogen. J. Phys. Chem. Ref. Data 6, 113.
26. Xu L, Nonaka H, Zhou HY, Ogino A, Nagata T, et al. (2007) Characteristics of surface-wave plasma with air-simulated N_2–O_2 gas mixture for low-temperature sterilization. J. Phys. D: Appl. Phys. 40: 803–808.
27. Kogelschatz U (2003) Dielectric-barrier Discharges: Their History, Discharge Physics, and Industrial Applications. Plasma Chem. Plasma Process.Vol. 23, No. 1, March.
28. Lu B, Zhang X, Yu X, Feng T, Yao S (2006) Catalytic oxidation of benzene using DBD corona discharges, Journal of Hazardous Materials 137: 633–637.
29. Guo YF, Ye DQ, Chen KF, He JC, Chen WL (2006) Toluene decomposition using a wire-plate dielectric barrier discharge reactor with manganese oxide catalyst in situ. Journal of Molecular Catalysis A: Chemical 245: 93–100.
30. Zhang YP, Ma PS, Zhu XL, Liu CJ, Shen YT (2004) A novel plasma-treated Pt/NaZSM-5 catalyst for NO reduction by methane, Catalysis Communications 5: 35–39.
31. Guo YF, Ye DQ, Chen KF, He JC (2007) Toluene removal by a DBD-type plasma combined with metal oxides catalysts supported by nickel foam, Catalysis.
32. Kim HH, Ogata A, Futamura S (2006) Effect of different catalysts on the decomposition of VOCS using flow-type plasma-driven catalysis, IEEE Transactions on Plasma Science 34: 984–995.
33. H. H Kim, A Ogata, S Futamura (2006) Effect of different catalysts on the decomposition of VOCs using flow-type plasma-driven catalysis, IEEE Trans. Plasma Sci. 34: 984–995.
34. Futamura S, Einaga H, Kabashima H, Hwan LY (2004) Synergistic effect of silent discharge plasma and catalysts on benzene decomposition. Catal. Today 89: 89–95.
35. Kim HH, Oh SM, Ogata A, Futamura S (2005) Decomposition of gas-phase benzene using plasma-driven catalyst (PDC) reactor packed with Ag/TiO2 catalyst. Appl. Catal. B: Environ.56: 213–220.
36. Malik MA, Kolb JF, Sun Y, Schoenbach KH (2011) Comparative study of NO removal in surface-plasma and volume-plasma reactors based on pulsed corona discharges. J. Hazard. Mater. 197: 220–228.
37. X Tu, J.C Whitehead (2012) Plasma-catalytic dry reforming of methane in an atmospheric dielectric barrier discharge: Understanding the synergistic effect at low temperature. Appl. Catal. B. 125: 439–448.
38. Fan HY, Shi C, Li XS, Zhao DZ, Xu Y, et al. (2009) High-efficiency plasma catalytic removal of dilute benzene from air. J. Phys. D: Appl. Phys. 42. 225105.
39. Ding HX, Zhu AM, Lu FG, Xu Y, Zhang J, et al. (2006) Low-temperature plasma-catalytic oxidation of formaldehyde in atmospheric pressure gas streams. J. Phys. D: Appl. Phys. 39: 3603–3608.
40. Mok YS, Koh DJ, Kim KT, Nam IS (2003) Nonthermal Plasma-Enhanced Catalytic Removal of Nitrogen Oxides over V_2O_5/TiO_2 and Cr_2O_3/TiO_2. Ind. Eng. Chem. Res., 42 (13): 2960–2967.
41. Chen Z, Mathur VK (2003) Nonthermal Plasma Electrocatalytic Reduction of Nitrogen Oxide. Ind. Eng. Chem. Res. 42 (26): 6682–6687.
42. Lin H, Huang Z, Shangguan W, Peng X (2007) Temperature-programmed oxidation of diesel particulate matter in a hybrid catalysis–plasma reactor. Proc. Combust. Inst 31: 3335–3342.
43. Zhang PY, Liang FY, Yu G, Chen Q, Zhu WP (2003) A comparative study on decomposition of gaseous toluene by O_3/UV, TiO_2/UV and $O_3/TiO_2/UV$. J. Photochem. Photobiol. A: Chemistry 156: 189–194.
44. Vandenbroucke AM, Morent R, Geyter ND, Leys C (2011) Decomposition of Trichloroethylene with Plasma-catalysis: A review. J. Hazard. Mater. 195: 30–54.
45. Masuda S, Hosokawa S, Tu X, Sakakibara K, Kitoh S, et.al (1993) Destruction of gaseous pollutants by surface-induced plasma chemical process (SPCS). IEEE Trans. Ind. Applicat. 29: 781–786.
46. Yan K, Kanazawa S, Ohkubo T, Nomoto Y (1999) Oxidation and Reduction Processes During NO_x Removal with Corona-Induced Nonthermal Plasma. Plasma Chem. Plasma Proc. 19: 421–443.

10

Use of Surface Enhanced Blocking (SEB) Electrodes for Microbial Cell Lysis in Flow-Through Devices

Abdossamad Talebpour*, Robert Maaskant, Aye Aye Khine, Tino Alavie

Qvella Incorporation, Richmond Hill, Ontario, Canada

Abstract

By simultaneously subjecting microbial cells to high amplitude pulsed electric fields and flash heating of the cell suspension fluid, effective release of intracellular contents was achieved. The synergistic effect of the applied electric field and elevated temperature on cell lysis in a flow-through device was demonstrated for Gram-negative and Gram-positive bacteria, and *Mycobacterium* species. The resulting lysate is suitable for downstream nucleic acid amplification and detection without requiring further preparation. The lysis chamber employs surface enhanced blocking electrodes which possess an etched micro-structured surface and a thin layer of dielectric metal oxide which provides a large effective area and blocks transmission of electrical current. The surface enhanced blocking electrodes enable simultaneous suppression of the rapid onset of electric field screening in the bulk of the cell suspension medium and avoidance of undesired electrochemical processes at the electrode-electrolyte interface. In addition the blocking layer ensures the robustness of the cell lysis device in applications involving prolonged flow-through processing of the microbial cells.

Editor: Lyle L. Moldawer, University of Florida College of Medicine, United States of America

Funding: The authors have no support or funding to report.

Competing Interests: Qvella is a development stage (pre-product) molecular diagnostics company. Qvella was founded in 2009 and has not sold any products or services to date. The authors declare that they are employed by Qvella Corporation and have a financial interest in the company. The authors have filed patent applications which include some aspects of the methods and devices described in this manuscript: METHODS AND DEVICES FOR ELECTRICAL SAMPLE PREPARATION, US20140004501. There are no further patents, products in development or marketed products to declare.

* Email: samadt@qvella.com

Introduction

The unprecedented advances in detecting and identifying microorganisms using nucleic acids have not been adequately matched with corresponding progress in pre-analytical sample preparation techniques required for efficiently and rapidly providing an inhibitor and contamination free nucleic acid suspension. Consequently, despite its specificity and sensitivity, polymerase chain reaction (PCR) has not replaced the much slower standard microbial culture-based techniques as the frontline diagnostic test in the clinical microbiology laboratory. For microfluidic applications, often favoured by developers of automated molecular based platforms, the difficulties are enhanced in the adaptation of traditional cell lysis techniques such as mechanical, chemical and enzymatic lysis methods. These generally require the addition of reagents for the lysis step and some form of extraction and purification of the target nucleic acids.

Over a decade ago, seemingly inspired by the extensive studies on microorganism inactivation by pulsed electric fields (PEF) [1], irreversible electroporation was suggested as a convenient method of microbial cell lysis for molecular assay chips [2–3]. This approach, involving the formation of electrically induced nano-scale pores in the cell membrane, has been successfully demonstrated for the lysis of mammalian cells and a range of microbial species [4–6]. In many previous studies involving the application of electric fields, the term "lysis" has been used to refer to both the inactivation of microbial cells and the permanent rupturing of the

cell wall. However, a careful review of the literature shows that in most cases, lysis efficiency has only been indirectly inferred based on microbial cell survival rates and not based on direct evidence of the release of intracellular contents such as nucleic acids. The use of indirect metrics for inferring microbial cell lysis can lead to misinterpretation of the effect of the electric fields on the microbial cells because, even though a microbial cell may be rendered inactive by the application of an electric field, it may not be sufficiently lysed to achieve release of nucleic acids.

According to a recent report, the thickness of a cell wall may serve as a barrier against the formation of large pores which are necessary for the release of larger cell contents such as ribosomes [7]. This may be the reason why only a limited number of studies have reported the success of electrical lysis methods to release nucleic acids [2,4], and success in this regard is restricted generally to Gram-negative bacteria such as *E. coli*. This may also be the reason why electrical lysis methods for efficient release of nucleic acid contents have not yet been demonstrated for microorganisms with bilayer membranes surrounded by a tough cell wall, as in the case of Gram-positive bacteria and fungi.

An alternative approach for achieving reagent-free inline microbial cell lysis is thermal lysis which employs external heating of capillary channels. A recent study [8] has concluded that thermal lysis of *E. coli* cells achieves DNA yields similar to lysing by bead beating only for heating times greater than 10 minutes. This may not be fast enough for typical applications and furthermore the method has not been demonstrated for bacteria with tougher cell walls. Therefore, a platform based on heating alone is not expected to offer a rapid cell lysing module for implementation on microfluidic devices.

In this article, a novel hybrid approach is described that employs both electrical and thermal lysis mechanisms to achieve effective lysis and release of intracellular contents for a wide range of bacterial cells types, including Gram-positive and Gram-negative bacterial and *Mycobacterium* microbial cells. The method involves exposing microbial cells to relatively strong electrical fields in the presence of rapidly increasing temperatures that arise during the application of the electric field. This electrical lysis method, which is termed "E-Lysis" in this manuscript, utilizes an applied electric field to generate a high cellular trans-membrane voltage while simultaneously inducing flash heating due to Joule heating from the ionic current in the cell suspension fluid, and the combined effect of the electric field and the flash heating results in the lysis of the suspended cells. A phenomenological rationale for performing electrical lysis at elevated temperatures is provided by the phase transition model of electro-permeabilization in an external electric field [9]. According to this model, bilayer membranes have several stable states each corresponding to a local minimum of the molecular free energy. At a sufficiently elevated temperature the local minima vanish and the membrane dissolves in the surrounding water forming small micelles. Similarly, in a strong enough external field the free energy minima cease to exist leading to the breakdown of the membrane. Therefore, if a trans-membrane voltage is applied to cells suspended in a high temperature medium, the electric field thresholds needed for the disintegration of the cell membrane and subsequent release of the macromolecules can be substantially lowered [9].

The E-Lysis method was implemented by exposing microbial cells to relatively high electric fields in a microfluidic channel and applying a train of bipolar square pulses through "surface enhanced blocking electrodes". These electrodes have a finely micro-structured surface on which a thin dielectric layer is formed. It is demonstrated that this approach enables the design of a robust microbial cell lysis device capable of exploiting the synergistic effects of electrical field and heat for efficient release of nucleic acid contents from the cells and delivering assay-ready lysate for performing downstream nucleic acid assays.

Materials and Methods

Fabrication of surface enhanced blocking (SEB) electrodes

The SEB electrodes are fabricated by electrochemical etching of the surface of 100 μm thick aluminium foils in an aqueous solution containing hydrogen chloride and appropriate additives referring to the recipes disclosed in the US patent 4,276,129 [10]. The etched foils are then anodized in an aqueous electrolytic solution containing boric acid and ammonium borate and a formation current of 0.05 mA/cm^2 at 100 V [11]. A scanning electron microscopy (SEM) picture of the electrode surface, presented in Figure 1, portrays the enhanced oxidized surface of the aluminium electrode.

Fabrication of cell lysis device

A microfluidic lysis device was fabricated by bonding the electrodes to two flat Lexan plates and then sandwiching a dielectric spacer element between the electrodes. The spacer serves to define the side walls of the electrical chamber, provides the fluid seal, and electrically insulates the electrodes from each other. Figure 2 shows an illustration of the chamber device and voltage controller. Two types of electrical chambers were constructed. The

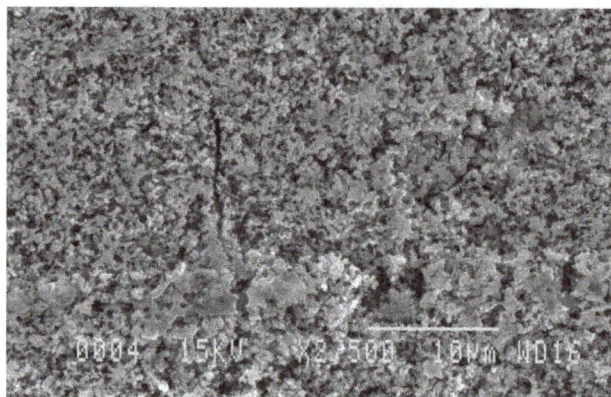

Figure 1. SEM image of the electrode surface. The microstructure profile of Al_2O_3 layer on aluminium surface, created by etching and then anodizing an aluminium foil, was revealed by an SEM image with 2500x magnification.

standard electrical chamber used in lysis testing had dimensions of 15 mm(L)×6.4 mm(W)×0.2 mm(H). For electrical characterization of the SEB electrodes an electrical chamber with the dimensions of 28 mm(L)×3.2 mm(W)×0.1 mm(H) was used. In this case the smaller height allowed the application of electrical fields with twice higher amplitudes.

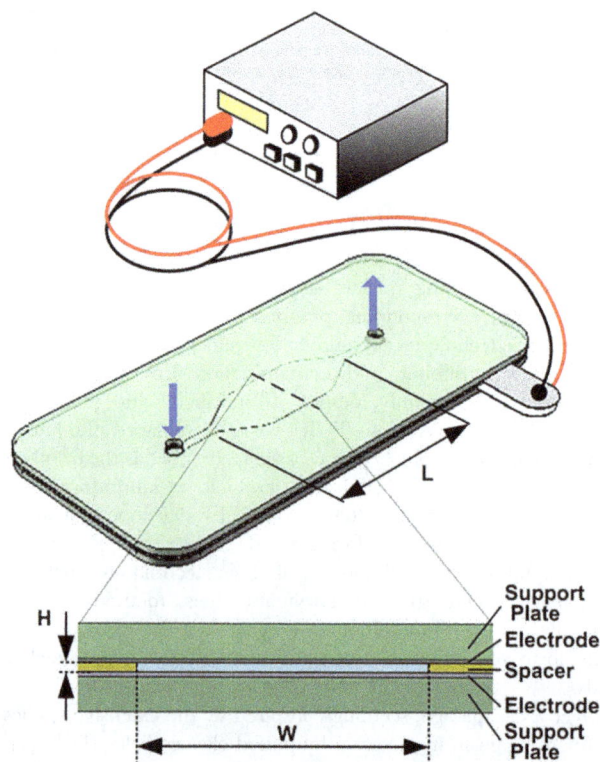

Figure 2. The schematic view of the flow-through electrical lysis device. A microfluidic chamber was fabricated by bonding the electrodes to two flat Lexan plates and then sandwiching a dielectric spacer element between the electrodes. The electronic system supplies bipolar square electric pulses whose amplitude or frequency can be controlled in response to chamber temperature obtained in real-time from chamber conductivity measurements.

Cell culture

Each single colony of Gram-negative bacteria (*Escherichia coli, Klebsiella pneumonia, Pseudomonas aeruginosa, Enterobacter cloacae and Acinetobacter baumanii*) grown on LB agar (G401, Hardy Diagnostics) was respectively cultured in LB broth (CG51, Hardy Diagnostics) overnight at 37°C. The cells were centrifuged at 7000 rpm for 5 min. The cell pellet was washed twice and re-suspended in 0.1 to 0.5 mM pre-filtered sodium phosphate buffer pH 7.4.

Each single colony of Gram-positive bacterial cells (*Streptococcus pneumonia, Streptococcus sanguis, Streptococcus pyogenes, Staphylococcus aureus and Enterococcus faecalis*) grown on Trypticase Soy agar with 5% sheep blood (A10, Hardy Diagnostics) was respectively cultured in Tryptic Soy broth (U63, Hardy Diagnostics) overnight at 37°C. The cultured cells were centrifuged at 10000 rpm for 5 min. The cell pellet was washed twice and re-suspended in 0.5 mM pre-filtered sodium phosphate buffer pH 7.4.

A single colony of acid-fast bacteria, *Mycobacterium smegmatis*, was grown with agitation for 2–3 days at 37°C in 5 ml of Middlebrook 7H9 (VWR,), containing sterile albumin dextrose complex (ADC) (0.5% BSA, 0.2% glucose, 3 µg of catalase per ml) supplement as well as Tween 90 (0.05%). The cultured cells were centrifuged at 10000 rpm for 5 min. The cell pellet was washed twice and re-suspended in 0.5 mM pre-filtered sodium phosphate buffer pH 7.4.

The concentrations of the bacterial suspensions were estimated by optical density measurement at 600 nm in Bio-Rad SmartSpec 3000 spectrometer, using sodium phosphate buffer as a blank and the conversion factor of 1 $OD = 5 \times 10^8$ cells/mL. The bacterial suspensions for the respective experiments were prepared at a concentration of 1×10^9 CFU/mL, 5×10^8 CFU/mL or 200 CFU/mL in sodium phosphate buffer, as indicated in respective experiments. The high concentration cell suspension was used for verification of verification of lysis efficiency with the Bradford protein assay and the low concentration cell suspension was used for testing with the real-time reverse transcription PCR assay.

Cell lysis

Electrical lysis was performed by passing a 60 µL aliquot of cell suspension through the electrical lysis chamber in increments of 10 µL and applying a voltage pulse train at each increment. The voltage pulse train consisted of bipolar square pulses which, prior to initialisation of feedback control, had amplitude of 195 V and a frequency of 10 kHz.

In some experiments, cells were also subjected to mechanical lysis using glass beads for the purpose of comparison. To perform glass bead lysis (GB), an equal volume of glass beads (<106 µm, G4649 Sigma) was added to 60 µL aliquot of cell suspension and mechanically lysed by vortexing at a high speed for 5 minutes. The GB cell lysate was centrifuged at 14,000 rpm for 1 minute to separate the beads and the supernatant was collected for the assay.

Bradford protein assay

Total protein released into the cell lysate was assayed using Bradford Reagent (B-6916, Sigma). The cell lysate was centrifuged at 7000 rpm for 5 minutes and 50 µL of the supernatant was collected and mixed with an equal volume of Bradford protein assay reagent. The protein concentration was determined by measuring light absorbance at 595 nm and referring to a dose response curve previously prepared by running the Bradford assay on different concentrations of Bovine Serum Albumin (BSA).

Real time RT-PCR assay

A real-time reverse transcription PCR (RT-PCR) assay was performed to detect a specific target region in 16S rRNA of the bacteria. The primers were designed by sequence alignment software (Bioedit, Ibis Biosciences, USA) and primer design software (Primer3, National Institute of Health). RT-PCR reaction volume of 20 µL was prepared by mixing 5 µL of lysate sample, 10 µL of KAPA2G Robust HotStart 2X PCR reaction mix (kk5515, Kapa Biosystems), 1.2 µL of GoScript reverse transcriptase (A5004 Promega), 1 µL of forward primer (10 µM), 1 µL of reverse primer (10 µM), 1 µl of 100 nM SYTO-9 (S34854, Invitrogen) and 0.8 µL of RNAase free water. One-step real time RT-PCR was performed by reverse transcription at 55°C for 5 min, and activation of HotStart DNA polymerase and inactivation of reverse transcriptase at 95°C for 2 min, followed by 35 cycles of cDNA amplification (denaturation at 95°C for 3 sec, annealing at 63°C for 3 sec, and extension at 72°C for 3 sec) in an Eco Real Time PCR system (Illumina).

For the detection of Gram-negative bacteria, a forward primer 5'-GTTACCCGCAGAAGAAGCACCG-3' and a reverse primer, 5'-ATGCAGTTCCCAGGTTGAGCC-3' were used to amplify the 16S rRNA gene fragment of 151 base pairs at a hypervariable region of all Gram-negative bacterial species (nucleotides 484 to 635 using *Escherichia coli* IaI1 as a reference).

For the detection of Gram-positive bacteria, a forward primer 5'-GACAGGTGGTGCATGGTTGTC-3' and a reverse primer, 5'-ACGTCATCCCCACCTTCCTC-3' were used to amplify the 16S rRNA gene fragment of 170 bases pair at a hypervariable region of all Gram-positive bacterial species (nucleotides 1035 to 1185 using *Streptococcus pneumoniae G54* as a reference).

For the detection of *Mycobacterium smegmatis*, a forward primer 5'-CCACACTGGGACTGAGATACGGC-3' and a reverse primer, 5'- CGTATCGCCCGCACGCTCAC-3' were used to amplify the 16S rRNA gene fragment of 202 base pairs at a hypervariable region of all *Mycobacterium* species (nucleotides 333 to 673 using *Mycobacterium bovis* BCG M140 as a reference).

Results and Discussion

Electrical characteristics of the lysis device

The E-Lysis method is performed using a fluidic device having a chamber formed between two opposing electrodes. During the process, electrochemical reactions can take place at the electrode-electrolyte interface, by products of which can create undesirable effects for the application of microfluidic reagent-free cell lysis. For example, gas generation and bubble formation due to electrolysis of the cell suspension fluid [12] can obstruct the fluidic channel. In addition, the products of redox reactions at the electrode interface may damage the electrodes, and degrade target biological molecules in the resulting lysate. In addition, the redox reaction may drastically change the ionic composition of the lysate rendering the liquid unsuitable for the downstream assays without further processing.

The electrolysis products can be partially avoided by operating the device with a high frequency bipolar pulse train. The ions produced at the electrode-electrolyte interface are then significantly neutralized in alternating cycles before diffusing away into the bulk medium [12]. The frequency should be chosen in consideration of the ionic strength of the chamber fluid so as to maximise the effective duration of the applied electric field, which results in higher electroporation efficiency [1].

The potential degradation of target molecules at the interfacial zone by the electrolysis products can be alleviated by providing the electrode surface with a protective permeation layer [2]. Such layers allow the movement of solute ions to the electrode while preventing the macromolecules from reaching the electrode-electrolyte interface. This approach, while reducing ill effects of electrolysis to some extent, does not prevent it entirely and the fabrication of the layer is a difficult multistep process. A preferred method for countering interfacial reactions is insulating the electrodes from the electrolyte with a thin layer of dielectric coating thus forming "blocking electrodes" which avoid direct electrical contact with the liquid within the chamber. However, the presence of the dielectric layer introduces an electrode surface capacitance which accelerates the formation of electric double layers in the ionic solutions near the electrodes and, consequently, screens the bulk fluid in the chamber from the applied electric field within a so-called charging time [13]. Accordingly, the effective electric field experienced by the suspended cells may drop to a small fraction of the nominally applied electric field shortly after the field is applied [14]. This problem can be circumvented by forming the dielectric layer on a conductive substrate with a finely micro-structured surface which substantially increases the surface area of the electrode. The large capacitance that is thereby achieved enables a charging time close to that of nominally smooth non-blocking electrodes, while concurrently avoiding the generation of electrolysis products. This type of electrode is henceforth referred to as a surface enhanced blocking (SEB) electrode.

The electrical characteristics of the chamber can be modelled by the equivalent electrical circuit presented in Figure 3 [15]. The capacitance C_{DL} corresponds to the dynamic double-layer capacitance near the interface of dielectric layer and the liquid in the chamber. R_{DL} is the parallel (in the direction of the chamber thickness) resistance corresponding to leakage current in the double layer. In general, values of C_{DL} for flat metal electrode surfaces fall in the range 5–50 $\mu F/cm^2$ depending on the type of electrode, ionic strength and composition of the solution, temperature and voltage [16]. However, electrode surface enhancement will increase this capacitance to higher values. Capacitance C_{DE} is the capacitance of the dielectric layer whose value depends on the layer thickness and the effective area of the electrode. For the electrodes used in the experiments described here, based on the empirically established relation between capacitance and the formation current and voltage [11], C_{DE} is estimated to be about 3 $\mu F/cm^2$. Resistance R_{DE} is the equivalent parallel resistance of the dielectric layer and accounts for leakage current in the capacitor. This resistance is in the order of 10 MΩ due to very low conductivity of the Al_2O_3 layer. R_{CH} represents the bulk solution resistance and C_{CH} the bulk capacitance. The value of the bulk solution capacitance $C_{CH} = \varepsilon\varepsilon_0 WL/H$, where $\varepsilon = 80$ is the dielectric constant of water, is estimated from the geometry of the chamber to be on the order of 1 pF. This value is so small that it can be approximated by an open circuit. R_{LOAD} is the sum of the power supply output resistance and the input resistance of the electrodes.

For the purposes of electrical characterisation the chamber of dimensions 28 mm(L)×6.4 mm(W)×0.1 mm(H) was filled with 0.25 mM NaCl solution. The resistance R_{CH} is estimated to be 220 Ω for this chamber configuration based on the fluid conductivity reported by McClesky [17]. Impedance spectrum measurements were obtained in the frequency range of 2 kHz to 4 MHz, and are presented in Figure 4. The chamber resistance, estimated from these results [18], gives a value of 210 Ω which agrees well with the estimated value. All the electrical parameter values, with the exception of R_{LOAD}, R_{DE} and C_{DE} are dependent on the ionic strength of the carrier solution. The load resistance modifies the voltage division among the circuit components and is more important at higher ionic strengths.

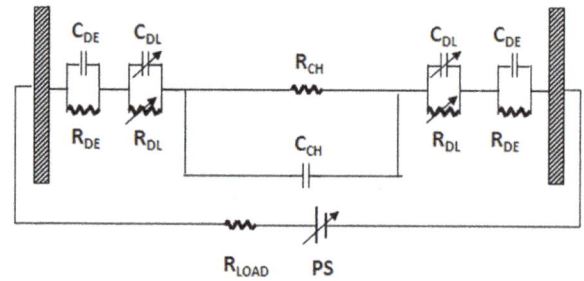

Figure 3. The equivalent circuit model for the electrical lysis device. The electrical lysis chamber modelled by an equivalent electrical circuit following the suggestions of reference [15].

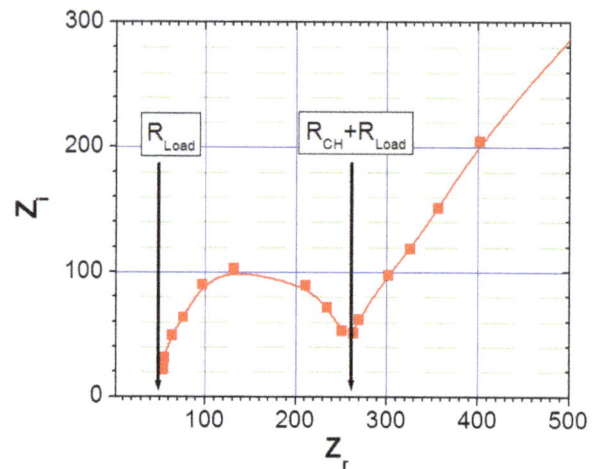

Figure 4. The impedance spectrum of the electrical lysis chamber. An electrical lysis chamber, with dimensions 28 mm(L)×6.4 mm(W)×0.1 mm(H), was filled with 0.25 mM NaCl solution and its impedance spectrum was recorded in the 2 kHz to 4 MHz frequency range.

The electrical response of chambers can be investigated by adopting the one dimensional model described by Bazant et al [14] for which a dilute 1:1 electrolyte such as NaCl is suddenly subjected to a DC voltage, V_A, by bare parallel-plate electrodes separated by a distance H. Due to the build-up of a double layer at the electrode-electrolyte interface the bulk electric field, $E = V_A/H$, starts to lose uniformity in the thickness (H) direction. Ions migrate in the bulk field and accumulate on electrodes. The resulting double layer eventually screens the electric field in the bulk and restricts the full potential drop in immediate vicinity of the electrode-electrolyte interface. Bazant et al. [14] have provided convincing arguments that the rate at which the bulk field decays is characterized by time constant $t_{scr} = \lambda H/2D$ where D is the diffusivity of the ions and λ (nm) is the Debye screening length, which at room temperature is related to the ionic strength of the medium, IS (M), by $IS = 10.75\lambda^2$. The switching model of Bazant et al. is valid at voltages of order $V_T = kT/e$ (k is Boltzmann's constant, T absolute temperature, and e electron's charge), which amounts to 26 mV at room temperature. In contrast, Beunis et al. [19] have offered a qualitative model that is applicable for arbitrarily high voltages.

An E-Lysis chamber with dimensions 28 mm(L)×6.4 mm(W)×0.1 mm(H) was filled with 0.25 mM

NaCl solution and a fast rising step voltage with an amplitude of either 4 or 95 Volts was applied, the electrical responses for which are shown in Figure 5. The response at the lower applied voltage (4 V) was fitted to the models proposed by Bazant et al. [14] and Beunis et al. [19]. The good agreement with the models which do not account for an electrode surface layer capacitance indicates that the presence of SEB electrodes does not appreciably reduce the switching response of the chamber in spite of the presence of a relatively thick dielectric (Al_2O_3) layer on the electrodes. The dielectric layer thickness d is about 140 nm since the thickness of the anodic Al_2O_3 layer is related to the formation voltage (100 V) by a proportionality factor of 1.4 nm/V [20]. In addition, the dielectric constant of Al_2O_3 layer is about 9 [21]. Therefore, without surface enhancement, these parameters would predict a capacitance $C_{DE} = \varepsilon\varepsilon_0 WL/d = 0.1\ \mu F$, with a corresponding $R_{CH}C_{DE}$ switching time of under 1.75×10^{-5} s, a value 35 times shorter than the observed screening time for the SEB electrodes given in Figure 5.

The switching behaviour at the higher applied voltage (95 V) cannot be easily explained. The attempts to tackle the high voltage regimes [22–23] have not yielded satisfactory analytical approximations due to the onset of phenomena such as steric effects [24]. There is an indication from these studies that the charging time in high voltage regime should be shorter than its corresponding value at low voltage regime. However, this is not observed in Figure 5 where the screening time for the high applied voltage has increased by a large factor. This phenomenon could be due to the involvement of what may be identified as fractal surface charging. According to Sakaguchi and Baba [25] the effect of fractals on the electrical response can be accounted for by assuming that the temporal behaviour of current is given by $I = I_0 \exp(-(t/t_{scr})^p)$ ($p<1$). In contrast, a qualitative inspection of Figure 5 shows that the temporal trend of the measured current is more akin to an exponential function of second order. We do not know how to explain this behaviour, but it should be noted that the longer charging time is advantageous for efficient cell lysis.

In order to demonstrate the blocking behaviour of SEB electrodes, two electrical chambers with dimensions of 15 mm(L)×6.4 mm(W)×0.2 mm(H) were fabricated using SEB and copper electrodes respectively. The chambers were filled with sodium phosphate buffer of ionic strength 0.5 mM and were driven by bipolar square pulses with amplitude of 195 V. The electric currents of two chambers are presented in Figure 6 for the first four cycles. As it is observed, in contrast to the case of copper electrode, there is an asymmetry in the charging and discharging behaviour for the SEB electrodes. The maximum amplitudes of the chamber currents during the charging cycle (positive current) are nearly identical for both cases. On the other hand, during the discharging cycle (negative current) the amplitude of current for SEB electrode is higher. This is due to the fact that SEB electrodes store electrical charge without electrochemical reactions at the electrode-electrolyte interface. The copper electrodes do not display this charging behaviour because of the presence of Faradic current involving electrolysis at the electrode-electrolyte interfaces.

Thermal characteristics of the lysis device

When a voltage pulse train is applied to the chamber electrodes, the chamber fluid is exposed to an alternating electric field and substantial ionic current flows across the chamber causing rapid Joule heating. The peak temperature which can be reached by the fluid in the chamber is limited to the boiling temperature of the fluid, a property which is influenced by bubble nucleation conditions and pressure. For the case of chambers with ports open to atmospheric pressure, the maximum attainable temperature is approximately 100°C. The high heating rates of the electrical lysis device (on the order of 5000°C/s) may delay boiling by a few degrees [26]. However, the large surface to volume ratio of the chamber and the presence of numerous inhomogeneous bubble nucleation sites on the electrode surface may offset this effect, and therefore it is assumed herein that the onset of the liquid to vapour phase transition takes place at 100°C.

In general, the conductivity of an aqueous solution is a linearly increasing function of temperature [17] with a rate of 2–3% per °C, and therefore, when applying a pulse train the chamber

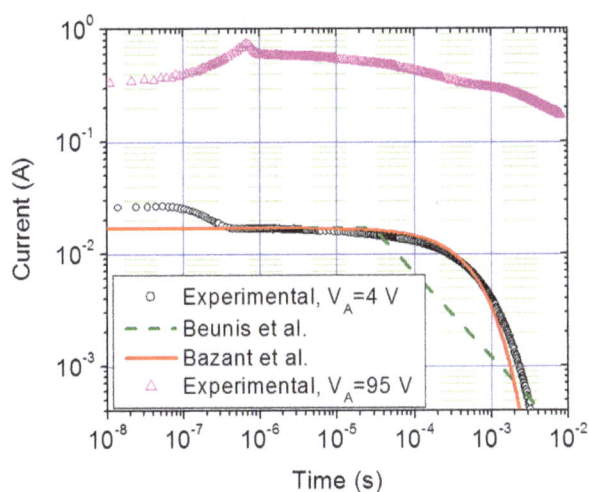

Figure 5. The electrical response of an electrical lysis chamber. An electrical lysis chamber, with dimensions 28 mm(L)×6.4 mm(W)×0.1 mm(H), was filled with 0.25 mM NaCl solution and a fast rising step voltage with an amplitude of 4 and 95 volts was applied to the surface enhanced blocking electrodes. The response at the lower applied voltage was fitted to the models proposed by Bazant et al. [14] and Beunis et al. [19].

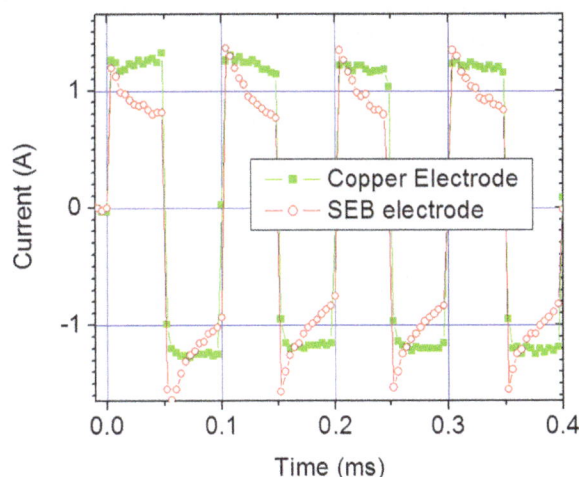

Figure 6. The effect of electrode type on current response of a lysis chamber. Two electrical lysis chambers, made respectively with copper and SEB electrodes, were filled with the microbial cell suspensions having ionic strength of 0.5 mM. Bipolar square pulses with amplitude of 195 V were applied and the electric current was recorded.

current will rise due to Joule heating of the fluid in the chamber. Further energy input will result in the inception of a liquid to vapour phase transition in the region of the chamber which has reached this temperature. Subsequently the mixed phase in the central region undergoes rapid expansion, displacing the liquid out of the chamber ports. Accordingly, as the mixed phase region expands further, the ionic current within the chamber decreases substantially as net conductivity of the chamber falls. The displacement of the sample from the chamber is generally undesirable since it limits the exposure of the cell suspension to the electric field and elevated temperatures.

The lysis performance can be improved by controlling the voltage pulse train applied to the chamber so as to maximise the fluid temperature but prevent evaporation. The suspended cells are then exposed to the applied electric field and elevated temperature for an extended time. This requires shaping the pulse train by dynamically varying the pulsing rate or pulse amplitude with reference to a feedback signal related to the instantaneous temperature of the suspension medium. The E-Lysis device conveniently provides such a feedback signal from the current flowing across the chamber. Active feedback control of the voltage pulse train applied to the chamber is depicted by the control diagram shown in Figure 7. The temperature of the fluid in the chamber is indicated by the electrical impedance of the chamber, $Z(t)$, which is obtained from measured voltage $V(t)$ and current $I(t)$. The instantaneous value of this impedance is used as the feedback signal. An error parameter $e(t)$ is obtained from the impedance relative to a predetermined setpoint impedance Z_{SP} and provided to the controller. A wide range of choices exist for controlling the voltage pulse train in response to the error signal $e(t)$. One example implementation is presented in the following paragraph.

An initial voltage pulse train of constant amplitude and frequency is applied and upon reaching a predetermined chamber impedance setpoint, active feedback control is activated. Thereafter, the voltage controller rapidly lowers the voltage amplitude to a fraction of the initial amplitude and adjusts the amplitude of the remainder of the voltage pulse train in response to measured electrical impedance in order to maintain the setpoint impedance level. A typical current response corresponding to this scheme is presented in Figure 8 for initial voltage pulse train amplitude of 195 V and a frequency of 10 kHz. The setpoint current, chosen to be approximately equal to the peak current experienced by the same chamber for a constant voltage pulse train amplitude, was reached after t = 16 ms after which the voltage was adjusted downward to approximately 60 V. Thus a peak chamber temperature near 100°C is reached at the setpoint and as can be seen from the near constant current response, the temperature is

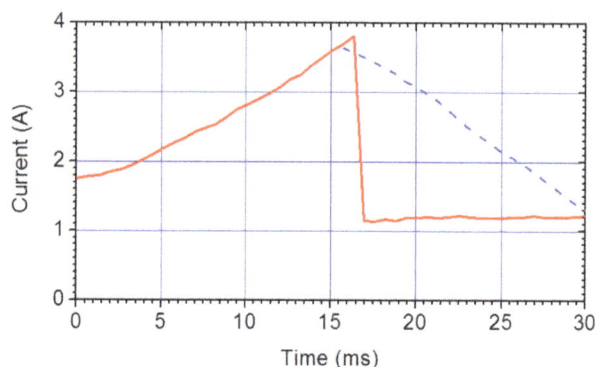

Figure 8. A typical current envelope of a lysis chamber operated under active feedback control of voltage pulse train amplitude. Feedback control of the voltage pulse amplitude in response to real-time chamber impedance measurements maintains the lysis chamber at a near constant temperature after reaching the setpoint at a time of about 16 ms. The dashed (blue) curve indicates the trend of current envelope in the absence of active feedback.

approximately maintained at this level for the remainder of the voltage pulse train. Without this feedback control a liquid to gas phase transition would have occurred as represented by the dashed curve in Figure 8.

Combined electric field and thermal lysis performance

The experiments described in this section are intended to demonstrate the synergistic effects of heat and electrical field on the efficiency of the device for lysing microbial cells. The performance was first demonstrated with Gram-negative bacteria *E. coli*. A bacterial suspension of 5×10^8 CFU/mL in 0.1 mM sodium phosphate buffer pH 7.4 was lysed in an electrical lysis chamber, with dimensions of 28 mm(L)×3.2 mm(W)×0.1 (H)mm. The 10 kHz pulse train of duration τ consisted of 50 μs bipolar pulses with amplitude V_A. The resulting electric field strengths across the chamber are indicated on Figure 9 which also shows the resulting current envelop. To ensure nearly equivalent electrical energy delivery to the chamber for purposes of comparison of lysis effectiveness, pulse train duration was chosen such that the quantity τV_A^2 was kept approximately constant ($=540$ sV^2) for all cases. Based on electrical conductivity changes, which are manifested by the rising currents in Figure 9, the temperature for all 0.1 mM cases increased a small amount, although not beyond the normal biological temperature range. The highest peak temperature reached was 42°C, which is obtained from the current rise for the applied field 19 kV/cm by applying the thermal conductivity factor of 2.5%/°C. This factor was obtained from the accompanying curve of the 0.4 mM suspension medium and applied field 15.2 kV/cm which reached boiling as indicated by the current drop near the end of the pulse train. A bacterial cell suspension of 5×10^8 CFU/mL in 0.4 mM sodium phosphate buffer pH 7.4 was electrically lysed at an applied voltage of 152 V for a duration of 23.5 ms, electrically equivalent to the 15.2 kV/cm condition of the 0.1 mM case in Figure 9. However, due to the higher ionic strength the medium is rapidly heated and prior to the end of the pulse train the temperature reaches the boiling point. The results of a total protein assay are presented in Figure 10. Although the electric field amplitude is correlated to the amount of protein released from the cells during electrical lysis, the lysing effectiveness as judged by total protein release is much lower for electrical lysis using the 0.1 mM phosphate buffer as compared to GB lysis. By plating lysate on culture plates, we verified that no

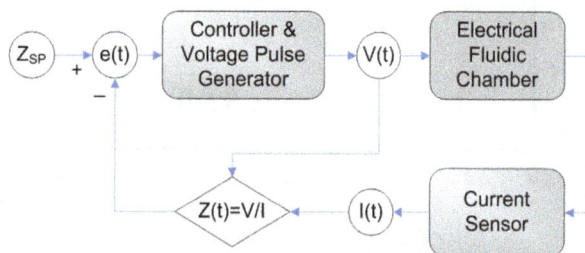

Figure 7. The block diagram of the impedance based temperature regulation scheme. The electric current supplied to the electrodes is measured and used as a proxy for the temperature of the chamber liquid. A feedback loop controls the voltage pulse train and maintains the chamber impedance at a pre-set value.

bacteria survived the injury inflicted by exposure to multiple electric pulses, as expected from published results [1,3,27–29]. However, the protein assay results indicate that employing an electric field alone for lysis of bacteria cells, with the goal of releasing macromolecules, is not an efficient procedure. However, when electrical lysis is combined with flash heating, as in the case for the example of the 15.2 kV/cm field in a 0.4 mM sodium phosphate buffer, the percentage of the released protein increased by over a factor of 10, and was about 80% as compared to GB lysis. Therefore, the simultaneous exposure of the bacteria to heat and electric field significantly enhances lysis efficiency.

The synergistic effect of the applied electric field and elevated temperature on cell lysis was further demonstrated with Gram-positive bacteria which are otherwise resistant to lysing by electric field alone due to their tough cell wall. The suspensions of 1×10^9 CFU/mL *S. pneumoniae* in 0.4 mM sodium phosphate buffer pH 7.4 were lysed in a chamber with different parameter pairs (τ, V_A) under nominally similar energy input defined by the quantity $\tau V_A^2 = 10^3$ sV^2. In all cases the peak temperature momentarily reached 100°C prior to the end of the voltage pulse train. The results of total protein assay are presented in Figure 11, and it is observed that the lysis efficiency varies from 50% to over 80% for the range of electric field strengths tested.

The effect of electrode type on lysis performance

The experiment described in this section is intended to demonstrate the advantages of using SEB electrodes for preserving the integrity of target macromolecules. *S. pneumoniae* cells, suspended in 0.5 mM sodium phosphate buffer pH 7.4, were lysed in two geometrically similar standard chambers, one chamber having SEB electrodes and the other having copper electrodes. The pulse amplitude was initially 195 V with feedback and a current response as shown in Figure 8. The corresponding electric field strength was about 10 kV/cm. The cell suspension intended for the Bradford protein assay and the real time RT-PCR assay respectively had a concentration of 1×10^9 CFU/mL and CFU/mL. The result of protein assay is presented in

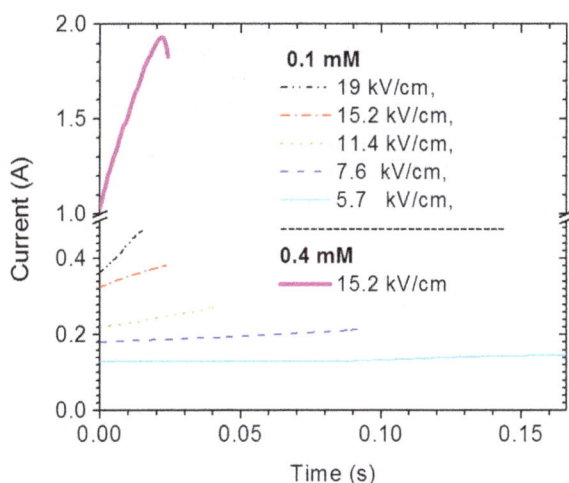

Figure 9. **The current response envelope of lysis chambers operated under different applied voltages.** Bipolar square pulses with different pulse amplitudes, V_A, and pulse train durations, τ, were applied to a lysis chamber containing a cell suspension with ionic strength of 0.1 mM and the current envelopes were measured. Also presented is the current envelope for the case of $V_A = 152$ V applied to a cell suspension having ionic strength of 0.4 mM. The suspension in this case reaches boiling temperature prior to the end of the pulse train.

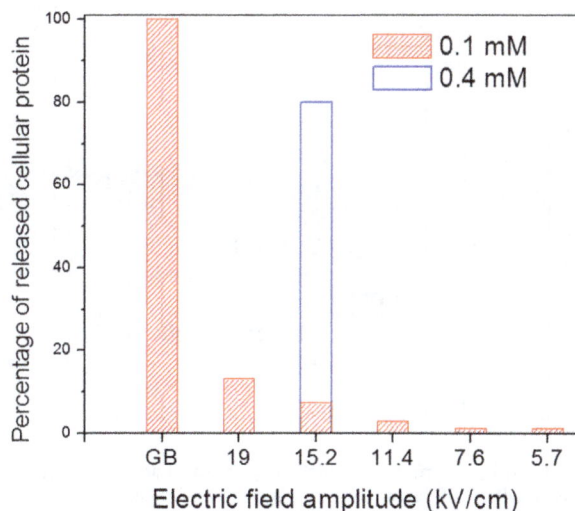

Figure 10. **The lysis efficiency of *E. coli* as a function of electric field strength.** The electrical lysis efficiency of *E. coli* cells for the cases presented in Figure 9 was determined by measuring the released protein as compared to glass bead beating. Only the case which involved flash heating gave comparable results indicating the synergy of electric field and heat lysis mechanisms.

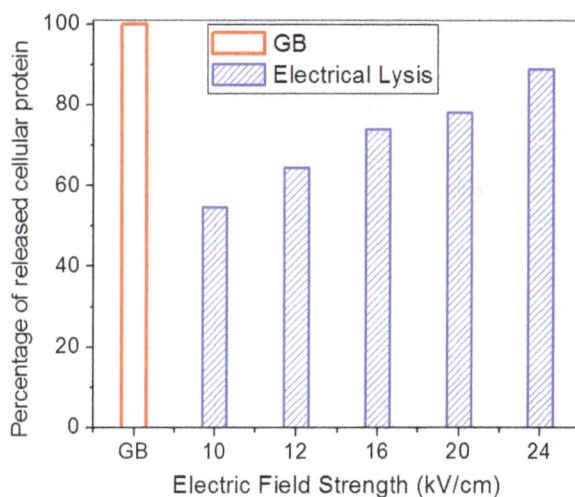

Figure 11. **The lysis efficiency of *S. pneumoniae* as a function of electric field strength.** The lysis efficiency of *S. pneumoniae* cells subjected to electrical lysis which combined electric field with flash heating to 100°C was determined by measuring the released protein. The amount of released protein is reported as a percentage of protein released by glass bead lysis.

Figure 12. The two chambers are similar in terms of protein release and their lysis performance is comparable to the GB beating. Therefore the electrode type does not affect lysis performance of the chambers for protein release.

With respect to performance in relation to nucleic acid assays the efficient release of molecular contents is not the only relevant criterion. The treated sample should not contain intolerable levels of PCR inhibitors and the target molecules must be intact. A real time RT-PCR assay was performed on the lysate of 200 CFU/mL cell suspension after being subjected to electrical lysis in chambers

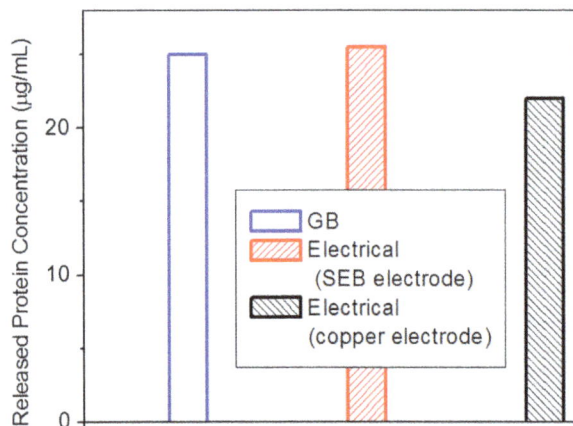

Figure 12. E-Lysis efficiency, with respect to protein release, in channels with SEB and copper electrodes. An electric field of 10 kV/cm was applied until a peak temperature near 100°C was reached, followed by a constant temperature dwell with voltage feedback control. Both SEB and plain copper electrodes reach protein release concentrations similar to glass bead (GB) lysis.

constructed with SEB and copper electrodes, respectively. The flash heating phase utilised a field strength of 10 kV/cm and voltage amplitude feedback control as above. A volume of 5 μL of the cell lysate, representing lysate from a single cell, was subjected to RT-PCR. The results, presented in Figure 13, indicate that the PCR amplification for the lysate from the chamber with SEB electrodes performs as well as GB lysis. The copper electrodes yield results which are delayed by 4 cycles relative to the SEB electrodes, implying an equivalent reduction in PCR products by a factor of 16. Since the protein assay result of Figure 12 suggests nearly similar macromolecule release efficiency, this implies that while using copper electrodes either the target nucleic acid has been degraded, possibly by the oxidizing species such as OH^- ions from the interfacial electrochemical reactions, or some ions released into the lysate are inhibiting the nucleic acid assay. In

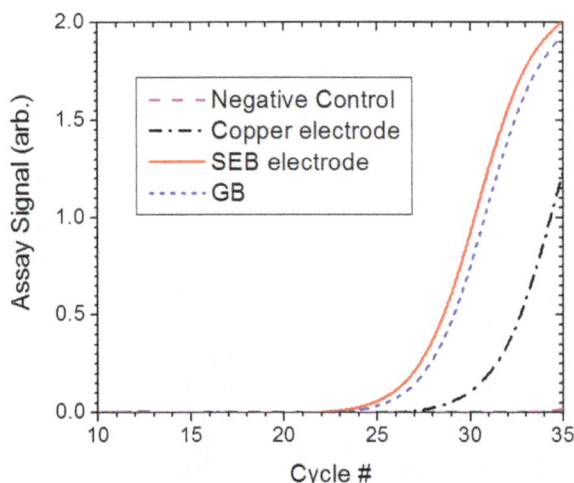

Figure 13. E-Lysis efficiency, with respect to nucleic acid assay, in channels with SEB and copper electrodes. The electrical lysis efficiencies of E-Lysis chambers using SEB and copper electrodes were compared by determining the amount of released ribosomal RNA from RT-PCR assay performed on the cell lysate.

either case, the ability of SEB electrodes to provide assay-ready lysate is evident.

By monitoring the electrical current it was observed that the chamber made with copper electrodes degrades following application of pulse trains, indicated by a reduction in the current amplitude which could be achieved. After 10 actuations, the maximum current decreases by about half and the chamber may be judged as ineffective for lysing of volumes greater than approximately 50 μL. This is accompanied by discoloration of the initially polished surface which is a manifestation of an electrochemical reaction with the chamber fluid. In contrast, the chamber made from SEB electrodes is much more robust, not showing any evidence of degradation after lysing of as much as 100 mL of sample.

Electrical lysing capability across different pathogenic bacteria

Further electrical lysing test results are presented to demonstrate the broad lysing capability of the device. Lysing was performed with the SEB electrode device for five pathogenic Gram-negative and five pathogenic Gram-positive bacteria species. Cell suspensions of each species were prepared at a concentration 200 CFU/ml in 0.5 mM sodium Phosphate buffer pH 7.4. A 60 μL each of cell suspension of each species and a 0.5 mM sodium phosphate buffer as a negative control was subjected to electrical lysis with an initial electric field of 10 kV/cm and voltage feedback control. RT-PCR was performed on 5 μL of lysate in each case. The detectability criterion was defined as follows. The standard deviation, σ, of the RT-PCR signal was calculated over the first 10 cycles, where the signal is predominately background noise. The cycle number for which the recorded real time PCR signal exceeds the threshold level of 6σ is defined as C_T. If the C_T value of a sample differs by more than 5 cycles (a signal to noise ratio of over 32) from the C_T value of the corresponding negative control running in parallel with the sample, the detection is considered to be unambiguous.

The C_T values for the different Gram-negative and Gram-positive bacteria species over five independent runs, each run with two replicates, are presented in Figure 14. The detection limit criterion as described above was satisfied for all bacterial species samples tested. This demonstrates the effectiveness of E-Lysis as a rapid and reagent-free inline lysis method for sensitive detection of bacterial species.

The performance of the E-Lysis device was verified for the case of a polymicrobial sample. A cell suspension, containing 200 CFU/ml of Gram-negative bacteria *E. coli* and 200 CFU/ml of Gram-positive bacteria *S. pneumoniae* cells was prepared in 0.5 mM sodium phosphate buffer pH 7.4. A 60 μL aliquot of the suspension was electrically lysed. Simultaneous real time RT-PCR assays were performed on two 5 μL samples of the lysate in two separate wells, one containing Gram-negative bacterial specific primer set and the other containing Gram-positive bacterial specific primer set. The results are presented in Figure 15. Comparing the observed C_T values with the case of single microbial samples (Figure 14) indicated that the lysis performance of the device for a given bacterial species is not being influenced by the presence of other species in the sample. Therefore, the device is easily applicable in situations requiring multiplexed detection of bacterial species using the respective specific primers.

Finally, the lysis performance of the E-Lysis device was tested for the *Mycobacterium* species, well known to be very difficult to lyse by conventional microbial lysis techniques. A cell suspension, containing 200 CFU/ml of the *Mycobacterium smegmatis* was prepared in 0.5 mM sodium phosphate buffer pH 7.4. A 60 μL

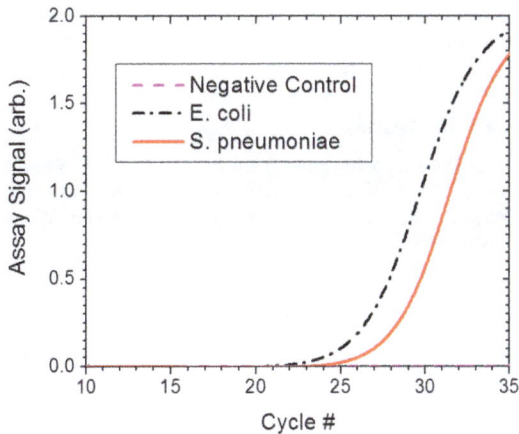

Figure 14. The real time RT-PCR results for detection of a variety of Gram-negative and Gram-positive bacterial species. Electrical lysis was demonstrated for a variety of Gram-negative and Gram-positive bacterial species using an E-Lysis chamber with combined electric field and flash heating to 100°C. This was demonstrated by performing real time RT-PCR assays directly on the lysate. Each data point represents 5 independent runs with two replica samples in each run.

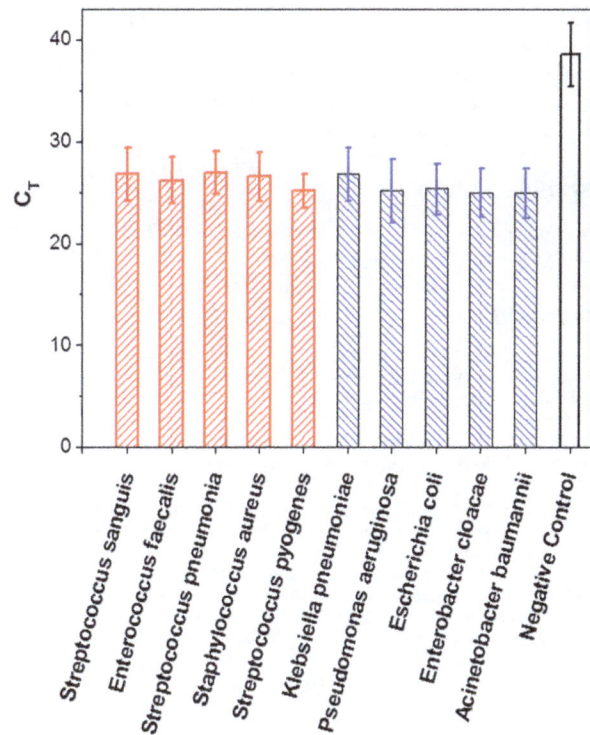

Figure 15. The electrical lysis efficiency of a polymicrobial sample. The performance of the electric lysis device for lysing a polymicrobial sample was demonstrated by lysing a cell suspension containing Gram-negative bacteria *E. coli* and Gram-positive bacteria *S. pneumoniae* cells and determining the amount of released ribosomal RNA from RT-PCR assay performed on the cell lysate.

aliquot of the suspension was subjected to electrical lysis using the E-Lysis device and another 60 μL aliquot was subjected to 5 minutes of glass bead beating as in previously discussed tests. Simultaneous real time RT-PCR assays were performed on 5 μL samples of the lysate in wells containing *Mycobacterium*-specific primer set. From the results presented in Figure 16 it is observed that the E-Lysis performance for the *Mycobacterium* is much more effective than the glass bead beating method used, yielding a C_T value about 10 cycles lower. Also, this result yields a C_T value similar to the values of the other bacteria species reported herein, indicating a similar lysis efficiency.

Conclusions

Owing to the disadvantages of electrolysis from unprotected electrodes in microfluidic cell lysis devices, blocking electrodes are preferred for electrical lysis of microbial cells. However, the selection of blocking electrodes must mitigate the main drawback of such approach, namely the reduction in electrical screening time which can severely limit the effectiveness of electroporation by shielding the target cells from the applied electric field. This effect is largely eliminated by employing surface enhanced blocking electrodes with a finely micro-structured surface and a thin dielectric coating which increases the screening time to levels expected from bare electrodes. The use of such electrodes in thin fluidic chambers and the application of a bipolar voltage pulse train allow high electric fields of the order of 10 kV/cm to be generated in the cell suspension for a sufficient length of time to cause irreversible electroporation of cellular membranes. This configuration also causes rapid Joule heating from the ionic current generated in the cell suspension fluid and, by applying feedback control from measured electrical current across the chamber, temperatures in the neighbourhood of 100°C can be maintained during the pulse train. While electrical lysis of Gram-negative bacteria such as *E. coli* has been previously demonstrated, the current configuration and voltage regulation scheme provides a cell lysis device suitable for microfluidic applications and which can be used for a wide range of microorganisms, including Gram-positive bacteria and *Mycobacteria* whose cell envelope includes a

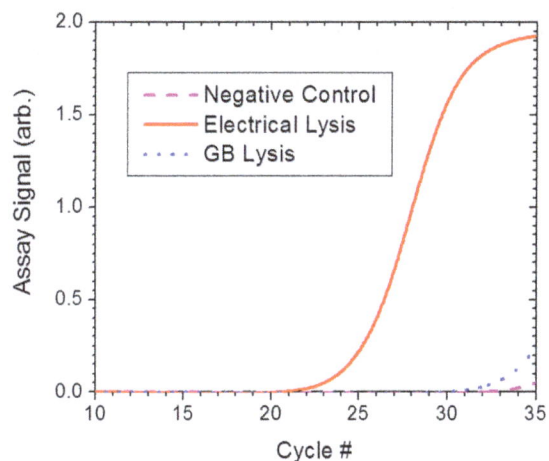

Figure 16. The electrical lysis efficiency of *Mycobacterium*. The lysis efficiency of *Mycobacterium smegmatis* cells subjected to electrical lysis which combined electric field with flash heating to 100°C was verified by determining the amount of released ribosomal RNA from RT-PCR assay performed on the cell lysate. The lysis efficiency of the electrical lysis method is higher than the lysis efficiency of glass bead beating technique by a factor of about 1000 (~10 cycles).

tough cell wall. The effectiveness of this technique was demonstrated by lysing a range of pathogenic Gram-negative and Gram-positive bacterial species and directly using the lysate for performing real time RT-PCR without any further processing. The high performance of this lysis method is attributed to the synergistic effect of electrical field and high temperature on the bilayer membrane and cell wall.

Acknowledgments

We would like to acknowledge the contributions of Dr. Tony Mazzulli of Mount Sinai Hospital for providing the clinical isolates we used in our experiments and for fruitful discussions. We thank Dr. Christopher Stone, Dr. Jooeun Lee and Binal Shah for their assistance with performing assays, and K. K. Chan and Sanjesh Yasotharan for their technical support.

Author Contributions

Conceived and designed the experiments: AT RM AAK TA. Performed the experiments: AT RM AAK TA. Analyzed the data: AT RM AAK TA. Contributed reagents/materials/analysis tools: AT RM AAK TA. Wrote the paper: AT RM AAK TA.

References

1. Htilsheger J, Potel J, Niemann EG (1983) Electric field effects on bacteria and yeast cells. Radiation and Environmental Biophysics 22: 149–162.
2. Cheng J, Sheldon EL, Wu L, Uribe A, Gerrue LO, et al. (1998) Preparation and hybridization analysis of DNA/RNA from E. coli on microfabricated bioelectronic chips. Nature Biotechnology 16: 541–546.
3. Lee SW, Tai YC (1999) A micro cell lysis device. Sensors and Actuators. 73: 74–79.
4. Morshed B, Shams M, Mussivand T (2013) Brief Review on Electrical Lysis: Dynamics Revisited and Advancements in On-chip Operation. Critical Reviews in Biomedical Engineering. 41: 37–50.
5. Geng T, Lu C (2013) Microfluidic electroporation for cellular analysis and delivery. Lab on a Chip, 13: 3803–3821.
6. Wang HY, Bhunia AK, Lu C (2006) A microfluidic flow-through device for high throughput electrical lysis of bacterial cells based on continuous dc voltage. Biosensors and Bioelectronics. 22: 582–588.
7. Mitchell GJ, Wiesenfeld K, Nelson DC, Weitz JS (2013) Critical cell wall hole size for lysis in Gram-positive bacteria. Journal of The Royal Society Interface, 10: 20120892.
8. Packard MM, Wheeler EK, Alocilja EC, Shusteff M (2013) Performance Evaluation of Fast Microfluidic Thermal Lysis of Bacteria for Diagnostic Sample Preparation. Diagnostics. 3: 105–116.
9. Pavlin M, Kotnik T, Miklavčič D, Kramar P, Lebar AM (2010) Chapter Seven, Electroporation of Planar Lipid Bilayers and Membranes; Advances in Planar Lipid Bilayers and Liposomes. 6: 165–226.
10. Kanzaki N, Toyama K, Nakano K, Shimatani R, Kudo F (1981) Method for producing foil electrodes for electrolytic capacitor. U.S. Patent No. 4,276,129. Washington, DC: U.S. Patent and Trademark Office.
11. Sarang PG, Maduskar SV (1986) A low-cost circuit for fabricating anodic oxide films. Journal of Physics E: Scientific Instruments. 19: 1063–1065.
12. Persat A, Suss ME, JG Santiago (2009) Basic principles of electrolyte chemistry for microfluidic electrokinetics. Part II: Coupling between ion mobility, electrolysis, and acid–base equilibria. Lab on a Chip. 9: 2454–2469.
13. Han ZJ, Morrow R, Tay BK, McKenzie D (2009) Time-dependent electrical double layer with blocking electrode. Applied Physics Letters. 94: 043118.
14. Bazant MZ, Thornton K, Ajdari A (2004) Diffuse-charge dynamics in electrochemical systems. Physical review E. 70: 021506.
15. Wu J (2008) Interactions of electrical fields with fluids: laboratory-on-a-chip applications. Nanobiotechnology, IET. 2: 14–27.
16. Chazalviel JN (1999) Coulomb Screening by Mobile Charges, Birkhauser, Boston, p. 72.
17. McCleskey RB (2011) Electrical conductivity of electrolytes found in natural waters from (5 to 90) C. J. Chem. Eng. Data. 56: 317–327.
18. Barsoukov E, Macdonald JR (Editors) (2005) Chapter 1, Impedance Spectroscopy, Theory, Experiment and Applications, Whiley Interscience.
19. Beunis F, Strubbe F, Marescaux M, Beeckman J, Neyts K, et al. (2008) Dynamics of charge transport in planar devices. Physical review E. 78: 011502.
20. Zeković LD, Urošević VVB, Jovanić B (1983) Investigation of anodic alumina by a photoluminescence method: II. Thin solid films. 105: 217–223.
21. Dyer CK, Alwitt RS (1978) Ellipsometric measurements of the barrier layer in composite aluminum oxide films. Electrochimica Acta. 23: 347–354.
22. Kilic MS, Bazant MZ, Ajdari A (2007) Steric effects in the dynamics of electrolytes at large applied voltages. I. Double-layer charging. Physical review E. 75: 021502.
23. Kilic MS, Bazant MZ, Ajdari A (2007) Steric effects in the dynamics of electrolytes at large applied voltages. II. Modified Poisson-Nernst-Planck equations. Physical review E. 75: 021503.
24. Borukhov I, Andelman D, Orland H (1997) Steric effects in electrolytes: a modified Poisson-Boltzmann equation. Physical review Letters. 79: 435–439.
25. Sakaguchi H, Baba R (2010) Electric double layer on fractal electrodes. Phys. Rev. E. 75: 051502.
26. Elias E, Chambre PL (2007) Limit of superheat in uniformly heated fluid. Heat Mass Transfer. 43: 957–963.
27. Zhong K, Chen F, Wu J, Wang Z, Liao X, et al. (2005) Kinetics of inactivation of Escherichia coli in carrot juice by pulsed electric field. Journal of food process engineering. 28: 595–609.
28. de la Rosa C, Tilley PA, Fox JD, Kaler KV (2008) Microfluidic Device for Dielectrophoresis Manipulation and Electrodisruption of Respiratory Pathogen Bordetella pertussis. IEEE Transactions on Biomedical Engineering. 55: 2426–2432.
29. Aronsson K, Rönner U, Borch E (2005) Inactivation of Escherichia coli, Listeria innocua and Saccharomyces cerevisiae in relation to membrane permeabilization and subsequent leakage of intracellular compounds due to pulsed electric field processing. International journal of food microbiology. 99: 19–32.

11

Theory of Electric Resonance in the Neocortical Apical Dendrite

Ray S. Kasevich[1], David LaBerge[2]*

1 Stanley Laboratory of Electrical Physics, Great Barrington, Massachusetts, United States of America, **2** Department of Cognitive Sciences, University of California Irvine, Irvine, California, United States of America

Abstract

Pyramidal neurons of the neocortex display a wide range of synchronous EEG rhythms, which arise from electric activity along the apical dendrites of neocortical pyramidal neurons. Here we present a theoretical description of oscillation frequency profiles along apical dendrites which exhibit resonance frequencies in the range of 10 to 100 Hz. The apical dendrite is modeled as a leaky coaxial cable coated with a dielectric, in which a series of compartments act as coupled electric circuits that gradually narrow the resonance profile. The tuning of the peak frequency is assumed to be controlled by the average amplitude of voltage-gated outward currents, which in turn are regulated by the subthreshold noise in the thousands of synaptic spines that are continuously bombarded by local circuits. The results of simulations confirmed the ability of the model both to tune the peak frequency in the 10–100 Hz range and to gradually narrow the resonance profile. Considerable additional narrowing of the resonance profile is provided by repeated looping through the apical dendrite via the corticothalamocortical circuit, which reduced the width of each resonance curve (at half-maximum) to approximately 1 Hz. Synaptic noise in the neural circuit is discussed in relation to the ways it can influence the narrowing process.

Editor: Eshel Ben-Jacob, Tel Aviv University, Israel

Funding: The authors have no support or funding to report.

Competing Interests: The authors have declared that no competing interests exist.

* E-mail: dlaberge@earthlink.net

Introduction

The pyramidal neuron, with its relatively long dendrite that extends upward from the top of the soma, is the main excitatory neuron and most numerous type of neuron found in the neocortex [1]. Its electrical activity is influenced by the thousands of other neurons, both excitatory and inhibitory, which make synaptic contacts along the membrane surface [2,3]. In the waking state, a large proportion of these synaptic contacts are continuously bombarded by local circuits, which maintains a low level of noisy subthreshold oscillations in the neural membrane [4,5].

Electrical brain rhythms, particularly synchronized oscillations generated by the pyramidal neurons, are measured as encephalograms (EEGs), and are believed to be related to a wide variety of cognitive functions [6,7]. The mechanisms of these synchronized oscillations have been studied under conditions in which the dendrite membrane voltage lies below the threshold for generating action potentials. Under these subthreshold conditions, an injected current typically induces the neuron to respond by transferring voltage along the dendritic membrane, and in some neurons, the magnitude of this response is frequency dependent. When the neural membrane responds preferentially to inputs within a narrow range of frequencies, the membrane is said to be exhibiting resonance [8,9]. As it is most generally used, the term *resonance* denotes the ability of a system to oscillate most strongly at a particular frequency.

Resonance activity apparently can arise from noise in neural networks in a variety of ways, ranging from the stochastic resonance produced by a moderate level of synaptic noise that enhances subthreshold signals, e.g., [10], to the coherent resonance produced by combining synaptic noise with network coupling, e.g., [11]. Resonance activity also arises from synaptic plasticity, for example in the regulation by synaptic plasticity of thalamocortical connections of gamma (30–100 Hz) oscillations in the primary visual cortex [12] and in filtering by short term depression and facilitation at synapses [13,14]. Plasticity has been combined with a coupling of 100 neurons to produce transient stochastic resonance [15]. The present approach regards resonance activity in the neocortical apical dendrite as arising from repeated operations of electrical circuit elements of the apical dendrite membrane (mostly capacitive) on current pulse trains injected into the distal segment of the apical dendrite. These pulse trains arise from the thalamus, and are sustained over time by the corticothalamic loop.

Apparently, transient alterations in synaptic function on the dendrites of the waking animal are continually being initiated by local circuit background activity on dendritic spines, which produce subthreshold membrane activity. Alterations are produced by locally produced action potentials and backpropagating action potentials. The finding [13,14] that temporal filtering of synaptic transmission arises from short-term synaptic depression and facilitation following synaptic events indicates that recent synaptic activity produces a transient context, and this context can affect local signal processing [16], and perhaps can affect the more global tuning of resonance at the neural or circuit levels.

The purpose of the present paper is to increase our understanding of the mechanisms by which the dendritic membrane can become tuned to a specific narrow band of frequencies by describing a model of the transfer of electrical energy along the apical dendrite of a neocortical pyramidal neuron. As current pulses move down the apical dendrite, they are influenced by the conductive and capacitive elements of the anisotropic membrane (in which capacitance is not the same in radial and longitudinal directions). The progressive narrowing of resonance in the present model of the neocortical apical dendrite is based on the surface impedance theory formulated by Delogne [17]. A standard measurement for describing the electrical properties of the membrane is impedance, which is the ratio of the voltage output to the current input [8,18]. The relationship of impedance amplitude to resonance characteristics has been investigated in several articles [[8,19–23]. The relationship of impedance phase to membrane resonance has received considerably less attention in the literature, but a recent study [18] highlights the potential role of membrane channel activity on time delays in hippocampal neurons. For purposes of analyzing possible determinants of resonance in the apical dendrite of the neocortical pyramidal neuron, the present study will use impedance amplitude as the principal measurement. In the particular circuit model of this paper, a peak in the transfer impedance amplitude, measured here over the frequency range 10 to 100 Hz, indicates the presence of resonance in the apical dendrite membrane [8]. We will investigate the characteristics of resonance tuning at four frequencies: 20, 40, 60, and 80 Hz.

Results

Simulated Model

The vertical shaft of the typical human neocortical apical dendrite is modeled here as a series of 6 compartments, as shown in Figure 1, following the reconstruction of the apical dendrite by [24]. The 6 compartments constitute the mid region of the apical dendrite, and these 6 compartments, each of a length of 200 μm, will serve as the main compartment model. An additional dendritic segment, located between the 6th compartment and the soma, has a length here of approximately 240 microns, and may be regarded as a transitional region. The diameter of this segment gradually decreases to accommodate the large difference in diameters of the soma and the mid region of the dendrite, while the 6 compartments maintain a relatively constant diameter [25]. Therefore the present model does not include the transitional segment between the soma and the 6 compartments. The synapses on the initial 40 μm part of the transitional segment, like the soma, are inhibitory and excitatory spines are absent [3]. As the distance from the soma increases the inhibitory synapses decrease in number while the spines increase in number. At the distance from the soma where the 6 model compartments begin, the synapses are mostly excitatory [26]. Membrane conductances are known to fluctuate somewhat over the middle section of the apical dendrite shaft, e.g., [27–29]. One of these changes is incorporated into the model, while other changes are tentatively assumed to produce relatively small perturbations in the behavior of the model. In some pyramidal neurons the main shaft of the neocortical apical dendrite bifurcates into two shafts (and in some it bifurcates more than once). The voltage output from each branch will depend on the transfer impedance of that branch and its input conductance. The input voltage of Êeach branch is identical and is derived from the input current at the initial segment multiplied by the segments transfer impedance. When the ends of the branches converge to a common segment leading to the soma, the total voltage experienced by the common branch may combine in a manner that depends on the transfer impedance of each branch. The

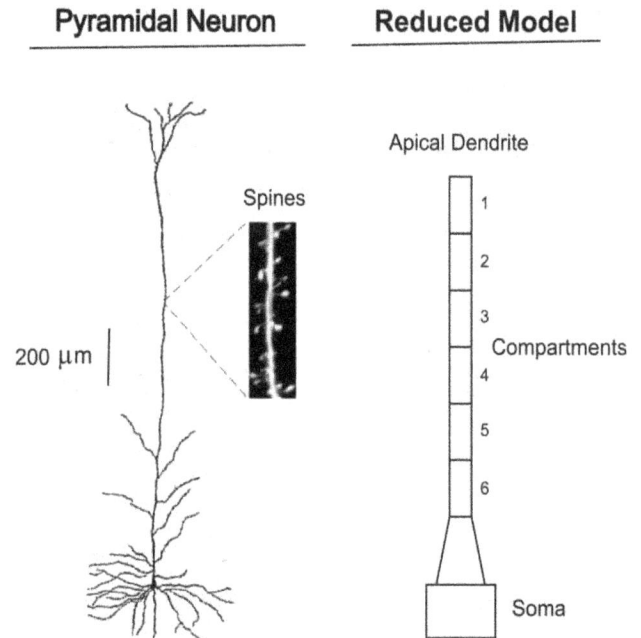

Figure 1. A neocortical layer 5 pyramidal neuron: camera lucida drawing and reduced compartment model. Human Layer 5 neocortical pyramidal neuron from [32], and a simplified compartment model of the apical dendrite. Also shown, on a 18-micron length of dendrite, is a sample from the hundreds of spines that dot the apical dendrite [33] (Figure 8). The schematic drawing of the apical dendrite in the reduced model contains 6 compartments, which provide the present model framework. Between the 6th compartment and the soma is a segment of the apical dendrite that contains distributions of electrical elements that differ from those in the 6 compartments, and this segment is regarded as a transitional region.

transfer impedance of a branch will, in turn depend on the number of compartments defined for each branch.

The input from the thalamic principal neuron to the first compartment is located at the distal region of the Layer 5 apical dendrite for all neocortical minicolumns except for minicolumns located in the primary sensory areas, where the input from the thalamic principal neuron is mainly in the mid region of the Layer 5 pyramidal neuron, but with some inputs at the distal region [30,31]. Relatively thin branches of the apical dendrite that appear most frequently near the soma do not receive direct thalamic inputs, and their exact function in the present model is not specified at this time. It appears that the spines are tonically activated by the same local circuitry that activates spines on the major shafts, and this activation could produce the voltage level needed to open local K^+ channels in a manner that supports the particular peak frequency resonance of the local minicolumn. In any case, the net contribution of the minor branches is presumed to be relatively weak.

Figure 2 shows a schematic drawing of a typical longitudinal cross-section of the pyramidal apical dendrite, in which the current-carrying core is surrounded by the lipid bilayer membrane (additional electrical detail of the membrane structure is shown in Figure 3). While the diameter of an apical dendrite has a range of approximately 0.30 to 8.5 microns [34], the thickness of the membrane is considerably smaller, and varies in the range of 3–5 nanometers [35,36]. Hence the scale of the membrane in Figure 2 is considerably expanded relative to the scale of the cytoplasmic core.

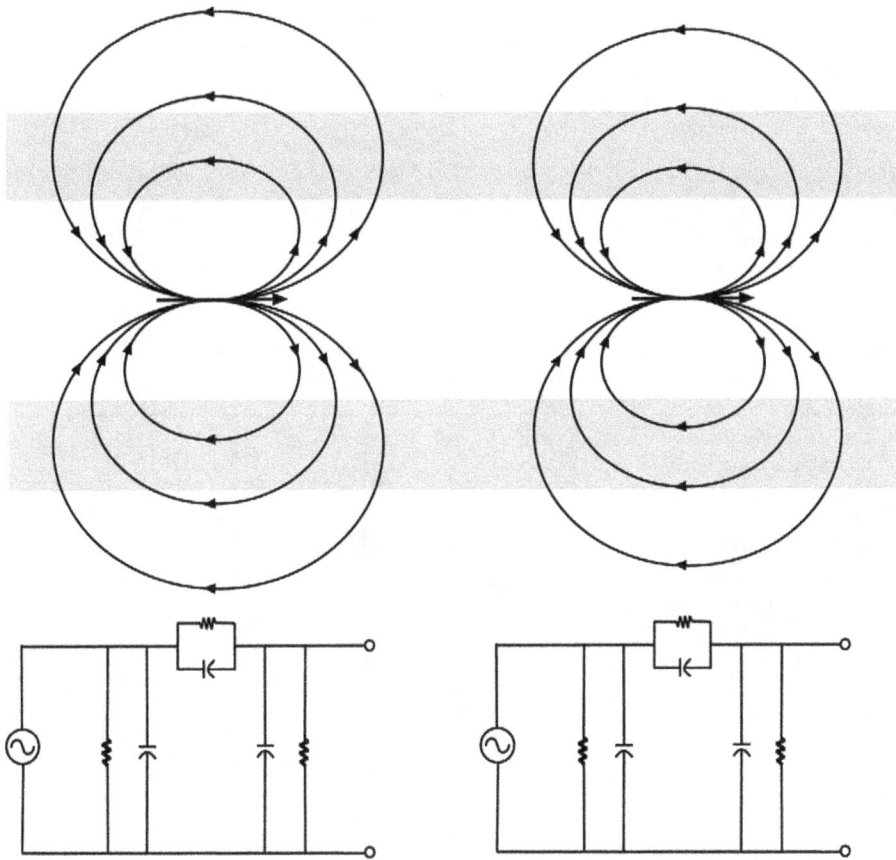

Figure 2. Electric field dipoles and corresponding circuits. Frequency-dependent lumped-element equivalent circuit diagrams of dendritic compartments.

The instantaneous movement of charges is represented in Figure 2 by a dipole arrow. The currents in the dendritic core generate electromagnetic fields, with the electric field being the dominant one [36]. As the energy field travels along the membrane, it acts upon the charges to move them along the axis of the conducting core. This view of the fields as electric dipoles associated with moving charges in a conducting medium follows from Maxwellian concepts, and some of the earliest articulations of this view were given by [37,38]. The

electric fields of the idealized dipole loop outward and back through the membrane dielectric and extracellular fluid to terminate on the dipole, thus completing the circuit. A steady-state frequency lumped-element representation of a neuron compartment suitable for frequency analyses of voltages and currents associated with the moving dipole field is shown in Figure 3.

Each circuit compartment contains an active voltage-gated outward conductance, g_{vg}, a passive leak conductance, g_{leak}, and

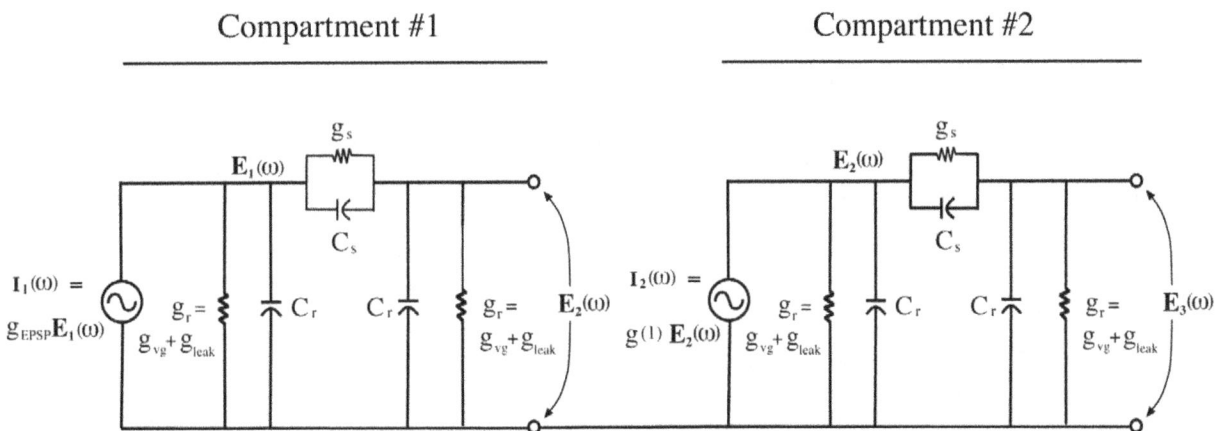

Figure 3. Circuit diagram of two adjacent connected circuit compartments of the membrane. The circuit can be viewed as a two-stage RC coupled amplifier circuit [42].

the capacitances, C_r and C_s, which represent anisotropic membrane electrical properties. The first compartment contains, in addition to these circuit elements, a synaptic input conductance, g_{EPSP}. The compartment equivalent circuit is a series-parallel RC circuit with voltage-dependent current sources. The preceding compartment output voltage defines the voltage-dependent input circuit source when multiplied by the compartment input conductance.

The present formulation may be considered a variation of traditional cable theory [39]. Traditional cable theory is based on electromagnetic energy transmission carried by two concentric conductors separated by a dielectric. The inner or center conductor is of positive polarity and carries current and voltage from the source to the end point of the cable or to a load. The outer conductor or shield of negative polarity carries the return current to the source. The shield is assumed to be continuous or uninterrupted. The electric circuit theory of the cable is completely described by series connected resistance (R) and inductance (L) elements, along with shunt connected capacitance (C) and conductance (G), together forming a continuous ladder-like electric network. The presence of inductive components normally enables the high-pass filtering in resonance behavior. The major problem that arises when classical cable theory is applied to neurons is to find a component in the membrane that acts like an inductor. One candidate for an inductive-like component is the temporal activity of h-channels [18]. The time delays in opening and closing of h-channels produce inductive-like effects at frequencies below 20 Hz (and therefore are appropriate for models of hippocampal activity in the 6–12 Hz theta range). However, because the opening and closing of h-channels is not sufficiently fast to account for inductive effects at frequencies above 20 Hz (18), another way must be found that produces high-pass filtering if resonance behavior is to be revealed in the neocortical apical dendrite, which produces EEGs in the 30–100 Hz range, and sometimes higher.

The approach taken here modifies the circuit of the traditional cable model by replacing the inductance (L) element with a capacitive element that is directed along the outer surface, longitudinal to current, so that the circuit now contains two kinds of capacitive elements: the non-linear surface capacitance, C_s, and the radial capacitance, C_r, which is the capacitive element in traditional cable theory that is directed perpendicularly across the membrane (see Figure 4). Energy flow along the outside of the membrane activates the longitudinal surface capacitive elements, and this energy flow is produced by holes or apertures in the shield of the traditional cable. The result is a leaky cable with a dielectric coating, which is a variation of the cable model that includes possibilities of external energy flow coupled to internal energy flow. The leaky feature may involve radiation, and is used in many practical applications such as tunnel communications (e.g. in the English Channel Tunnel and the Ted Williams Tunnel in Boston). The external current flows through a surface impedance associated with the boundary between the outer surface of the perforated sheath and external medium which is air for the engineering application and extracellular fluid for the dendritic application. The electromagnetic theory of leaky cables in air was developed by Delogne [17].

Delogne's theory [17] has guided our description of surface current energy flow along the outer membrane of the apical dendrite. The apertures or channels are represented by a nonlinear capacitance and conductance coupled to the internal RC parameters of the cable. Unlike the radial capacitance, the surface capacitance and conductance of the dendrite has a frequency, geometry, and wave number dependence. This new

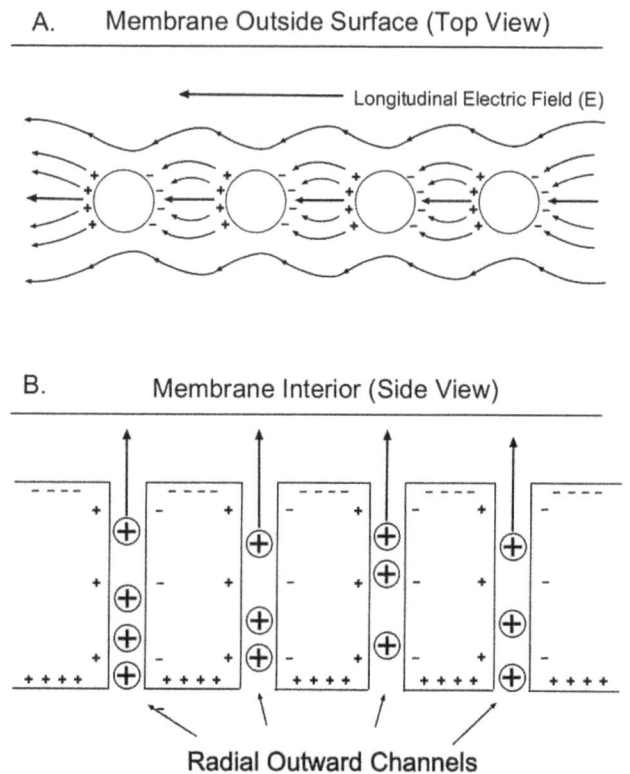

Figure 4. Longitudinal and radial capacitances. (A) Longitudinal surface capacitance, acting across the membrane segments between channel openings. (B) Radial capacitance, acting across the dielectric segments between the intracellular and extracellular charges.

combination of electric circuit elements leads in a straight-forward manner to band-pass frequency behavior for the dendrite.

The apical dendrite model we have proposed has a single peak value in the band-pass characteristic not because of any energy exchanges back and forth between inductance L and capacitance C as in electric circuit theory. A resonance or peak amplitude in the transfer impedance of the neuron is created by the way the electric energy divides between the surface and membrane capacitive and conductive elements as a function of frequency. At low frequencies, capacitance tends to block current flow and at high frequencies capacitance tends to act as a short.

The horizontal capacitance C_s and horizontal conductance g_s are derived from the surface wave electrical properties of the membrane and extracellular fluid interface. Delogne [17] derived the general surface impedance of the leaky cable with dielectric coating by using electromagnetic field theory. The surface impedance component of the circuit is capacitive in nature owing to dielectric coating of the leaky cable. Surface impedance is a fundamental electrical property of the external energy flow. According to Barlow and Brown (see p. 5 in [40]), "A surface wave is one that propagates along an interface of two different media without radiation; such radiation being construed to mean energy converted from the surface wave field to some other form". In the present case, the surface wave includes both the instantaneous capacitive stored energy associated with moving charges along the interface between the outer membrane surface and extracellular fluid, represented by C_s, and the dissipation of energy associated primarily with the internal energy loss of the dendritic core and extracellular fluid, represented by g_s. The surface impedance concept provides an anisotropic membrane model of the neuronal dendrite, which in this respect, contrasts with the compartment model of [20].

For the single compartment, the derived steady-state complex transfer impedance is:

$$E_2(\omega)/I_1(\omega) = \frac{k(s+\omega_1)}{(s+\omega_2)(s+\omega_3)}, \quad (1)$$

where

$$s = j\omega, \quad (2)$$

$$\omega_1 = g_s/C_s, \quad (3)$$

$$\omega_2 = \frac{g_r+2g_s}{C_r+2C_s}, \quad (4)$$

$$\omega_3 = g_r/C_r, \quad (5)$$

and

$$k = \frac{C_s}{(C_r+2C_s)(C_r)}. \quad (6)$$

The parameters, ω_1, ω_2, and ω_3 have units of radian frequency and are referred to here as relaxation frequencies, following classical circuit theory concepts. The relaxation frequencies and k completely characterize the bandpass frequency behavior of the dendrite model.

Equation (1) contains expressions for two different capacitances: C_s, and C_r, corresponding, respectively, to the holding of charge in the longitudinal direction of the membrane and the holding of charge in the radial direction of the membrane. The operation of C_r is crucial for producing the low-pass filtering at frequencies below the peak frequency, while the operation of C_s is crucial for producing the high-pass filtering at frequencies above the peak frequency.

The dynamic characteristics of this network model can be specified by either its voltage transfer function, $E_2(\omega)/E_1(\omega)$, or its impedance transfer function, $E_2(\omega)/I_1(\omega)$. Equation (1) is the typical form for the frequency response of an RC coupled electronic amplifier with bandpass behavior governed by frequency, ω, and the values of ω_1, ω_2, ω_3, and constant k [41].

The sheath geometry for the classical leaky cable involves a cylindrical outer surface defined by a two-layer construction: a conducting layer on one side and a dielectric layer on the other side. Holes (channels) exist through the surface allowing current flow from inside the cable to outside or in the opposite direction. The surface impedance representation of the current and voltage boundary condition for a leaky cable is assumed to be approximately equivalent to the dendrite outer surface impedance boundary in the range of neuronal resonant frequencies of our paper by symmetry. The effect of attenuation or resistivity (conductivity) in Equation (7), is given by the propagation constant or wave number for the classical neuron cable model [36]. The ratio of the longitudinal electric field E to the circumferential

magnetic field H evaluated at the boundary separating conductor and dielectric gives the result of Delogne's [17] surface admittance as shown in Equation (7), separated into capacitive and conductive components.

The surface transfer admittance derived from the Delogne leaky cable theory [17] is approximately

$$\frac{j\omega\epsilon}{\gamma^2 b(\ln b/a)} \cong j\omega C_s + g_s, \quad (7)$$

where ϵ is the membrane dielectric constant, γ is the complex propagation constant given by [36], b is the outer radius of the membrane, and a is the inner radius of the membrane. The membrane dielectric constant includes the conductive channels, as described by [42] in artificial dielectric theory.

Using the definition of the classical cable propagation constant,

$$\gamma = \sqrt{\frac{z_a(f)}{z_m(f)}} \quad [36,p.42] \quad (8)$$

and the well-known formulas for the intracellular impedance, $z_a(f)$, and for the membrane impedance, $z_m(f)$, the horizontal capacitance, C_s, and associated conductance, g_s, are from Equation (7):

$$C_s = \frac{C_0\lambda^2(0)}{2\pi b(1+\omega^2\tau^2)} \quad (9)$$

and

$$g_s = \frac{\omega^2 C_0\tau\lambda^2(0)}{2\pi b(1+\omega^2\tau^2)} \quad (10)$$

and

$$C_0 = \frac{2\pi\epsilon}{ln(b/a)} \quad \text{(per unit length)}, \quad (11)$$

where λ is the neuronal space constant, τ is the neuronal time constant, b is the outer radius of the membrane, and a is the inner radius of the membrane.

The electrical theory of leaky cables applied here to the dendritic segment provides a way to consider the effect of coupling between longitudinal currents (along the outer surface of the membrane) and radial membrane currents (across the membrane). If we consider the charge per unit length on each of two concentric cylindrical coaxial capacitor surfaces, the per unit length capacitance of a coaxial cable is given by Equation (11) [43]. The numerator of C_0 is an expression of area and the denominator is an expression of thickness, which together correspond to the basic physical description of a capacitor. The surface (longitudinal) capacitance, C_s, is proportional to C_0, and the parameter of proportionality is dependent on frequency. The radial capacitance, C_r, is also proportional to C_0, but is not dependent on frequency. If C_s were set to zero (implying that g_s would be changed to represent the core conductance of the dendrite), the present model would reduce to the traditional cable model.

The many radial leak paths of the membrane containing high conductivity fluid greatly increase the membrane conductivity over that of the axon membrane. Although many kinds of conductances exist in the dendritic membrane [44], the present model is based mainly on the outward potassium (K^+) conductances, which are of two kinds: passive leak conductance, g_{leak} and active voltage-gated conductance g_{gv}. For convenience these two K^+ radial conductances are combined under the symbol, g_r. The passive membrane of the dendrite with its radially oriented leak channels constitutes an anisotropic dielectric (see Figure 4). The electrical conductivity in the radial direction is made up of conductive channels with an intracellular resistivity on the order of 29.7 ohm-cm (see p. 481 in [36]). Axon membrane resistivity, in contrast, is on the order of 10^{11} ohm-cm (see p. 483 in [36]). This anisotropic condition strongly influences the shape and location of the membrane model passband as a function of frequency and amplitude.

To observe the operation of the one-compartment circuit over the 10–100 Hz range where most of the EEG measurements in the literature are located, we estimated the transfer impedances that showed peak values at 20, 40, 60, and 80 Hz. Using the circuit parameters assumed for the present model, we obtained maximum impedance amplitude values for these 4 frequency locations of 2.5369, .46009, .13223, and .04952 $M\Omega$, corresponding, respectively, to g_r values of 105, 310, 627, and 1056 nS.

Impedance transfer function for amplitude and phase in the single compartment

The four transfer impedance amplitude curves shown in Figure 5 were calculated from Equation (12). Because the expression of Equation (1) involves the imaginary number $j\omega$, ($s = j\omega$), Equation (1) was converted to the real number form:

$$E_2(\omega)/I_1(\omega) = \frac{k\sqrt{(\omega^2 + \omega_1{}^2}}{\sqrt{(\omega^2 + \omega_2{}^2)}\sqrt{(\omega^2 + \omega_3{}^2)}}, \quad (12)$$

where $\omega_1, \omega_2, \omega_3$, and k are defined by Equations (3–6).

To illustrate the amplitude calculation of the impedance values shown in Figure 5, we arbitrarily select the 50 Hz point on two curves: the 20 Hz and the 40 Hz peak frequency curves. Entering into Equation (12) the values for $\omega, \omega_1, \omega_2$, and ω_3 of 314, 2957.88, 1680.82, and 1112.52, respectively, with $k = 1.6315 \times 10^9$, we obtain for the 20 Hz peak frequency curve an impedance of 2.45 $M\Omega$ at 50 Hz. For the 40 Hz peak frequency curve we change only the values of ω_2 and ω_3 to 3183.98 and 3284.59, respectively, to obtain an impedance of 0.460 $M\Omega$ at 50 Hz (see Figure 5).

The corresponding phase angles calculated from Equation (1) are shown in Figure 6. The net phase angle for each compartment derived from Equation (1) is:

$$\phi = \arctan(\omega/\omega_1) - [\arctan(\omega/\omega_2) + \arctan(\omega/\omega_3)]. \quad (13)$$

When we insert into Equation (13) the same ω parameter values we just used to calculate the two impedance amplitude points at 50 Hz, we obtain ϕ values of -20.28 degrees and -5.03 degrees for the 20 Hz and 40 Hz peak frequency curves, respectively (see Figure 6).

The impedance amplitude curves, shown in Figure 6, indicate that current injected into the circuit of one compartment produces a voltage output that shows moderate variation for the 20 Hz peak curve, for which the radial conductance value is relatively low

($g_r = 105$ nS). The variation in impedance amplitude shows a slight narrowing around the peak frequency of 20 H (particularly by decreases in amplitude at lower neighboring frequencies), which enables the circuit to pass input current at frequencies near 20 Hz, while exerting a filtering effect on input current at other, more outlying frequencies. However, as the radial conductance increases, and the frequency of the peak impedance amplitude increases with it, the shape of the impedance curve flattens, until, at a peak amplitude frequency of 80 Hz a narrowing of the curve around the 80 Hz peak is no longer detectable by visual inspection. Therefore, the progressive flattening of the curve as radial conductance increases results in a lowering of the ability of the single compartment circuit to operate effectively as a bandpass filter.

The impedance phase curves, shown in Figure 6, depart slightly from resonance phase curves that are based on simple RC-coupled circuits, because the frequencies of the zero-crossings do not exactly correspond to the frequencies of the peak amplitudes for each curve. These deviations at the zero crossings are small for the 20 Hz peak amplitude, and increase toward the 80 Hz peak amplitude. Therefore, the maximum energy transfer for a given set of circuit parameter values occurs at a frequency where the voltage and current are slightly out of phase. An explanation for this deviation is that the longitudinal or coupling capacitance is frequency-dependent in the present circuit, while it is frequency-independent in the simple RC-coupled circuit.

The general characteristics of the impedance amplitude and phase curves of the present model, shown in Figures 5 and 6, can be compared with the empirical curves obtained by an in vitro experiment with rats [45], in which chirps of frequencies of 0.1 to 25 Hz were injected into a Layer 5 apical dendrite either at the soma or at distances of 120–270 μm from the soma, while a second electrode recorded the voltage response at the other location. The measured impedance amplitudes and phase are reproduced here in Figure 7. The obtained curves of amplitude and phase shift appear to be independent of the direction of excitation along the apical dendrite. The impedance amplitude curve shows a peak at approximately 6 Hz, and the zero-crossing of the phase at approximately 3 Hz. The maximum impedance level is higher than the level obtained in the present analysis by a factor of approximately 12. This difference in maximum impedance may be due to the lower frequency of the obtained peak resonance and to the passive state of the dendritic membrane in the preparation used by [45].

Simulated Narrowing of the Resonance Curve

The amplitude curves shown in Figure 5 indicate that the present anisotropic membrane model generates impedance transfer functions that show a peak value at a specific frequency in the range 10–100 Hz. Therefore the transfer functions appear to operate as filters which select a particular frequency or band of frequencies of pulse inputs while blocking other frequencies. However, the effectiveness of selecting or filtering frequencies appears to decrease as the peak frequency of the curves move from 20 to 80 Hz because the slopes of the curves on each side of the peak frequency become progressively smaller. Clearly, the present single compartment circuit model is not adequate for producing effective filtering of input frequencies over the 10–100 Hz range, nor is the model adequate for supporting other activities related to the existence of a sharp resonance at a specific frequency (e.g., [9]).

To achieve a narrow, peaked resonance curve, we extend the present single compartment model to a 6-compartment model, with the added feature that the pulse train is recycled repeatedly through the 6 compartments by way of the thalamus within the

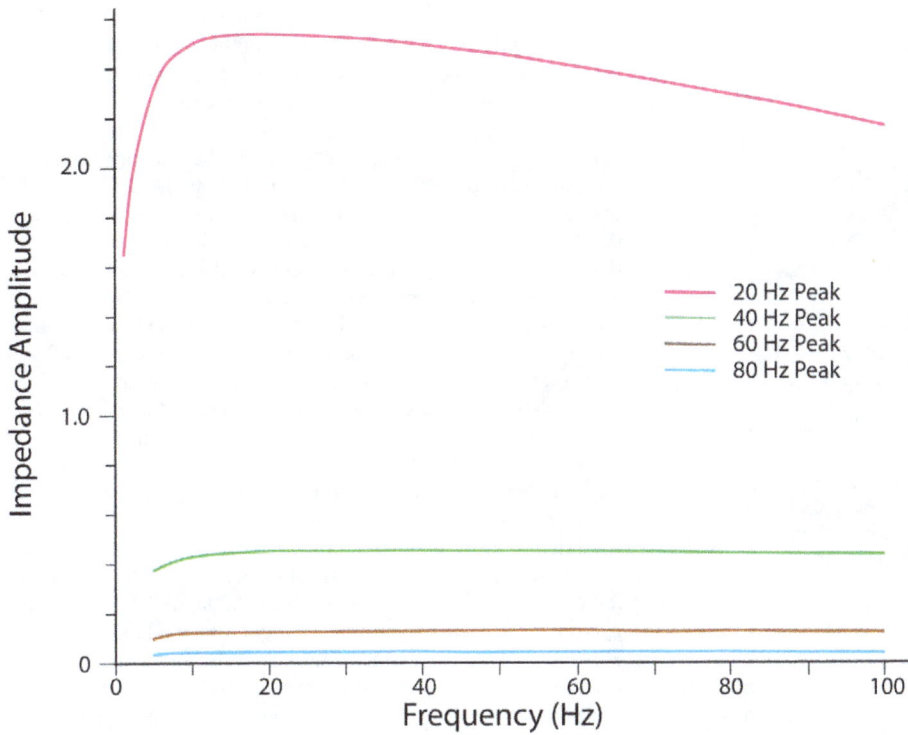

Figure 5. Impedance amplitude curves. Impedance amplitude curves produced from four outward conductances resulting from four levels of internal voltage produced by subthreshold activity of local dendritic spines. Peak amplitudes of the curves are at 20, 40, 60, and 80 Hz.

corticothalamic circuit. Figure 8 shows a schematic diagram of the corticothalamic circuit within which a pulse train repeatedly traverses the compartments of the apical dendrite of the neocortical Layer 5 pyramidal neuron. Each time the pulse train travels through the 6 compartments of the dendrite it produces a cascade of circuit operations which successively narrow the impedance transfer function around its peak value. The thalamus sends current pulses not only to the distal part of the Layer 5 apical

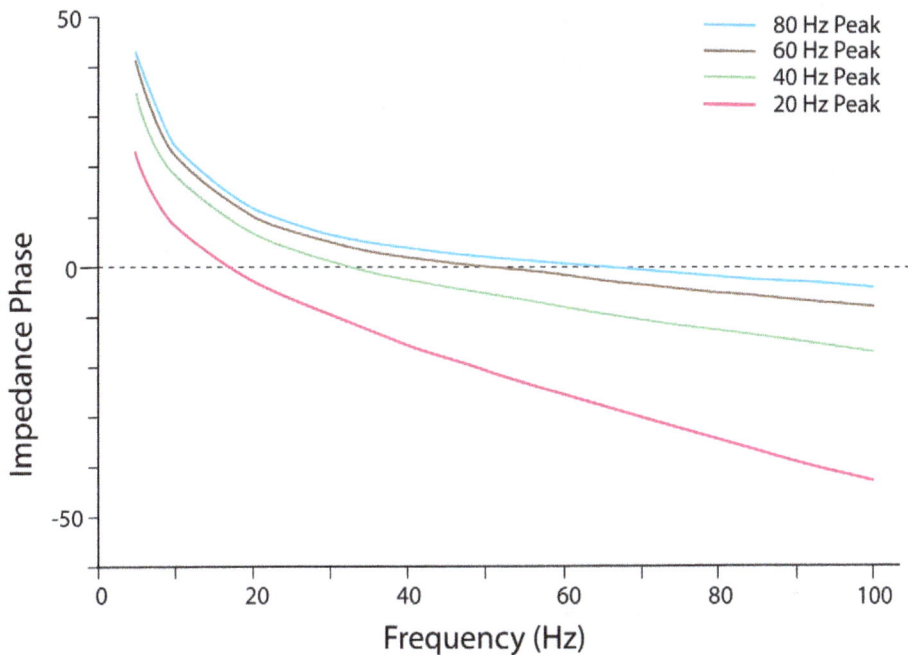

Figure 6. Impedance phase curves. Impedance phase shift curves produced from four outward conductances resulting from four levels of internal voltage produced by subthreshold activity of local dendritic spines.

A

B

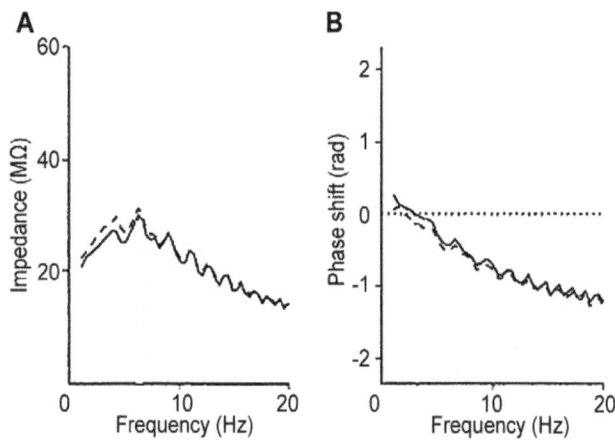

Figure 7. Empirical impedance amplitude and phase curves. (A) Impedance amplitude profile. (B) Impedance phase shift curve. Recording electrode inserted in the soma (filled line), or inserted into the dendrite (dashed line)120–270 μm from the soma. Data from [45].

pulse train as it moves from the pyramidal neuron through the thalamus and back to the distal region of the pyramidal neuron. In the Discussion Section the issue of synaptic noise in the apical dendrite spines and in the corticothalamic circuit will be addressed.

Both the single compartment and the looping multiple-compartment models assume that the average level of the radial conductance, g_r, determines the frequency location of the peak in the transfer function. Figure 9 shows, in a local sector of the dendrite, a schematic representation of three levels of voltage-gating which are produced by varying the number of active channels in the membrane. This figure also shows the predicted narrowed shape of the impedance transfer function after a moderate number of loops through the 6 compartments have successively operated on the original current input. It is assumed that the same level of g_r (allowing for slight changes in the leak current for purposes of the model; see Equation 15) operates across all of the compartments of the model, although the level of g_r is assumed to vary for apical dendrites across cortical locations.

Although transmembrane ion channels are distributed non-uniformly over a segment of dendritic membrane [27], we assume for simplicity that the present derivation of the radial conductance, g_r, can be reasonably based on averages over the large number of channel openings in a typical segment of the apical dendrite. The activation of the voltage-gated channels is presumed to be produced through the many local spines in a segment of dendrite. This subthreshold activation is, in turn, activated in a tonic manner by the bombardment of local axons, which is commonly

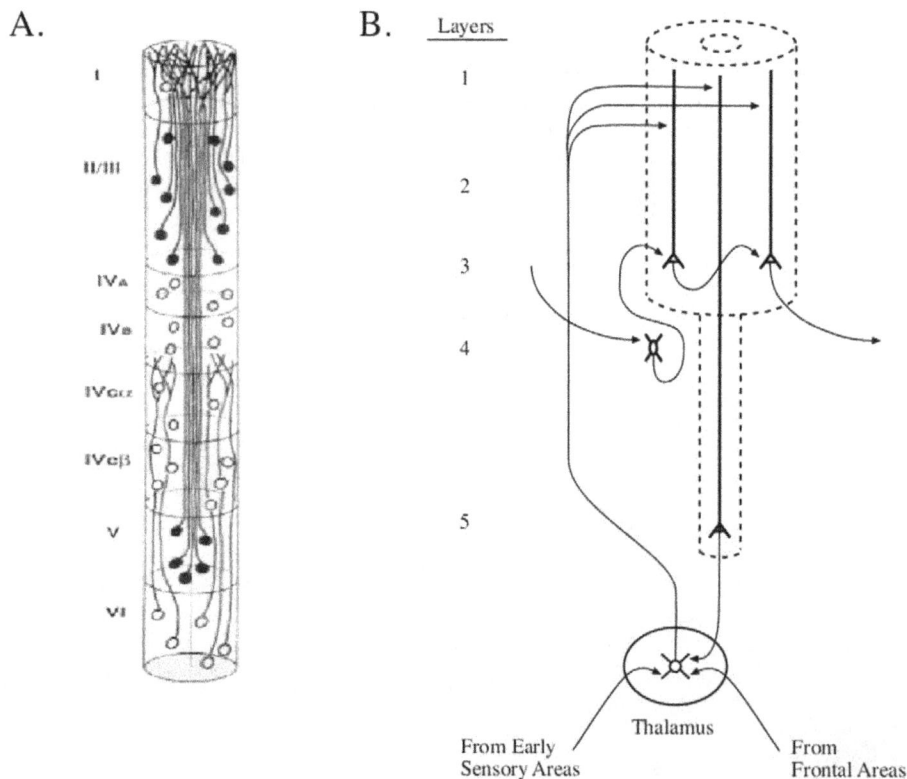

dendrite, completing the loop activity, but it also sends current pulses to the distal part of the layer 2/3 apical dendrite, which may influence the input-output activity of the minicolumn within corticocortical circuits that interconnect regions of the cortex [32]. For simplicity in analyzing the general properties of the cascade-loop model, it is assumed initially that no noise is added to the

Figure 8. Cortical minicolumn and corticothalamic circuit. (A) Schematic diagram of a minicolumn in monkey visual cortex (see Figure 19 in [46]). Shown are somas and apical dendrites of pyramidal neurons and somas of stellate neurons. (B) Schematic diagram showing the corticothalamic loop. A specific resonance profile in a Layer 5 apical dendrite is projected, via the thalamus, back to the first compartment of the same apical dendrite. The profile is also projected to apical dendrites of Layer 2/3 pyramidal neurons where it spreads to the basal dendrites and may influence the selection of inputs arriving from other cortical minicolumns. For clarity, the reticular nucleus, with its inhibitory projections to the thalamic principal neurons, is not shown. Adapted from [30,31].

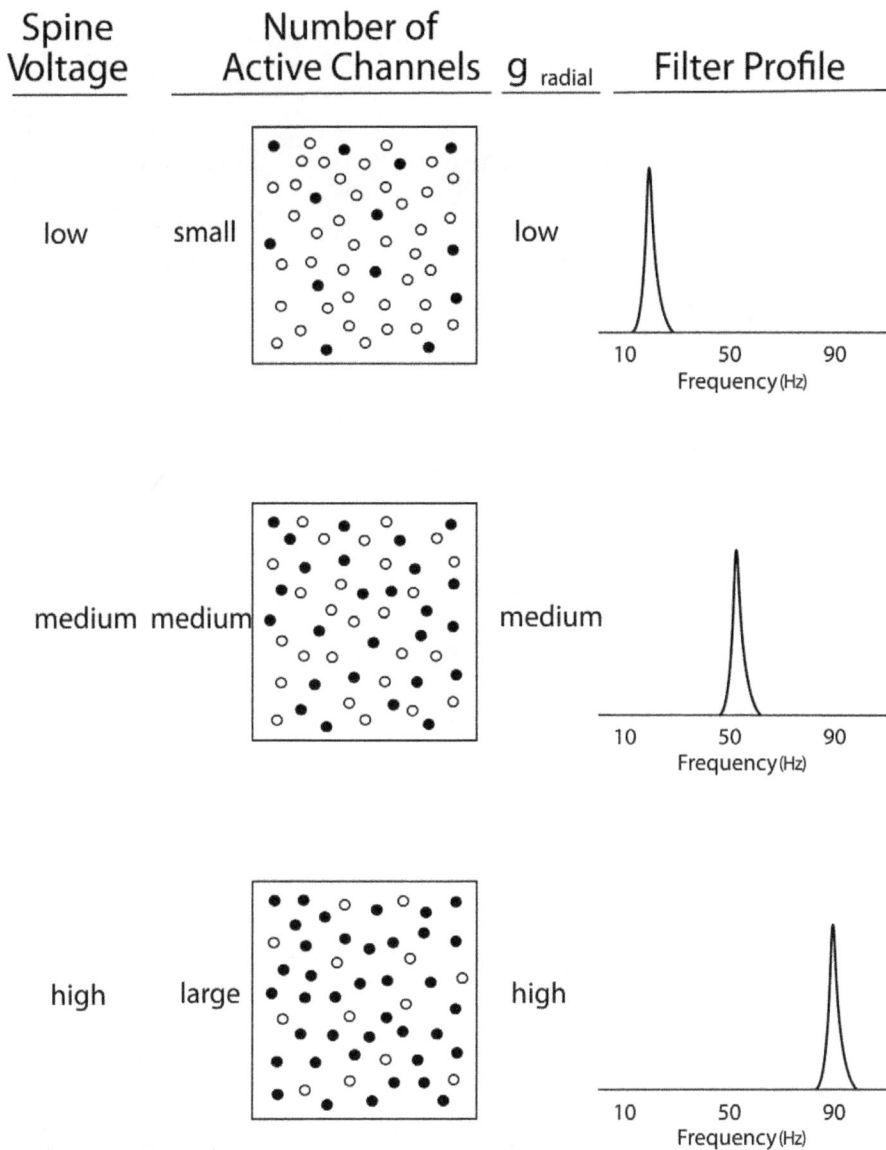

Figure 9. Resonance peak frequency and activity of local spines. Schematic diagram of the hypothesized relationship between peak frequency of a resonance curve and the internal voltage level produced by subthreshold activity of local dendritic spines.

termed background noise. Cortical neurons, including the pyramidal neurons, receive a large majority of their synaptic inputs from neurons in local circuits that are less than a millimeter away [2]. The average local bombardment level at a cortical minicolumn is presumed to be constant, on average, but the level is presumed to vary among the minicolumns across the neocortex. Hence, the level of local subthreshold synaptic activation for a cluster of apical dendrites in a particular minicolumn can determine the resonant frequency of those apical dendrites.

Impedance transfer function magnitude with repeated cascades across the 6 compartments

The magnitude of the impedance transfer functions of the cascade-loop model, plotted as percent of maximum voltage for 20, 40, 60, and 80 Hz peak frequencies, are shown in Figure 10. The results show two main tuning effects: The adjustment of the peak frequency of the impedance transfer function, shown here

more definitively than in the single compartment model results of Figure 5, and the sharpening of the resonance curve around the peak frequency. The resonance curves corresponding to successive loops through the 6 compartments of an apical dendrite gradually narrows toward an asymptote described by a single vertical line. Visual inspection of the curve adjacent to the vertical line indicates that, at half the height of the curve, the width of the curve is approximately 1 Hz. It would seem that adding additional loops would narrow the width even further than the width obtained in the successions of curves observed in Figure 10. However, the presence of synaptic noise in the components of the cascade-loop model, particularly at linkages in the circuit loop, presumably opposes the narrowing process, and a balance between total synaptic noise and the compartmental narrowing process would then be expressed by a width value that conceivably could lie within a range of a fraction of a Hz to several tens of Hz.

According to the model, the peak frequency of the impedance transfer function is shifted upward or downward by varying the

A.

B.

C.

D.

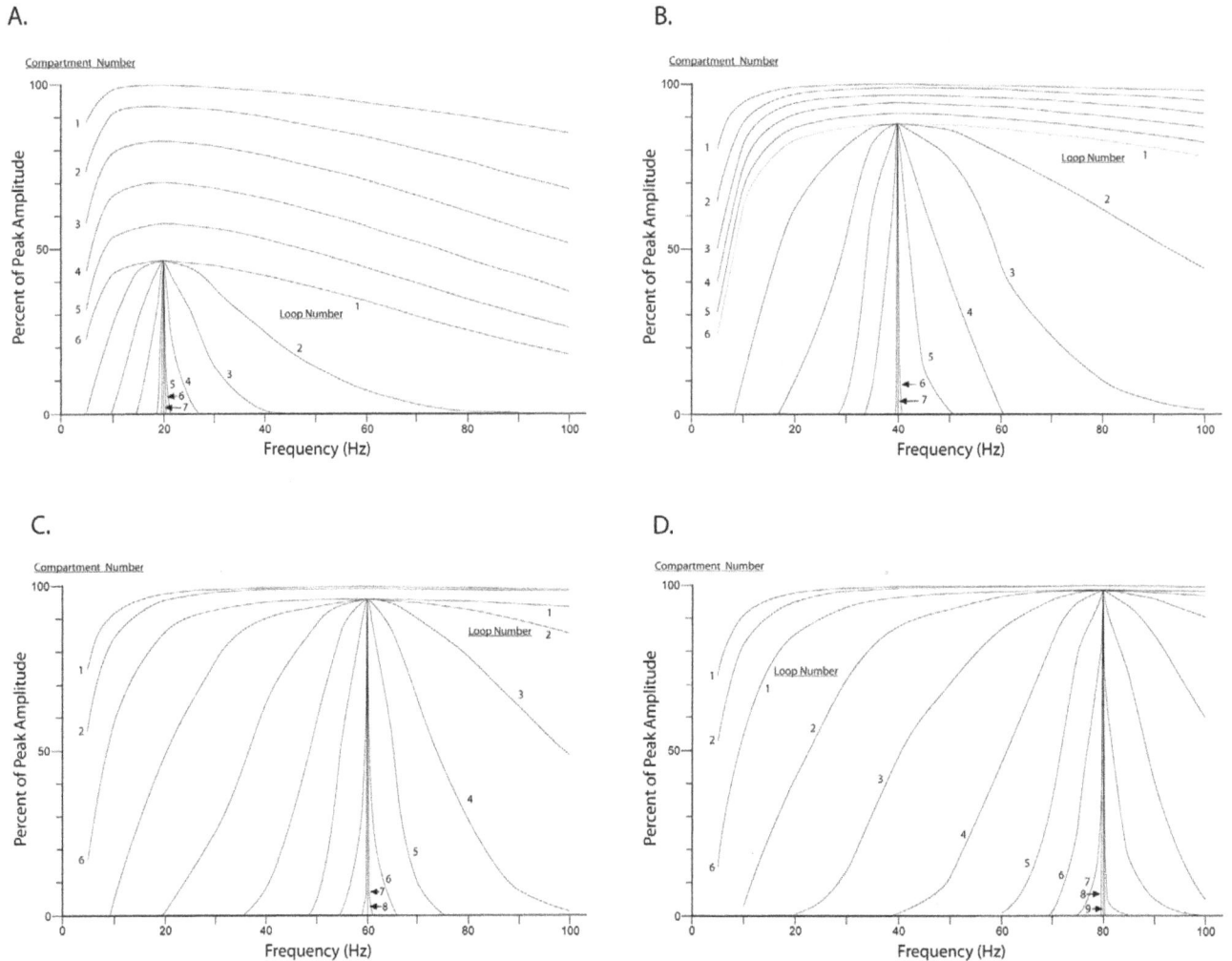

Figure 10. Narrowing of resonance curves. Resonance curves with peaks at 20, 40, 60, and 80 Hz, are narrowed by repeated cascades through 6 compartments via circuit loops through the thalamus. The resonance curves for the first 6 compartments are shown at the top of each of the 4 sets of curves (for clarity, the resonance curves for the 3rd, 4th, and 5th compartments are omitted for the 60 and 80 Hz conditions). The remaining curves in each of the 4 conditions show the effects of subsequent loops through the 6 compartments. The curve representing a loop is actually the output of the 6th compartment of each successive loop. Radial conductances = 105, 310, 627, and 1056 nS for the curves with peaks 20, 40, 60, and 80 Hz, respectively.

level of the total output conductance, which in turn is influenced by the tonic voltage level maintained inside the dendrite by the subthreshold synaptic activity at the hundreds of local spines. The area of the resonance curve around the peak is narrowed by successive electrical operations in the compartments through which the pulse travels as it makes its way toward the soma. As more compartments are traversed the resonance peak becomes more sharply defined.

Discussion

Resonance Tuning

Two aspects of resonance tuning are treated in this paper: (a) the selection of the peak frequency of the resonance curve, and (b) the sharpening of the resonance curve around the peak frequency.

(a) Selective activity is indicated by the finding that the impedance transfer function exhibits a peak, which suggests that only one compartment of the apical dendrite model is needed to exhibit some degree of selective filtering over the total range of

frequencies of an injected input. Although the peak in the single-compartment case is weakly defined here, the peak becomes more noticeably narrowed when the compartment operations are repeated through more compartments. Thus when the current containing a range of frequencies is passed to the next compartment, the voltage of some frequencies will be larger than that of other frequencies. As mentioned in the previous section, the physical basis hypothesized for the selection of the (peak) resonance frequency is the outward conductance of potassium ions. A particular value of outward conductance is produced by a corresponding level of voltage on the inside surface of the membrane. This level of internal voltage is maintained by the ongoing subthreshold synaptic activity in local spines, which in turn is produced by pulses from axons of the local cortical circuitry. The local circuitry of the cortex is assumed to be noisy, owing to its complex circuitry and the firing characteristics of the constituent neurons. The source or sources that regulate the level of this background noise apparently are not yet known in detail. However, it would seem that the noise level for a specific minicolumn (or for a column cluster of minicolumns) should differ from the noise level of minicolumns serving other functions. For example, the ambient background noise level for minicolumns involved in processing the color red should exhibit a different noise level than the minicolumns involved in processing the square shape of an object. In this way the corticocortical circuits that connect Layer 2/3 pyramidal neurons of a group of minicolumns can function somewhat independently of the noise level of minicolumn local circuits, once the local noise level selects the resonance frequency of the participating pyramidal neurons. The amplitude of the noise of a particular minicolumn will, of course, exhibit a variance, and the size of this variance is expected to affect the sharpness of the resonance function.

(b) Sharpening activity, shown in Figure 10, is indicated by the increase in slope on each side of the resonance peaks of the curves as the number of participating compartments is increased. Thus it would appear that when cascades of electrical operations are performed across an appropriately chosen number of compartments (using the looping circuit to add compartments), the dendrite can exhibit a similar resonance curve, both in amplitude and in sharpness (often defined as Q), for any impedance peak in the range 10-100 Hz. However, when synaptic noise is taken into account, the predicted sharpness of a resonance curve will be attenuated. In general, the amount that noise attenuates the sharpness of the resonance peak depends upon the balance between number of compartments in the apical dendrite and the amount of noise in the corticothalamic loop.

In the search for experimental evidence supporting the existence and function of surface capacitance, C_s, in our model we have analyzed the transfer impedance data shown in Figure 7 from the double-patching of neurons by Ulrich [45] from the viewpoint of classical electric network theory. The obtained impedance magnitude shape and zero phase angle condition would seem to require both low-pass and high-pass filtering effects that are only possible when the network model contains both series and shunt capacitances (assuming the absence of an inductance element). The series capacitance is defined as C_s in our model. Future experiments that are designed to measure the complex transfer impedance (phase angle and magnitude) combined with known values of the membrane conductance and capacitance would enable a determination of the g_s/C_s ratio in our model.

The Effects of Synaptic Noise

While it is recognized that noise is added to the resonance processing within the pyramidal neuron itself, we assume here that the level of that noise is small relative to the level of noise added to the resonance processing by synapses. The issue of synaptic noise in the present model can be separated into the noise produced in two different places in the cascade-loop model of resonance tuning: the ambient noise in the thousands of spines along the apical dendrite shaft which produce subthreshold activation of neighboring conductance channels (chiefly potassium, in this model), and the noise at the synapses in the corticothalamic circuit which enables output current pulses from the dendritic compartments to loop back to the apical dendrite. Many hundreds of spines dot the apical dendrite (with a density of about one spine per dendritic length of 2 microns [33]), and the shapes of the spines apparently vary widely, without any clear indication of subtypes having specific functional properties [47]. Some spines apparently produce action potentials (from external axon contacts or from backpropagating action potentials) while others operate at subthreshold levels, which affect local conductance gating without producing action potentials. Whenever synaptic activity occurs in a spine it is expected that short-term depression and facilitation

will follow and the balance between these two effects will serve as a transient context for the next synaptic event [16]. At present it is not known exactly how the operations underlying the tuning process in the dendrite may be affected by the transient context effect at the spines, including spines that produce subthreshold activity as well as those that produce suprathreshold spiking activity. One important consequence of altered synaptic context is that local conductances may be changed. In the present model of trains of current pulses, this contextual effect may be frequency dependent, and at higher frequencies, in particular, the effect may be repeated in a manner that prolongs the context over extended periods of time in which resonance is tuned.

The second location of synaptic noise of interest to the present model is in the part of the circuit loop that lies outside of the compartments of the dendrite. Included are synapses on the thalamic principal cell and on the inhibitory cells of the reticular nucleus (and synapses on the principal cell from the reticular nucleus and the interneurons), as well as the synapses of the principal cell axon on the apical dendrite of the cortical pyramidal cell. Additional synapses to be considered contact the corticothalamic loop from other sources. They include the important synapses by which ascending afferent fibers contact the thalamus en route to the primary sensory cortical areas. For the higher cortical areas, both ascending fibers from lower areas and descending fibers from frontal cortical areas make synaptic contacts with the thalamus (see Figure 8). Synaptic activities here initiate and help to maintain the loop activity over varying periods of time, The main synaptic activity is presumed to be the suprathreshold production of action potentials, and transient context effects are expected to influence spiking activity [15,16]. However, it is not yet known the manner of and extent to which short-term resonance-like activities at these synapses influence the more global resonance tuning in the apical dendrite that is described by the present cascade-loop model.

If the traditional view of noise is applied to the workings of the present model, then all of the synapses under consideration, especially the synapses in the corticothalamic circuit, produce a significant level of noise that tends to flatten the shape of the resonance curve. While every cycle of the corticothalamic circuit adds noise to flatten the shape of the profile of the resonance curve, the next cascade through the apical dendrite compartments narrows the shape of the profile of the resonance curve. If the total noise in the corticothalamic loop is sufficiently large relative to the number of dendritic compartments, then the resonance curve cannot be narrowed to the maximum asymptotes shown in Figure 5. Therefore, a relatively large number of compartments in the apical dendrite provides a means of effectively offsetting the noise in the corticothalamic circuit. Hence, longer apical dendrites (e.g., in primates) that contain a larger number of compartments should produce more narrow resonance profiles, while very short apical dendrites (e.g., in mice) should show a limit on the narrowness of the resonance profile. For an illustrative comparison of apical dendrite lengths of Layer 5 and Layer 2/3 pyramidal neurons across 5 mammalian species see Figure 1 in [32].

Methods

Simulation method for a single compartment

The purpose of the single compartment model simulation is to demonstrate resonance in the apical dendrite by generating four resonance curves, corresponding to peak resonant frequencies of 20, 40, 60, and 80 Hz. To simulate the operations of a segment of the Layer 5 apical dendrite, we employ here the assumed equivalent circuit in a single compartment (see Figure 3). The choices of parameter values for the circuit are based on the following rationale: The diameter of the apical dendrite tapers from approximately 8.5 μm near the soma to a range of .30 to 1.33 μm at the distal dendrite [33]. Here we select a dendritic segment lying in the middle region between the soma and distal endpoints, and we assume the dendrite diameter, d, to have a constant value of 5.0 μm throughout the selected segment (following the compartment assumptions of [48] for the apical dendrite). The radial capacitance C_r is assumed to be .09438 $\mu F/cm^2$, based on a diameter value of 5.0 μm, a compartment length of $200\mu m$, and a total membrane capacitance which is which is at the top of the range of capacitance values estimated by [25] for the Layer 5 apical dendrite. The relatively large capacitance value for C_r takes into account the presence of spines that increase the area of the membrane surface. The surface capacitance, C_s, is frequency dependent and, following Equation (9), depends on the capacitance value of $C_0 = 9.4nS$, which is based on the minimum value of membrane capacitance given by [25]. The voltage-gated conductance, g_r, is estimated for each of the 4 resonance curves, corresponding to peak resonances of 20, 40, 60, and 80 Hz. The estimated g_r is assumed to be approximately constant over the the 6 compartments of the model, following the results of [25], who found that the voltage-gated K^+ density gradient changes rapidly near the soma, while maintaining a relatively flat slope in the mid region of the apical dendrite. The time constant of the membrane, τ, is assumed to be $30ms$, and the space constant, λ, is frequency-dependent, with the steady state component, $\lambda(0) = .97mm$ (based on a membrane resistance, $R_m = 50,000\Omega cm^2$, internal resistance, $R_i = 200\Omega cm$, $d = 1.5$ μm). The frequency dependence of the space constant is defined as [36]:

$$\lambda(f) = \frac{\lambda(0)}{Re\sqrt{1 + i2\pi f\tau}}. \tag{14}$$

To calculate the transfer impedance, E_2/I_1, for the circuit of one compartment (see Figure 3), the expressions for C_s and g_s, along with a value for the outward radial conductance, which is the sum of the voltage-gated conductance and the leak conductance, were condensed into the ω parameters and the constant k (via Equations (3–6). The ω parameters and the constant k were then entered into Equation (12). The choice of the g_r conductance value determines the location of the peak frequency and influences the shape of the transfer impedance curve plotted as a function of frequency. In particular, the maximum value, or peak, of the transfer impedance function will appear at a frequency that increases monotonically (but not linearly, owing to the frequency-dependent parameters) with g_r.

Simulation method for multiple compartments

We now consider the way that the multiple compartment model could tune an apical dendrite to narrow its resonance curve around a particular peak frequency. For the model, we select a segment of the apical dendrite 1200 μm in length, which is assumed to have a constant diameter, d of $5\mu m$. Each compartment is assumed to have a length of 200 μm, so that the total number of compartments of the model is 6. Following the work of [27], we assume that g_r decreases linearly from the current injection site at the first compartment, and reaches a constant conductance level after the 6th compartment. Thereafter, a constant level of g_r is assumed for each compartmental circuit.

The decrease in g_r at each compartment is applied only to the g_{leak} portion of the radial conductance. At the beginning of the first

compartment the value of g_{leak} is assumed to be 30 nS (which is slightly larger than the 25 nS value assumed for g_{leak} in the model of [20]), and this initial leak current is decreased by 5 nS to obtain the amount of leak current for the second compartment, which is 25 nS. After the 6th compartment the leak current is zero. Therefore,

$$g_{leak}(n) = [30 - 5(n-1)], \tag{15}$$

The excitatory input conductance to the nth compartment $(1 \leq n \leq 6)$ is denoted as $g_e(n)$, and it is related to the input of the previous compartment and to the leak conductance by the recursive equation

$$g_e(n+1) = g_e(n) - g_{leak}(n). \tag{16}$$

At the end of the first compartment,

$$E_2/E_1 = g_e(1)E_2/I_1, \tag{17}$$

which states that the transfer impedance of that compartment circuit, E_2/I_1, is multiplied by the output conductance of that compartment circuit, $g_e(1)$, to obtain the voltage transfer function. For the second compartment output,

$$E_3/E_2 = g_e(2)E_2/I_1, \tag{18}$$

so that, in general,

$$_{n+1}/E_1 = (E_2/I_1)^n \prod_{i=1}^{n} g_e(i). \tag{19}$$

The cascading operations across the n compartments is initiated by choosing a value for $g_{EPSP} = g_e(1)$. In order to plot the energy transfer values as a percentage of the maximum energy value, the value of $g_e(1)$ was calculated from the reciprocal of the peak impedance of each of the four curves, keeping in mind that the impedance is given at the end of the first compartment (which is the input to the second compartment):

$$g_e(1) = g_e(2) + 30nS. \tag{20}$$

A cascade through the 6 compartments produces an array of new impedances across the frequency range. These impedance values are sent back to the first compartment via the corticothalamic loop (see Figure 8), and processed again through the 6 compartments. As a first approximation, the model assumes that no noise is added at synapses in the corticothalamic loop, nor within the pyramidal neuron itself. The impedance value at the peak frequency is assumed to be constant at the beginning of the first compartment across all loops, owing to the enhancement of the EPSP amplitude by the thalamic circuitry [49,50]. Then the impedances of the other frequencies, above and below the peak frequency, are renormalized relative to the amplitude of the peak frequency. The looping sequence continues until the narrowing of the resonance curve reaches a width on the x-axis of approximately 1 Hz. Specifically, when the percentage values were less than 0.10 for each of the two frequencies bordering the peak frequency (e.g., for a peak frequency of 20 Hz, the bordering frequencies were 19 Hz and 21 Hz) the resonance curve was plotted as a single vertical line positioned at the peak frequency value.

Acknowledgments

The authors express their appreciation to Professor Robert Matthews of the University of Puget Sound for his programming assistance in the calculation of the values shown in Figure 10. We also thank three anonymous reviewers for their helpful comments. The second author expresses his appreciation to the University of Washington, where he was a Visiting Scholar for the past two years, and to UCIrvine, where he is Professor Emeritus.

Author Contributions

Conceived and designed the experiments: DL RSK. Performed the experiments: RSK DL. Analyzed the data: RSK DL. Contributed reagents/materials/analysis tools: RSK DL. Wrote the paper: RSK DL.

References

1. Mountcastle VB (1998) The cerebral cortex. Cambridge, Massachusetts: Harvard University Press.
2. Binzegger T, Douglas RJ, Martin KA (2004) A quantitative map of the circuit of cat primary visual cortex. J Neurosci 24: 8441–8453.
3. DeFelipe J, Farinas (1992) The pyramidal neuron of the cerebral cortex: morphological and chemical characteristics of the synaptic inputs. Prog Neurobiol 3: 563–607.
4. Heider B, McCormick DA (2009) Rapid neocortical dynamics: cellular and network mechanisms. Neuron 62: 171–189.
5. Steriade, M. , Timofeev, I. , Grenier, F (2001) Natural waking and sleep states: a view from inside neocortical neurons. J Neurophysiol 85: 1969–1985.
6. Kahana MJ (2006) The cognitive correlates of human brain oscillations. J Neurosci, 26: 1669–1672.
7. Sejnowski TJ, Paulsen O (2006) Network oscillations: emerging computational principles. J Neurosci 26: 1673–1676.
8. Hutcheon B, Yarom Y (2000) Resonance, oscillation and the intrinsic frequency preferences of neurons. Trends Neurosci 23: 216–222.
9. Higgs MH, Spain WJ (2009) Conditional bursting enhances resonant firing in neocortical Layer 2-3 pyramidal neurons. J Neurosci 29: 1285–1299.
10. Longtin A, Bulsara A, Moss F (1991) Time-interval sequences in bistable systems and the noise-induced transmission of information by sensory neurons. Phys Rev Lett 67: 656–659.
11. Stacey WC, Lazarewicz MT, Litt B (2009) Synaptic noise and physiological coupling generate high-frequency oscillations in a hippocampal computational model. J Neurophysiol 102: 2342–2357.
12. Paik S-B, Glaser DA (2010) Synaptic plasticity controls responses through frequency-dependent gamma oscillation resonance. PLOS Comput Biol e1000927.
13. Izhikevich EM, Desai NS, Walcott EC, Hoppensteadt FC (2003) Bursts as a unit of neural information: selective communication via resonance. Trends in Neurosci 26: 161–167.
14. Zucker RS, Regehr WG (2002) Short-term synaptic plasticity. Annu Rev Physiol 64: 355–405.
15. Volman V, Levine H (2008) Activity-dependent stochastic resonance in recurrent neuronal networks. Phys Rev E 77: 060903(1–3).
16. Volman V, Levine H, Ben-Jacob E, Sejnowski TJ (2009) Locally balanced dendritic integration by short-term synaptic plasticity and active dendritic conductances. J Neurophysiol 102: 3234–3250.
17. Delogne P (1982) Leaky feeders and subsurface radio communication. New York: Peter Peregrinus, LTD.
18. Narayanan R, Johnston D (2008) The h channel mediates location dependence and plasticity of intrinsic phase response in rat hippocampal neurons. J Neuroscience 28: 5846–5860.
19. Llinas RR (1988) The intrinsic electrophysiological properties of mammalian neurons: Insights into central nervous system function. Science 242: 1654–1664.
20. Gutfreund Y, Yarom Y, Segev I (1995) Subthreshold oscillations and resonant frequency in guinea-pig cortical neurons: physiology and modeling. J Physiol (Lond.) 483: 621–640.
21. Pike FG, Goddard RS, Suckling JM, Ganter P, Kasthuri N, et al. (2000) Distinct frequency preferences of different types of rat hippocampal neurons in response to oscillatory input currents. J Physiol (Lond.). pp 205–213.
22. Hu H, Vervaeke K, Storm JF (2002) Two forms of electrical resonance at theta frequencies, generated by M-current, h-current and persistent Na⁺ current in rat hippocampal pyramidal cells. J. Physiol (Lond.) 545: 783–805.
23. Narayanan R, Johnston D (2007) Long-term potentiation in rat hippocampal neurons is accompanied by spacially widespread changes in intrinsic oscillatory dynamics and excitability. Neuron 56: 1061–1075.
24. Destexhe A (2001) Simplified models of neocortical pyramidal cells preserving somatodendritic voltage attenuation. Neurocomput 38-40: 167–173.
25. Keren N, Bar-Yehuda D, Korngreen A (2009) Experimentally guided modeling of dendritic excitability in rat neocortical pyramidal neurons. J Physiol (Lond.) 587.7: 1413–1437.
26. Jones EG, Powell TPS (1970) Electron microscopy of the somatic sensory cortex of the cat. I. Cell types and synaptic organization, Philos. Trans R Roc London B 257: 1–11.
27. London M, Meunier C, Segev I (1999) Signal transfer in passive dendrites with nonuniform membrane conductance. J Neurosci 19: 8219–8233.
28. Reyes AD (2001) Influence of dendritic conductances on the input and output properties of neurons. Annu Rev Neurosci 24: 653–675.
29. Williams SR, Stuart G (2003) Role of dendritic synapse location in the control of action potential output. Trends Neurosci 26: 147–154.
30. Jones EG (2002) Thalamic circuitry and thalamocortical synchrony. Phil Trans Royal Soc London, B 357: 1659–1673.
31. Jones, EG (2007) The Thalamus (2nd ed.). Cambridge, United Kingdom: Cambridge University Press.
32. LaBerge D (2005) Sustained attention and apical dendrite activity in recurrent circuits. Brain Res Reviews 50: 86–99.
33. Duan H, Wearne SL, Rocher AB, Macedo A, Morrison JH, et al. (2003) Age-related dendritic and spine changes in corticocortically projecting neurons in macaque monkeys. Cerebral Cortex 13: 950–961.
34. Van Elburg RA, Ooyen A (2010) Impact of dendritic size and dendritic topology on burst firing in pyramidal cells. PLoS Comp Biol 6: e1000781.
35. Doyle DA, Cabral JM, Pfuetzner RA, Kuo A, Gulbis JM, et al. (1998) The structure of the potassium channel. Molecular basis of K⁺ conduction and selectivity. Science 280: 69–77.
36. Koch C (1999) Biophysics of computation: Information processing in single neurons. New York: Oxford University Press.
37. Heaviside O (1892) Electrical Papers (2 vols). London: Macmillan.
38. Fitzgerald GF In: Larmor J, ed (1902) The scientific writings of the late George Francis Fitz, Gerald. Dublin: Hodges and Figgis.
39. Rall W (1989) Cable theory for dendritic neurons. In: C Koch C, Segev I, eds. Methods in neuronal modeling. Cambridge, MA: MIT Press. pp 9–62.
40. Barlow HM, Brown J (1962) Radio Surface Waves. Oxford, UK: Oxford University Press.
41. Angelo EJ, Jr, Papoulis A (1964) Pole-zero patterns in the analysis and design of low-order systems. New York: McGraw-Hill.
42. Collins RE (1960) Field theory of guided waves. Artificial dielectrics,Section 12.4 528, 529, York, PA: The Maple Press.
43. Plonus MA (1978) Applied Electromagnetics, McGraw-Hill Book Co., New York.
44. Hille, B (2003) Ion channels of excitable membranes (3rd Edition). Sunderland, MA: Sinauer Associates.
45. Ulrich D (2002) Dendritic resonance in rat neocortical pyramidal cells. J Physiol (Lond.) 87: 2753–2759.
46. Peters A, Sethares C (1991) Organization of pyramidal neurons in Area 17 of monkey visual cortex. J Comp Neurol 306: 1–23.
47. Arellano JI, Benevides-Piccione R, DeFelipe J, Yuste R (2007) Ultrastructure of dendritic spines: correlation between synaptic and spine morphologies. Front. Neurosci 1: 131–143.
48. Keren N, Peled N, Korngreen A (2005) Constraining compartmental models using multiple voltage recordings and genetic algorithms. J Neurophysiol 94: 3730–3742.
49. LaBerge D, Carter M, Brown V (1992) A network simulation of thalamic circuit activity in selective attention. Neural Computation, 41: 318–331.
50. Noesselt T, Tyll S, Boehler CN, Budinger E, Heinze HJ, et al. (2010) Sound-induced enhancement of low-intensity vision: multi-sensory influences on human sensory-specific cortices and thalamic bodies relate to perceptual enhancement of visual detection sensitivity. J Neurosci 230: 13609–13923.

12

Rapid and Label-Free Separation of Burkitt's Lymphoma Cells from Red Blood Cells by Optically-Induced Electrokinetics

Wenfeng Liang[1,2,3], Yuliang Zhao[3], Lianqing Liu[1]*, Yuechao Wang[1], Zaili Dong[1], Wen Jung Li[1,3]*, Gwo-Bin Lee[4], Xiubin Xiao[5], Weijing Zhang[5]

1 State Key Laboratory of Robotics, Shenyang Institute of Automation, Chinese Academy of Sciences, Shenyang, China, 2 University of Chinese Academy of Sciences, Beijing, China, 3 Department of Mechanical and Biomedical Engineering, City University of Hong Kong, Kowloon, Hong Kong, 4 Department of Power Mechanical Engineering, National Tsing Hua University, Hsinchu, Taiwan, 5 Department of Lymphoma, Affiliated Hospital of Military Medical Academy of Sciences, Beijing, China

Abstract

Early stage detection of lymphoma cells is invaluable for providing reliable prognosis to patients. However, the purity of lymphoma cells in extracted samples from human patients' marrow is typically low. To address this issue, we report here our work on using *optically-induced dielectrophoresis* (ODEP) force to rapidly purify Raji cells' (a type of Burkitt's lymphoma cell) sample from red blood cells (RBCs) with a label-free process. This method utilizes dynamically moving virtual electrodes to induce negative ODEP force of varying magnitudes on the Raji cells and RBCs in an *optically-induced electrokinetics* (OEK) chip. Polarization models for the two types of cells that reflect their discriminate electrical properties were established. Then, the cells' differential velocities caused by a specific ODEP force field were obtained by a finite element simulation model, thereby established the theoretical basis that the two types of cells could be separated using an ODEP force field. To ensure that the ODEP force dominated the separation process, a comparison of the ODEP force with other significant electrokinetics forces was conducted using numerical results. Furthermore, the performance of the ODEP-based approach for separating Raji cells from RBCs was experimentally investigated. The results showed that these two types of cells, with different concentration ratios, could be separated rapidly using externally-applied electrical field at a driven frequency of 50 kHz at 20 V_{pp}. In addition, we have found that in order to facilitate ODEP-based cell separation, Raji cells' adhesion to the OEK chip's substrate should be minimized. This paper also presents our experimental results of finding the appropriate bovine serum albumin concentration in an isotonic solution to reduce cell adhesion, while maintaining suitable medium conductivity for electrokinetics-based cell separation. In short, we have demonstrated that OEK technology could be a promising tool for efficient and effective purification of Raji cells from RBCs.

Editor: Masaya Yamamoto, Institute for Frontier Medical Sciences, Kyoto University, Japan

Funding: The work was supported by National Natural Science Foundation of China (Project No. 61107043) http://www.nsfc.gov.cn; CAS-Croucher Joint Lab Scheme (Project No. 9500011) http://www.croucher.org.hk/. The funders had no role in study design, data collection and analysis, decision to publish, or preparation of the manuscript.

Competing Interests: The authors have declared that no competing interests exist.

* E-mail: lqliu@sia.cn (LL); wenjli@cityu.edu.hk (WJL)

Introduction

B-cell lymphomas are a species of lymphomas derived from the carcinogenesis of B lymphocytes in the human lymphatic system. They are generally classified into two categories: 1) indolent lymphomas – cancerous cells that are under control and patients have a long-term survival rate even without treatments; and 2) malignant lymphomas – which are cancerous cells that could spread rapidly and cause a rapid deterioration of the health and even death of patients, and hence, need timely and thorough treatments. Burkitt's lymphoma [1], one of the fourteen kinds of B-cell lymphomas, is a type of malignant lymphoma and propagates quickly inside a patient's body, often to the bone marrow, blood, and central nervous system. Without timely treatment, Burkitt's lymphoma could cause death rapidly. However, this kind of malignant lymphoma can be cured, depending on the histology, type, and stage of the disease [2]. Thus, early stage detection of this type of lymphoma cell is essential and invaluable for achieving

a favorable prognosis, as well as for potentially improving the patient's quality of life. However, different patients may exhibit varying degrees of drug resistance to the same drugs commonly used in targeted therapy for the clinical treatment of lymphomas. Thus, it is necessary to explore the clinicopathological characteristics of these cancerous cells from human lymphoma patients in order to better understand the relationship between cell histology and disease pathology in patients. Correlating data of cell histology and disease pathology to improve the accuracy of an early patient diagnosis will assist doctors in choosing the best treatments for patients. However, there are typically many red blood cells (RBCs) in a solution sample of Raji cells (a type of Burkitt's lymphoma cell) extracted from patients. Thus, a rapid and efficient technique is required to enable the identifying, discriminating, and purifying of target Raji cells in a mixed cell population from RBCs that may interfere with later detection and research protocols. For this purpose, technologies with a high degree of sensitivity, specificity, and reproducibility are required to separate Raji cells from RBCs. Existing technologies are broadly categorized using specific biological markers or differential biomechanical and electromechanical schemes. Of these schemes, biomechanical and electromechanical methods are known as "label-free" techniques as no biomarkers are required to implement them. For example, the density gradient centrifugation method [3–4] is a label-free method commonly used to remove the RBCs or plasma for isolating the cancerous cells in peripheral blood, using the density variation mechanism of cells with the assistance of commercial available liquid kits (e.g., using Ficoll as given in [5]). This technique, however, simultaneously contaminates all of the isolated RBCs. Another label-free technique is using microfluidic systems, i.e., based on purely hydrodynamic forces. This technique has already been demonstrated to be capable of isolating cancerous cells with a recovery rate of over 90% [6]. However, a strong drawback of this method is that separation of cells of similar inertia (i.e., similar sizes) is very difficult. *Acoustophoresis* is another separation method based on biomechanical mechanism, and has been employed to separate cancer cell lines from leukocyte fractions by acoustic standing-wave forces with a purity of over 79.6% [7]. Nevertheless, acoustophoresis seems to be a low throughput method, since separation performance is determined by retention time in the acoustic field (i.e., slow flow rate in the microfluidic channel is desired), the size of the cells, and the acoustic properties of the cells relative to the suspending medium. Alternatively, *immunomagnetic-based cell separation* approaches are commonly used to identify and discriminate cancerous cells by coupling magnetic beads with cell surface antigen-specific antibodies with an auxiliary of applied magnetic field [8–9]. This approach is based on the identification of selected proteins on the tumor cells' surface, including epithelial cell adhesion molecule, cytokeratins, and other proteins. However, these proteins are not expressed in all tumors [10] and, hence, some cells cannot be labeled and targeted, which limits the applications of this technique for cell separation. Moreover, the preparation of cell samples is also more complicated and time-consuming than other methods. In addition, dielectrophoresis (DEP) force, generated by the interaction between an applied non-uniform electric field and the dielectric objects in the field, can be used to induce motions of different directions on the objects, reflecting the differential inherent properties (e.g., dielectric properties and sizes) of the objects. Accordingly, this technique is a label-free manner and has been applied to separate cancerous cells [11–13]. Nevertheless, the DEP-based separation technique requires unique conductive electrodes fabricated by micro-lithographic techniques to be integrated with a microfluidic system, and, hence, this technology lacks the flexibility of allowing for dynamic and reconfigurable manipulation once the electrodes are fabricated. Moreover, integrated micro pumps/valves are needed to enhance the separation performance, which complicates the fabrication process of the experimental system.

More recently, *optically-induced dielectrophoresis* (ODEP), or *optical electronic tweezer* (OET) [14], similar in its operational principle to DEP, permits researchers to dynamically alter the virtual electrodes that serve the function of generating the DEP force in a real-time, programmable, and reconfigurable approach by controlling the optically-projected patterns that are digitally generated by a computer. This relatively new ODEP force-based micro-/nano-manipulation technique has already been demonstrated in a large variety of applications in the bioengineering field, including cell separation (separation of live and dead human B-cells) [15], discrimination of normal oocytes [16], manipulation of DNA molecules [17], cell counting and cell lysis [18], sperm

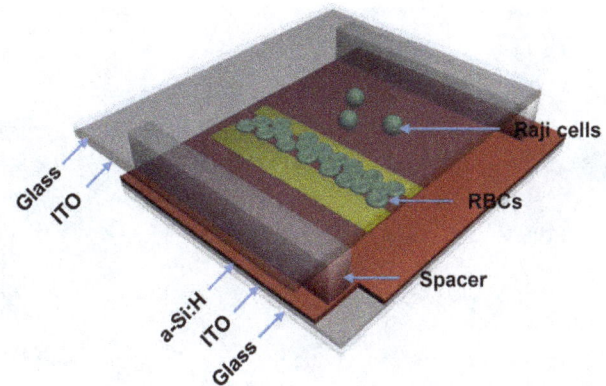

Figure 1. Three-dimensional schematic illustration of the OEK chip.

diagnostic manipulation [19], mouse embryo selection [20], circulating tumor cell isolation [21–22], and cancerous cell identification [23]. We present in this paper our experimental validation of rapidly separating Raji cells from RBCs by employing a *dynamic optically-induced DEP force* in an OEK (*optically-induced electrokinetics*) chip. The polarization models for the two types of cells were established, and then the finite element method (FEM) numerical solutions of their velocities induced by the ODEP force were obtained with the purpose of demonstrating the feasibility of using two optically-projected lines (with one dynamically moving and one stationary) of different widths to separate the two types of cells. We note that, in addition to the ODEP force, there could simultaneously exist two other types of electrokinetics forces, namely AC electroosmosis (ACEO) and AC electrothermal (ACET) flows, during the cell manipulation and separation process. These two forces, ACEO and ACET, are due to the interaction of the electric double layers with the tangential component of the electric field and the presence of the non-uniform electric field resulting in Joule heating, respectively. Consequently, we herein define the microfluidic device used in our experiments as an "*optically-induced electrokinetics*" (OEK) chip, instead of an "ODEP chip" or and "OET chip". Moreover, within an applied electric field across the OEK device, Raji cells tended to randomly adhere to the a-Si:H surface and, thus, influence cell separation performance. To reduce this cell-adhesion phenomenon, the composition of the isotonic solution in which the two types of cells are suspended was experimentally optimized and the results are reported there. With a proper percentage of bovine serum albumin (BSA) in an isotonic solution, the affinity force between the cells and the bottom substrate surface of the OEK chip (i.e., the a-Si:H layer) could be decreased significantly, and, hence, enhance the success rate of the ODEP-based manipulation and separation of cells. Most importantly, the separation of Raji cells and RBCs with two different concentration ratios were experimentally explored using an OEK chip, and the results are presented in this paper.

Theory and FEM modeling

Generation of "ODEP" force

Figure 1 is a diagrammatic sketch of our OEK chip, which consists of a top glass substrate coated with a transparent and conductive indium tin oxide (ITO) film that is used as an electrode; a microfluidic chamber that is the working area of the chip (the volume of this chamber is 3 cm ×1 cm ×60 μm); and a bottom

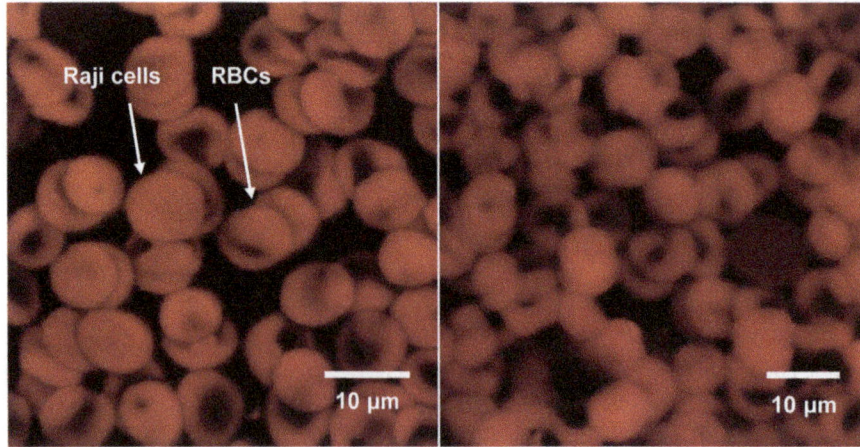

Figure 2. Atomic force microscope-scanned images of Raji cells and RBCs.

substrate with a thin photoconductive film of hydrogenated amorphous silicon (a-Si:H) deposited onto another ITO glass substrate. The a-Si:H layer behaves as an insulator due to its inherent lower conductivity when not illuminated. When an optical pattern is projected onto a particular area of the a-Si:H layer via a commercial digital projector, the electron-hole pairs are excited by the migration of electrons from the valence band to the conduction band of this corresponding area of the a-Si:H layer and, thus, locally increases the conductivity of the illuminated area of the a-Si:H layer via the photoconductive effect. Then, the electric field across the liquid chamber dramatically increases above the locally illuminated a-Si:H area since most of the externally applied voltage is shifted to the liquid chamber; thus, a non-uniform electric field can be created in the liquid chamber. Additionally, any dielectric particles suspended in the liquid at locations near this optically-induced non-uniform electric field will experience a force caused by an interaction between the electrically-polarized dipole moments of both of the particles and

the liquid solution, known as the "ODEP force" in this OEK chip. Unlike conventional DEP chips, no metal electrodes are required to create the non-uniform electric field; hence, ODEP provides a dynamic and flexible manipulation technique. The ODEP force can be either positive or negative under specific conditions, which means that the particles can be either attracted to (pDEP), or repelled from (nDEP), the illuminated areas, respectively. Our prior work presented in [24] provides a more detailed treatment of how optical wave-lengths, applied AC wave forms, and frequency can affect the optically-induced DEP force.

Polarization models for Raji cells and RBCs

The time-averaged DEP force exerted on a dielectric particle in a fluidic medium is given as [25]:

$$\langle \vec{F}_{DEP} \rangle = 2\pi\varepsilon_m abc \mathrm{Re}[K(\omega)]\nabla|\vec{E}_{rms}|^2 \qquad (1)$$

where a, b, and c are the semi-axes of the particle radii, respectively; ε_m denotes the permittivity of the liquid medium; E_{rms} is the root-mean-square value of the electric field; and $\mathrm{Re}[K(\omega)]$ is the real part of the Clausius-Mossotti (CM) factor, which is further expressed as [25]:

$$\mathrm{Re}[K(\omega)] = \mathrm{Re}\left[\frac{\varepsilon_p^* - \varepsilon_m^*}{3\left(\varepsilon_p^* - \varepsilon_m^*\right)A_j + 3\varepsilon_m^*}\right] \qquad (2)$$

In the above equation, $\varepsilon^* = \varepsilon - j\sigma/\omega$, where ε and σ are the permittivity and conductivity, respectively; $\omega = 2\pi f$, where f is the applied voltage frequency across the liquid medium; the subscripts p and m denote the properties of the particle and liquid medium, respectively; and A_j is the depolarization factor along the j-axis ($j = x, y, z$). Intuitively, based on Eq. (1), the DEP force depends on a particle's geometry and inherent dielectric property. We therefore speculated that the separation of Raji cells and RBCs is feasible because these cells have very different geometric shapes and dielectric properties. Figure 2 shows atomic force microscope (AFM) scanned images of both Raji cells and RBCs, indicating their large differences in shape. The remaining parts of this paper will show, both theoretically and experimentally, that our intuitive conjecture was correct.

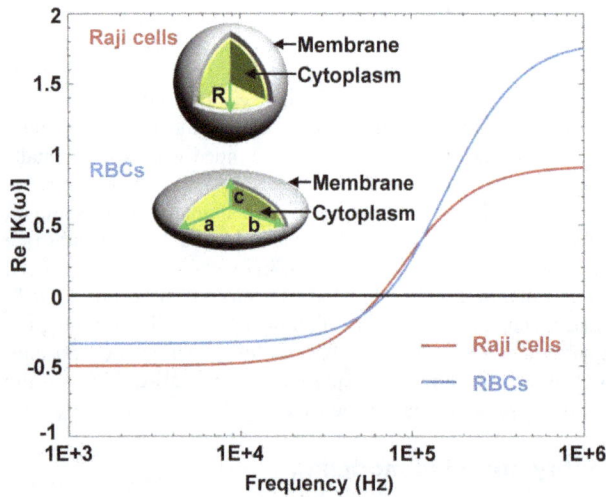

Figure 3. Re[K(ω)] for the two types of cells as a function of the applied frequency. The two types of cells are suspended in an isotonic solution with a conductivity of 1.3×10^{-2} S/m. Positive values represent an attractive (positive) DEP force, and negative values represent a repulsive (negative) DEP force.

Table 1. Dielectric properties of Raji cells and RBCs.

Type	σ_c (S/m)	ε_c (F/m)	σ_{mem} (S/m)	ε_{mem} (F/m)	C_{mem} (F/m²)	D (nm)	R (μm)
Raji [28]	0.58	$60\varepsilon_0$	8.2e-6	$8.8\varepsilon_0$	1.07e-2	7	6
RBCs [29]	0.5	$60\varepsilon_0$	1.6e-6	$8.5\varepsilon_0$	8.7e-3	8	$a=3.5$

a) Raji cells. Raji cells can be modeled as a spherical particle ($R = a = b = c$, $A_j = 1/3$), as shown in the inset of Figure 3. The shell and interior refer to the cell membrane and to the cytoplasm, respectively. Consequently, a single layer of the shell-core model can be used as an approximation for Raji cells, and the effective complex permittivity for Raji cells is given by [26]:

$$\varepsilon_p^* = C_{mem} R \frac{j\omega\tau_c + 1}{j\omega(\tau_c + \tau_m) + 1} \tag{3}$$

where $\tau_c = \varepsilon_c/\sigma_c$, $\tau_m = C_{mem}R/\sigma_c$; the subscripts c and mem represent the cell cytoplasm and membrane, respectively; C_{mem} is the capacitance of the cell membrane; and R is radius of the cell. The real part of the CM factor for Raji cells is further given by [26]:

$$\text{Re}[K(\omega)] = \text{Re}\left[-\frac{\omega^2(\tau_1\tau_m - \tau_c\tau_m') + j\omega(\tau_m' - \tau_1 - \tau_m) - 1}{\omega^2(\tau_c\tau_m' + 2\tau_1\tau_m) - j\omega(\tau_m + 2\tau_1 + \tau_m) - 2} \right] \tag{4}$$

where $\tau_1 = \varepsilon_m/\sigma_m$, $\tau_m' = C_{mem}R/\sigma_m$.

b) RBCs. RBCs are modeled as single-shell oblate ellipsoids having semi-axes of a, b, and c, which satisfy the relationships where $a = b > c$ and $e = a/c = 4$, as shown in the inset of Figure 3. The shell and interior also refer to the cell membrane and the cytoplasm, respectively. The effective complex permittivity for RBCs is given by [27]:

$$\varepsilon_p^* = \varepsilon_{mem}^* \left(\frac{\varepsilon_{mem}^* + \left(\varepsilon_c^* - \varepsilon_{mem}^*\right)\left(A_j + v(1 - A_j)\right)}{\varepsilon_{mem}^* + \left(\varepsilon_c^* - \varepsilon_{mem}^*\right)\left(A_j - vA_j\right)} \right) \tag{5}$$

where v is the volume fraction of the cell interior; $v = (c-d)(a-d)^2/ca^2$; and A_j is the depolarization factor along the long axis, $A_j = 0.5(e^2\arctan(e^2-1)^{0.5} - (e^2-1)^{0.5})/(e^2-1)^{1.5}$ [27]. The polarization properties for Raji cells and RBCs, i.e., the real part of their respective CM factors, as a function of the applied frequency, can be obtained by using (2), (4), and (5), as shown in Figure 3. Herein, the negative value for each of the two curves in Figure 3 represents a nDEP force under the given frequency range; therefore, a pDEP force will be exerted on the cells with an applied frequency higher than the crossover frequency. Typical dielectric parameters for the two types of cells, as shown in Table 1, are used to calculate the polarization for each cell type, which is suspended in an isotonic solution with a conductivity of 1.3×10^{-2} S/m.

Raji cells and RBCs velocities induced by the ODEP force

Once a cell begins to move due to the DEP force, an opposing Stokes' drag force will be produced to retard the cell's motion. The Stokes' equation for estimating the drag force on a cell moving in a fluid with velocity v is given as [30]

$$F_d = f_r v \tag{6}$$

where f_r is the friction factor of the cell in the fluid. Considering the geometrical configurations of the two types of cells, the friction factor for Raji cells and RBCs are $6\pi\eta R$ and $32/(3\eta a)$ [25], respectively, where η is the dynamic viscosity of the fluid. Since the Stokes' drag force is directly proportional to the particle's moving velocity, the magnitude of the DEP force acting on a particle can be experimentally inferred. However, in order to theoretically obtain the magnitude of the DEP force exerted on the two types of cells, the vector $\nabla|\vec{E}_{rms}|^2$ in (1) should be calculated. As an example,

Unit: V²/m³   25µm optical line   15µm optical line

(a)

(b)

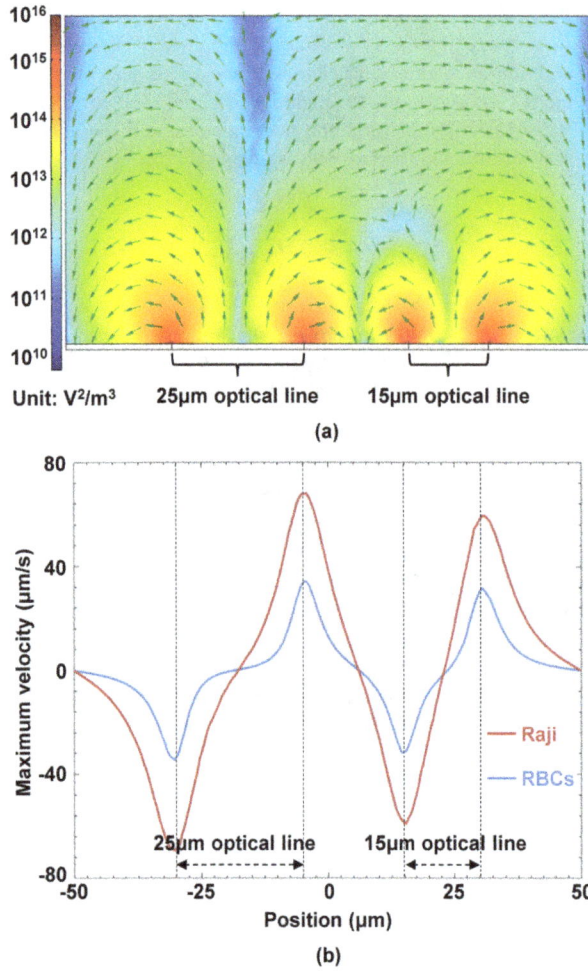

Figure 4. FEM simulation results of the DEP force. (a) Cross-sectional view of the x-component of the vector $\nabla|\vec{E}_{rms}|^2$. The arrows indicate the direction of the nDEP force induced by two optical lines of different widths, i.e., one moving line with a width of 25 µm and the other line being stationary with a width of 15 µm. (b) Velocities of Raji cells and RBCs induced by the ODEP force. The velocity of the Raji cells caused by the DEP force is higher than that of the RBCs, and the DEP force induced by the 25 µm-wide optical line for cells is also higher than that by the 15 µm optical line for cells.

Figure 4(a) shows the cross-sectional distribution of the x-component of the vector $\nabla|\vec{E}_{rms}|^2$(denoted as $\nabla|\vec{E}_{rms}|^2_x$) calculated using a commercial FEM software package (Multiphysics, COMSOL AB, Sweden). The FEM simulation method is the same as discussed in our prior work [31]. In this example, two optically-projected lines with a width of 25 µm and 15 µm, respectively, serve as virtual electrodes and the applied frequency is 50 kHz with a voltage of 20 V_{pp}. These virtual electrode geometries and electric field parameters reflect the actual experimental conditions used in our Raji cells and RBCs separation experiments, as will be discussed in the 'Results and discussions' section. As shown in Figure 4(a), the DEP force sharply decreases along the vertical direction and has a maximum value around the illuminated areas. As illustrated and described in our prior work [24], the maximum DEP force exerted on the two types of cells can be determined at the position of the height of cells' radii above the bottom of the a-Si layer, defined as

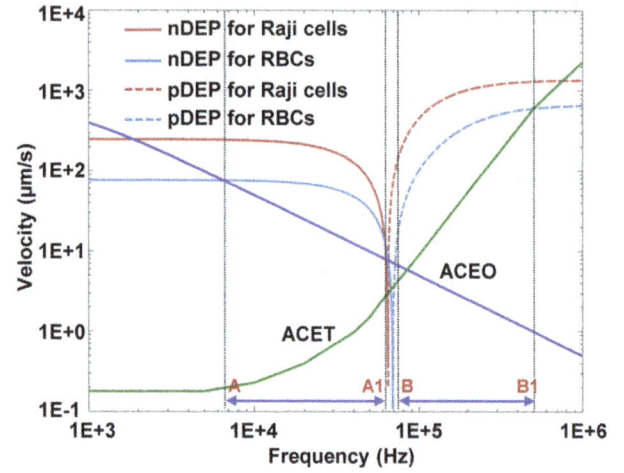

Figure 5. Contribution of various optically-induced electrokinetics forces on the velocities for Raji cells and RBCs as a function of the applied frequency. The velocity generated by the DEP, ACEO, and ACET were at 6 µm and 3.5 µm height above the a-Si:H film surface for Raji cells and RBCs, respectively. The width of the optical line that served as the virtual electrode to generate the electric field was 25 µm.

$$F_{DEPx} = \pi\varepsilon_m R^3 \text{Re}[K(\omega)]\nabla|\vec{E}_{max}|^2_x \qquad (7)$$

Then, the velocity for Raji cells and RBCs induced by the maximum DEP force can be obtained by balancing (6) and (7), expressed as

$$v = \frac{\pi\varepsilon_m R^3 \text{Re}[K(\omega)]\nabla|\vec{E}_{max}|^2_x}{f_r} \qquad (8)$$

As shown in Figure 4(b), the maximum velocity of the Raji cells, as induced by the DEP force, is higher than that of the RBCs; additionally, the DEP force induced by the 25 µm-wide optical line is higher than that induced by the 15 µm optical line. This means that the DEP force induced by two different optical line widths can be employed to perform the separation of Raji cells from RBCs.

Comparison of AC- and optically-related electrokinetics forces

As noted in the previous section, there exist several optically-related electrokinetic forces in an OEK chip, mainly including the DEP force and forces on a particle caused by ACEO- and ACET-induced flows. To enhance the experimental performance of the separation of the two types of cells, experimental conditions should be set up such that the DEP force will be dominant in the separation process with a specific applied voltage frequency across the upper and lower substrates of an OEK chip. Using the equations in Table 2, the numerical solution showing the velocities for the two types of cells caused by the AC electrokinetics forces (i.e., DEP, ACEO, and ACET), as a function of the applied frequency, were obtained by using the FEM method with same simulation parameters and simulation modules as those described in [31], except that a liquid conductivity of 1.3×10^{-2} S/m was used to reflect the experimental conditions for the work reported

Table 2. Equations of the velocities caused by the optically-induced electrokinetics forces exerted on the two types of cells.

Forces	Governing Equations [30]
DEP	*Equation 8*
ACEO	$-\varepsilon_m\varsigma E_t/\eta$
ET flow	$0.5\varepsilon_m[(\alpha-\beta)(\nabla T\cdot E)E^*/(1+(\omega\tau)^2)-$ $0.5\alpha|E|^2\nabla T]$

Here, ς is the zeta potential; η is the dynamic viscosity of DI water; E_t is the tangential electric field; T is the temperature of the liquid solution; and $\tau = \varepsilon_m/\sigma_m$. For DI water at room temperature, approximately, $k=0.6$ J·m^{-1} s^{-1}·K^{-1}; ρ_m $=1$ g·cm^{-3}, $\alpha=-0.4\%$ K^{-1}; $\beta=2\%$ K^{-1}; and $\partial\rho_m/\partial T/\rho_m = 10^{-4}$ K^{-1} [30].

in this paper. Figure 5 shows the simulated and calculated results of the cellular velocities induced by various forces. The results indicate that within specific frequency range, the ACEO or ACET may affect the DEP force-based separation process, respectively. Furthermore, for these two types of cells, when the frequency is from ~7 kHz to ~60 kHz, the nDEP force will be the dominate force; when the frequency is from ~70 kHz to ~500 kHz, the pDEP force will be the dominant force on both the RBCs and the Raji cells. However, the ACET force in the frequency range of ~70 kHz to ~500 kHz is much higher than that in the frequency range of ~7 kHz to ~60 kHz, and thus may affect the cellular properties due to the increased temperature in the fluidic medium. Hence, a frequency in the range of ~7 kHz to ~60 kHz, which induces an nDEP force, is selected to conduct the separation experiments in this study.

Materials and Methods

Fabrication of the OEK chip

The fabrication process of the OEK chip employed in this paper was described in detail in our prior work [32]. In order to apply an electric field across the chip, part of the a-Si:H thin film on the bottom ITO-glass substrate was etched to establish an electrical

connection, as shown in Figure 1. Specifically, a 5 mm ×8 mm area of a-Si:H was patterned through standard photolithography and dry-etching (using the Oxford Plasma Lab 80 etching system) with 2% oxygen, 12.5% CF4 gas, with a 30mTorr etching chamber pressure, and 6-minute plasma exposure. The chip was then rinsed and cleaned with acetone and DI water before being dried by nitrogen gas. The microfluidic chamber, into which the cells and solution were injected, has a height of ~60 μm. This microfluidic chamber between two ITO-glass substrates was constructed by using a patterned polydimethylsiloxane (PDMS) thick film or a double-sided tape as a spacer.

Experimental setup

The experimental setup for our OEK platform is shown in Figure 6 (illustration) and Figure S1 (picture). The OEK chip was fixed on a three-dimensional digital translation platform (Leetro Automation Co. Ltd, China), which would accurately regulate the spatial movement of the OEK chip, as well as span the working areas on the chip. To create the optically-projected patterns, a commercial graphics software package (Flash 11, Adobe, U.S.A.) was employed to generate the virtual electrodes with any desired geometrical configurations, which were projected onto the lower surface of the OEK chip via a commercial LCD projector (VPL-F400X, Sony, Japan) coupled with a computer. The manipulation and separation processes of the cells were observed and recorded using a charged coupled device (DH-SV1411FC, DaHeng Image, China) mounted on a microscope (Zoom 160, OPTEM, U.S.A.). In addition, a condenser lens (Nikon, MS plan, 50×), fixed between the LCD projector and the OEK chip, was used to focus and collimate the optical pattern onto the OEK chip. In order to power the OEK chip to perform cell separation process, an AC bias potential, supplied by a function generator (Agilent 33522A, U.S.A.), was applied to the transparent ITO glasses, which are located at the top and bottom of the OEK chip as shown in Figure 1.

Cell preparation and counting

The RBCs were obtained from volunteers as previously described in [33]. The samples were centrifuged at 1000 rpm

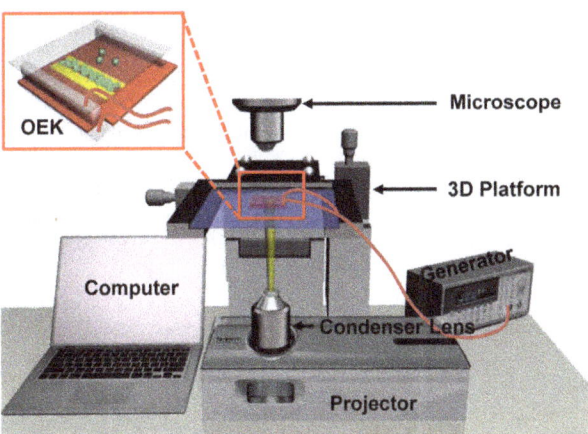

Figure 6. Schematic illustration of the experimental setup for the OEK chip. The experimental system consists of an image acquisition system to observe the cell separation process, and a pattern generation system to generate the virtual electrodes which are projected onto the lower surface of the OEK chip by a LCD projector.

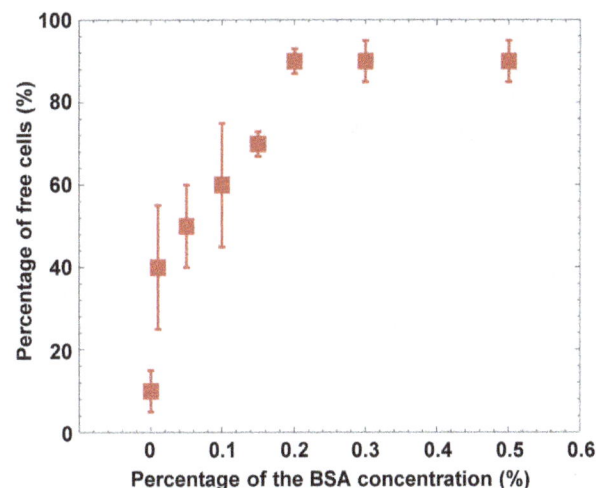

Figure 7. Percentage of free cells as a function of the percentage of the BSA concentration. Raji cells to substrate adhesion could be significantly decreased when 0.2% BSA was added to the conventional isotonic solution. The error bars indicate the standard deviation of the measurements.

for 5 min at 4°C (Sigma 3–30K, Germany). Then, the supernatant was discarded and the remained RBCs were resuspended into 1 mL of the isotonic solution for a second round of centrifugation using the same parameters. Then, the collected RBCs were resuspended into 1mL of isotonic solution for further experiments. The purpose of using the isotonic solution was to maintain an appropriate osmotic condition for the viability of the cells.

The Raji cell line [34] was a gift from Dr. Xiubin Xiao of the Affiliated Hospital of Military Medical Academy of Sciences, Beijing, China. The Raji cell line was cultured in Roswell Park Memorial Institute (RPMI-1640) culture medium supplemented with 10% (v/v) fetal calf serum, 1% penicillin (v/v) (100 U/mL), and 1% streptomycin (v/v) (100 μg/mL) at 37°C in a humidified atmosphere of 5% CO_2 (Model 371, Thermo Scientific). Before each experiment, 1 mL of Raji cell suspension was taken directly out of the culturing flask, and centrifuged at 1000 rpm for 5 min at 4°C with supernatant discarded. The collected Raji cells were resuspended into 1 mL of RPMI-1640 medium and centrifuged again using the same parameters to remove the residual culture medium. Then, the resulting Raji cells were resuspended into 1 mL of isotonic solution for further experiments.

After the cell suspensions were prepared, cells counts were performed for both types using a commercial hemocytometer (Qiujing Co. Ltd., China) to control the cell concentration. In order to achieve standardized and comparable cell separation results using the proposed OEK chips, 1 mL of Raji cells suspension with a concentration of 1×10^6 cells/mL was spiked into 1 mL of RBCs suspension to obtain various cell ratios. For our experiments, RBCs concentrations of 1×10^7 cells/mL and 4×10^7 cells/mL were used to achieve cells ratios (Raji cells: RBCs) of 1:10 and 1:40, respectively. Finally, the resulting cellular mixture of RBCs and Raji Cells was loaded into the OEK chip for further ODEP force based cell separation experiments.

Results and Discussion

Investigation of Raji cellular adhesion phenomenon inside an OEK chip

In a typical experimental procedure to investigate the cell-to-substrate adhesion phenomenon, Raji cells (at a concentration of 5×10^5 cells/mL) suspended in an isotonic solution consisting of 8.5% (w/v) sucrose and 0.3% (w/v) glucose were introduced into

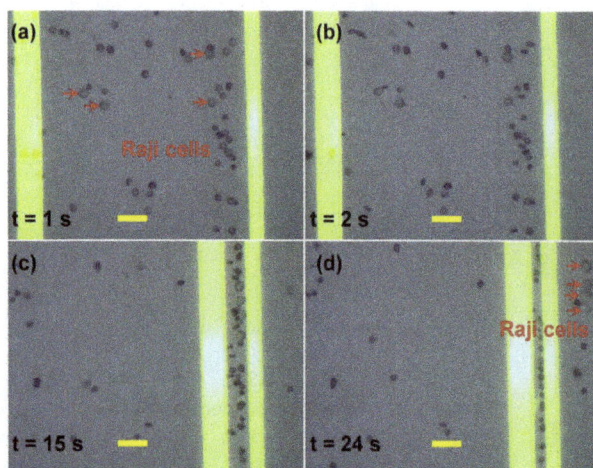

Figure 8. Microscopic images showing the dynamic separation of Raji cells from RBCs with a concentration ratio of 1:10. All scale bars are 25 μm.

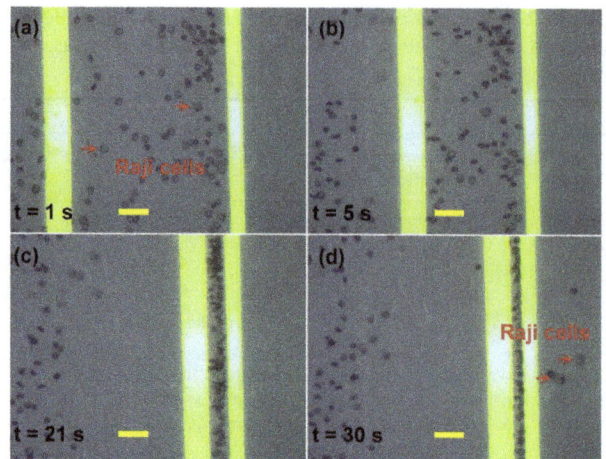

Figure 9. Microscopic images showing the dynamic separation of Raji cells from RBCs with a concentration ratio of 1:40. All scale bars are 25 μm.

the OEK chip (with a microfluidic chamber that holds ~18 μL solution). All Raji cells within the microscope's field of view (280 μm ×200 μm) were then investigated. For the purpose of data analysis, the cells that could be moved by the optically projected patterns (i.e., virtual electrodes) are defined as "free cells". And, the cells that did not show any movement when ODEP force was applied around them are defined as "adherent cells". For a typical experiment, ~90 cells in the field of view of the microscope were investigated. In addition, these tests were performed in ten different OEK chips. Moreover, ten different "field of view areas" (FOVA) were investigated in each chip. Two experiments were performed for each of the FOVA in each of the ten OEK chips used in this study. Accordingly, 200 experiments carried out. Therefore, we have observed the adhesion behavior of ~18, 000 cells (90 cells ×2×10 FOVA ×10 OEK chips) in obtaining each data point shown in Figure 7. The experimental results showed that only ~10% "free cells" of the Raji cells could be freely transported by ODEP force if only the isotonic solution is used, since the a-Si:H has a native oxide present at the a-Si:H/ liquid interface, as described by N. K. Lau, et al., [35]. According to them, the adhesion force between a-Si:H surface and mammalian cells were in the order of nanonewtons, while the ODEP force was in the order ranging from tens to hundreds of piconewtons. Consequently, in order to facilitate ODEP-based cell separation, this adhesion force between the Raji cells and the a-Si:H surface should be minimized. N. K. Lau, et al., have reported chip-modification methods that can avoid cell adhesion by surface treatments; however, the processes for fabricating those modified chips are time-consuming and complicated. In this study, the method of adding different concentrations of bovine serum albumin (BSA) into the isotonic solution was experimentally investigated and characterized. The BSA was added into the solution with the purpose of decreasing the affinity force between the cells and the a-Si:H substrate of the OEK chip, and thereby enhancing the performance of the ODEP manipulation and separation process of the two types of cells. We have investigated cell-to-substrate adhesion behavior as a function of the BSA's concentration in the isotonic solution. As mentioned above, for each of the BSA concentration tested, Raji cells were observed in ten distinct field of views of the microscope for each of the ten different OEK chips. Results shown in Figure 7 revealed that the cell-to-substrate adhesion could be quite significantly decreased, i.e., with $90\pm3\%$ of the cells being easily transported when 0.2% (wt) BSA was added to the conventional isotonic solution. Furthermore, the measured liquid conductivity of the optimized isotonic solution was 1.3×10^{-2} S/m (obtained by using a Cond 3110 conductivity meter, Germany) and also met the liquid conductivity requirement necessary to enable the OEK chip to work well [36].

Performance of the separation of Raji cells from RBCs with different concentration ratios

In this study, an electrical field of 20 V_{pp} with frequency of 50 kHz (based on the theoretical calculations as discussed in the 'Theoretical and FEM modeling' section) was employed to power the OEK chip. Two optical lines of different widths, i.e., one moving line of 25 μm width and one stationary line of 15 μm width, were adopted as virtual electrodes to generate a DEP force in the OEK chip to separate the Raji cells from the RBCs. Five attempts were done for separating Raji cells from RBCs with two different Raji cells-to RBCs concentration ratios of 1:10 and 1:40, respectively. For each time, the Raji cells were successfully manipulated and separated from RBCs.

a) A concentration ratio of 1:10. Figure 8 and Video S1 show an example of the experimental process for separating the Raji cells from the RBCs with a concentration of 1:10 (i.e., approximately 1 Raji cell for every 10 RBCs in the isotonic solution). Initially, when the optical patterns were projected onto the a-Si:H surface and there was no voltage applied, the two types of cells were randomly suspended in the liquid solution as shown in Figure 8(a). Once the voltage was switched on, the two types of cells were repelled, and the RBCs would change their orientations because they would align their longest axis to be parallel to the direction of the electric field (Figure 8(b)). Then, the 25 μm-wide optical line was set with a translational velocity of 12 μm/s from left to right by using an animation software (Flash), and this motion continuously pushed the two types of cells toward the location of the 15 μm optical line, due to the larger nDEP force induced by the 25 μm-wide optical line. We have observed that, when driven by the 25 μm-wide optical line, both of the two types of cells simultaneously have self-rotational and translational behaviors (see Video S1), which are the same observation obtained in our prior experiments with other types of cells as reported in [23]. After 15 s, the two types of cells are aligned and located within the gap between the two lines (Figure 8(c)). When the 25 μm-wide optical line was moved further to the right, all of the Raji cells were pushed past the 15 μm-wide optical line, as shown in Figure 8(d). Whereas, there were also some RBCs pushed towards the stationery line because Raji cells' movement affected those RBCs. The Raji cells could be further separated and purified by employing subsequent virtual electrodes with a similar function explained in this section.

b) A concentration ratio of 1:40. In order to experimentally explore and estimate the throughput of the OEK chip, we further execute the separation experiment of the two types of cells with a lower Raji cells-to-RBCs concentration ratio of 1:40. Figure 9 and Video S2 present the performance of the separation of Raji cells from the cellular mixture with the same experimental procedure as discussed for the 1:10 concentration ratio above. As shown in Figure 9(c), there were a few Raji cells and many RBCs trapped within the gap between the two lines. Then, when the gap was further narrowed, the Raji cells were pushed past the stationery line (Figure 9(d)). However, there were some RBCs that were also simultaneously separated. Again, the Raji cells could be further separated and purified from these RBCs by employing subsequent virtual electrodes with a similar function as described above.

The separation efficiency for Raji cells and RBCs of the two different concentrations ratios of 1:10 and 1:40 were 67% and 50%, respectively. This efficiency could be easily improved by projecting subsequent virtual electrodes configured as the two optical lines of different widths discussed in this paper to further separate the cells. This process can be repeated in the OEK chip until the cells are completely separated.

The purpose of this study is to verify the capability of using the ODEP technique to separate Raji cells from RBCs, and potentially provide a new automated and label-free technology for cell separation and purification. Since there are much more RBCs in a blood sample than white blood cells, we have focused the current study on only Raji cells and RBCs. We simulated the blood sample of lymphoma patients by spiking the Raji cells (from the cell line described in [34]) into the RBCs sample from volunteers and then perform the cell separation experiment proposed in this study. Although the size of the Raji cells is similar to the white blood cells, theoretically ODEP force can be still applied to discriminate them due to their inherent different dielectric parameters (i.e., membrane/cytoplasm/nucleus permittivity and conductance), which will result in the different direction and/or magnitude of ODEP force exerted on them. We are currently investigating the required electrokinetic parameters for the separation of white blood cells, red blood cells, and Raji cells; we will report our findings in the future.

Conclusion

We experimentally demonstrated rapid separation of Raji cells from red blood cells (RBCs) by employing dynamic *optically-induced dielectrophoretic* (ODEP) force on the cells. The separation of these cells with two different concentration ratios was achieved by using two optically-projected lines (one dynamically moving and one fixed) that generate negative ODEP force of different magnitudes on the cells. Inside an *optically-induced electrokinetics* (OEK) chip, these projected line images act as virtual electrodes to generate the negative ODEP force on the cells under a non-uniform electric field produced by a 20 V_{pp} at frequency of 50 kHz across the fluidic medium. We also report the selection of an appropriate isotonic solution with the purpose of addressing the problems of cellular adhesion behavior and proper fluid conductivity for cell separation in an OEK chip. In addition, an FEM simulation of the ODEP force acting on the cells was carried out to explain the separation phenomenon. This simulation includes the usage of different polarization models for the Raji cells and RBCs. Further work is required to investigate the viability of cells using the separation technique reported in this paper. However, our initial work showed that both Raji cells and RBCs do survive for several hours after the separation experiments in an OEK chip. Pending on the results of more extensive cell viability experiments, cell separation using dynamic ODEP forces could prove to be a unique method that is capable of rapidly separating and purifying cells using relative simple procedures.

Supporting Information

Figure S1 A picture of the actual ODEP system setup used to manipulate and separate cells in our experiments.

Video S1 Video of the experimental results of the dynamic separation of Raji cells from RBCs with a concentration ratio of 1:10.

Video S2 **Video of the experimental results of the dynamic separation of Raji cells from RBCs with a concentration ratio of 1:40.**

Author Contributions

Conceived and designed the experiments: WJL LL WL. Performed the experiments: WL. Analyzed the data: WL YZ. Contributed reagents/materials/analysis tools: YW ZD G-BL XX WZ. Wrote the paper: WL.

References

1. Burkitt D (1958) A sarcoma involving the jaws in african children. Brit J Surg 46(197): 218–223.
2. Janeway CA, Travers P, Walport M, Shlomchik MJ (2001) Immunobiology, 5th edition. New York: Garland Science. 414 p.
3. Byum A (1976) Isolation of lymphocytes, granulocytes and macrophages. Srand J Immunol 5: 9–15.
4. Rosenberg R, Gertler R, Friederichs J, Fuehrer K, Dahm M, et al. (2002) Comparison of two density gradient centrifugation systems for the enrichment of disseminated tumor cells in blood. Cytometry 49(4): 150–158.
5. Vettore L, Concetta MD, Zampini P (1980) A new density gradient system for the separation of human red blood cells. Am J Hematol 8(3): 291–297.
6. Sun J, Li M, Liu C, Zhang Y, Liu D, et al. (2012) Double spiral microchannel for label-free tumor cell separation and enrichment. Lab Chip 12(20): 3952–3960.
7. Augustsson P, Magnusson C, Nordin M, Lilja H, Laurell T (2012) Microfluidic, label-free enrichment of prostate cancer cells in blood based on acoustophoresis. Anal Chem 84(18): 7954–7962.
8. Ryu BY, Kubota H, Avarbock MR, Brinster RL (2005) Conservation of spermatogonial stem cell self-renewal signaling between mouse and rat. Proc Natl Acad Sci 102(40): 14302–14307.
9. Talasaz AH, Powell CC, Huber DE, Berbee JG, Roh KH, et al. (2009) Isolating highly enriched populations of circulating epithelial cells and other rare cells from blood using a magnetic sweeper device. Proc Natl Acad Sci 106(10): 3970–3975.
10. Mikolajczyk SD, Millar LS, Tsinberg P, Coutts SM, Zomorrodi M, et al. (2011) Detection of EpCAM-negative and cytokeratin-negative circulating tumor cells in peripheral blood. J Oncol 2011: 1–10.
11. Huang CJ, Chien HC, Chou TC, Lee GB (2011) Integrated microfluidic system for electrochemical sensing of glycosylated hemoglobin. Microfluid Nanofluid 10(1): 37–45.
12. Moon HS, Kwon K, Kim SI, Han H, Sohn J, et al. (2011) Continuous separation of breast cancer cells from blood samples using multi-orifice flow fractionation (MOFF) and dielectrophoresis (DEP). Lab Chip 11(6): 1118–1125.
13. Gao J, Riahi R, Sin LY, Zhan S, Wong PK (2012) Electrokinetic focusing and separation of mammalian cells in conductive biological fluids. Analyst 137(22): 5215–5221.
14. Chiou PY, Ohta AT, Wu MC (2005) Massively parallel manipulation of single cells and microparticles using optical images. Nature 436: 370–372.
15. Ohta AT, Chiou PY, Phan HL, Sherwood SW, Yang JM, et al. (2007) Optically controlled cell discrimination and trapping using optoelectronic tweezers. IEEE J Sel Top Quant 13(2): 235–243.
16. Hwang H, Lee DH, Choi W, Park JK (2009) Enhanced discrimination of normal oocytes using optically induced pulling-up dielectrophoretic force. Biomicrofluidics 3(1): 014103.
17. Lin YH, Chang CM, Lee GB (2009) Manipulation of single DNA molecules by using optically projected images. Opt Express 17(17): 15318–15329.
18. Lin YH, Lee GB (2010) An integrated cell counting and continuous cell lysis device using an optically induced electric field. Sensor Actuat B Chem 145(2): 854–860.
19. Ohta AT, Garcia M, Valley JK, Banie L, Hsu HY, et al. (2010) Motile and non-motile sperm diagnostic manipulation using optoelectronic tweezers. Lab Chip 10(23): 3213–3217.
20. Valley JK, Swinton P, Boscardin WJ, Lue TF, Rinaudo PF, et al. (2010) Preimplantation mouse embryo selection guided by light-induced dielectrophoresis. PloS ONE 5(4): e10160.
21. Huang SB, Chen J, Wang J, Yang CL, Wu MH (2012) A new optically-induced dielectrophoretic (ODEP) force-based scheme for effective cell sorting. Int J Electrochem Sci 7: 12656–12667.
22. Huang SB, Wu MH, Lin YH, Hsieh CH, Yang CL, et al. (2013) High-purity and label-free isolation of circulating tumor cells (CTCs) in a microfluidic platform by using optically-induced-dielectrophoretic (ODEP) force. Lab Chip 13(7): 1371–1383.
23. Chau LH, Liang W, Cheung FWK, Liu WK, Li WJ, et al. (2013) Self-rotation of cells in an irrotational AC E-field in an opto-electrokinetics chip. PLoS ONE 8(1): e51577.
24. Liang W, Wang S, Dong Z, Lee GB, Li WJ (2012) Optical spectrum and electric field waveform dependent optically-induced dielectrophoretic (ODEP) micro-manipulation. Micromachines 3(2): 492–508.
25. Morgan H, Green NG (2003) AC Electrokinetics: colloids and nanoparticles, 1th edition. UK: Research Studies Press. 39 p.
26. Jones TB (1995) Electromechanics of Particles. UK: Cambridge Univ Press. 46 p.
27. Kakutani T, Shibatani S, Sugai M (1993) Electrorotation of non-spherical cells: Theory for ellipsoidal cells with an arbitrary number of shells. Bioelectrochem Bioenerg 31(2): 131–145.
28. Polevaya Y, Ermolina I, Schlesinger M, Ginzburg BZ, Feldman Y (1999) Time domain dielectric spectroscopy study of human cells II. Normal and malignant white blood cells. BBA Biomembranes 1419(2): 257–271.
29. Yang J, Huang Y, Wang XB, Becker FF, Gascoyne RC (1999) Cell separation on microfabricated electrodes using dielectrophoretic gravitational field-flow fractionation. Anal Chem 71(5): 911–918.
30. Castellanos A, Ramos A, Gonzalez A, Green NG, Morgan H (2003) Electrohydrodynamics and dielectrophoresis in microsystems: scaling laws. J Phys D Appl Phys 36(20): 2584–2597.
31. Liang W, Liu N, Dong Z, Liu L, Mai JH, et al. (2013) Simultaneous separation and concentration of micro- and nano-particles by optically-induced electrokinetics. Sensor Actuat A-Phys 193: 103–111.
32. Wang S, Liang W, Dong Z, Lee GB, Li WJ (2011) Fabrication of micrometer-and nanometer-scale polymer structures by visible light induced dielectrophoresis (DEP) force. Micromachines 2(4): 431–442.
33. Li M, Liu L, Xi N, Wang Y, Dong Z, et al. (2012) Atomic force microscopy imaging and mechanical properties measurement of red blood cells and aggressive cancer cells. Sci China Life Sci 55(11): 968–973.
34. Li M, Liu L, Xi N, Wang Y, Dong Z, et al. (2011) Imaging and measuring the rituximab-induced changes of mechanical properties in B-lymphoma cells using atomic force microscopy. Biomech Bioph Res Co 404 (2): 689–694.
35. Lau NK, Ohta AT, Phan HL, Hsu HY, Jamshidi A, et al. (2009) Antifouling coatings for optoelectronic tweezers. Lab Chip 9(20): 2952–2957.
36. Hsu H, Ohta AT, Chiou PY, Jamshidi A, Neale SL, et al. (2010) Phototransistor-based optoelectronic tweezers for dynamic cell manipulation in cell culture media. Lab Chip 10(2): 165–172.

13

Extracellular Neural Microstimulation May Activate Much Larger Regions than Expected by Simulations: A Combined Experimental and Modeling Study

Sébastien Joucla[1,2], Pascal Branchereau[1,2], Daniel Cattaert[1,2], Blaise Yvert[1,2]*

1 Université Bordeaux, Institut des Neurosciences Cognitives et Intégratives d'Aquitaine, UMR5287, Bordeaux, Talence, France, **2** CNRS, Institut des Neurosciences Cognitives et Intégratives d'Aquitaine, UMR5287, Bordeaux, Talence, France

Abstract

Electrical stimulation of the central nervous system has been widely used for decades for either fundamental research purposes or clinical treatment applications. Yet, very little is known regarding the spatial extent of an electrical stimulation. If pioneering experimental studies reported that activation threshold currents (TCs) increase with the square of the neuron-to-electrode distance over a few hundreds of microns, there is no evidence that this quadratic law remains valid for larger distances. Moreover, nowadays, numerical simulation approaches have supplanted experimental studies for estimating TCs. However, model predictions have not yet been validated directly with experiments within a common paradigm. Here, we present a direct comparison between experimental determination and modeling prediction of TCs up to distances of several millimeters. First, we combined patch-clamp recording and microelectrode array stimulation in whole embryonic mouse spinal cords to determine TCs. Experimental thresholds did not follow a quadratic law beyond 1 millimeter, but rather tended to remain constant for distances larger than 1 millimeter. We next built a combined finite element – compartment model of the same experimental paradigm to predict TCs. While theoretical TCs closely matched experimental TCs for distances <250 microns, they were highly overestimated for larger distances. This discrepancy remained even after modifications of the finite element model of the potential field, taking into account anisotropic, heterogeneous or dielectric properties of the tissue. In conclusion, these results show that quadratic evolution of TCs does not always hold for large distances between the electrode and the neuron and that classical models may underestimate volumes of tissue activated by electrical stimulation.

Editor: Eleni Vasilaki, University of Sheffield, United Kingdom

Funding: This work was supported by the French ministry for research and technology (RMNT Neurocom project No 03J489 and ACI Neurosciences Intégratives et Computationnelles No 2003541), the French National Research Agency (ANR – Programme Blanc No ANR06BLAN035601 and Programme TecSan No ANR07TECSAN01404), the Institut pour la Recherche sur la Moelle Epinière et l'Encéphale (IRME), the Région Aquitaine (No 20030301201A and 20040301202A), and the European Union (EU) FP7 program. The funders had no role in study design, data collection and analysis, decision to publish, or preparation of the manuscript.

Competing Interests: The authors have declared that no competing interests exist.

* E-mail: blaise.yvert@u-bordeaux1.fr

Introduction

Extracellular electrical stimulation of neural tissues has been used for decades, for either fundamental research purposes or clinical treatment applications. Macroscopic stimulation using millimeter-scale electrodes is used in a number of clinical applications, including suppression of tremors in Parkinson's disease using deep brain stimulation (DBS) [1], restoration of auditory perception using cochlear implants [2,3], and alleviation of chronic pain [4] or restoration of motor functions [5,6] using spinal cord stimulation. Over the past decade, smaller electrodes with typical size of the order or below 100 microns have been assembled into microelectrode arrays (MEAs). These multichannel probes allow interfacing large neural networks with hundreds of recording and stimulating sites. These new devices have triggered a surge towards finding pertinent paradigms of extracellular microstimulation to modify and even control the dynamics and plasticity of neural networks [7,8,9], and also, in clinical applications, to restore visual perception using retinal implants

[10,11] or to create bidirectional brain-machine interfaces [12,13,14]. For all these applications, a key step is to control the spatial extent of an electrical stimulation, which often remains an open question.

Two types of studies have been carried out previously to estimate the spread of activation of an electrical stimulation: either experimentally or using simulations. Pioneering experimental studies carried out several decades ago have suggested that the threshold current required to elicit an action potential in a cell or a fiber is proportional to the square of the distance to the electrode [15,16,17,18,19,20]. This current-distance relationship has been verified for electrode-neuron/fiber distances smaller than a few hundreds of microns, and there is no experimental evidence that this quadratic law remains valid for larger distances. Moreover, in these studies, the exact position of the electrode with respect to the full arborization of the cell is often difficult to determine for obvious experimental constraints. More recently, modeling studies have been used to numerically estimate the neural response to an electrical stimulation. The particular advantage of these ap-

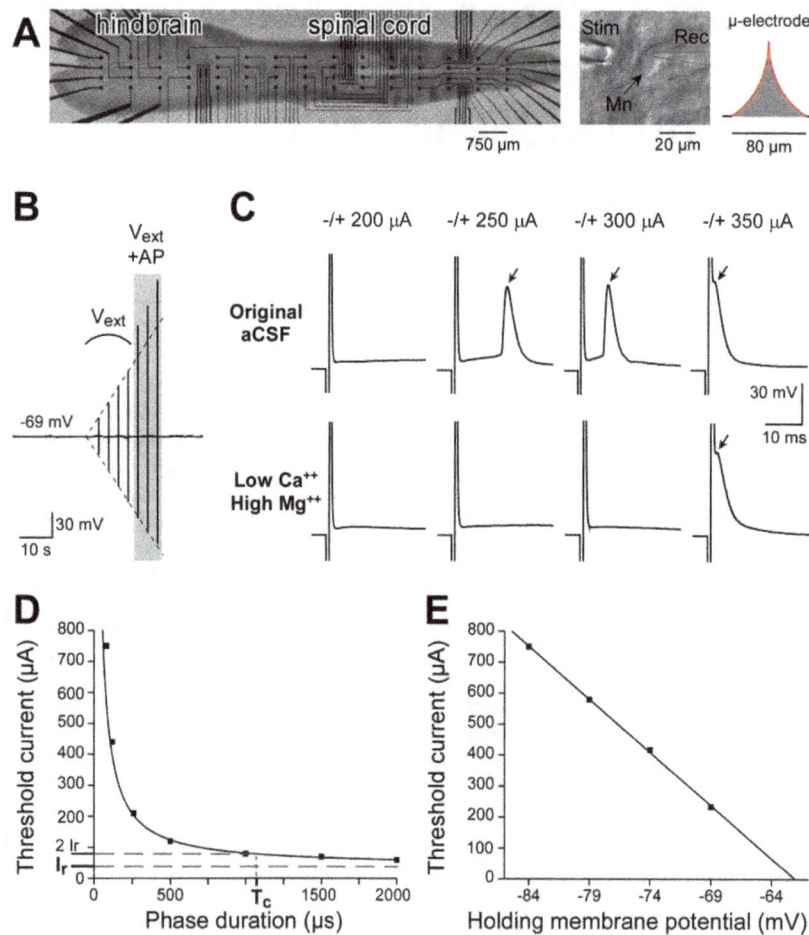

Figure 1. Experimental protocol of extracellular stimulations. A: Dorsally-opened hindbrain-spinal cords of ($n = 9$) E14.5 mouse embryos were positioned on MEAs, with the side of the central canal facing the 4×15 microelectrodes of the array (left). In each preparation, one motoneuron was patch-clamped in whole cell configuration, and stimulated using both a glass pipette positioned close to its cell body (middle) and the microelectrodes of the MEA. The theoretical shape of the 3D microelectrodes (height: 80 µm, base diameter: 80 µm) is shown on the right (the conductive part is in red, the insulated part is in black). **B**: Current-clamp recording during a series of stimuli with increasing intensities. In order to perform all trials in the same conditions, a slow holding current (time constant: 5 s) was injected to maintain V_m at a corrected value of about -69 mV. The dashed line highlights the fact that the amplitude of the signal increases linearly with the pulse intensity (here steps of 50 µA) until an action potential is elicited which superimposes on top of the extracellular potential (3 shaded events). **C**: Threshold currents (TCs) were determined by increasing regularly the intensity of the cathodic-first biphasic current-controlled pulse. Under standard aCSF, indirect activations of the neuron were observed (top row), while only direct activations were achieved using a low Ca^{2+}/high Mg^{2+} aCSF solution (bottom row). **D**: The motoneuron chronaxie (T_c) was calculated by determining TCs for different phase durations and fitting the hyperbolic Weiss's law [32]. On average, we found $T_c = 913 \pm 132$ µA ($n = 3$). Based on this value, we chose a phase duration of 1 ms for all experiments. **E**: Activation thresholds decreased linearly with holding membrane potential values.

proaches is to offer a very flexible way to predict the volumes of activated tissue (VAT) for various electrode configurations, neuronal morphologies and conductive media. It has been shown that VATs estimated with simulation approaches fit with previous experimental recordings of the literature for distances below about 200–300 microns [21,22]. Indirect validation of simulation results have also been reported by others, by correlating modeling predictions with clinical data [23]. Based on these results, simulation approaches are expected to accurately predict VATs over large distances of several millimeters [24]. However, to our knowledge, there has been no direct comparison, within the same study, between experimental and modeling prediction of the spatial extent of extracellular electrical stimulation. Moreover, the validation of computational approaches over distances of several millimeters has not been performed. This is of primary importance for instance in light of DBS paradigms aiming at stimulating

millimeter-scale regions of the central nervous system (CNS) using currents of the order of the mA [25].

The purpose of this study was thus to confront, in a common paradigm, experimental and modeling determination of the spread of extracellular neural stimulation over distances encompassing several millimeters. First, we determined experimentally the direct activation thresholds of patch-clamp-recorded spinal motoneurons subject to electrical microstimulations using MEAs. Distances up to 3 millimeters were considered. Second, we built a model mixing finite elements and compartmentalized neurons (as introduced in a previous study [26]), corresponding to the experimental paradigm. We found that, while experimental and simulation thresholds closely match for short electrode-neuron distances, computational models strongly overestimate (by two orders of magnitude) these thresholds at large distances.

Materials and Methods

1. Experimental protocols

Using whole-cell patch-clamp recordings of spinal cord motoneurons and microelectrode arrays dedicated to *in vitro* experiments, we characterized the spatial extent of current-controlled extracellular electrical microstimulations. All experimental protocols conformed to recommendations of the European Community Council Directive of November 24, 1986 (86/609/EEC) and local French legislation for care and use of laboratory animals.

a. Microelectrode array (MEA). Extracellular electrical stimulations were delivered to whole hindbrain-spinal cord preparations of mouse embryos using MEAs, which were composed of a 8 2D electrodes and a grid of 4×15 3D microelectrodes (width spacing: 250 μm, length spacing: 750 μm, see Figure 1A, left), all made of Pt (*Ayanda Biosystems – now Qwane Biosciences –*, Lausanne, Switzerland). The 2D microelectrodes had a rectangular shape (60 μm×250 μm), and their impedance, measured at 1 kHz in a saline solution, was in the range 60–100 kΩ. The 3D microelectrodes, obtained by standard isotropic photolithography [27], had a height of about 80 μm and a base diameter of about 80 μm, all the electrode being in electrical contact with the tissue (see red line in Figure 1A, right). According to this shape, these 3D microelectrodes had a theoretical surface of 7285 μm². Their impedance was in the range 100–150 kOhm. The array was surrounded by a home-made Sylgard square chamber (side length: 2 cm, height: 3 mm), and the bottom part, including electrode leads, was insulated from the extracellular medium by a 5-μm-thick SU-8 epoxy layer [27] (black line in Figure 1A, right). An external cylindrical Ag/AgCl ground electrode pellet (diameter: 2 mm, height: 4.3 mm, *World Precision Instruments*, Aston, England) was used for the return of the stimulation current.

b. Tissue preparation. Hindbrain-spinal cord preparations were dissected as described previously [26,28,29,30]. E14.5 embryos were surgically removed from pregnant OF1 mice (*Charles River Laboratories*, L'Arbresle, France) previously killed by cervical dislocation. The whole spinal cord and medulla were dissected in a cool (6–8°C) artificial CSF (aCSF) solution (pH 7.5) composed of (in mM): 113 NaCl, 4.5 KCl, 2 CaCl$_2$2H$_2$O, 1 MgCl$_2$6H$_2$O, 25 NaHCO$_3$, 1 NaH$_2$PO$_4$H$_2$O and 11 D-Glucose) gassed with carbogen (95% O$_2$ and 5% CO$_2$). The spinal cord and hindbrain were opened dorsally (open-book preparation) and meninges were removed. This preparation was then placed on the MEA (Figure 1A left), with the side of the central canal facing the microelectrodes of the array, and stabilized by a plastic net with small holes (70×70 μm²) in order to achieve a tight and uniform contact with the microelectrodes. The neural tissue was continuously perfused at a rate of 1.5 ml/min. Experiments were performed at room temperature (22±3°C).

c. Whole-cell current-clamp recordings. Spinal motoneurons ($n = 9$) were identified as previously reported [31], according to their morphological features (pear-shaped large cell body) and their disposition in columns in the lumbar ventral horn. An Olympus BX51WI microscope (*Micro Mécanique*, Évry, France) equipped with differential interference contrast (DIC) and a CCD camera (SPOT RT-SE6, *Diagnostic Instruments*, Sterling Heights, MI, USA) were used to visualize motoneurons. Patch-clamp electrodes were constructed from thin-walled single-filamented borosilicate glass (outer diameter: 1.5 mm; *Harvard Apparatus*, Les Ulis, France) using a two-stage vertical microelectrode puller (PP-830; *Narishige*, Tokyo, Japan; first step: 60.1, second step: 49.6). Their resistances ranged from 4 to 6 MΩ. Electrodes were filled

with Neurobiotin (0.4%) (*CliniSciences*, Montrouge, France) diluted in the following medium (mM): 5 NaCl, 130 K-gluconate, 1 CaCl$_2$2H$_2$O, 10 Hepes, 2 ATP Mg^{2+}, 10 EGTA (pH adjusted to 7.4 with KOH). Patch-clamp electrodes were positioned on visually identified motoneurons (Figure 1A middle) using motorized micromanipulators (*Luigs & Neumann*, Ratingen, Germany). Whole-cell recordings were performed after a high-resistance seal was achieved (between 10 and 16 GΩ). Recordings were made using an Axon Multiclamp 700B amplifier (*Molecular Devices*, Sunnyvale, CA, USA), with reference to the external cylindrical ground electrode. Data were low-pass filtered (3 kHz), sampled at 50 kHz (Micro 1401, *Cambridge Electronic Design (CED)*, Cambridge, UK) and acquired with the Spike2 software (v5.14, CED). Measurements of the somatic membrane potential (V_m) were corrected for liquid junction potentials (-14 mV). In order to perform all trials in the same conditions, a slow holding current (time constant: 5 s) was injected to maintain V_m at a corrected value of about -69 mV (maximal variations recorded at rest: ± 2 mV). This value was chosen so as to maintain V_m close enough to the neuron spike threshold (about 10 mV below threshold) in order to be able to activate the neuron at low currents, but far enough to avoid spontaneous firing. We verified that activation thresholds decreased almost linearly for increasing holding membrane potentials (Figure 1E).

d. Direct neuronal activation with extracellular electrical stimulations. After establishment of the whole-cell configuration, the original aCSF was replaced by a low calcium/high magnesium solution (composed of (in mM): 111.85 NaCl, 4.5 KCl, 0.1 CaCl$_2$2H$_2$O, 5 MgCl$_2$6H$_2$O, 25 NaHCO$_3$, 1 NaH$_2$PO$_4$H$_2$O and 11 D-Glucose), in order to block synaptic transmission and avoid network-induced activation (see Figure 1C). Current-controlled monopolar stimulations were applied between the 3D or 2D microelectrodes of the array and the external ground electrode. A glass pipette with a tip diameter of about 10-20 μm was also used to deliver electrical stimulations as close as possible to the recorded motoneuron cell body (32±3 μm, $n = 5$, see Figure 1A middle). Stimuli were delivered using the STG2008 stimulator controlled by the MC_Stimulus II v2.1.4 software (*Multi Channel Systems*, Reutlingen, Germany). We used trains of cathodic-first biphasic current pulses separated by 3 sec (phase duration: 1 ms, chosen accordingly to motoneuron chronaxie, see below). For each stimulation site, activation thresholds were determined by delivering stimulations of increasing intensities, from 50 μA to 800 μA (maximal value allowed by the stimulator) in steps of 50 μA (see Figure 1B). The sequence of stimuli was stopped when three consecutive action potentials were triggered (see shaded area in Figure 1B), and the activation threshold current (TC) was defined as the current for which the neuron triggered the first action potential. If the TC was less than 200 μA, a more precise threshold value was determined with steps of 10 μA. Stimulation sites of the MEA were systematically explored, from the closest to the farthest. At the end of the stimulation procedure (which generally lasted between one and two hours), we verified that the neuron excitability had not changed by determining again a TC with a stimulation site close to the neuron.

e. Motoneuron chronaxie. In order to determine the appropriate phase duration of the pulses, extracellular stimulations of variable durations were initially delivered to 3 motoneurons. TCs were determined for different durations (60, 120, 250, 500, 1000, 1500 and 2000 μs), and the threshold-duration curve was fitted according to the hyperbolic Weiss's law [32]:

$$I_{th}(T_{stim}) = I_r \times \left(1 + \frac{T_c}{T_{stim}}\right), \qquad (1)$$

where T_{stim} is the stimulation duration, I_{th} is the current activation threshold, I_r is the neuron rheobase and T_c is the neuron chronaxie. This is further illustrated in Figure 1D, with a neuron for which we found a chronaxie of 1037 μs. On average, we determined a chronaxie of 913±132 μA (mean ± s.e.m., $n = 3$). Based on this value, a 1-ms phase duration was chosen for all experiments.

f. Activation threshold analysis. Our main goal was to determine the evolution of TCs with distance from the stimulation electrode. The signal recorded by the patch-clamp pipette is the sum of two components: the extracellular potential V_{ext} due to the stimulation and the motoneuron membrane potential (see Figure 1B). During the 2-ms-long stimulus, the signal mainly reflected the extracellular electrical potential (namely, the stimulation artifact), which was characterized by a return to baseline with a short time constant, of the order of hundreds of μs (see for instance Figure 1C, $I = -/+$ 200 μA). After the end of the stimulus, action potentials elicited by suprathreshold stimulations were undoubtedly detected, even when action potential peaks occurred before the end of the stimulus (see arrows in Figure 1C). Furthermore, the value of V_{ext} corresponding to the TC was also measured at the end of the cathodic phase of the stimulus. The distance between the motoneuron cell body and each stimulation electrode was calculated using 2D images taken in the focal plane using the CCD camera and the ×10 microscope objective. The 3D distance between the cell body and the tip of the electrode was then calculated from its 2D position and altitude above the MEA substrate. The TCs and threshold V_{ext} values determined for 9 motoneurons were analyzed together, data being binned into distance classes. In each class, averages and standard errors were determined.

g. Immunolabeling. After each experiment, immunostaining was processed on the whole spinal cord in order to reveal the morphology of the recorded E14.5 neuron ($n = 9/9$) as well as the location of voltage-gated sodium channels ($n = 3/9$). The spinal cord was fixed in 2% paraformaldehyde for 1 h at 4°C, and then rinsed three times with 0.1 M PBS (composed of (in mM): 77 Na2HPO4, 23 NaHPO, 154 NaCl, pH: 7.4). It was then immersed in 0.1 M PBS containing 10% goat serum and 0.4 Triton X-100 (T 8787, *Sigma*, St Louis, MO, USA), and incubated for 12–24 h at 4°C with a mouse monoclonal anti-sodium channel antibody (Pan Nav antibody, S 8809, Sigma) and Cy3-Streptavidin (1:400, *Invitrogen SARL*, Cergy Pontoise, France) to reveal the Neurobiotin-injected motoneuron. After three rinses (at least 30 minutes each), the spinal cord was incubated for 1 h at room temperature in a solution containing the FITC-coupled secondary antibody (Goat anti-mouse, Alexa Fluor 1:400, *Molecular Probes*) and Cy3-Streptavidin. After three rinses in 0.1 M PBS (at least 10 minutes each) and a last rinse in distilled water, the spinal cord was mounted in a mixture containing 90% glycerol and 10% PBS to which 2.5% 1,4-diazabicyclo[2,2,2]octane (DABCO, *Sigma*) was added, in order to reduce the rate of fluorescence quenching. All incubations and rinses were performed sheltered from light.

h. Confocal microscopy. Preparations were imaged with a BX51 Olympus Fluoview 500 confocal microscope equipped with blue argon (488 nm) and green helium–neon (546 nm) laser sources to visualize the pan/FITC and Neurobiotin/Cy3 labelings, respectively. Neuron morphologies and sodium channels immunolabelings were acquired with a ×60 oil-immersion objective (serial optical section thickness: 0.2 μm). Images were averaged over 2 scannings (Kalman filter).

i. Neuron morphology reconstruction. One motoneuron was fully reconstructed to build a compartmental neural model (Figure 2B₁, see modeling part of this study described below). For this purpose, 2 overlapping stacks of 246 confocal images were acquired at ×60 magnification (serial optical section thickness: 0.2 μm), each stack covering more than half of the whole neuron morphology. These stacks were then merged into a single stack (using the *vias* software, v2.1, freely downloadable at http://www.mssm.edu/cnic/tools-vias.html), which was then loaded into the Neurolucida software (v 6.02.2, *MBF Bioscience*, Williston, USA) to interactively reconstruct the neuron arborization. Finally, the morphology was converted to a.hoc file compatible with the NEURON software, v7.2 [33] using the cvapp software (v1.2, freely downloadable at http://www.compneuro.org/CDROM/docs/cvapp.html).

2. Modeling approach: General approach

We further performed simulations to compare theoretical and experimental estimations of the spatial extent of microstimulation. For that purpose, we used a classical 2-stage model made of a finite element model (FEM) for the calculation of the electrical potential field in the tissue, and a compartmentalized neuron model for the computation of the response of a neuron stimulated by this field (see [34] for a review of the biophysical and mathematical background of these models). The calculation of the potential field was performed either in the case of an homogeneous, isotropic and purely resistive tissue, or in the case of an anisotropic, heterogeneous, or leaky dielectric tissue.

3. Calculation of the potential field in the case of an homogeneous, isotropic and purely resistive tissue

The electrical potential field generated in the neural tissue by an electrical stimulation was computed using a 3D FEM using COMSOL Multiphysics 3.4 (COMSOL AB, Stockholm, Sweden) interfaced with Matlab 7.7 (The Mathworks, Natick, USA), under Linux (Fedora 14).

a. Model geometry. The 3D model geometry corresponded to the experimental MEA, including the chamber, the neural tissue, one 3D stimulation electrode and the ground electrode. The outer limits of the model corresponded to the inner geometry of the MEA square chamber (side length: 2 cm, height: 3 mm). This volume was subdivided into two regions, representing the aCSF solution and the neural tissue. The neural tissue had a parallele-pipedical shape whose dimensions fitted the E14.5 embryonic mouse hindbrain-spinal cord preparation (length: 13 mm, width: 2 mm, height: 200 μm, see Figure 2A₁). We verified that modeling a more realistic shape did not alter the distribution of the electrical potential within the tissue. A single stimulation electrode was modeled on the bottom surface of the chamber, represented by a 3D conical boundary, whose dimensions corresponded to the actual MEA used experimentally (base diameter: 80 μm, height: 80 μm, see Figure 2A₂). The external ground electrode was modeled by a cavity inside the domain representing its immersed part in contact with the aCSF solution (diameter: 2 mm, height: 2.7 mm).

b. Volume equation. The finite element model solved the homogeneous Poisson equation:

$$-div(\sigma \nabla V_{ext}) = 0, \qquad (2)$$

where V_{ext} is the extracellular electrical potential. The electrical

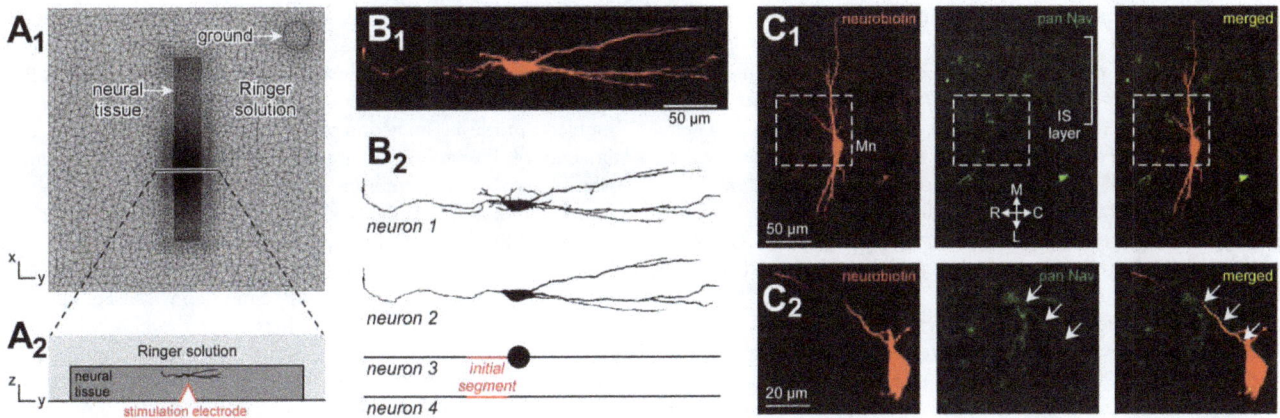

Figure 2. Modeling approach. Simulations were performed to compare theoretical and experimental estimations of the spatial extent of electrical microstimulation. **A**: A 3D Finite Element Model (FEM) was used to compute the electrical potential field generated in the neural tissue by an electrical stimulation. The model geometry corresponded to the experimental MEA, including the chamber, the neural tissue, one 3D electrode (which shape was approximated by a cone with a height of 80 μm and a base diameter of 80 μm) and the cylindrical ground electrode (height = 2.7 mm, diameter = 2 mm). This volume was subdivided into two regions, representing the aCSF solution (side length: 2 cm, height: 3 mm, conductivity $\sigma = 1.65$ S/m) and the neural tissue (length: 13 mm, width: 2 mm, height: 200 μm, $\sigma = 0.10$ S/m). **B**: The FEM-calculated extracellular potential field was then applied to compartmentalized motoneuron models in order to determine theoretical activation thresholds. Four different neurons were considered (B2), the most complex of them (1895 compartments) being a 3D reconstruction of a Neurobiotin-injected motoneuron considered in the experimental study (B1). These neuron models were equipped with passive properties in the soma, dendrites and most of the axon. Voltage-dependent Na$^+$ and K$^+$ channels were located in the initial segment of the axon (length = 34 μm), leaving from the soma, in accordance with the immunolabelings (C). **C**: Confocal projection of a neurobiotin-injected motoneuron (red) and anti-sodium channel labeling (pan Nav, green). C1: Global view of the neuron and the layer of initial segments (IS), oriented in the medio-lateral direction (60 optical sections). C2: Close-up in the perisomatic region (6 optical sections), revealing the localization of the sodium channels in the IS of the axon (merged image, yellow, see arrows). In $n = 3$ experiments, the IS had a length of 34 +/− 3 μm. [C: caudal; M: median; L: lateral; R: rostral].

conductivities σ of the aCSF and neural tissue were supposed homogeneous and isotropic in each region. The conductivity of the aCSF was set to $\sigma_{aCSF} = 1.65$ S/m, a value measured with a conductimeter at room temperature. The conductivity of the spinal cord was set to $\sigma_{tissue} = 0.10$ S/m, a value close to that estimated in a previous study [26].

c. Boundary conditions. Insulating boundary conditions (BCs) were assigned to the circumference of the chamber, the air-aCSF solution interface (top part of the chamber), and the floor of the chamber:

$$\sigma \nabla V_{ext} \cdot \mathbf{n} = 0. \qquad (3)$$

The conductive elements (ground and stimulation electrodes) were assigned Robin BCs (derived from Ohm's law at the electrode-electrolyte interface), as previously validated by comparing experimental measurements and modeling calculations of the potential field [26]:

$$\sigma \nabla V_{ext} \cdot \mathbf{n} + g V_{ext} = g V_{metal}, \qquad (4)$$

where g is the surface conductance of the electrode-electrolyte interface. The metal voltage in Equation 4 was set to $V_{metal} = 0$ for the ground and $V_{metal} = V_{stim}$ for the stimulation electrode, with V_{stim} adjusted to have a normalized current of 1 μA. The surface conductances of the electrodes were set respectively to $g_{stim} = 338$ S/m^2 and $g_{ground} = 975$ S/m^2 [26].

d. Mesh and solver. The 3D geometry of the model was densely meshed with 450628 tetrahedral Lagrange P2 elements and 629034 degrees of freedom (Figure 2A$_1$), in order to minimize the numerical error on the electrical potential V_{ext} and its

derivatives. The problem was solved with the SSOR-preconditioned conjugated gradient algorithm.

4. Calculation of the potential field in the case of an anisotropic, heterogeneous, or leaky dielectric tissue

The influence of different electrical properties of the neural tissue on the potential field were further evaluated separately by modeling the spinal cord either with a globally anisotropic conductivity, or with local variations of the (isotropic) conductivity, or by taking into account dielectric properties of the tissue. Each of these models was extended from the original model described above, as follows:

a. Anisotropy. The originally isotropic conductivity of the tissue was replaced by an anisotropic model, made of a longitudinal value (in the x direction) equal to 0.33 S/m and a transverse value (in the y and z directions) equal to 0.083 S/m, according to a former model of the spinal cord white matter [22].

b. Heterogeneity. In the original model geometry, we added two series of ten 20-μm-side cubes, with an inter-cube space of 20 microns, located respectively at 500 microns and 3 millimeters from the stimulation electrode. These small domains were assigned conductivities equal to 0.1 and 10 times the original tissue conductivity (0.1 S/m), in alternation. In this model, the neuron model was positioned on the edge of these heterogeneities in order to maximize their effect on the neural response.

c. Dielectric effects. Dielectric properties, neglected in the quasi-static approximation leading to the Poisson equation (Eq. 2), were finally included in the neural tissue domain, which led to the resolution of the Helmholtz equation:

$$\nabla^2 V_{ext} + k^2 V_{ext} = 0, \qquad (5)$$

where $k^2 = -j\omega\mu(\sigma + j\omega\varepsilon)$ is the propagation constant, $\omega = 2\pi f$ is the angular frequency of the stimulation, μ is the magnetic permeability of the tissue and ε its dielectric permittivity [34,35]. The spectrum of the cathodic-anodic pulse used in this study displays a main component at $f_1 = 500$ Hz and secondary harmonics at $f_3 = 1500$ Hz, $f_5 = 2500$ Hz, and every f_{2p+1}. Their amplitude decreased in a cardinal sine so that 99% of the energy of the pulse is contained in the harmonics below 10500 Hz (81% being already contained in the main frequency 500 Hz). For that reason, we focused on the field generated at 500 Hz and 10500 Hz. The magnetic permeability of the spinal cord was set to that of the vacuum ($\mu = 4\pi \times 10^{-7}$ H/m) and its dielectric permittivity was set to $\varepsilon = \varepsilon_0 \times \varepsilon_r$, with $\varepsilon_0 = 8.854 \times 10^{-12}$ F/m (dielectric permittivity of the vacuum) and $\varepsilon_r = 3 \times 10^5$ at 500 Hz and 2×10^4 at 10500 Hz [35]. In a subsequent model, we also modeled an heterogeneity of the dielectric permittivity by adding an 80-μm-sided cube at 500 microns or 3 millimeters from the stimulation electrode, in which ε_r was set to 10 or 100 times its original value. In this model, the neuron was positioned on the edge of this heterogeneity in order to maximize its effect on the neural response.

5. Computation of the neuronal response

We used a cable equation framework to compute the membrane response of a non-myelinated compartmentalized neuron (indeed, motoneurons are not myelinated at embryonic stages) embedded in the extracellular potential field calculated as above. Simulations were run using the NEURON software, v7.2 [33].

a. Neuron morphologies. Four neuron morphologies of decreasing dendritic complexities were used for the compartment model (Figure 2B$_2$). The most complex morphology (neuron 1, $N = 1895$ compartments) was obtained from a full reconstruction of stacks of confocal images (Figure 2B$_1$), as detailed in the Experimental Methods. Neuron 2 was an approximation of neuron 1, with only its main morphological features (soma, axon and the main dendrites, $N = 1250$ compartments). Neuron 3 was a straight neuron displaying the same basic morphological characteristics as Neuron 2 (soma diameter, axon and dendrite length, 300 1-μm-long compartments). Finally, Neuron 4 was a straight fiber of same length as Neuron 3 (300 1-μm-long compartments).

b. Electrical characteristics. Each compartment was assigned identical passive characteristics: the surface capacitance was set to its classical value ($c_m = 1$ μF/cm^2), the intracellular resistivity was set to $\rho_i = 100$ Ω.cm, which was within the range of values used in the literature (33–300 Ω.cm) [36,37,38,39,40]. Using these parameters, the surface leakage conductance was then estimated to $g_l = 3.10^{-5}$ S/cm^2 by comparing the experimental and modeling responses of a neuron to a hyperpolarizing pulse of −50 pA. Active Hodgkin-Huxley-like [41] sodium and potassium conductances were set to the initial segment of the axon ($g_{Na} = 0.24$ S/cm^2, $g_K = 0.036$ S/cm^2). This initial segment, which corresponded to a 34-μm-long portion of the axon starting from the soma, was chosen in accordance with the experimental anti-sodium channel immunolabelings (see Figure 2C). In this active region, the leakage potential was calculated so that the passive leakage current compensated the non-zero sodium and potassium active currents at rest, hence imposing a uniform value of the resting potential over the neuron ($E_l = +13.88$ mV, $E_{Na} = +50$ mV, $E_K = -77$ mV, $V_{rest} = -69$ mV).

c. Extracellular stimulations. The extracellular potential field V computed with the FEM for a nominal current of $I = +1$ μA was interpolated at the center of each compartment j (V_j) and assigned with the extracellular mechanism in NEURON. As in the experimental protocol, cathodic-first biphasic stimulations

with phase duration of 1 ms were used, that is the extracellular potential of compartment j was 0 before the stimulus, $-I \times V_j$ during the cathodic phase, $+I \times V_j$ during the anodic phase, and 0 after. Starting from $I = 1$, the amplitude I of the stimulus was recursively doubled until firing an action potential in the soma (detected after the end of the anodic phase, as follows: An action potential occurred when the membrane potential V_m at time t_k was greater than V_m at times t_{k-1} and t_{k+1}, and above a given threshold, set to −20 mV). Then, the current I was successively decreased and increased by dichotomy until finding the threshold current, with a relative precision of 0.1%. A 50-μs time step was used, allowing a reduced error on the activation threshold estimation (using a 10 times smaller time step led to a threshold difference of less than 1%).

d. Threshold-distance curves. In these simulations, the neurons were oriented parallel to the y-axis in order to mimic the position of experimentally recorded motoneurons, which laid perpendicular to the central canal of the spinal cord (confocal observations). They were moved along the x-axis (in the direction of the spinal cord) at $z = 130$ μm, in order to determine the evolution of the activation thresholds as a function of the electrode-neuron distance. The height of 130 μm (about 50 μm above the electrode tip) corresponded to the actual depth of the experimentally stimulated motoneurons.

Results

1. Experimental determination of threshold current-distance curves

Extracellular electrical stimulations were applied to spinal cords of mouse embryos (embryonic age E14.5, $n = 9$) using MicroElectrode Arrays (MEAs). In each experiment, a single motoneuron was patch-clamped in whole-cell configuration, and the threshold current necessary to elicit an action potential was determined for as many electrodes of the array as possible. Figure 3A$_1$ gives an example of thresholds obtained for most electrodes of an array and a stimulating pipette located 20 microns from the soma (see Figure 1A) of the recorded neuron. Surprisingly, nearly all electrodes of the array could be used to elicit a spike with currents below 800 μA. Moreover, threshold values were almost independent of the electrode-neuron distance when the electrode was beyond about 1 millimeter from the neuron. This phenomenon was observed across all experiments, as shown in Figure 3A$_2$ reporting the average threshold-distance curve obtained from all neurons considered in this study ($n = 9$). On average, we found that the threshold at 2 millimeters (550 μA) was only 2.75 times stronger than that at 250 microns (200 μA). This unexpected result raises a strong issue for achieving focal stimulation of all cells in the vicinity of a stimulating electrode. Indeed, as shown in Figure 1E, thresholds depend linearly on the resting potential of the neuron prior stimulation, and in our case, the slope was found to be about 35 μA/mV. Thus, while a current of 200 μA is required for an electrode to trigger a neuron located at 250 microns with a resting potential of −69 mV, the same stimulus would be sufficient to elicit a spike in a neuron located at 2 millimeters with a resting potential of −59 mV.

Quadratic threshold-distance curves have been introduced based on an initial intuition that the quantity that determines a threshold is the value of the gradient of the extracellular potential at the neuron location. This hypothesis implies that the gradient of the extracellular potential should reach the same value each time the neuron reaches its firing threshold. That is $|\beta \nabla V_{ext}|_{threshold} = C$, where the constant C does not depend on the stimulating electrode. However, here, we did not find that the gradient of the

A₁

A₂

B

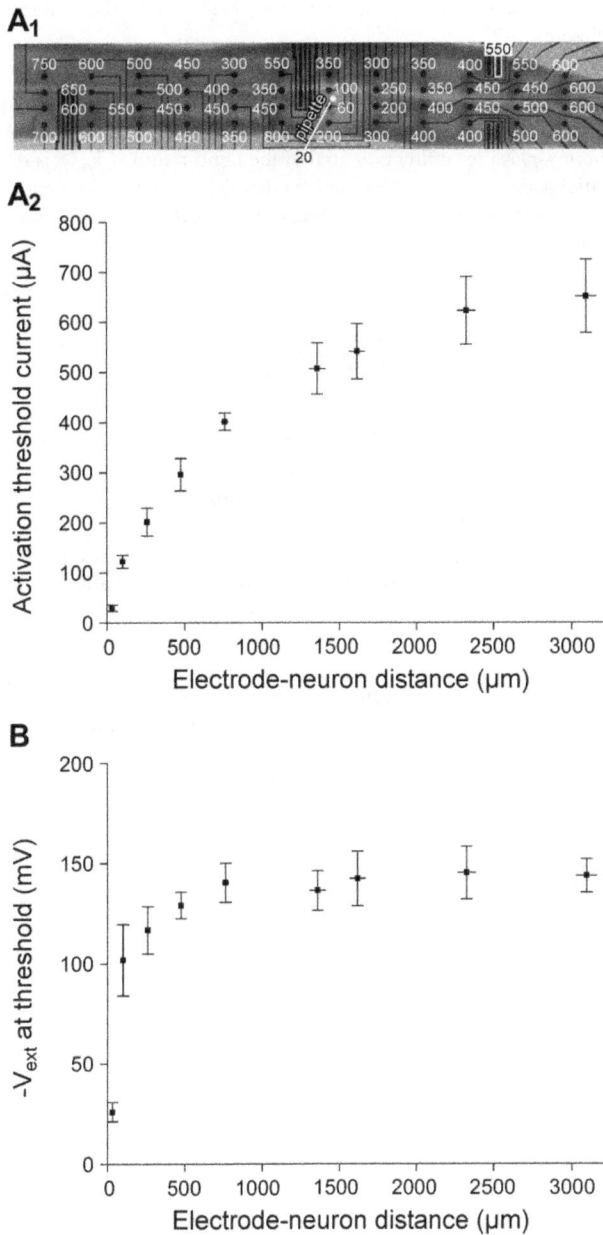

Figure 3. Experimental results. A: Experimental determination of direct activation threshold currents. For each patch-clamped motoneuron, direct activation threshold currents (TCs) were determined by systematically scanning the electrodes of the MEA. A1: Example of thresholds obtained for most electrodes of an array (ranging from 60 to 800 μA) and a stimulating pipette located 20 μm from the soma of the recorded neuron (20 μA). A2: The threshold-distance curve, obtained by pooling the results of 9 experiments, displays a non-quadratic increase of TCs with distance. For instance, on average, TCs reached 200 μA at 250 μm, and 550 μA at 1.6 mm, which is only 2.75 higher. **B: Evolution of the extracellular potential values at threshold with distance.** The amplitude of the extracellular potential field was measured at the end of the cathodic phase of the threshold stimulation. The average potential-distance curve (n = 9) shows that this value increases sharply in the first hundred microns and then stagnates at about 140 mV, meaning that action potentials are elicited at the moment when V_{ext} reaches this value.

extracellular potential was constant at threshold. Rather, we found

that beyond 750 microns, the value of the extracellular potential itself was nearly constant around $V_{ext} \approx -140$ mV (Figure 3B).

2. Modeling results

We next compared these experimental results with modeling predictions. For this purpose, we developed a modeling approach, where we estimated activation thresholds for compartmentalized neurons placed in a potential field calculated using finite element models similar to those already used in the literature [26,38,42].

a. Discrepancy between experimental and simulated activation thresholds. Threshold currents were computed for different positions of model neuron 1 along a line passing over a stimulation electrode (Figure 4A). We determined TCs for electrode-neuron distances ranging from 50 to 3200 microns (Figure 4B). We found that modeled TCs were close to experimental ones for short distances, in the first hundreds of microns. However, for larger distances (above 1 millimeter), TCs reached and stabilized to values of the order of 10^5 μA well above experimental thresholds. Thus, simulated thresholds were dramatically over-estimated at these distances (more than 100 times) compared to experimental ones.

b. Influence of model parameters on simulated thresholds. We then tested whether some model parameters could be responsible for this strong discrepancy between experiments and modeling, and found that this was not the case (Figure 5). First, we tested the influence of the complexity of the neuron geometry. We replaced the original neuron morphology (model 1) with simpler morphologies made of a reduced number of dendrites (neurons 2, 3 and 4, see Figure 2B₂). As shown in Figure 5A, this did not affect strongly the simulated TCs. Second, we modified the electrical conductivity σ_{tissue} of the neural tissue,

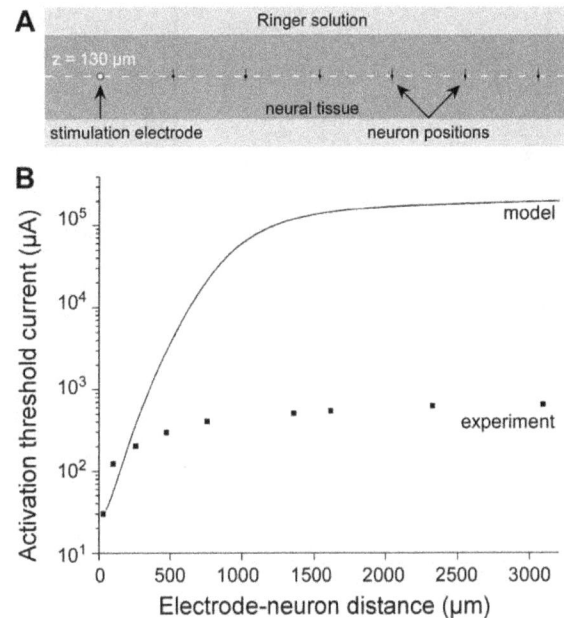

A

B

Figure 4. Modeling prediction of activation threshold currents. **A**: Simulated TCs were computed for different positions of Neuron 1 (see Figure 2) along a line passing 50 μm above the tip of the stimulation electrode. **B**: Theoretical TCs (continuous line) were found to be close to experimental TCs (square symbols) for short electrode-neuron distances (typically below 250 μm). For increasing distances, theoretical TCs reached levels several orders of magnitude above experimental ones (note the logarithmic scale).

with values ranging from 0.10 S/m (the original value) to 1.65 S/m (the conductivity of the aCSF solution). Increasing σ_{tissue} increased TC values for short electrode-neuron distances (Figure 5B), due to the fact that a greater current delivered to the electrode was necessary to achieve similar amplitudes of the electrical potential field. However, TCs were not affected for high distances (above about 1 millimeter). Since neither the neuron morphology nor the tissue conductivity substantially influenced TCs at large distances, we finally tested the influence of the passive and active properties of the neuron. We found that increasing (respectively decreasing) the leakage conductance by one order of magnitude globally scaled the thresholds by a factor comprised between +16% and +18.5% (respectively −3% and −4%) depending on the electrode-neuron distance (Figure 5C$_1$, note the vertical shift in logarithmic scale). Similarly, decreasing (respectively increasing) the maximum sodium conductance by a factor 2 globally scaled the thresholds by a factor comprised between +36% and +40% (respectively −26.8% and −27.5%, see Figure 5C$_2$). Therefore, modifying the electrical properties of either the neuron or the surrounding medium did not change simulated TCs substantially enough as to explain the strong discrepancy with the experimental TCs.

c. Link between the membrane response and the extracellular potential field. To understand why such large currents are necessary to activate modeled neurons at large distances, we determined the link between the profile of the membrane potential V_m and the profile of the extracellular potential V_{ext} along the neuronal geometry. In a previous study, we showed theoretically that the membrane potential may often be

inferred directly from the potential field using the "mirror estimate" [42]:

$$V_m(s) = V_{rest} - V_{ext}(s) + \langle V_{ext} \rangle, \tag{6}$$

where V_{rest} is the resting potential of the neuron and $<V_{ext}>$ is the spatial average of the extracellular potential field over the neuron morphology, weighted by the local diameters $d(j)$ of the N neuron compartments:

$$\langle V_{ext} \rangle = \sum_{j=1}^{N} d(j)V_{ext}(j) \bigg/ \sum_{j=1}^{N} d(j) \tag{7}$$

Here, we verified that the mirror estimate was a good predictor of the membrane polarization in the present case. For sake of simplicity, and because the neuronal geometry did not influence noticeably the TCs at large distances, we considered only Neuron 3, which displays the main elements of the reconstructed motoneuron. Figures 6A shows the profile of V_{ext} values along the neuron model located at a distance of 500 microns, for a current of −/+442 μA that is large enough to trigger an action potential experimentally (see Figure 3). Figure 6B shows the membrane potential predicted by the mirror estimate and the actual membrane potential at the end of the cathodic phase of the stimulus (1 ms), calculated by solving the full cable equation with the NEURON software. It can be seen that both curves are almost

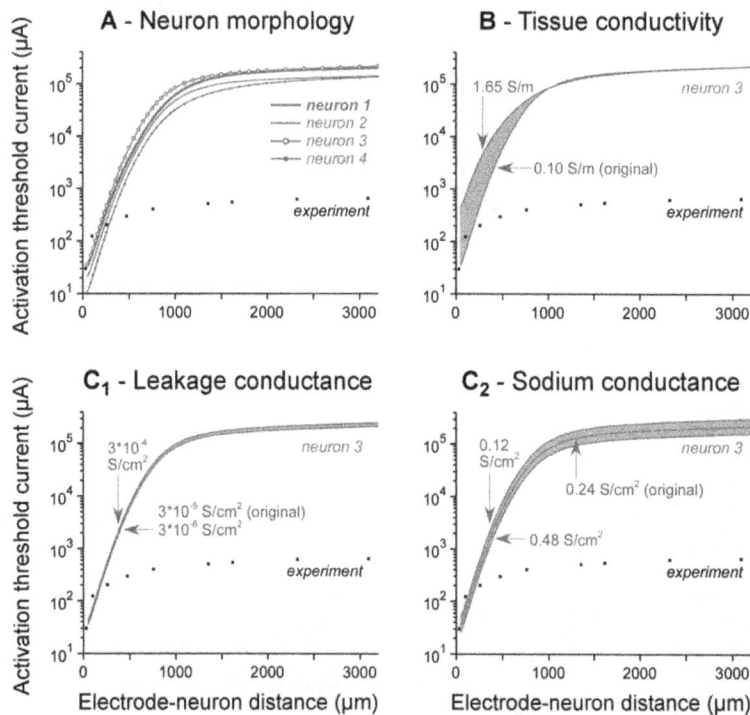

Figure 5. Influence of model parameters on modeling thresholds. A: Different morphologies were considered, from the most complex one (model neuron 1) to the simplest straight fiber (model neuron 4, see Figure 2B). The resulting TC-distance curves were not substantially altered by the choice of the morphology. **B:** Increasing the conductivity of the neural tissue to that of the aCSF (1.65 S/m) increased TCs for short electrode-neuron distances only (below about 1 mm). **C.** Increasing or decreasing the leakage conductance by one order of magnitude only altered the amplitude of the TCs by a factor comprised between −4% and +18.5% (C1). Finally, increasing or decreasing the maximum sodium conductance by a factor of 2 only altered the amplitude of the TCs by a factor comprised between −27.5% and +40% (C2). Overall, modifying any of these model parameters did not change TC values down to levels comparable to experimental ones at large distances (over 1 mm).

superimposed, assessing the validity of the mirror estimate in this case.

The *mirror estimate* states that the membrane polarization is directly related to how much the potential field fluctuates along the neuron, rather than to the absolute value of the potential itself. At large distances, the potential field around the neuron is nearly constant with only weak spatial fluctuations along the neuron (Figure 6A). Consequently, the membrane potential was also nearly constant. Indeed, as can be seen in Figure 6B, a 442-µA-stimulation induced a fluctuation of the membrane potential of only about 2mV with respect to the resting potential. This explains why reaching the action potential threshold in the neuron model could only be achieved with very high currents (about 6 mA at 500 microns).

d. Influence of anisotropic, heterogeneous and dielectric properties of the tissue.   Because the mirror estimate holds in the present situation, understanding how the potential field is affected by modifications of the FEM model helps predicting the consequences of these modifications on the neural response. We thus further modified the original FEM model to explore whether anisotropy, heterogeneities or dielectric properties could lead to more ample fluctuations of V_{ext} along the neuron and thus to smaller TCs, as observed experimentally.

First, we replaced the isotropic conductivity of the spinal cord with an anisotropic model of the spinal white matter (see Methods). We found that while the induced fluctuations of V_{ext} were large at 500 microns (Figure 7A, left), these were very small at 3 millimeters from the electrode (Figure 7A, middle) as with the isotropic model. Accordingly, TCs were reduced for small distances but unchanged at large distances (Figure 7B).

Second, we evaluated the influence of local heterogeneities of the tissue, modeled by embedding small cubic domains with variable conductivity in the spinal cord model (see Methods and Figure 8A). As illustrated in Figure 8B, this induced fluctuations of V_{ext} along the neuron geometry (red curves) compared to the smooth profile obtained with the homogeneous model (blue curves). However, these variations were small and led to a modest reduction of the threshold currents as compared to the homogeneous model. This reduction was more pronounced at short than at large distances: 2.1 mA vs 3.8 mA at 500 microns from the electrode (Figure 8B, left) and 87.4 mA vs 100.0 mA at 3 millimeters (Figure 8B, right). It should be noted that V_{ext} variations were maximum along a line passing on the edges of the heterogeneities (Figure 8A) and rapidly vanished away from these edges. Yet, this effect remained far too limited to explain the much smaller threshold currents observed experimentally.

Third, we tested if the discrepancies between the experimental and modeled TCs could stem from the absence of propagation, capacitive, and inductive effects in the model of neural tissue. To test this hypothesis, we modified the FEM equation to take into account the dielectric permittivity ($\varepsilon = \varepsilon_0 \times \varepsilon_r$) of the tissue, which led to the resolution of the frequency-dependent Helmholtz equation (see Methods). This equation was initially solved at a frequency of 500 Hz, which is the main component of the spectrum of the biphasic pulse used in this study. As shown in Figure 9B, the resulting potential field had a magnitude very close to that obtained with the resistive model, both at small and large distances (compare the blue and black lines). Moreover, as illustrated in Figure 9C, the phase shift was also extremely limited, with values ranging between −0.5° and +0.5° and remaining almost uniform along the neuron geometry, meaning that the time courses of V_{ext} were in phase for all compartments (an anti-phase configuration would correspond to a phase shift of 90°). These similarities in magnitude and phase between the resistive

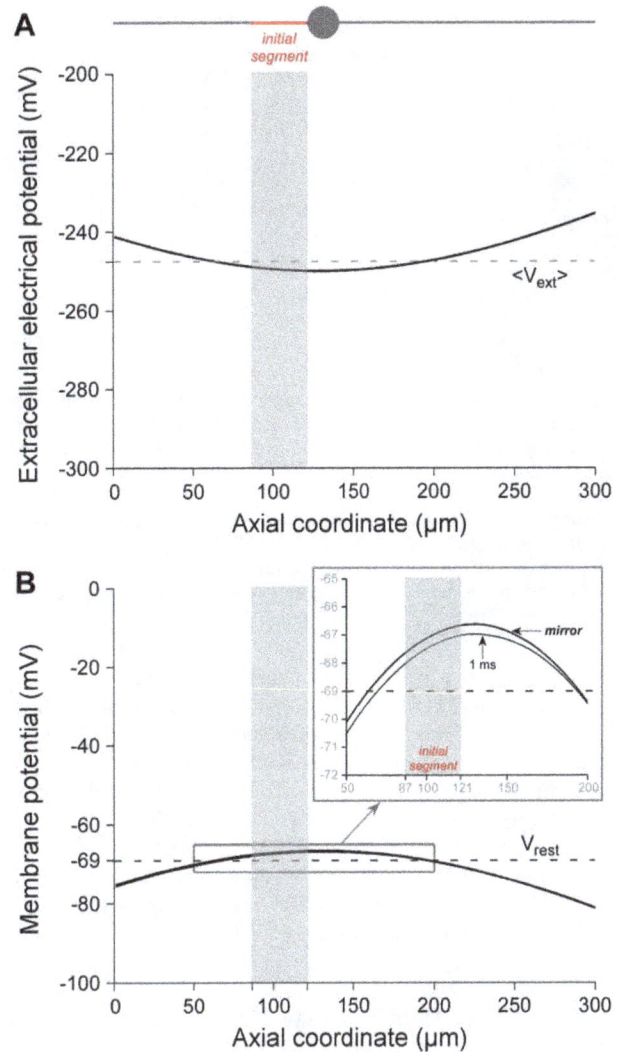

Figure 6. Membrane polarization is well predicted by the *mirror* estimate, and is small at large distances. Longitudinal profiles of extracellular potential V_{ext} (**A**) and membrane potential V_m (**B**) are plotted for the neuron model 3 (illustrated at the top of the figure), located at 500 µm from the stimulation electrode, for a current of −/+ 442 µA, a level that elicits a spike in experiments. Both profiles are the *mirror* image of each other. The inset in B shows that the membrane potential at the end of the cathodic phase of the stimulus is well predicted by the *mirror* estimate (Eq. 6). It can further be noted that the extracellular potential plotted in panel A displays very small variations around its spatial mean (dashed line). Similarly, the variations of V_m around the resting potential are only of a few mV (B), which is not enough to reach the activation threshold and elicit a spike. At this distance, a current of 6080 µA is actually required to elicit an action potential in the modeled neuron, about 15 times higher than the experimental thresholds.

and dielectric models were also observed for higher frequencies up to 10500 Hz (not shown).

Finally, we evaluated the effect of possible heterogeneity of the dielectric permittivity, by including a cubic domain in the spinal cord model (see Figure 9A and dark gray bands in Figure 9B), in which ε_r was set to either 10 or 100 times its initial value (3×10^5 at 500 Hz). The first model ($\varepsilon_r \times 10$) resulted in electrical potential profiles (thin red curves) close to those obtained with the homogeneous dielectric model, both at short and large distances,

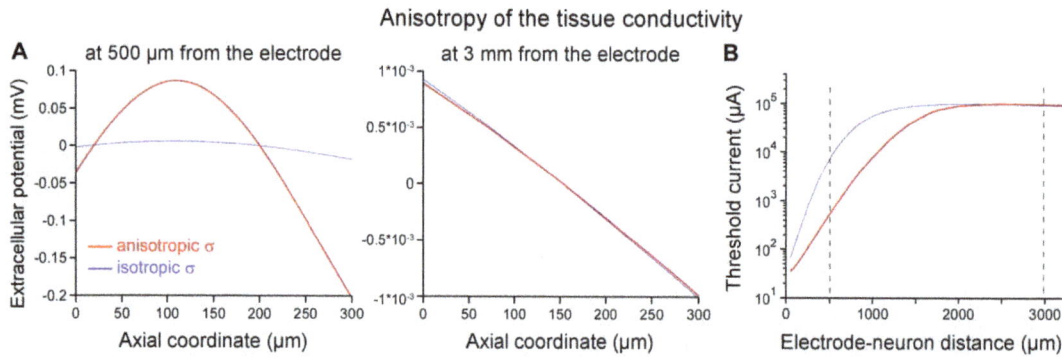

Figure 7. Influence of the anisotropy of the tissue conductivity on the potential field (A) and threshold-distance relationship (B). The originally isotropic tissue model was replaced with an anisotropic volume conductor with a longitudinal (along the x direction) conductivity of 0.33 S/m and a transverse (along the y and z directions) conductivity of 0.083 S/m. **A:** The resulting profiles of the potential field V_{ext} (centered on its spatial mean) along Neuron 3 (oriented along the y direction) are plotted at small (500 μm, left) and large (3 mm, right) distances from the electrode. Large variations of V_{ext} are observed at small distances, but not at large distances, where the isotropic and anisotropic profiles are almost indistinguishable (compare red and blue lines). **B:** Activation thresholds currents as a function of the electrode-neuron distance for an isotropic (blue) and anisotropic (red) tissue. Consistently with the V_{ext} profiles shown in A, TCs in the anisotropic case were smaller than in the isotropic case at small distances (541 μA vs. 7280 μA at 500 μm), but almost identical at large distances (96.1 mA vs. 93.4 mA at 3 mm). The 2 vertical dashed lines correspond to the neuron positions at which the V_{ext} profiles are plotted in A.

and characterized by a small phase shift. Similarly to the homogeneous dielectric model, this was observed at all frequencies. The second model ($\varepsilon_r \times 100$) induced larger modifications, essentially on the potential field magnitude and at short distance from the stimulation electrode (Top left, thick red curve). However, at a large distance of 3 millimeters, the amplitude of the potential field along the neuron remained again within its original range, and the phase shift was only very slightly modified.

Overall, taking into account anisotropy, heterogeneities, or dielectric properties of the tissue did not bring noticeable

modifications to the potential field that could explain the discrepancies between experimental and modeled TCs.

Discussion

A major issue in CNS extracellular microstimulation using MEAs is to control the spatial extent of tissue activation. Ideally, each electrode of an array should be considered as an individual stimulation pixel acting exclusively on cells located in its vicinity but not on cells located in the vicinity of neighboring electrodes. However, the spatial influence of an electrode remains largely

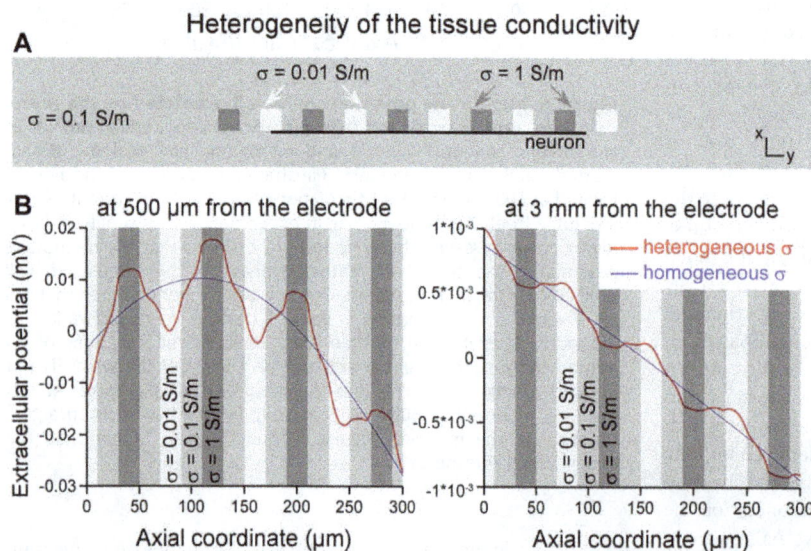

Figure 8. Influence of the heterogeneity of the tissue conductivity on the potential field and threshold currents. A: In the originally homogeneous tissue (with $\sigma = 0.1$ S/m), ten 20-μm-sided cubes were added, in which the conductivity was alternatively set to 0.01 S/m and 1 S/m. The neuron model 3 was positioned at the edge of these cubes to maximize the effect of the heterogeneity. **B:** The extracellular potential field (centered on its spatial mean) is plotted along the neuron model for distances of 500 μm (left) and 3 mm (right) from the stimulation electrode. At 500 μm, the profile of V_{ext} (in red) displays larger variations than that obtained with the homogeneous model (in blue), with large gradients in the regions of low conductivity (light gray bands) and small gradients in the regions of high conductivity (dark gray bands). These variations result in smaller threshold currents than in the homogeneous model (2.1 mA vs 3.8 mA, respectively). At 3 mm, these field variations are even smaller leading to little decrease of the threshold currents (87.4 mA vs 100.0 mA).

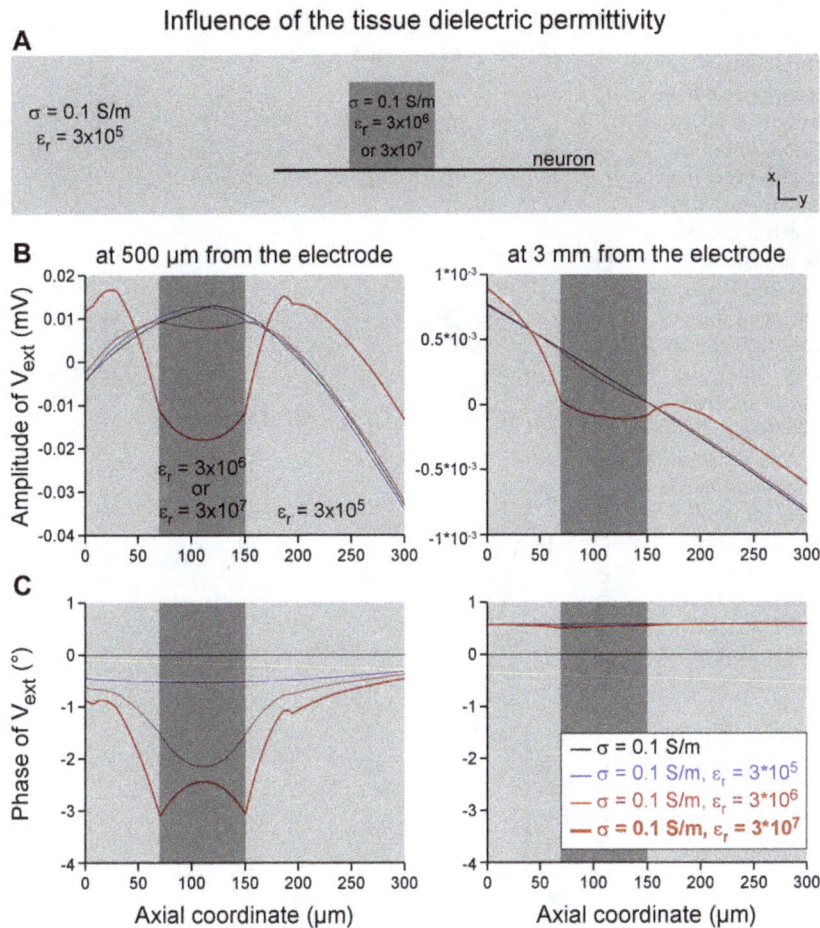

Figure 9. Influence of the dielectric permittivity of the tissue on the potential field. The original purely resistive description of the neural tissue was modified to incorporate its dielectric permittivity ε ($= \varepsilon_0 \times \varepsilon_r$), leading to the resolution of the Helmholtz equation instead of the Poisson equation (see Methods, Eq. 5). **A**: The influence of a dielectric anisotropy was also tested by introducing a 80-μm-sided cube with a different relative permittivity value ε_r (either 10 or 100 times higher). The neuron model 3 was positioned at the edge of this cube to maximize the effect of the heterogeneity. **B**: The magnitude of the potential field along the neuron (centered on its spatial mean) is plotted for distances of 500 μm (left) and 3 mm (right). Very few differences appear between the homogeneously dielectric (blue) and resistive (black) models. When introducing dielectric anisotropy (red curves), the profile of the field magnitude was modified more importantly for small than for large distances, yet remaining overall within the same range of values as with the homogenous resistive model. **C**: Phase shifts generated by the isotropic dielectric model were uniform along the neuron geometry and very small in amplitude (less than 1°). In the case of the anisotropic dielectric tissue, the phase shift was unchanged at long distance and only slightly modified at short distance.

unknown. Here, we characterized experimentally the spatial extent of monopolar stimulations using MEAs, and compared these results to modeling predictions.

From our experimental results, it appears that, depending on electrode-neuron distance, average currents required to elicit an action potential ranged between 20 μA (at about 20 microns) and 650 μA (at 3 millimeters). These current values fell within the wide range of current values reported in the literature, from a few μA to several mA (for review, see [43]). As a rule of thumb, we found that intensities of about 100–200 μA were required to activate cells within 100–250 microns, which corresponded to an activation spread of the order of 1 μm/μA.

Previous experimental studies have reported that current-distance curves could be well approximated by a quadratic law over distances of a few hundreds of microns [15,16,17,18,19,20]. However other work has not confirmed these results. For instance, Abzug and colleagues reported non quadratic but rather linear current-distance curves when spanning distances up to almost 1 millimeter [44]. More recently, fMRI and optical imaging

techniques have also been used to determine the spread of activation due to a microstimulation on millimeter-scaled areas [45,46,47]. These approaches generally report larger extent of CNS activation than electrophysiological experiments, which can be attributed to the fact that these imaging techniques equally reflect direct and indirect trans-synaptic as well as sub-threshold responses, and not only direct supra-threshold responses. Here, we found that the experimental current-distance relationship for direct activation was not quadratic over distances up to 3 millimeters, and that thresholds actually varied much more slowly than expected.

Although experimental studies provide the only possible way to determine directly the evolution of threshold currents with electrode-neuron distance, such approaches remain difficult to achieve technically, and do not allow an exhaustive screening of different stimulation paradigms or electrode geometries and configurations. For these reasons, computational modeling approaches have been developed to numerically estimate the spatial extent of electrical stimulations [37,38,48,49]. However, to our knowledge, these approaches have not been validated with experimental data directly in a common situation. Here we thus confronted theoretical estimation of activation thresholds with experimental results. We found that simulated TCs corresponded well to experimental TCs for distances smaller than about 250 microns (see Figure 4), with values similar to those found by others (e.g., 200 µA at 150 microns, see [21]). However, an unexpected discrepancy between modeling and experiments was found for distances above about 250 microns, for which simulated thresholds were strongly overestimated (by 2 orders of magnitude at 3 millimeters). We verified that experimental thresholds were not underestimated because of stimulation of distant dendritic or axonal arborization extending in the vicinity of distant stimulating electrodes, as reported in a recent *in vivo* study [47]. Here, this hypothesis could be discarded since the morphology of recorded neurons was found to be much more compact (usually about 300-µm-long, see Figure 2B) than the distances under consideration (several millimeters). Moreover, neurons were chemically isolated from one another using a low-Ca^{2+} aCSF solution, preventing network-induced activation. We also ruled out the possible influence of the neuron morphology and electrical properties (leakage and sodium conductances). Indeed, overestimated simulated thresholds only weakly depended on these model parameter values (Figure 5), a result consistent with other modeling work also reporting robust threshold estimation with respect to model parameters [21].

In the experimental study, motoneurons were activated with different types of electrodes, including 3D microelectrodes and one larger 2D rectangular electrode. At large distance, the activation threshold currents were similar for these two types of electrodes, which is consistent with the current-controlled mode of stimulation (in Figure 3A$_1$, compare the threshold values for the planar rectangular electrode and the 3D electrode on its right on the top row of the array). Accordingly, the modeled thresholds obtained with these two types of electrodes only differed by less than 4% for distances beyond 300 microns, and less than 1% beyond 800 microns. We also further checked that threshold currents obtained with disk-shaped stimulation electrodes of increasing radius reached identical values for distances greater than about twice the electrode's radius (Figure 10). Hence, the electrode size does not influence threshold currents at large distances. This is consistent with a previous modeling study showing that microelectrodes could be well approximated by point sources for distances above a few hundreds of microns [50].

The high values of threshold currents at large distance can be explained by the small variations of the extracellular potential along the neuron geometry. Indeed, we found that the spatial profile of the membrane potential at the end of the cathodic phase was well predicted by the "mirror estimate" (Figure 6), which states that the neuron response relative to the resting potential is equal to the opposite of the extracellular potential centered on its spatial average along the neuron morphology [42]. The relative uniformity of V_{ext} at large distances, due to the dispersion of the current in all directions of the experimental chamber, thus produced small variations of V_m and hence led to high TCs.

A possible cause of the discrepancy between the experimental and the modeled thresholds could be an oversimplification of the potential field model, based on the Poisson equation and characterized by isotropic, homogeneous and resistive conductivities. We thus further modified this model to embed either anisotropy (Figure 7), heterogeneity (Figure 8) or dielectric properties (Figure 9) in the neural tissue. Some of these features could modify substantially the extracellular potential profiles at small distances (500 microns), and thus (based on the above

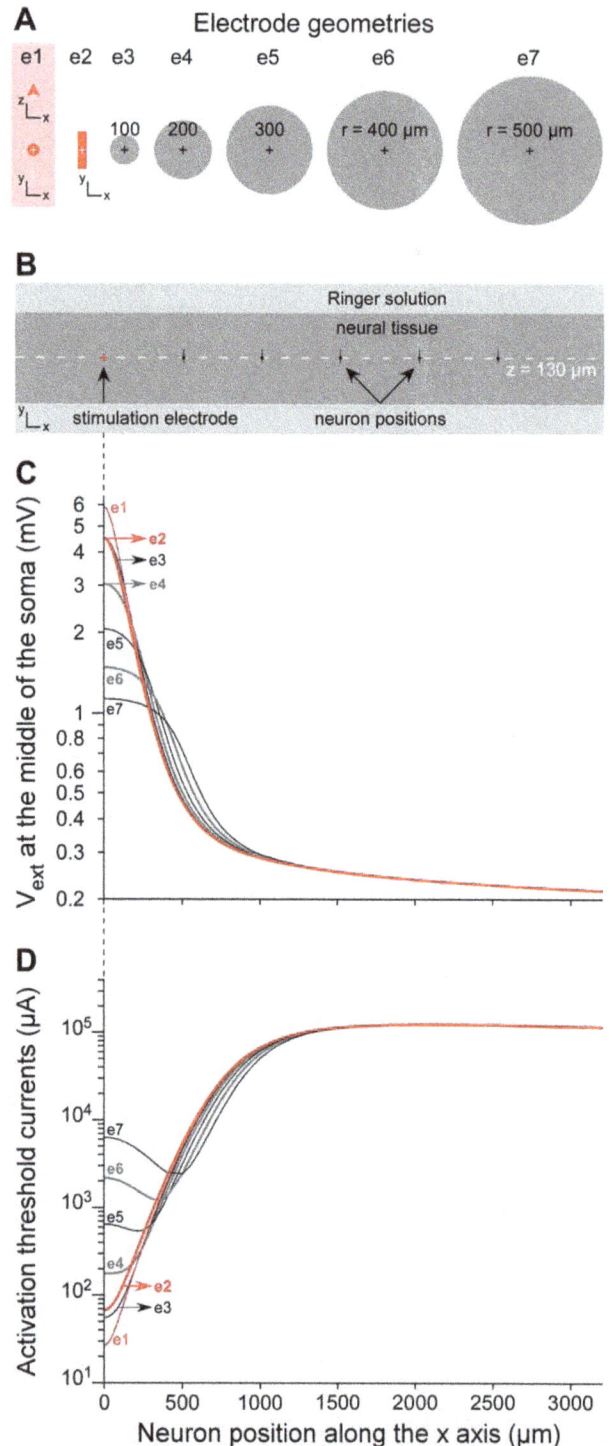

Figure 10. Influence of the electrode shape and size on the potential field and the threshold-distance curves. Seven stimulation electrode geometries were compared (**A**): The original 3D conical electrode used for all simulations of this study (e1, pink area, represented in *x–z* and *x–y* planes), a 2D rectangle electrode (e2, width = 60 µm, length = 250 µm) similar to that used in the experimental study, and 5 disk-shaped electrodes of increasing radius (from *r* = 100 µm for electrode e3 to *r* = 500 µm for electrode e7). For each stimulation electrode (centered on *x* = 0), the neuron model 3 was positioned at different locations along the *x*-axis, at *z* = 130 µm (**B**). The extracellular electrical potential V_{ext}, obtained for a 1-µA-stimulus, was interpolated at the middle of the soma (**C**) and activation threshold currents were computed as a function of the neuron position (**D**). It can

be seen that both the potential field and threshold currents were influenced by the electrode shape only for distances smaller than twice the size of the electrode.

"mirror" considerations), the neuron polarization and activation threshold currents. However, this was not the case at large distance (3 millimeters), where the model shows important robustness to the different parameters. In particular, the potential field was hardly modified by the dielectric model, which is consistent with a recent modeling study assessing the validity of the quasi-static approximation in the case of extracellular neural stimulation [35]. In this work, the authors indeed found that the propagation and inductive effects were negligible over the considered range of distances (below 1 cm), and that the capacitive effects, although being the most significant, only changed TCs by less than 20% for an homogeneous dielectric tissue. Here, we also tested the influence of local variations of the relative permittivity ε_r, and did not found substantial changes of the potential field, at least way behind explaining the two orders of magnitude between experimental and modeled TCs at large distances. The most influencing factor was the anisotropy of the spinal cord model, yet insufficient to explain the overestimation of TCs at large distances. Moreover, the embryonic spinal cord being not yet myelinated at the considered embryonic stage (E14.5), such anisotropy is unlikely to take place in our experimental study.

The amplitude of the pulses used in the present study was usually above the safe charge injection limit of Pt microelectrodes (about 25 μA for a 1000-μs pulse and the array that we used, see http://www.qwane.com/Documents/ safe_charge_injection_limit.pdf), which could have influenced the neural activity. In this study, every experiment was performed on an acute preparation for only a few hours during which we only delivered a few tens of pulses through each microelectrode. Moreover, although we used stimulations above the safe charge injection limit, we did not observe changes of the threshold currents between the beginning and the end of the experiment (see Methods section 1.d). To further verify the stability of the stimulation, we measured the actual current delivered through the electrode for a given current command, and how this current evolved after repetitive stimulation. We performed such experiments in a Ringer solution, delivering series of −800/+800 μA 1-ms-phase-long biphasic stimuli, and recorded simultaneously the delivered current and the voltage of the electrode. These stimuli corresponded to the highest amplitudes used in our experiments. After 1000 stimuli, we observed no variation of the injected current. Thus, because the extracellular potential field is proportional to the current (and not to the voltage) applied to the electrode [34], stimulations were very stable across the experiments. When delivering these pulses, the potential applied to the stimulating electrode was of the order of 4 V, which was over the safe charge injection limit for platinum (about 1 V). These high voltage values could locally induce water hydrolysis and alter the neural tissue around the stimulation electrode. However, these degradations would mostly influence the response of neurons located close to the electrode. Yet, modeling predictions differed from experimental results only for large electrode-neuron distances but not when the electrode was close to the neuron. Thus, possible overstimulations above the safe charge injection limit could not explain the mismatch between model and experimental results.

In conclusion, two main results are reported in this study. First, we found current-distance curves that were not quadratic over large distances beyond 1 millimeter. Second, we found that simulations predicted threshold currents in accordance with experiments for neuron-electrode distances below 250 microns, but largely overestimated at larger distances. Although the reason for this discrepancy remains to be determined, the present results suggest that activated extents may be underestimated by conventional simulation paradigms. This finding obtained *in vitro* on an embryonic preparation should further be investigated in other types of tissue and also *in vivo* with either MEA microstimulation or more macroscopic deep brain stimulation paradigms.

Acknowledgments

The authors wish to thank Serge Korogod and Lionel Rousseau for 'stimulating' discussions, Philippe Chauvet and Gilles Bouchard ("Or et Façon") for skilful technical work, and François Couraud for providing the pan Nav antibody samples.

Author Contributions

Conceived and designed the experiments: SJ BY. Performed the experiments: SJ PB BY. Analyzed the data: SJ BY. Contributed reagents/materials/analysis tools: SJ DC BY. Wrote the paper: SJ BY.

References

1. Benabid AL, Pollak P, Gervason C, Hoffmann D, Gao DM, et al. (1991) Long-term suppression of tremor by chronic stimulation of the ventral intermediate thalamic nucleus. Lancet 337: 403–406.
2. Clark GM, Tong YC, Black R, Forster IC, Patrick JF, et al. (1977) A multiple electrode cochlear implant. J Laryngol Otol 91: 935–945.
3. Wilson BS, Finley CC, Lawson DT, Wolford RD, Eddington DK, et al. (1991) Better speech recognition with cochlear implants. Nature 352: 236–238.
4. Winfree CJ (2005) Spinal cord stimulation for the relief of chronic pain. Curr Surg 62: 476–481.
5. Courtine G, Gerasimenko Y, van den Brand R, Yew A, Musienko P, et al. (2009) Transformation of nonfunctional spinal circuits into functional states after the loss of brain input. Nat Neurosci 12: 1333–1342.
6. Harkema S, Gerasimenko Y, Hodes J, Burdick J, Angeli C, et al. (2011) Effect of epidural stimulation of the lumbosacral spinal cord on voluntary movement, standing, and assisted stepping after motor complete paraplegia: a case study. Lancet 377: 1938–1947.
7. Shahaf G, Marom S (2001) Learning in Networks of Cortical Neurons. J Neurosci 21: 8782–8788.
8. Eytan D, Brenner N, Marom S (2003) Selective Adaptation in Networks of Cortical Neurons. J Neurosci 23: 9349–9356.
9. Wagenaar DA, Madhavan R, Pine J, Potter SM (2005) Controlling Bursting in Cortical Cultures with Closed-Loop Multi-Electrode Stimulation. J Neurosci 25: 680–688.
10. Weiland JD, Liu W, Humayun MS (2005) Retinal prosthesis. Annu Rev Biomed Eng 7: 361–401.
11. Djilas M, Olès C, Lorach H, Bendali A, Degardin J, et al. (2011) Three-Dimensional Electrode Arrays for Retinal Prostheses: Modeling, Geometry Optimization, and Experimental Validation. J Neural Eng 8(4): 046020.
12. Nicolelis MA (2001) Actions from thoughts. Nature 409: 403–407.
13. Jackson A, Mavoori J, Fetz EE (2006) Long-term motor cortex plasticity induced by an electronic neural implant. Nature 444: 56–60.
14. O'Doherty JE, Lebedev MA, Ifft PJ, Zhuang KZ, Shokur S, et al. (2011) Active tactile exploration using a brain-machine-brain interface. Nature 479: 228–231.
15. Adrian ED (1936) The spread of activity in the cerebral cortex. J Physiol 88: 127–161.
16. Brooks VB, Enger PS (1959) SPREAD OF DIRECTLY EVOKED RESPONSES IN THE CAT'S CEREBRAL CORTEX. J Gen Physiol 42: 761–777.
17. Stoney SD Jr, Thompson WD, Asanuma H (1968) Excitation of pyramidal tract cells by intracortical microstimulation: effective extent of stimulating current. J Neurophysiol 31: 659–669.
18. Bagshaw EV, Evans MH (1976) Measurement of current spread from microelectrodes when stimulating within the nervous system. Exp Brain Res 25: 391–400.
19. Nowak LG, Bullier J (1996) Spread of stimulating current in the cortical grey matter of rat visual cortex studied on a new in vitro slice preparation. J Neurosci Methods 67: 237–248.
20. Gustafsson B, Jankowska E (1976) Direct and indirect activation of nerve cells by electrical pulses applied extracellularly. J Physiol 258: 33–61.
21. McIntyre CC, Grill WM (2000) Selective microstimulation of central nervous system neurons. Ann Biomed Eng 28: 219–233.

22. McIntyre CC, Grill WM (2002) Extracellular stimulation of central neurons: influence of stimulus waveform and frequency on neuronal output. J Neurophysiol 88: 1592–1604.

23. Struijk JJ, Holsheimer J, Barolat G, He J, Boom HBK (1993) Paresthesia thresholds in spinal cord stimulation: a comparison of theoretical results with clinical data. IEEE Trans Biomed Eng 1: 101–108.

24. Butson CR, McIntyre CC (2006) Role of electrode design on the volume of tissue activated during deep brain stimulation. J Neural Eng 3: 1–8.

25. Benabid AL, Chabardes S, Mitrofanis J, Pollak P (2009) Deep brain stimulation of the subthalamic nucleus for the treatment of Parkinson's disease. The Lancet Neurology 8: 67–81.

26. Joucla S, Yvert B (2009) Improved focalization of electrical microstimulation using microelectrode arrays: a modeling study. PLoS ONE 4: e4828. doi:4810.1371/journal.pone.000482.

27. Heuschkel MO, Fejtl M, Raggenbass M, Bertrand D, Renaud P (2002) A three-dimensional multi-electrode array for multi-site stimulation and recording in acute brain slices. J Neurosci Methods 114: 135–148.

28. Branchereau P, Chapron J, Meyrand P (2002) Descending 5-hydroxytryptamine raphe inputs repress the expression of serotonergic neurons and slow the maturation of inhibitory systems in mouse embryonic spinal cord. J Neurosci 22: 2598–2606.

29. Yvert B, Branchereau P, Meyrand P (2004) Multiple Spontaneous Rhythmic Activity Patterns Generated by the Embryonic Mouse Spinal Cord Occur Within a Specific Developmental Time Window. J Neurophysiol 91: 2101–2109.

30. Yvert B, Mazzocco C, Joucla S, Langla A, Meyrand P (2011) Artificial CSF motion ensures rhythmic activity in the developing CNS ex vivo: A mechanical source of rhythmogenesis? J Neurosci 31: 8832–8840.

31. Delpy A, Allain AE, Meyrand P, Branchereau P (2008) NKCC1 cotransporter inactivation underlies embryonic development of chloride-mediated inhibition in mouse spinal motoneuron. J Physiol 586: 1059–1075.

32. Weiss G (1901) Sur la possibilité de rendre compatible entre eux les appareils servant à l'excitation électrique. Arch Ital Biol 35: 416–446.

33. Hines ML, Carnevale NT (1997) The NEURON simulation environment. Neural Comput 9: 1179–1209.

34. Joucla S, Yvert B (2011) Modeling extracellular electrical neural stimulation: From basic understanding to MEA-based applications. J Physiol Paris. In press. doi:10.1016/j.jphysparis.2011.10.003.

35. Bossetti CA, Birdno MJ, Grill WM (2008) Analysis of the quasi-static approximation for calculating potentials generated by neural stimulation. J Neural Eng 5: 44–53.

36. Plonsey R, Barr RC (1998) Electric field stimulation of excitable tissue. IEEE Eng Med Biol Mag 17: 130–137.

37. McIntyre CC, Grill WM (1999) Excitation of central nervous system neurons by nonuniform electric fields. Biophys J 76: 878–888.

38. McIntyre CC, Grill WM, Sherman DL, Thakor NV (2004) Cellular Effects of Deep Brain Stimulation: Model-Based Analysis of Activation and Inhibition. J Neurophysiol 91: 1457–1469.

39. Manola L, Holsheimer J, Veltink P, Buitenweg JR (2007) Anodal vs cathodal stimulation of motor cortex: a modeling study. Clin Neurophysiol 118: 464–474.

40. Mainen ZF, Joerges J, Huguenard JR, Sejnowski TJ (1995) A model of spike initiation in neocortical pyramidal neurons. Neuron 15: 1427–1439.

41. Hodgkin AL, Huxley AF (1952) A quantitative description of membrane current and its application to conduction and excitation in nerve. J Physiol 117: 500–544.

42. Joucla S, Yvert B (2009) The "Mirror" Estimate: An Intuitive Predictor of Membrane Polarization during Extracellular Stimulation. Biophysical Journal 96: 3495–3508.

43. Tehovnik EJ (1996) Electrical stimulation of neural tissue to evoke behavioral responses. J Neurosci Methods 65: 1–17.

44. Abzug C, Maeda M, Peterson BW, Wilson VJ (1974) Cervical branching of lumbar vestibulospinal axons. J Physiol 243: 499–522.

45. Seidemann E, Arieli A, Grinvald A, Slovin H (2002) Dynamics of depolarization and hyperpolarization in the frontal cortex and saccade goal. Science 295: 862–865.

46. Tolias AS, Sultan F, Augath M, Oeltermann A, Tehovnik EJ, et al. (2005) Mapping cortical activity elicited with electrical microstimulation using FMRI in the macaque. Neuron 48: 901–911.

47. Histed MH, Bonin V, Reid RC (2009) Direct Activation of Sparse, Distributed Populations of Cortical Neurons by Electrical Microstimulation. Neuron 63: 508–522.

48. Struijk JJ, Holsheimer J, van der Heide GG, Boom HB (1992) Recruitment of dorsal column fibers in spinal cord stimulation: influence of collateral branching. IEEE Trans Biomed Eng 39: 903–912.

49. Struijk JJ, Holsheimer J, Boom HB (1993) Excitation of dorsal root fibers in spinal cord stimulation: a theoretical study. IEEE Trans Biomed Eng 40: 632–639.

50. McIntyre CC, Grill WM (2001) Finite element analysis of the current-density and electric field generated by metal microelectrodes. Ann Biomed Eng 29: 227–235.

14

Negative Refraction Angular Characterization in One-Dimensional Photonic Crystals

Jesus Eduardo Lugo*, Rafael Doti, Jocelyn Faubert

Visual Psychophysics and Perception Laboratory, School of Optometry, University of Montreal, Montreal, Quebec, Canada

Abstract

Background: Photonic crystals are artificial structures that have periodic dielectric components with different refractive indices. Under certain conditions, they abnormally refract the light, a phenomenon called negative refraction. Here we experimentally characterize negative refraction in a one dimensional photonic crystal structure; near the low frequency edge of the fourth photonic bandgap. We compare the experimental results with current theory and a theory based on the group velocity developed here. We also analytically derived the negative refraction correctness condition that gives the angular region where negative refraction occurs.

Methodology/Principal Findings: By using standard photonic techniques we experimentally determined the relationship between incidence and negative refraction angles and found the negative refraction range by applying the correctness condition. In order to compare both theories with experimental results an output refraction correction was utilized. The correction uses Snell's law and an effective refractive index based on two effective dielectric constants. We found good agreement between experiment and both theories in the negative refraction zone.

Conclusions/Significance: Since both theories and the experimental observations agreed well in the negative refraction region, we can use both negative refraction theories plus the output correction to predict negative refraction angles. This can be very useful from a practical point of view for space filtering applications such as a photonic demultiplexer or for sensing applications.

Editor: Matteo Rini, Joint Research Centre - European Commission, Germany

Funding: This work was supported by NSERC-Essilor Research Chair and an NSERC operating grant. The funders had no role in study design, data collection and analysis, decision to publish, or preparation of the manuscript.

Competing Interests: The authors have declared that no competing interests exist.

* E-mail: je.lugo.arce@umontreal.ca

Introduction

Photonic crystals can be considered as multidimensional periodic gratings, in which the features of refraction at flat surfaces are dominated by Bragg diffraction effects. The refraction angle from positive to negative can be tailored based on photonic band theory [1]. Numerous studies on diffraction gratings and periodic planar waveguides, essentially the one-dimensional counterparts for the photonic structures, led to the observation of a vast variety of anomalous refraction effects, including "birefringence" [2–6]. These systems have undergone extensive and systematic study based on the wave vector diagram formalism. This formalism has proven to be an excellent tool in explaining the unusual refractive properties for the one-dimensional diffraction grating system. In the late 1990s, diffraction characteristics that appeared to be negative refraction were explained in terms of the dispersion surfaces of photonic bands and prism, lens, and collimation effects based on refraction were predicted [7–9]. Specifically, it has been demonstrated that light propagation in strongly modulated 2D/3D photonic crystals becomes refraction-like in the vicinity of the photonic bandgap, even in the presence of strong multiple diffraction [4]. In these conditions, it is possible to define an effective phase refractive index to explain the propagation inside the photonic crystal using the conventional Snell's law. Since such effective index is determined by the photonic band structure, it can be negative and less than unity, which leads to negative refraction [9].

This behavior can be understood by using the effective-mass model in electron-band theory. In the photonic case a Bloch photon, near the bandgaps, can be considered as free, and be regarded as a refracted photon inside of a medium with an effective refractive index. These particular index states only appear close the photonic bandgap in a similar way as the effective mass states in a semiconductor. The same conclusion has been reached by others groups [10]. For instance, the effective dielectric constant of a 2D photonic crystal in all optical bands, for both TE and TM polarizations, was calculated. It has been found that near the gamma point (center of the Brillouin zone), the dispersion relationship for the TM mode is independent of the propagation direction, while the TE mode in general depends on the electromagnetic waves propagation direction. Therefore, for a 2D photonic crystal, there always exists an effective dielectric index for the TM mode near the gamma point. However, it cannot be defined as an effective refractive index for TE mode unless the photonic crystal is highly symmetric. By using similar arguments presented in [9], Kavokin theoretically explored negative refraction in one-dimensional photonic crystals (1D PCs) [11]. By using the dispersion of the photonic bands, he inferred negative refraction zones from frequency regions where the effective mass is negative. Recently, we have simulated a lossless 1D PC structure and showed that negative refraction could be present near the low frequency edge of at least the second, fourth and sixth bandgaps [12]. The same conclusion was reached by other groups [13–15]. Furthermore, we experimentally demonstrated negative refraction in strongly modulated porous silicon 1D-PC in the visible and near infrared regions. However, in [12] negative refraction was explored with only one angle of incidence. Therefore, a complete angular characterization is still missing.

Moreover, in regards to the theory of negative refraction in 1D PCs, the existence of antiparallel energy and phase velocity has been thoroughly analyzed in [16]. The existence of negative refraction in 2D PCs is substantially different from the one-dimensional case because 2D PCs with a negative slope band demonstrates negative refraction beam propagation. This is not true for 1D PCs because the correctness propagation condition needs to be fulfilled. The "correctness" of propagation in 1D PCs implies that the correct physical conditions, required to observe negative refraction, are met. The analysis presented in [16], for 2DPCs, only tackles negative refraction for on-plane propagation where the crystal is periodic. The exact analogy for negative refraction propagation between 2D PCs and 1D PCs is the normal incidence case, where the 1D PCs are periodic in that particular direction. Nonetheless the aforementioned correctness condition should also be applied in the 2D PCs case when you have off-plane propagation, a point that we will discuss later.

In this paper, we experimentally completed the angular characterization of negative refraction in a 1D PC structure, near the low frequency edge of the fourth photonic bandgap. We compared the experimental results with current negative refraction theory in 1D PCs [11] and with a theory developed here, based on the group velocity. We confronted both negative refraction theories and found good agreement between them with differences up to 4 degrees, within the explored incidence angle interval. We analytically derived the correctness condition and showed that for the experimental conditions we used, the correctness condition is fulfilled up to an incidence angle of 15 degrees. We also theoretically verified the correctness condition near the second bandgap edge (1350 nm) and found that it is fulfilled up to an incidence angle of 20 degrees. In order to compare the experiments with theory we developed an approximation that accounts for the positive refraction that the negative refraction beam suffers at the structure output. The correction uses Snell's law and an effective refractive index, based on two effective dielectric constants [17]. We found good agreement between experimental observations and the theory developed here for the whole incidence angle interval explored. The agreement between current theory and experimental results was good for incidence angles smaller than 15 degrees because the effective mass approximation begins to fail for incidence angles larger than 15 degrees and so does its consequent correction approximation. Since both theories and experimental results agreed well in the negative refraction region, given by the correctness condition, we can use both negative refraction theories with the addition of the output correction given herein to predict negative refraction angles.

Results and Discussion

Sample preparation and negative refraction angle for the output measurement

Porous silicon (Psi) multilayers (Fig. 1) were prepared by electrochemical anodization of crystalline silicon (c-Si) [18]. Porous silicon was fabricated by wet electrochemical etching of highly boron-doped c-Si substrates with orientation (100) and electrical

resistivity of 0.001–0.005 Ohm-cm (room temperature = 25°C, humidity = 30%). On one side of the c-Si wafer, an aluminum film was deposited and then heated at 550°C during 15 minutes in nitrogen atmosphere to produce a good electrical contact. In order to have flat interfaces, an aqueous electrolyte composed of HF/ethanol/glycerol was used to anodize the silicon substrate. It is well known that the Psi refractive index increases by decreasing the electrical current applied during the electrochemical etching. However, reducing the porosity too much might stop the electrolyte flow through the porous and limit the subsequent high porosity layer that makes the contrast. One way to allow the electrolyte to flow is by increasing the ethanol fraction in the solution. For this reason, an electrolyte composition of 3:7:1 was used. In addition, the HF concentration was maintained constant during the etching process using a peristaltic pump to circulate the electrolyte within the Teflon^{TM} cell. Anodization begins when a constant current is applied between the c-Si wafer and the electrolyte by means of an electronic circuit controlling the anodization process. To produce the multilayers, current density applied during the electrochemical dissolution was alternated from 3 mA/cm^2 (layer a) to 40 mA/cm^2 (layer b) and eighty periods (160 layers) were made. Psi samples were partially oxidized at 350°C for 10 minutes. The best refractive index values we found that fit the experimental photonic bandgap structure studied here are $n_a = 1.1$ and $n_b = 2$ [12]. We have experimentally measured the refractive indices of single Psi layers made with the same electrochemical conditions as the multilayers [18] and we found that $n_a = 1.40 \pm 0.07$ and $n_a = 2.20 \pm 0.11$. The refractive indices were measured by using interference fringes from reflectance measurements [18–19]. Nevertheless, it is known that the refractive index and etching rate for a single layer are modified in the presence of a multilayer structure up to approximately 14%, a phenomenon that has been systematically observed [19]. This result might have the consequence of compromising the mechanical stability of the structure. Indeed, in certain regions seen in Fig. 1 layers appear to be collapsed. Nevertheless, negative refraction was observed in all our experiments where several regions were scanned. Scanning electron microscopy (SEM) was used to measure the films thicknesses which were 326 ± 11 nm (a) and 435 ± 11 nm (b).

Once the samples were ready, we investigated the relationship between the negative refraction angle and the incidence angle at 633 nm (TE polarization) for the 1D PC structure. Figure 2 shows the experimental setup we used. The apparatus consists of a plate on which we find a *curved support with a sliding base (4)*, a *turning bar (7)* and a *turning platform (8)*. There is a *light source of 633 nm (5)* that can slide on the *curved support (4)* that points towards the turning

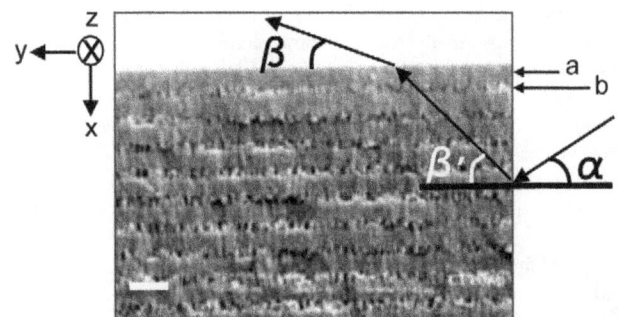

Figure 1. 1D PC structure. SEM picture showing the layers a and b, angle of incidence α and negative refraction angle β' inside the structure and corrected negative refraction angle β at the output, which can be measured experimentally. The light impinges at the right interface (The white line on the left represents 1 micron).

center. On top of the *turning bar (7) we* find a *xyz platform (2)* that holds the *1D PC (1)* under test. The *turning platform (8)*, which holds the video camera (10), has *two movement axes (9)*. These materials were placed on a standard optical table. We illuminated the sample edge with a light source at the desired incidence angle α and, by exploring the sample side with the video camera; we found the output refracted beam (corrected negative refraction angle β). Once the beam was detected, its direction was confirmed by means of the beam spot luminance on the image monitor (not shown) that was measured with a luminance meter. As the refracted beam gets weaker for higher incidence angles, we explored angles up to 25 degrees in order to have enough discrimination of the spot luminance in reference to the monitor image background luminance. More details are given in the methods section.

Negative refraction theory

In order to compare the experimental corrected negative refraction angles at the output (angle β) with theory, first we need

to discuss negative refraction theory that allows us to calculate negative refraction angles β' as a function of incidence angles α (see Fig. 1). Negative refraction theory for 1D PCs has been presented in reference [11], where the condition for negative refraction uses the notion that, if in a given direction the effective mass is negative, the corresponding components of group and phase velocities of light have different signs. This seems to be true for 1D PCs because they are strongly anisotropic, so that the effective masses have different signs in on-plane and normal-to-plane directions [11]. However, in order to fully warrant the occurrence of negative refraction the correctness condition needs to be fulfilled [16].

It is well known that there are significant differences between the properties of 1D PCs and 2D PCs. In 2D PCs, when the plane of incidence is chosen to be the periodic plane, the entire wave vector is confined in the first Brillouin zone. In the theory of wave propagation through a crystal lattice, the Brillouin zone is a fundamental region of wavevectors; every vector outside this region is tantamount to some other vector inside it. In contrast, in the 1D PC, only the component of the wave vector along the direction of

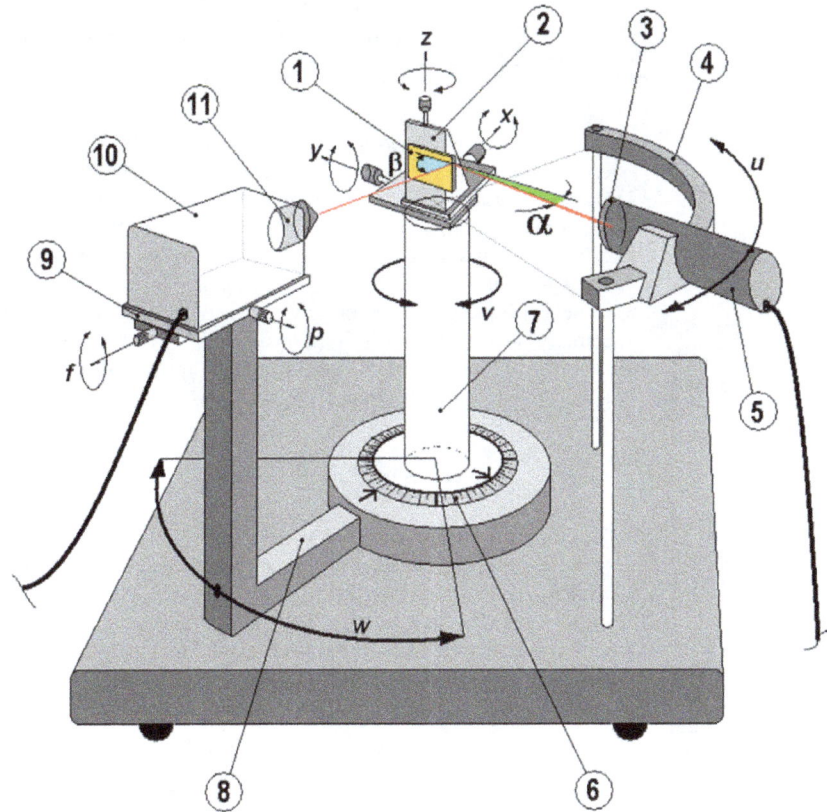

1- P-Silicon specimen
2- x-y-z, platform
3- Pin-hole + polarizer
4- Sliding support
5- Source, λ 633 nm (movement: u)
6- Goniometer (fixed, degree scale)
7- Turning bar (movement: v)

8- Turning platform (movement: w)
9- Camera movement:
     Parallax correction (p),
     Focus correction (f)
10- Video camera
11- Camera objective: X100
α - Incidence angle
β - Corrected Neg-Refraction angle

Figure 2. Experimental set up. The eleven components of the experimental setup for negative refraction observation.

periodicity is restricted within the first BZ. This has a very important implication. In 2D PCs, a band with negative slope corresponds to a negative refraction beam. However, this is not true for 1DPCs. We have chosen x to represent the direction of the periodicity (Fig. 1). The slope of a certain band will then be given by $V_{gper}k_{per}$, where V_{gper} and k_{per} are the group velocity and wavevector components in the normal-to-plane direction respectively. Since $V_{gpar}k_{par}$ is always positive, where V_{gpar} and k_{par} are the group velocity and wavevector components in the on-plane direction respectively, and for a band with positive slope $V_{gper}k_{per} > 0$, then $\vec{S} \cdot \vec{k} > 0$, where $\vec{S}$ and $\vec{k}$ are the Poynting vector and the wavevector respectively. For a band with negative slope $V_{gper}k_{per} < 0$, then the correctness condition for negative refraction to occurs $\vec{S} \cdot \vec{k} < 0$ gives

$$V_{gpar}k_{gpar} < |V_{gper}k_{per}|. \tag{1}$$

Where we have used the fact that the Poynting vector is proportional to the group velocity. The group velocity components V_{gper} and V_{gpar} can be obtained from the photonic bands' dispersion relationship as outlined in [13] as:

$$V_{gper} = \frac{1}{\dfrac{\partial k_{per}}{\partial \omega}}, \tag{2}$$

$$V_{gpar} = -\frac{\dfrac{\partial k_{per}}{\partial k_{par}}}{\dfrac{\partial k_{per}}{\partial \omega}}. \tag{3}$$

We have verified condition 1 (see the methods section), for our proposed 1D PC structure, close to the second (1350 nm) and fourth (633 nm) low frequency band edges (TE Polarization) and used refractive index values and layer thickness described in the experimental section and with n_0 equals one. Condition 1 is fulfilled for incidence angles up to 20 degrees and 15 degrees respectively. For 633 nm light, traveling in the structure, one should expect that for angles of incidence larger than 15 degrees there will be more than one beam travelling inside the structure. For the normal-to-plane direction the second and fourth allowed band ends at 1345 nm and 630 nm respectively and they are characterized by a negative parabolicity close to the band edge. On the other hand, for the on-plane direction it is also parabolic close to band edge but it is characterized by a positive effective mass. In the case where the relevant bands have different band slope signs, one can observe the simultaneous propagation of beams [16]. Condition 1 can be generalized as

$$\vec{V}_{gparT} \cdot \vec{k}_{parT} < |\vec{V}_{gperT} \cdot \vec{k}_{perT}|. \tag{4}$$

Where $\vec{k}_{parT}$ and $\vec{V}_{gparT}$ are the total wavevector and group velocity in the parallel direction. $\vec{k}_{perT}$ and $\vec{V}_{gperT}$ are the total wavevector and group velocity in the perpendicular direction. In the 1D PC case the vectors, according with figure 3, are given by $\vec{k}_{parT} = (k_{par1}, k_{par2}, 0)$, $\vec{V}_{gparT} = (V_{par1}, V_{par2}, 0)$, $\vec{k}_{perT} = (0,0,k_{per1})$, $\vec{V}_{gperT} = (0,0,V_{per1})$ and in the 2D PC case by $\vec{k}_{parT} = (k_{par1},0,0)$, $\vec{V}_{gparT} = (V_{par1},0,0)$, $\vec{k}_{perT} = (0,k_{per2},k_{per1})$ and $\vec{V}_{gperT} = (0,V_{per2},V_{per1})$.

It is clear that if we are in a band with a negative slope where $\vec{V}_{gperT} \cdot \vec{k}_{perT} < 0$ is always true and since in the parallel direction

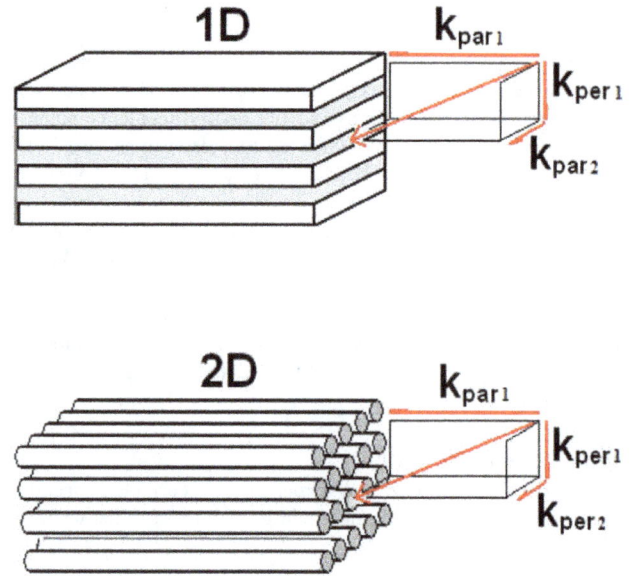

Figure 3. Correctness condition. Correctness condition generalization from 1D PCs to 2D PCs.

$\vec{V}_{gparT} \cdot \vec{k}_{parT}$ is always positive. Therefore inequality 4 gives the negative refraction correctness condition either for 1D PCs or 2D PCs. The parallel direction represents the direction where there are no periodic dielectric regions to coherently scatter the light. For instance, in a 1D PC is the on-plane direction (known as off-axis as well) and for a 2D PC is the off-plane direction.

The expression for the negative refraction angle β' for the geometry showed in figure 1 is obtained in [11] by using the continuity of the electric and magnetic fields at the boundary and the effective mass approximation as:

$$\sin^2(\beta') = \frac{n_0^2 \sin^2(\alpha)}{\Gamma^2}, \tag{5}$$

where n_0 is the air refractive index, α is the incidence angle, c is the speed of light, h is Planck's constant, λ is the working light wavelength, λ_0 is the wavelength associated with the top of the fourth subband, for instance and $\lambda > \lambda_0$. The effective mass approximation works fine only if the condition $(\lambda - \lambda_0)/\lambda \ll 1$ is fulfilled. The parameter Γ^2 is expressed as

$$\Gamma^2 = \left(1 - \frac{m_{per}}{m_{par}}\right) n_0^2 \sin^2(\alpha) + \frac{2m_{per}^2 c\lambda}{h m_{par}}\left[1 - \frac{\lambda}{\lambda_0}\right]. \tag{6}$$

The effective masses of light in the normal-to-plane direction m_{per} and on-plane direction m_{par} are calculated by using the expressions given in the methods section.

We can also calculate negative refractive angles by using the group velocity, which represents the direction of propagation inside the medium as follows:

$$\tan(\beta') = \frac{V_{gper}}{V_{gpar}}. \tag{7}$$

Equation (7) is given in the methods section. Figure 4 shows the comparison between Eqs. (5) and (7). We used a working wavelength of 633 nm (TE Polarization) which is close to the fourth low frequency band edge (λ_0-630 nm), and we used refractive index values and layer thickness described in the experimental section with n_0 equals one. Clearly both curves are similar with angle value differences up to 4 degrees, within the explored incidence angle interval. This result supports the use of the theoretical approach represented by Eq. (5) to predict negative refraction angles.

Comparison between experimental results and theory

Figure 5, shows the experimental and negative refraction results for the theories (eqs. (5) and (7)) for the behavior between angle of incidence versus angle of refraction. Since the experimental values represent corrected negative refraction angles at the output (angle β) we cannot compare them directly with the theories because they represent negative refraction angles β' inside the structure.

Reference [16] investigated light propagation in a 2D PC that consisted of dielectric rods in air with a hexagonal arrangement for the *H*-polarization case. They first performed a finite-difference time-domain (FDTD) simulation of light propagation along the Γ-K interface with an incidence angle of 8 degrees. The Γ-K interface goes from the center of the Brillouin zone to a vertex that joins two edges. Second, they supposed that their periodic structure could be described with an effective medium having an effective dielectric constant consistent with Maxwell-Garnett theory [16] and, therefore, an effective refractive index. In such a case, the field inside the PC is a plane wave. Third, by using the plane wave expansion method (PWE) they determined that for low dielectric contrasts between rods and air there is mainly one predominant component contributing to the Floquet-Bloch wave (FB). If the dielectric contrast between rods and air is bigger than 2, mixing between the different components in the FB sum starts to occur. This was corroborated by their FDTD simulations. Fourth, for both treatments, the effective homogeneous medium and the periodic medium with the PWE method gave almost the same angle for the propagating beam. This value is in excellent agreement with the FDTD simulation result. Given this, one might think that it is possible to describe a photonic crystal medium, for low dielectric contrast, as a homogeneous medium with an effective index. However, if you take the same angle of incidence, but choose a different interface such as Γ-M. The Γ-M interface goes from the center of the Brillouin zone to the middle of an edge. The propagation results are completely different to the precedent case and cannot be described by a homogeneuos medium approximation. From this we can infer that the wave is able to see the periodicity of the medium even when the index contrast is low.

Nonetheless, the fact the effective medium approach fails to generally describe beam propagation in some cases, this does not preclude the use such approximation to describe beam propagation in a particular direction if there is only one predominant FB wave travelling in that direction. We have done a finite element simulation of our structure (figure 6-top) where we can observe a single negative refraction beam (beam with angle β') that impinges towards the normal-to-plane interface where it is positively refracted as a single beam. This implies that we can use an effective medium approximation in such direction but we have to bare in mind that the effective refractive index does not represent the refractive index of the structure as if it were a homogeneous medium in all directions.

Since the negative refraction beam does not only propagate in the normal-to-plane direction we have to construct an effective medium approximation that takes into account the on-plane propagation direction as well. We can use the normal-to-plane $\varepsilon_{per} = \left(f_a n_a^2 + f_b n_b^2 \right)$ and on-plane $\varepsilon_{par} = \left(\dfrac{f_a}{n_a^2} + \dfrac{f_b}{n_b^2} \right)^{-1}$ effective dielectric constants, known to work fine in a multilayer system [17]. Therefore we can construct an effective medium approximation with an effective index of refraction given by

$$n_{output} = \sqrt{ \frac{|\beta'|}{\frac{\pi}{2}} \left(f_a n_a^2 + f_b n_b^2 \right) + \frac{\frac{\pi}{2} - |\beta'|}{\frac{\pi}{2}} \left(\frac{f_a}{n_a^2} + \frac{f_b}{n_b^2} \right)^{-1} }, \quad (8)$$

where f_j represents the fraction volume of each layer (43% for a-layers and 57% for b-layers), and n_j is the refractive index of each layer. The two angular prefactors multiplying each dielectric bound are necessary to account for the contribution of each component (normal-to-plane and on-plane). Then we can use Snell's law and the effective refractive index as:

$$\cos(\beta) = n_{output} \cos(\beta'). \quad (9)$$

Combining equations (5), (7), (8) and (9) the corrected negative refraction angles at the output can be calculated as:

$$\beta = \cos^{-1} \left(n_{output} \sqrt{1 - \frac{n_0^2 \sin^2(\alpha)}{\Gamma^2}} \right). \quad (10)$$

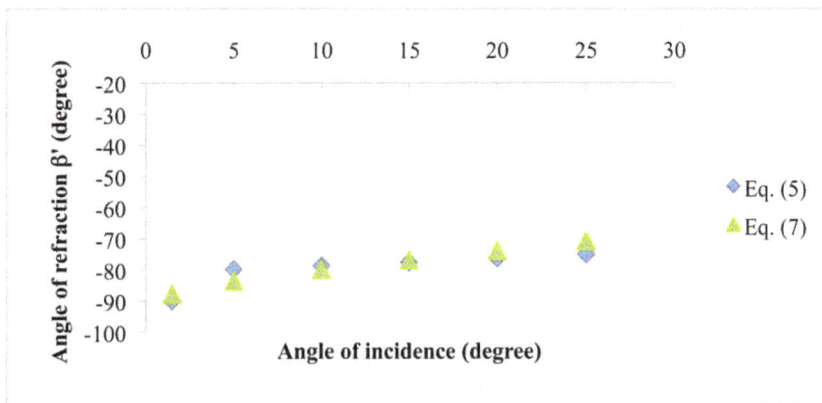

Figure 4. Negative refraction theories comparison. Angle of refraction β′ vs. angle of incidence for the 1D PC proposed structure. The theory presented in Kavokin [11] is compared against group velocity theory. The light wavelength is 633 nm (TE polarization) and we used refractive index values and layer thickness described in the experimental section.

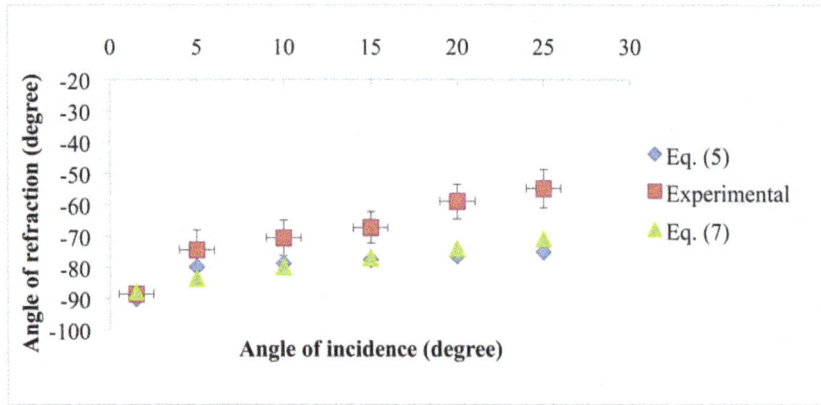

Figure 5. Comparison between negative refraction theories and experiments. Negative refraction experimental values compared against uncorrected theoretical values (Eqs. (5) and (7)).

Equation (10) is valid for $0 < \alpha_{min} \leq \alpha$, where α_{min} is given by

$$\left(1 - \frac{m_{per}}{m_{par}}\right) n_0^2 \sin^2(\alpha_{min}) \geq \left| \frac{2m_{per}^2 c\lambda}{hm_{par}} \left[1 - \frac{\lambda}{\lambda_0}\right] \right|. \qquad (11)$$

Inequality (11) assures us that the angle β' is real.

Now, by combining equations (7), (8), and (9) the corrected negative refraction angles at the output can be calculated as well as:

$$\beta = \cos^{-1}\left(\frac{n_{output}}{\sqrt{\tan^2(\beta') + 1}}\right), \qquad (12)$$

and the analytical expression for $\tan(\beta')$ is given in the methods section. Clearly, Eq. (5) values fit well with experimental values up to 15 degrees (fig. 6-middle). The angular difference $\delta\beta'$ between values predicted by eqs. (5) and (7) is the reason why the corrected refraction angles at the output, obtained by using eq. (10), differs from the experimental ones for angles of incidence larger than 15 degrees. This is understandable because as the incidence angle increases the effective mass approximation begins to fail. Indeed by increasing the incidence angle, the band edge is pushed towards small wavelengths making the separation $(\lambda - \lambda_0)$ increase. The results for the corrected negative refraction angles obtained with eq. (12) are shown in figure 6, (bottom). Notice that equation (12) predicts values that lie within the experimental accuracy obtained for all the angles of incidence. Notwithstanding, equation (10) is a good approximation to calculate corrected negative refraction angles at the output and it works well in the negative refraction region given by the correctness condition. All the experiments and calculations were done for TE polarization and a similar approach can be used for TM polarization where we expect to find analogous results as it was shown in reference [12].

Conclusion

In conclusion, we have experimentally completed the angular characterization of negative refraction in a 1D PC structure, near the low frequency edge of the fourth photonic bandgap and compared it with current theory and theory based on group velocity developed here. We have validated the current negative refraction theory approach with our theory. We found good agreement between both theories with differences within 4 degrees in the explored incidence angle interval. In order to know the negative refraction zone, we have analytically derived the

correctness condition and showed that for the experimental conditions we used, the correctness condition is fulfilled up to an incidence angle of 15 degrees. We also theoretically verified the correctness condition near the second bandgap edge (1350 nm) and found that it is fulfilled up to an incidence angle of 20 degrees. Finally, we corroborated the angular experimental values with negative refracted angular values obtained with both negative refraction theories by applying an output correction that uses Snell's law and an effective refractive index, based on the two effective dielectric constants. We found good agreement between experimental results and our theory in the entire incidence angle interval explored. The agreement between current theory and experimental results was good up to an incidence angle of 15 degrees because the effective mass approximation begins to fail for incidence angles larger than 15 degrees and the same is true for its consequent correction approximation. Since both theories and the experimental observations agreed well in the negative refraction region given by the correctness condition, we can use the combination of theory and output correction to predict negative refraction angles. This is very useful from a practical point of view. For instance, it could be useful for space filtering applications [20] such as a photonic demultiplexer or for sensing applications. A demultiplexer could be based on the fact that it is possible to have different wavelengths light impinging on the same incidence angle, since β' depends on the wavelength, light with different wavelengths is dispersed in different directions at the output. Equations (10) and (12) will consequently be useful to estimate the output angles. A (Bio)chemical sensor could instead exploit the fact that the multilayers are porous and we can change their refractive indices by infiltrating different chemical or biological compounds that again would shift the angles β' and β. Compound concentration should be proportional to this angular shift.

Materials and Methods

Determination of the refracted angle β: step sequence (see Fig. 2)

The first step consisted on choosing a convenient position of the light source on the sliding support. That was chosen in function of the free space needed for hand intervention. Once this position was determined, it was kept invariant along all the measurements. The light beam was kept as angular reference for zero degrees. So, through the *turning movement* v and the linear *displacement* x, the second step consisted in obtaining a regular tangent light beam

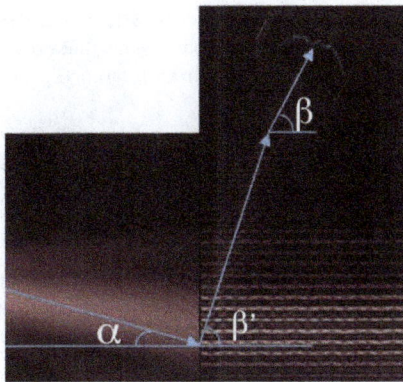

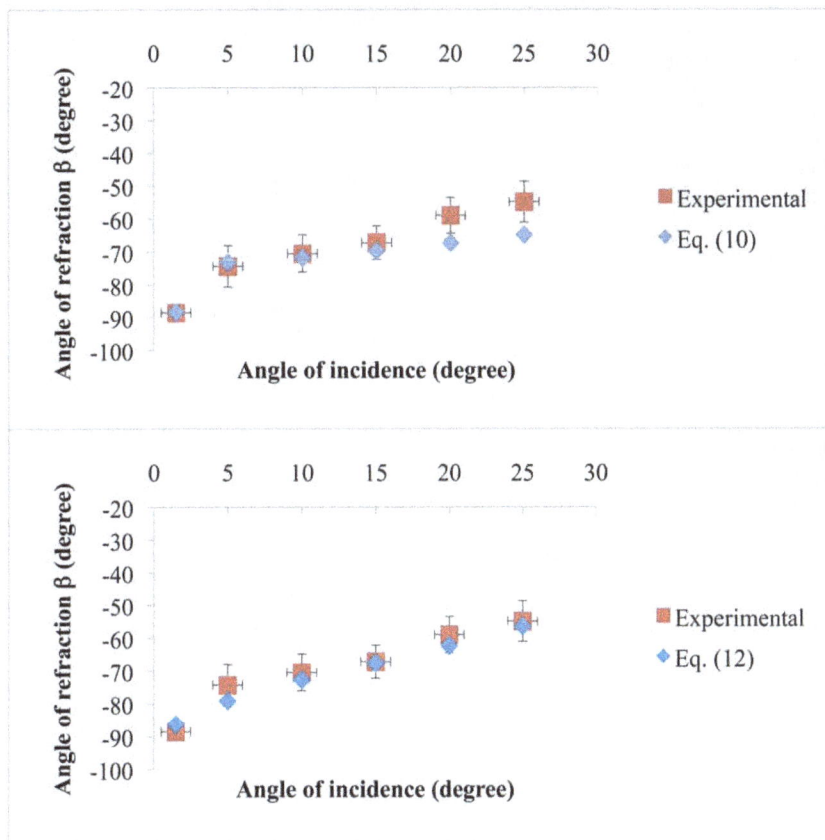

Figure 6. 1D PC results at a working wavelength of 633 nm (TE polarization). (Top) finite element negative refraction simulation showing beam propagation inside the structure and input and output interfaces. The angle of incidence α is 15 degrees, angles β' and β are -72 and -63 degrees respectively. (Middle) Comparison between negative refraction experimental values and corrected theoretical values (Eq. (10)). (Bottom) Comparison between negative refraction experimental values and corrected theoretical values (Eq. (12)). We used refractive index values and layer thickness described in the experimental section. Error bars represent systematic errors plus random errors (two standard deviations).

observed all along the lateral face of the specimen (parallel to y direction). The third step was to assure that the specimen illuminated edge was placed just over the turning center of the apparatus. This task was performed by acting the *movement y*. The fourth step was moving the *turning bar (7)* around to place the 1D-PC in the desired incident angle α in reference to the light beam. To achieve this, we acted the *movement v* and verified the angular position on the *goniometer (6)* scale. At this point, it was necessary to place the 1D-PC specimen in a way that assured us that the incident beam was totally contained in the illuminated edge, and without reaching the specimen normal face (x direction). This was

done by means of the *movement x* (sixth step). Then we explored the specimen side looking for a negative-refracted beam by means of the video camera, the turning *platform* and controlling the parallax error (by keeping the refracted light spot centered on the TV monitor and in focus). This seventh step involved the movements: w, p and f. After we found the light spot, we explored a narrow angle δw maintaining the light spot centered in the monitor, as we explained before. Using a luminance-meter (measure of the luminous intensity of light travelling in a given direction) we controlled the light level emitted by the monitor in the portion of the image containing the refracted light spot. With this procedure we found the angle β for which the luminance-meter gave the highest reading (*Lmax*), and then we checked the refracted intensity for points five degrees away from this last one, verifying that their intensity was less than *50% of Lmax*. This tedious procedure was repeated for incident angles ranging from 1.5 degrees up to 25 degrees. Each angle was measured four times, but we reported the average value as the negative refraction value and two standard deviations errors as random errors.

The transference from the refracted light intensity (that we expect to follow a Gaussian-like distribution according to our simulations [12]) to the monitor emitted light (measured with the luminance meter), cannot be considered proportional because of the energy conversions involved (all with their own non linearities and convergence limits). The narrow intensity per unit of area distribution of the refracted light and the acceptance angle of the camera suggested that the most important systematic error was due to two factors. First, the angular determination error: angular measurement through mechanic goniometers could reach without problems ± 1 degree error; but the *Lmax* reading gave us a non discernible reading along 3 degrees around the *Lmax* β angle. This effect is known as spatial filtering. Second, we explored the negative refracted spot light along a circumference centered in the same spot as if it where the center for the *w* movement. Unfortunately the real center (for the turning platform) and the refracted spot was several microns away (at least the distance from the spot to the specimen edge). Therefore, a further correction due to the parallax and eccentricity compensation is needed to solve this problem. Once more, as the refracted beam presented a narrow distribution and the *Lmax* gave us a 3 degree error, this was covered largely other systematic errors involved. We used the same light source and polarizer reported in [12] and the negative refractive transmitted light was captured by a CCD camera (KP-D50, Hitachi)) coupled with a singlet lens (focal length of 8 mm, NT-45114, Edmund Optics) placed at 8 mm from the sample. The signal from the camera was sent to a color analogical monitor and a luminance-meter (CS-100, Minolta) was placed at 50 cm from the monitor.

The correctness condition can be expressed as

$$\frac{\left[\begin{array}{l}\left(\dfrac{Ca^2}{A}+\dfrac{b^2}{B}\right)\cos(A)\sin(B)+\\[2mm]\left(\dfrac{a^2}{A}+\dfrac{Cb^2}{B}\right)\sin(A)\cos(B)-\\[2mm]\left(\dfrac{2(C^2-1)ab}{AB}\right)\sin(A)\sin(B)\end{array}\right]}{\left[\begin{array}{l}\sqrt{1-(\cos(A)\cos(B)-C\sin(A)\sin(B))^2}\times\\[2mm]\left|\cos^{-1}(\cos(A)\cos(B)-C\sin(A)\sin(B))\right|\end{array}\right]}<\frac{\lambda^2}{(2\pi n_0\sin(\alpha))^2}$$

where

$$A=\frac{2\pi a}{\lambda}\sqrt{n_a^2-n_0^2\sin^2(\alpha)},$$

$$B=\frac{2\pi b}{\lambda}\sqrt{n_b^2-n_0^2\sin^2(\alpha)},$$

$$C=\frac{1}{2}\left(\frac{\sqrt{n_a^2-n_0^2\sin^2(\alpha)}}{\sqrt{n_b^2-n_0^2\sin^2(\alpha)}}+\frac{\sqrt{n_b^2-n_0^2\sin^2(\alpha)}}{\sqrt{n_a^2-n_0^2\sin^2(\alpha)}}\right).$$

Effective mass approximation expressions

The effective mass expressions that only work close to the band-edge can be obtained from reference [11] and are given by:

$$m_{per}^{TE}=m_{per}^{TM}=\frac{\hbar}{(a+b)^2c}\left[\begin{array}{l}(bn_b+aC)\sin(B_0)\cos(A_0)+\\(an_a+bD)\sin(A_0)\cos(B_0)\end{array}\right],$$

$$m_{par}^{TE/TM}=\frac{(a+b)^2 2\pi m_{per}^{TE/TM}}{\lambda_0\left[\begin{array}{l}\left(\dfrac{b}{n_b}+\dfrac{aC}{n_a^2}\right)\sin(B_0)\cos(A_0)+\\[2mm]\left(\dfrac{a}{n_a}+\dfrac{bD}{n_b^2}\right)\sin(A_0)\cos(B_0)\mp\\[2mm]\dfrac{(n_a^2+n_b^2)(n_a-n_b)^2\lambda_0}{4\pi n_a^3 n_b^3}\sin(A_0)\sin(B_0)\end{array}\right]},$$

where

$$A=\frac{2\pi a}{\lambda_0}\sqrt{n_a^2-n_0^2\sin^2(\alpha)},$$

$$B=\frac{2\pi b}{\lambda_0}\sqrt{n_b^2-n_0^2\sin^2(\alpha)},$$

$$C=\frac{(n_a^2+n_b^2)}{2n_b},$$

$$D=\frac{(n_a^2+n_b^2)}{2n_a}.$$

The signs "$-$" and "$+$" in the expression for $m_{par}^{TE/TM}$ correspond to TE and TM polarized light respectively.

Equation (7) expression

$$\tan(\beta')=-\frac{(a+b)\sqrt{1-(\cos(A)\cos(B)-C\sin(A)\sin(B))^2}}{\dfrac{2\pi n_0\sin(\alpha)}{\lambda}\left[\begin{array}{l}\left(\dfrac{Ca^2}{A}+\dfrac{b^2}{B}\right)\cos(A)\sin(B)+\\[2mm]\left(\dfrac{a^2}{A}+\dfrac{Cb^2}{B}\right)\sin(A)\cos(B)-\\[2mm]\left(\dfrac{2(C^2-1)ab}{AB}\right)\sin(A)\sin(B)\end{array}\right]},$$

where

$$A=\frac{2\pi a}{\lambda}\sqrt{n_a^2-n_0^2\sin^2(\alpha)},$$

$$B=\frac{2\pi b}{\lambda}\sqrt{n_b^2-n_0^2\sin^2(\alpha)},$$

$$C=\frac{1}{2}\left(\frac{\sqrt{n_a^2-n_0^2\sin^2(\alpha)}}{\sqrt{n_b^2-n_0^2\sin^2(\alpha)}}+\frac{\sqrt{n_b^2-n_0^2\sin^2(\alpha)}}{\sqrt{n_a^2-n_0^2\sin^2(\alpha)}}\right).$$

Acknowledgments

The authors thank Dr. J. Antonio del Rio, Dr. Julia Tagüeña-Martinez for providing porous silicon samples, helpful discussions and comments to this manuscript. Dr. Alejandro Reyes for constructive comments on the manuscript. Dr. Rocio Nava and Dr. Beatriz de la Mora for sample preparation and useful comments. Gildardo Casarrubias and Rene Guardian for the sputtering system and SEM pictures and Patrick Perron for technical assistance.

Author Contributions

Conceived and designed the experiments: JEL RD. Performed the experiments: JEL RD. Analyzed the data: JEL. Contributed reagents/materials/analysis tools: JF. Wrote the paper: JEL RD JF.

References

1. Baba T, Asatsuma T, Matsumoto T (2008) Negative refraction in photonic crystals. MRS Bulletin 33: 927–930.
2. Russell PStJ (1986) Optics of Floquet-Bloch Waves in Dielectric Gratings. Appl Phys B: Photophys Laser Chem 39: 23.
3. Russell PStJ, Birks TA, Loyd-Lucas FD (1995) In: E Burstein, C Weisbuch, eds. Confined Electrons and Photons, New Physics and Applications, Vol. 340 of NATO Advanced Studies Institute, Series B: Physics, Plenum, New York. 585 p.
4. Russell PStJ, Birks TA (1996) In: Soukoulis M, ed. Photonic Band Gap Materials, Vol. 315 of NATO Advanced Studies Institute, Series E: Applied Sciences, Kluwer, Dordrecht. 71 p.
5. Russell PStJ (1986) Interference of integrated Floquet-Bloch waves. Phys Rev A 33: 3232.
6. Zengerle R (1987) Light propagation in singly and doubly periodic planar waveguides. J Mod Opt 34: 1589.
7. Kosaka H, Kawashima T, Tomita A, Notomi M, Tamamura T, et al. (1998) Superprism phenomena in photonic crystals. Phys Rev B 58: R10096.
8. Kosaka H, Kawashima T, Tomita A, Notomi M, Tamura T, et al. (1999) Self-collimating phenomensa in photonic crystal. Appl Phys Lett 74: 1212.
9. Notomi M (2002) Negative refraction in photonic crystals. Opt Quantum Electron 34: 133.
10. Zeng Y, Fu Y, Chen X, Lu W (2006) Effective dielectric constant of two-dimensional photonic crystals in optical bands. Solid state communications 138: 205–210.
11. Kavokin AV, Malpuech G, Shelykh I (2005) Negative refraction of light in Bragg mirrors made of porous silicon. Physics Letters A 339: 387–392.
12. Lugo JE, de la Mora B, Doti R, Nava R, Tagueña J, et al. (2009) Multiband negative refraction in one-dimensional photonic crystals. Optics Express 17: 3036–3041.
13. Yuan YC, Ming HZ, Long SJ, Fang LC, Wang Q (2007) Frequency bands of negative refraction in finite one-dimensional photonic crystals. Chin Phys 16: 173.
14. Srivastava R, Thapa BK, Pati S, Ojha SP (2008) Negative refraction in 1D photonic crystals. Solid State Communications 147: 157–160.
15. Boedecker G, Henkel C (2003) All-frequency effective medium theory of a photonic crystal. Optics Express 11: 1590.
16. Foteinopoulou S, Soukoulis CM (2005) Electromagnetic wave propagation in two-dimensional photonic crystals: A study of anomalous refractive effects. Phys Rev B 72: 165112.
17. Joannopoulos JD, Meade RD, Winn JN (1995) Photonic Crystals Molding the Flow of Light. New Jersey: Princenton University Press.
18. Nava R, de la Mora MB, Tagüeña-Martínez J, del Río JA (2009) Refractive index contrast in porous silicon multilayers. Phys Status Solidi C 6: 1721.
19. Pavesi L (1997) Porous silicon dielectric multilayers and microcavities. La Rivista del Nuovo Cimento 20: 1.
20. Gerken M, Miller DAB (2003) Wavelength demultiplexer using the spatial dispersion of multilayer thin-film structures. IEEE Photon Technol Lett 15: 1097–1099.

15

Fractional Calculus Model of Electrical Impedance Applied to Human Skin

Zoran B. Vosika[1], Goran M. Lazovic[2], Gradimir N. Misevic[3]*, Jovana B. Simic-Krstic[1]

1 Department of Biomedical Engineering, Faculty of Mechanical Engineering at University of Belgrade, Belgrade, Serbia, 2 Department of Mathematics, Faculty of Mechanical Engineering at University of Belgrade, Belgrade, Serbia, 3 Department of Research, Gimmune GmbH, Zug, Switzerland

Abstract

Fractional calculus is a mathematical approach dealing with derivatives and integrals of arbitrary and complex orders. Therefore, it adds a new dimension to understand and describe basic nature and behavior of complex systems in an improved way. Here we use the fractional calculus for modeling electrical properties of biological systems. We derived a new class of generalized models for electrical impedance and applied them to human skin by experimental data fitting. The primary model introduces new generalizations of: 1) Weyl fractional derivative operator, 2) Cole equation, and 3) Constant Phase Element (CPE). These generalizations were described by the novel equation which presented parameter (β) related to remnant memory and corrected four essential parameters ($R_0, R_\infty, \alpha, \tau_\alpha$). We further generalized single generalized element by introducing specific partial sum of Maclaurin series determined by parameters ($\beta*, \gamma, \delta \ldots$). We defined individual primary model elements and their serial combination models by the appropriate equations and electrical schemes. Cole equation is a special case of our generalized class of models for $\beta* = \gamma = \delta = \ldots = 0$. Previous bioimpedance data analyses of living systems using basic Cole and serial Cole models show significant imprecisions. Our new class of models considerably improves the quality of fitting, evaluated by mean square errors, for bioimpedance data obtained from human skin. Our models with new parameters presented in specific partial sum of Maclaurin series also extend representation, understanding and description of complex systems electrical properties in terms of remnant memory effects.

Editor: Boris Rubinsky, University of California at Berkeley, United States of America

Funding: This work is supported by the Ministry of Education and Science of the Republic of Serbia through Project No. III 41006 and is also partially supported by GNM private funds. The funders had no role in study design, data collection and analysis, decision to publish, or preparation of the manuscript.

Competing Interests: Author GNM declares that he is the owner of Gimmune GmbH. GNM declares that Gimmune GmbH is active in the field of tumor and viral research, as well in the field of nanotechnology, which are not overlapping with the field of science expressed in this publication. No support from other profitable organizations has been given to the authors affiliated with this research. Furthermore Gimmune GmbH does not have any financial and non-financial relationships with the University of Belgrade. It does not compromise any objectivity or validity of the research, the results and methods expressed in this paper.

* E-mail: gradimir@gimmune.com

Introduction

Application of the mathematics in biology and medicine requires interdisciplinary approach employing the high level of theoretical and experimentally based knowledge in all three disciplines [1],[2]. Here we will briefly introduce thematically relevant overview of the previous use of fractional calculus related to electrical impedance with application to human skin. Fractional calculus is a branch of mathematical analysis that generalizes the derivative and integral of a function to non-integer order [3],[4]. Application of fractional calculus in classical and modern physics greatly contributed to the analysis and our understanding of physico-chemical and bio-physical complex systems [5]. In the past two decades fractional calculus extended popularity in other natural science branches such as chemistry, biology and medicine. Living organisms are the most complex systems composed of over billons of different interconnecting entities at different spatial and temporal scales [6]. Therefore, our understanding of biological systems organization requires fractional calculus as a mathematical tool [5],[7],[8],[9],[10]. A large number of useful biophysical studies reported applications of fractional calculus; however, they were limited to relatively small number of biological model system examples such as: (1) electrical properties of neurons in neurobiology [11], (2) viscoelastic properties of muscles and bones in tissue bioengineering [12], [13], (3) kinetic properties of cell growth and differentiation during morphogenesis in developmental biology [14].

Fractional calculus is a mathematical field extending classical calculus for non-integer order of derivation thus dealing with derivatives and integrals of arbitrary and complex orders [3–5,8,15]. The fractional derivatives are non-local operators because they are defined using integrals. Consequently the fractional derivative in time contains information about the function at earlier points, thus it possesses a memory effect, and it includes non-local spatial effects [3–5,8,15]. In other words fractional derivatives are not a local property (point – quantity) and they consider the history and non-local distributed effects which are essential for better and more precise description and understanding of the complex and dynamic system behavior.

Fractional calculus applications in life sciences provides possibility to analytically focus on modeling of biological life processes where fractional order model will span multiple scales (nanoscale, microscale, mesoscale, and macroscale). Skin is the largest human organ with extremely high cellular and molecular complexity functioning as the protecting, communication and transfer interface between body and environment [16]. Therefore, human skin as highly ordered multilayer organ is particularly suitable model system for applying fractional calculus approach. Commonly, structural and functional studies of human skin employed measurements of bioelectrical and biochemical properties as well as simplified modeling. These approaches were incapable to provide mathematically precise analytical information and statistically significant predications on the electrical behavior of skin. Here we are extending fractional calculus application in biomedical field of natural sciences by modeling electrical properties of human skin which are based on its structural components and thus of its physiological state. Conductance and dielectric features of material, including biological tissues, are known to exhibit frequency dispersions [17], [18]. Impedance is therefore a complex resistivity (real and imaginary part) displayed under alternative current. We have used fractional calculus to model electrical impedance and applied derived models to describe bioimpedance properties of human skin as a test system.

Multifrequency measurements and modeling of electrical impedance is an important spectroscopy method in study of complex biological tissues and materials such as human skin. Passive electric properties of human skin were studied by Cole mathematical method employing bioimpedance measurements below 100 kHz [17], [19]. Cole model deals with both conductive and dielectric properties [20], whereas Cole-Cole approach primarily describes dielectric features (determined as permittivity) [21]. Since the human skin, as the complex organ, displays both conductive and dielectric behavior, neither of the two models can be applied to precisely describe and study bioimpedance properties of this organ.

Biological membranes show a high capacitance and a low but complicated pattern of conductivity [20]. Biological tissues as complex multi-layer systems behave as an anisotropic material due to the variable orientation of cells and their plasma membranes. As mentioned above in 1940 Cole formulated a mathematical model of electrical properties of cell membranes based on impedance measurements at multi frequency alternative current. Kenneth Cole and Robert Cole have conveyed another mathematical model of dielectric properties for materials in 1941 [21]. In 2001 El-Lakkani [22] attempted to analyze electric and dielectric properties of different types of human tissues either by Cole-Cole model [20] or Dissado model [23] in the alternative current frequency range from 20 Hz to100 kHz. This type of modeling was reviewed by Grimnes & Martinsen [17]. Such Multi Frequency Bioelectrical Impedance Analysis (MF-BIA) is a noninvasive and relatively new technique for studying biological systems. The complex impedance as a function of frequency of the external alternating voltage source (ω is frequency $\omega\in(0,\infty)$) is one of the powerful linear passive characteristic of materials in the frequency domain. One of the passive characteristics of materials in alternating current circuits models is a Constant Phase Element (*CPE*) which will be here further mathematically defined and generalized by impedance equations using fractional calculus approach.

We have used fractional calculus approach to construct simple models with unified principles. Without fractional calculus approach it would not be possible to make this generalized type of superior and more precise class of models where Cole model is a special case. Here we report modeling of bioimpedance using fractional calculus approach and experimental data fitting for human skin test system. We have derived new bioimpedance equations introducing one new parameter and corrections for four

parameters by employing generalized Weyl fractional derivative operator. Our model and results provide significant mathematical advance for solving complex biosystems when compared to the classical Cole model. Therefore, presented bioimpedance fractional calculus modeling may be useful for further fundamental research with applications in medicine which are related to physiological and pathological analysis of human skin.

Results

Our modeling strategy of the bioimpedance with application to human skin is based on generalization of Weyl fractional derivative operator, generalization of Cole equation, and generalization of *CPE*.

1. Generalized Weyl fractional calculus

Our primary idea was to generalize Weyl fractional calculus because this method is necessary for mathematical analysis of complex periodic functions describing characteristic values of alternating current in electric circuits which are employed in bioimpedance modeling.

1.1. Introduction to Weyl fractional calculus. In 1917 Weyl introduced his new approach of fractional calculus to analysis of periodic functions. In summary the α-th fractional integral and α-th fractional derivative of a 2π periodic adequate complex function of real variable are defined respectively by

$$(D^{-\alpha}\varphi)(t) := \sum_{l=-\infty}^{\infty} (j\cdot l)^{-\alpha}\cdot\varphi_f(l)\cdot e^{j\cdot l\cdot t},$$

$$(D^{\alpha}\varphi)(t) := \sum_{l=-\infty}^{\infty} (j\cdot l)^{\alpha}\cdot\varphi_f(l)\cdot e^{j\cdot l\cdot t}, j^2 = -1 \quad (1)$$

where $\alpha > 0$ and

$$\varphi_f(l) = \tfrac{1}{2\pi}\cdot\int_0^{2\pi} e^{-j\cdot l\cdot t}\cdot\varphi(t)\cdot dt, \, l=\pm 1,\pm 2,\pm 3,..., \varphi_f(0)=0 \quad (2)$$

This discrete Fourier transform can be viewed as the most usual way of defining fractional integrals and derivatives of periodic functions. The Equations (1) are correctly defined under the condition that $\varphi_f(0)=0$. Supplement to definition (1) is $D^{\alpha=0}=Id$ where *Id* is unit operator. In the case of considering formula for Weyl fractional integral

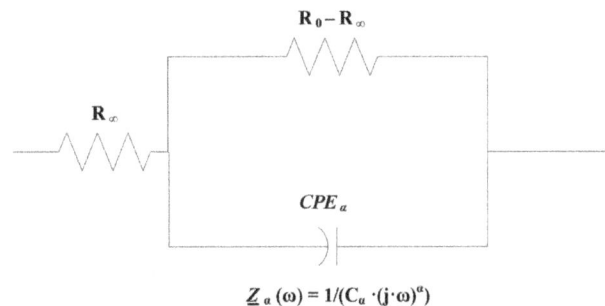

$$\underline{Z}_a(\omega) = 1/(C_a\cdot(j\cdot\omega)^a)$$

Figure 1. Schema of circuit for modified Cole element [20] according to Magin [7], [8].

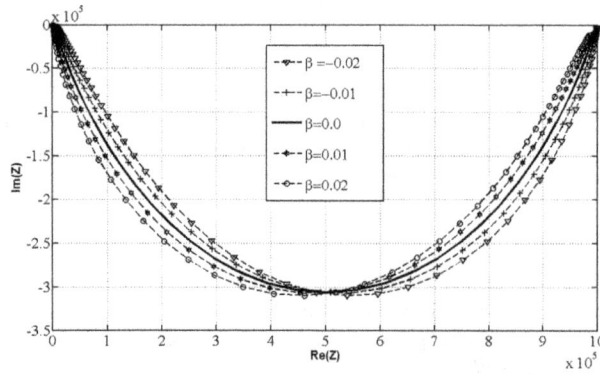

Figure 2. Cole Plot of Generalized Cole Model (GC1). Cole plot of GC1 model with five selected values of β parameter and suitable fixed parameters $R_0^* = 1 M\Omega$, $R_\infty^* = 1 k\Omega$, $\tau_{\alpha^*;\beta} = 1s$, $\alpha^* = 0.7$ in the frequency interval $\omega \in [0.1 Hz, 10^5 Hz]$.

$$_{-\infty}L_t^\alpha \varphi(t) := \frac{1}{\Gamma(\alpha)} \int_{-\infty}^{t} (t-t')^{\alpha-1} \varphi(t') \cdot dt', \ t > -\infty, \ 0 < \alpha < 1 \quad (3)$$

for

$$\psi_l(t) = e^{j \cdot l \cdot t} \quad (4)$$

then it follows (Butzer and Nessel, [9])

$$_{-\infty}L_t^\alpha \psi_l(t) = (j \cdot l)^{-\alpha} \cdot \psi_l(t), \ \forall l \in \mathbb{Z} : l \neq 0 \quad (5)$$

hence, if we define

$$_{-\infty}L_t^\alpha \varphi(0) = 0 \quad (6)$$

then it follows

$$\left(_{-\infty}L_t^\alpha \varphi\right)(t) = \sum_{l=-\infty}^{\infty} (j \cdot l)^{-\alpha} \cdot \varphi_f(l) \cdot \psi_l(t) \quad (7)$$

This operator $_{-\infty}L_t^\alpha$ is identical to the operator $D^{-\alpha}$ for $0 < \alpha < 1$. Then the Equation (7) can be written in the following form

$$\left(_{-\infty}L_t^\alpha \varphi\right)(t) \sim \sum_{l \neq 0} (j \cdot l)^{-\alpha} \cdot \varphi_f(l) \cdot \psi_l(t) \quad (8)$$

Previously defined operator is linear

$$\left(_{-\infty}L_t^\alpha \varphi\right)(t) \sim \sum_{l=-\infty}^{\infty} \varphi_f(l) \cdot _{-\infty}L_t^\alpha \psi_l(t) \quad (9)$$

1.2. Rigorous treatment of the Weyl approach to fractional calculus. The Weyl approach to fractional calculus, can be rigorously treated in the Banach spaces 2π periodic complex function of real variable $\varphi = \varphi(t) : \mathbb{R} \to \mathbb{C}$, defined by the interval $[0, 2 \cdot \pi]$

$$L_{2 \cdot \pi}^p := \left\{ \varphi | \|\varphi\|_{L_{2 \cdot \pi}^p} < \infty \right\} \quad (10)$$

with norms of this function defined by

$$\|\varphi\|_{L_{2 \cdot \pi}^p} := \left\{ \frac{1}{2 \cdot \pi} \cdot \int_0^{2 \cdot \pi} |\varphi(t)|^p dt \right\}^{1/p}, \ 1 \leq p < \infty \quad (11)$$

The factor $1/2\pi$ is characteristic for the usual Fouirer analysis. In the case $p = 2$ describes Hilbert space $L_{2 \cdot \pi}^2$. Dot product and norm are defined in the standard way (over line means complex conjugation).

$$(\varphi, \psi)_{L_{2 \cdot \pi}^2} := \frac{1}{2 \cdot \pi} \cdot \int_0^{2 \cdot \pi} \varphi(t) \cdot \overline{\psi(t)} \cdot dt, \|\varphi\|_{L_{2 \cdot \pi}^2} := \left[\frac{1}{2 \cdot \pi} \cdot \int_0^{2 \cdot \pi} |\varphi(t)|^2 dt \right]^{1/2} \quad (12)$$

If $\varphi, \psi \in L_{2 \cdot \pi}^2$ and φ, ψ are real functions the following relation holds

$$(\varphi, \psi)_{L_{2 \cdot \pi}^2} = \overline{(\psi, \varphi)}_{L_{2 \cdot \pi}^2} = (\psi, \varphi)_{L_{2 \cdot \pi}^2} \quad (13)$$

As described by Butzer and Westphal [24], Equation (8) can be considered as a motive for the definition of fractional integral $I^\alpha \varphi$, in the form of the convolution integral for $\alpha > 0$ and $\varphi \in L_{2 \cdot \pi}^p$. In order to more accurately describe the fractional integral of Weyl I^α for $\alpha > 0$ and $\varphi = \varphi(t) \in L_{2 \cdot \pi}^p$ the right side of Equation (8) will be rewritten as

$$\sum_{l=-\infty}^{\infty} \eta_{f\alpha}(l) \cdot \varphi_f(l) \cdot \psi_l(t) \quad (14)$$

with

$$\eta_{f\alpha}(l) := \begin{cases} (j \cdot l)^{-\alpha}, \ l \neq 0 \\ 0, \ l = 0 \end{cases} \quad (15)$$

The corresponding Fourier transformed function is

$$\eta_\alpha(t) := \sum_{l \neq 0} (j \cdot l)^{-\alpha} \cdot \psi_l(t) = \sum_{l=-\infty}^{\infty} \eta_{f\alpha}(l) \cdot \psi_l(t) \quad (16)$$

We now move to more precise definition of Weyl fractional integral. This function will be used as a kernel of the associated convolution integral, so-called Weyl fractional integral I^α, defined as follows

$$(I^\alpha \varphi)(t) := (\eta_\alpha * \varphi)(t) = \frac{1}{2 \cdot \pi} \cdot \int_0^{2 \cdot \pi} \eta_\alpha(t-t') \cdot \varphi(t') \cdot dt' \quad (17)$$

Then according to Butzer and Nessel, [9], I^α bounded linear operator, is actually a continuous operator over the space $L_{2\cdot\pi}^p$ for all $p\in[1,\infty)$

$$\|I^\alpha\varphi\|_{L_{2\cdot\pi}^p}\le\|\eta_\alpha\|_{L_{2\cdot\pi}^1}\cdot\|\varphi\|_{L_{2\cdot\pi}^p}\ \forall\varphi\in L_{2\cdot\pi}^p \qquad (18)$$

Moreover, the convolutional theorems for periodic functions (Butzer and Westphal, [24], Theorem 4.1.3), will be

$$(\eta_\alpha*\varphi)_f(l)=\eta_{f\alpha}(l)\cdot\varphi_f(l)\forall l\in\mathbb{Z}$$
$$=\begin{cases}(j\cdot l)^{-\alpha}\cdot\varphi_f(l)&l\ne0\\0&l=0\end{cases} \qquad (19)$$

Comparing Equations (1) and (8) for the linear operator I^α, it can be reasonably concluded that I^α is a fractional integral operator. According to the Equation (17) I^α is defined

$$(I^\alpha\varphi)(t):=\sum_{l\ne0}(j\cdot l)^{-\alpha}\cdot\varphi_f(l)\cdot\psi_l(t) \qquad (20)$$

Since $I^\alpha\varphi$ defines for all $\varphi(t)\in L_{2\cdot\pi}^p$ then $I^\alpha\varphi\in L_{2\cdot\pi}^p$. Same relation does not hold for Weyl fractional derivative

$$(D^\alpha\varphi)(t):=\sum_{l\ne0}(j\cdot l)^{\alpha}\cdot\varphi_f(l)\cdot\psi_l(t),\alpha>0 \qquad (21)$$

If the index $l=0$ then the corresponding member in Equation (21) is equal zero. For $\alpha>0$ the function $\eta_\alpha(t)\in L_{2\cdot\pi}^2$ because $L_{2\cdot\pi}^2\subset L_{2\cdot\pi}^1$. In the particular case when $p=2$, for the Hilbert space $L_{2\cdot\pi}^2$ Parseval Equation gives the maximum definitional domain for fractional derivative D^α. This domain is defined by

$$Dom(D^\alpha):=\{\varphi\in L_{2\cdot\pi}^2|D^\alpha\varphi\in L_{2\cdot\pi}^2\}$$
$$=\left\{\varphi\in L_{2\cdot\pi}^2|\sum_{l=-\infty}^{\infty}l^{2\alpha}\cdot|\varphi(l)|^2<\infty\right\} \qquad (22)$$

To eliminate the obvious deficiency, one of the ways is to define the operator D^α in the theory of distribution. Although we will not consider such general problem, we will use the most appropriate fractional operators and his domains. Later we will derive generalizations of fractional operators and their domains. Furthermore, we will consider the existence of a nontrivial, nonempty, common domain for all fractional integrals I^α and derivatives D^α for $\alpha>0$, which will be also codomain.

In this regard, we will review the maximal set $W_{all}\subset L_{2\cdot\pi}^2$ of complex functions of a real variable $\varphi=\varphi(t)$, so that following is valid.

$$(\forall\beta,\gamma>0)(\varphi\in W_{all}\Rightarrow D^\beta\varphi,I^\gamma\varphi\in W_{all}) \qquad (23)$$

It follows that the set of W_{all} will define Weyl fractional derivatives and fractional integrals, so that the W_{all} is domain and codomain. Previously mentioned nonempty set W_{all} (e.g., function

$e^{j\cdot t}$ belongs to such a set) is by construction closed under the operation of Weyl fractional integrals. The idea for the basic theorem of generalized Weyl fractionation derivatives and integrals is contained in the following theorem.

Theorem 1. For all $\alpha,\beta>0$

(i) $I^\alpha I^\beta\phi=I^{\alpha+\beta}\phi$, $\phi\in W_{all}$.

(ii) $D^\alpha I^\alpha\phi=I^\alpha D^\alpha\phi=\phi-\phi_f(0)$, $\phi\in W_{all}$.

Proof: We should bear in mind the sum member $\phi_f(0)$ must be excluded. The proof is analogous to that given in (Butzer and Westphal, [24], Proposition 4.1).

1.3. Generalization of Weyl fractional integral and derivative. Previous mathematical description is the basis for generalization of Weyl fractional integral $I^{\alpha;\beta}$, for $\alpha>0$ and $\beta\ge0$. Motivation for introducing generalization of Weyl fractional integral $I^{\alpha;\beta}$ is to use useful modification of Riemann –Liouville fractional integral $K^{\nu,\gamma}$ described by Nigmatullin [15]. Our basic idea for a new type of generalization is to use fractional integral operator acting on a periodic function so that the result of the operator action is a periodic function. For that purpose the set of Equations (14), (15) and (16) will be written in more general form in account member $\exp\left(-\nu\cdot\log(s)-\gamma\cdot(\log(s))^2\right)$ from Nigmatullins Equation [15].

$$\sum_{l=-\infty}^{\infty}\eta_{f\alpha;\beta}(l)\cdot\varphi_f(l)\cdot\psi_l(t),\alpha>0,\ \beta\ge0 \qquad (24)$$

and

$$\eta_{f\alpha;\beta}(l):=\begin{cases}(j\cdot l)^{-(\alpha+\beta\cdot\log(j\cdot l))},\ l\ne0\\0,\ l=0\end{cases} \qquad (25)$$

respectively,

$$\eta_{\alpha;\beta}(t):=\sum_{l\ne0}(j\cdot l)^{-(\alpha+\beta\cdot\log(j\cdot l))}\cdot\psi_l(t)=\sum_{l=-\infty}^{\infty}\eta_{f\alpha;\beta}(l)\cdot\psi_l(t) \qquad (26)$$

Lemma 1. $\eta_{\alpha;\beta}(t)\in L_{2\cdot\pi}^1$ for all $\alpha>0,\beta\ge0$.
Proof For $\beta=0$

$$\eta_{\alpha;\beta=0}(t)=2\cdot\sum_{l>0}\frac{\cos(l\cdot t-\alpha\cdot\pi/2)}{l^\alpha} \qquad (27)$$

which is a convergent series in $L_{2\cdot\pi}^1$, if $\beta>0$ then

$$\eta_{\alpha;\beta}(t)=2\cdot e^{\beta\cdot\frac{\pi^2}{4}}\sum_{l>0}\frac{\cos(l\cdot t-\alpha\cdot\pi/2-\beta\cdot\pi\cdot\log(l))}{l^{\alpha+\beta\cdot\log(l)}} \qquad (28)$$

This series is also convergent with respect to the previous, because there is a functional dependence of the exponent denominator fraction by the member of the sum.

Theorem 2. Generalization of Weyl fractional integral $I^{\alpha;\beta}$

$$(I^{\alpha;\beta}\varphi)(t):=(\eta_{\alpha;\beta}*\varphi)(t)=\frac{1}{2\cdot\pi}\int_0^{2\cdot\pi}\eta_{\alpha;\beta}(t-t')\cdot\varphi(t')\cdot dt' \qquad (29)$$

represents a linear and bounded, actually continuous operator in $L_{2\cdot\pi}^p$ for all $p\in[1,\infty)$.

Proof: The operator $I^{\alpha;\beta}$ is bounded because the following is valid.

$$\left\|I^{\alpha;\beta}\varphi\right\|_{L_{2\cdot\pi}^p} \leq \left\|\eta_{\alpha;\beta}\right\|_{L_{2\cdot\pi}^1}\cdot\|\varphi\|_{L_{2\cdot\pi}^p} \quad \forall\varphi\in L_{2\cdot\pi}^p \qquad (30)$$

Norm of the first element to the right is finite because of the previous lemma.

Therefore, if $\varphi(t)\in L_{2\cdot\pi}^p$ then $I^{\alpha;\beta}\varphi\in L_{2\cdot\pi}^p$. As with the Weyl fractional derivatives, there is a problem of completeness of the domain operator if other real value for α (non zero) and β are used. In general case instead of $I^{\alpha;\beta}$ we use $K^{\alpha;\beta}$. Therefore, we define the operator $K^{\alpha;\beta}$ acting on the functions $\varphi=\varphi(t)\in L_{2\cdot\pi}^2$ which are written

$$\varphi(t) := \sum_{l=-\infty}^{\infty} \varphi_f(l)\cdot\psi_l(t), \psi_l(t)=e^{j\cdot l\cdot t},$$

$$\varphi_f(l) = \frac{1}{2\cdot\pi}\cdot\int_0^{2\cdot\pi}\varphi(t)\cdot\overline{\psi_l}(t)\cdot dt, l=\pm1,\pm2,\pm3,...,\varphi_f(0)=0, \qquad (31)$$

In the case that the operator $\left(K^{\alpha;\beta}\varphi\right)(t)$ makes sense, analogous to (1) we define:

$$\left(K^{\alpha;\beta}\varphi\right)(t) := \sum_{l=-\infty}^{\infty}(j\cdot l)^{(\alpha+\beta\cdot\log(j\cdot l))}\varphi_f(l)\cdot\psi_l(t); \alpha\neq0,\beta\in\mathbb{R} \quad (32)$$

Supplement to the previous definition is $K^{0;0}=Id$.

The idea presented in the previous section on non-trivial and non-empty common domain and codomain for all operators that are defined by formula (32) will be used here. By analogy we define domain of the operator $K^{\alpha;\beta}$.

$$Dom\left(K^{\alpha;\beta}\right) := \left\{\varphi\in L_{2\cdot\pi}^2|K^{\alpha;\beta}\varphi\in L_{2\cdot\pi}^2, \alpha\neq0 \wedge \beta\in\mathbb{R}\right\} =$$
$$\left\{\varphi\in L_{2\cdot\pi}^2| \sum_{l=-\infty}^{\infty}|(j\cdot l)|^{2(\alpha+\beta\cdot\log(j\cdot l))}\cdot|\varphi(l)|^2<\infty\right\} \qquad (33)$$

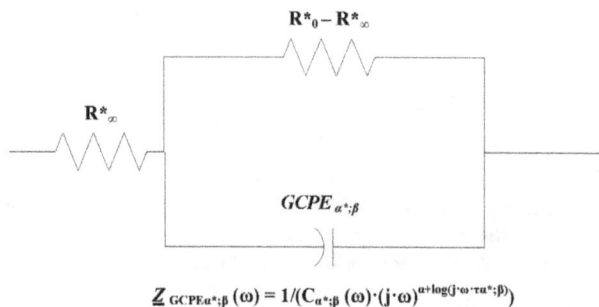

$$Z_{GCPE\alpha^*;\beta}(\omega) = 1/(C_{\alpha^*;\beta}(\omega)\cdot(j\cdot\omega)^{\alpha+\log(j\cdot\omega\cdot\tau\alpha^*;\beta)})$$

Figure 3. Schema of Circuit for Generalized Cole (GC1) Element with GCPE.

This domain is neither equal to $L_{2\cdot\pi}^p$ nor to $L_{2\cdot\pi}^2$. As indicated previously here we will define maximal set $W'_{all}\subset L_{2\cdot\pi}^2$ of complex functions of a real variable $\varphi=\varphi(t)$, with set properties.

$$(\forall\alpha\neq0)(\forall\beta\in\mathbb{R})\left(\phi\in W'_{all}\Rightarrow K^{\alpha;\beta}\phi\in W'_{all}\right) \qquad (34)$$

Described non-empty set W'_{all}, which is domain and codomain for all described operators $K^{\alpha;\beta}$, is by construction closed in relation to the generalization of Weyl fractional derivative or integral.

Theorem 3. The basic properties of operators in a given domain $(\phi\in W'_{all})$

$$\begin{aligned}&1.\ K^{\alpha;\beta=0}=D^\alpha,\ K^{-\alpha;\beta=0}=I^\alpha,\ \alpha>0,\\&2.\ K^{\alpha_1;\beta_1}K^{\alpha_2;\beta_2}=K^{\alpha_1+\alpha_2;\beta_1+\beta_2},\ \alpha_1,\alpha_2\in\mathbb{R}\backslash\{0\},\ (\alpha_1\cdot\alpha_2)>0,\\&3.\ K^{-\alpha;0}K^{\alpha;\beta}\phi=K^{0;\beta}\left(\phi-\phi_f(0)\right),\ \alpha\neq0,\ \beta\neq0\\&4.\ K^{-\alpha;-\beta}K^{\alpha;\beta}\phi=K^{\alpha;\beta}K^{-\alpha;-\beta}\phi=\phi-\phi_f(0),\ \alpha\in\mathbb{R}\backslash\{0\},\beta\in\mathbb{R}\end{aligned} \quad (35)$$

Proof: For 1–3 is obvious and for 4 it can be considered if $\phi\in W'_{all}\subset L_{2\cdot\pi}^2$ then this function can be written in a unique way $\phi=\sum_{l\in\mathbb{Z}}(\phi,\psi_l)_{L_{2\cdot\pi}^2}\cdot\psi_l(t)$.

Because $\chi=\chi(t)=K^{\alpha;\beta}K^{-\alpha;-\beta}\phi(t)\in W'_{all}$ for $\alpha\neq0$ it follows

$$\chi=\sum_{l\in\mathbb{Z}}(\phi,\psi_l)_{L_{2\cdot\pi}^2}\cdot\psi_l(t)$$

then for all $l\in\mathbb{Z}$

$$\left(K^{\alpha;\beta}K^{-\alpha;-\beta}\varphi,\psi_l\right)_{L_{2\cdot\pi}^2}=\left(\varphi,\left(-K^{-\alpha;-\beta}\right)\left(-K^{\alpha;\beta}\right)\psi_l\right)_{L_{2\cdot\pi}^2}$$
$$=\begin{cases}(\varphi,\psi_l)_{L_{2\cdot\pi}^2} & l\neq0\\0 & l=0\end{cases}$$

therefore

$$K^{\alpha;\beta}K^{-\alpha;-\beta}\phi=\sum_{l\in\mathbb{Z}}(\phi,\psi_l)_{L_{2\cdot\pi}^2}\cdot\psi_l(t)-(\phi,\psi_0)_{L_{2\cdot\pi}^2}\cdot\psi_0(t)$$
$$=\phi-\phi_f(0)$$

In a similar way proofs are provided for the other three equations.

Also, instead $[0,2\cdot\pi]$ it is possible to use a symmetrical segment $[-\pi,\pi]$. In the case the periodic function $\varphi=\varphi(t)T=2\cdot\pi/\omega,\omega>0$, appropriate formulas will be done by the transformation $2\cdot\pi\to2\cdot\pi/\omega$. Then instead $\psi_l(t)=\exp(j\cdot l\cdot t)$ we write $\psi_{l\omega}(t)=\exp(j\cdot l\cdot\omega\cdot t)$, etc. In this case, if operator $K^{\alpha;\beta}$ acts on $\psi_{1\omega}(t)=\exp(j\cdot\omega\cdot t)$, then following holds

$$K^{\alpha;\beta}(\exp(j\cdot\omega\cdot t))=(j\cdot\omega)^{(\alpha+\beta\cdot\log(j\cdot\omega))}\cdot\exp(j\cdot\omega\cdot t) \qquad (36)$$

The previous operator is analogous to the following one

$$D^{(\alpha + \beta \cdot \log(D))} \qquad (37)$$

in the case when the Weyl operator derivatives is true $D^{\alpha}(j\cdot\tilde{\omega})^{\alpha} \alpha \in \mathbb{R} \setminus \{0\}$.

If one introduces a non-negative relaxation time parameter $\tau_{\alpha^*;\beta}$, the modified operator $\left(\tau_{\alpha^*;\beta} \cdot K\right)^{\alpha^*;\beta} \alpha^* > 0$ for operator of fractional type derivative) acts as follows

$$\left(\tau_{\alpha^*;\beta} \cdot K\right)^{\alpha^*;\beta}(\exp(j\cdot\omega\cdot t+\theta)) = (j\cdot\omega\cdot\tau_{\alpha^*;\beta})^{\left(\alpha^* + \beta \cdot \log\left(j\cdot\omega\cdot\tau_{\alpha^*;\beta}\right)\right)} \\ \exp(j\cdot\omega\cdot t+\theta), \theta \in \mathbb{R} \qquad (38)$$

If $\alpha = \alpha^*$ and $\beta = 0$ then operator $\left(\tau_{\alpha^*;\beta} \cdot K\right)^{\alpha^*;\beta}$ is fractional derivative of Weyl type

$$(\tau_{\alpha}\cdot D)^{\alpha}(\exp(j\cdot\omega\cdot t+\theta)) = (j\cdot\omega\cdot\tau_{\alpha})^{\alpha}\cdot\exp(j\cdot\omega\cdot t+\theta), \tau_{\alpha} \geq 0 \qquad (39)$$

Using appropriate mathematical operations and weakened conditions described in Theorem 3 the Equation (38) can be reduced to form (36) due to specific functional dependence. Subsequently we will use operators $(\tau_{\alpha}\cdot D)^{\alpha}$ and $\left(\tau_{\alpha^*;\beta} \cdot K\right)^{\alpha^*;\beta}, \alpha, \alpha^* > 0$.

2. Generalized Cole element and Constant Phase Element (CPE)

Multi Frequency Bioelectrical Impedance Analysis (MF-BIA) is a noninvasive technique for studying biological systems. The complex impedance as a function of frequency of the external alternating voltage source (ω is frequency, $\omega \in (0,\infty)$) is one of the linear passive characteristic of materials in the frequency domain. In alternating current circuits the Constant Phase Element of capacitance type such as CPE_{α} ([7], [8], [25]) is defined by the impedance equation

$$\underline{Z}_{CPE_{\alpha}}(\omega) = \frac{1}{C_{\alpha}\cdot(j\cdot\omega)^{\alpha}}, C_{\alpha} > 0, \alpha \in (0,1] \qquad (40)$$

$$\underline{Z}_{CPE_{\alpha}}(\lambda\cdot\omega) = \frac{\lambda^{-\alpha}}{C_{\alpha}\cdot(j\cdot\omega)^{\alpha}} = \lambda^{-\alpha}\cdot\underline{Z}_{CPE_{\alpha}}(\omega)$$

C_{α} is capacitance of order α. For fractional index $\alpha = 0, \underline{Z}_{CPE_{\alpha}}(\omega) = R$ (resistance).

Mathematical model of electrical properties of cell membranes based on impedance measurements at multi frequency alternative

current was firstly formulated by Cole in 1940 [20]. Magin [7], [8] derived and generalized Cole equation using fractional calculus. His model comprises of three hypothetical circuit elements: a low-frequency resistor R_0, a high-frequency resistor R_∞ and CPE_{α}, arranged as shown in Figure 1. This circuit can be also visualized as two serial connected elements, where the first one is R_∞ and second one is $((R_0 - R_\infty)||CPE_{\alpha})$. Here we name the second element as reduced Cole type element.

The complex impedance described by Cole model is given by the following equation [20].

$$\underline{Z}_{\alpha}(\omega) = R_\infty + \frac{R_0 - R_\infty}{1 + (j\cdot\omega\cdot\tau_{\alpha})^{\alpha}} \qquad (41)$$

Here is the characteristic relaxation time according to Magin [7],[8], positive constant.

$$\tau_{\alpha} = \sqrt[\alpha]{C_{\alpha}\cdot(R_0 - R_\infty)} \qquad (42)$$

All of the four following parameters $R_0, R_\infty, C_{\alpha}, \alpha$ are material constants independent of frequency.

Here we derive Cole equation (41) for the circuit shown in Figure 1. Applied alternating voltage to the system is $V(t) = V_0\cdot\iota\varepsilon\iota\varkappa\iota\rho(j\cdot\omega\cdot t+\theta)$ (V_0 is the voltage amplitude, θ is the phase angle between the voltage and the current), current passing through the system is $i(t) = i_0\cdot\exp(j\cdot\omega\cdot t)$, and strength of current amplitude is i_0. Impedance of the system is given by the equation

$$V(t) = \underline{Z}_{\alpha}(\omega)\cdot i(t) \qquad (43)$$

Where the relations describing the system D^{α} (the Weyl fractional derivative, $D^{\alpha}(j\cdot\tilde{\omega})^{\alpha}$) is

$$(1 + (\tau_{\alpha})^{\alpha}\cdot D^{\alpha})V(t) = (R_\infty\cdot(\tau_{\alpha})^{\alpha}\cdot D^{\alpha} + R_0)i(t) \qquad (44)$$

In terms of fractional derivatives usage, our Eq. (2.44) has some analogy to the modified Zener model of a viscoelastic body [26] and Bohannan equation [27]. However, Bohannan applied fractional derivative according to Riemann-Liouville, whereas we have done according to Weyl. The second difference is that Bohannan [27] is deriving Cole-Cole equation [21], which is describing frequency dispersion of complex dielectric constant, and we have derived Cole equation [17,20], which is describing frequency dispersion of impedance. If one takes into account geometrical properties of the physical system, such as surface of electrode and distance between them, complex dielectric constant becomes complex capacitance ($\underline{C}$). Note that in the theory of alternative current the admittance $\underline{Y}$ is described by for-

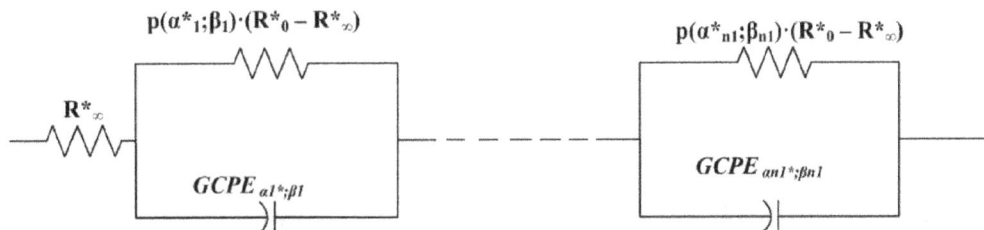

Figure 4. Schema of Circuit for Serially Linked Reduced Generalized Cole Elements.

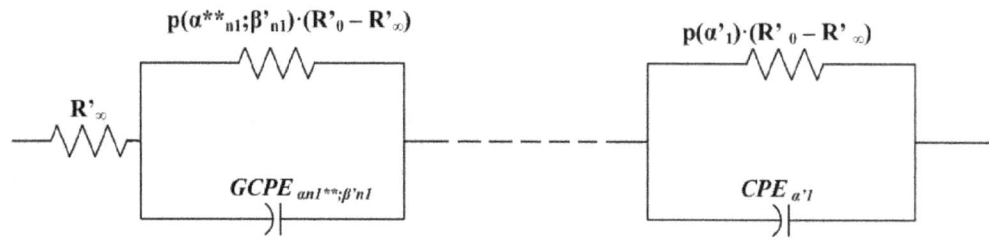

Figure 5. Schema of Circuit for all Permutation of Serially Linked Reduced Generalized Cole Elements and Reduced Cole elements.

mula $\underline{Y} = j \cdot \omega \cdot \underline{C}$. The Eq. (2.44) describes frequency dispersion of complex resistivity in the case when geometry is not taken into account.

In order to generalize our previous Eq. (2.44) we have used similar principal of mathematical approach as reported by Nigmatullin [15]. They described that for a strongly correlated fractal medium a generalization of the Riemann–Liouville fractional integral is obtained. In this paper we have modified generalization of Riemann-Liouville fractional integral and derivative for the case of periodic functions and obtained a new type of generalization of Weyl fractional integral and derivative.

Formally in Equation (44) replacement is done by

$$(\tau_\alpha)^\alpha \cdot (D^\alpha) \to (\tau_{\alpha^*;\beta} \cdot K)^{\alpha^*;\beta} \tag{45}$$

By introducing new parameters for resistance R_0^* and R_∞^* in Equation (44) we derive next equation

$$\left(1 + \left(\tau_{\alpha^*;\beta} \cdot K\right)^{\alpha^*;\beta}\right) V(t) = \left(R_\infty^* \cdot \left(\tau_{\alpha^*;\beta} \cdot K\right)^{\alpha^*;\beta} + R_0^*\right) i(t) \tag{46}$$

Then in (43) we perform following change $\underline{Z}_\alpha(\omega) \to \underline{Z}_{\alpha^*;\beta}(\omega)$ to obtain

$$\underline{Z}_{\alpha^*;\beta}(\omega) = R_\infty^* + \frac{R_0^* - R_\infty^*}{1 + \left(j \cdot \omega \cdot \tau_{\alpha^*;\beta}\right)^{\alpha^* + \beta \cdot \log\left(j \cdot \omega \cdot \tau_{\alpha^*;\beta}\right)}} \tag{47}$$

The Equation (47) describes our new generalized Cole model based on fractional calculus (our primary model). Five new physical and phenomenological parameters $\alpha^*, \beta, \tau_{\alpha^*;\beta}, R_0^*, R_\infty^*$ were introduced. In our work we have introduced new β parameter for modeling electrical impedance of complex systems. This parameter has formal mathematical analogy to γ parameter presented by Nigmatullin [15]. Parameter γ describes relaxation properties of dielectric phenomena of the medium and is derived as the generalized equation of the well-known Kohlrausch Williams Watts relaxation law. Constants R_0^* and R_∞^* are the corresponding resistances, therefore, we can write.

$$\lim_{\beta \to 0} R_0^* = R_0, \lim_{\beta \to 0} R_\infty^* = R_\infty \tag{48}$$

It should be mentioned that for small β, $\alpha^* \approx \alpha$, $R_0^* \approx R_0$ and $R_\infty^* \approx R_\infty$. The R_0^* and R_∞^* constants have different meaning to those described by the Cole model (Equation (41)), because the

leading members of Equation (47) $\beta \cdot \log\left(j \cdot \omega \cdot \tau_{\alpha^*;\beta}\right)$ and $\underline{Z}_{\alpha^*;\beta}(\omega)$ have different asymptotic behavior when $\omega \to 0$ and $\omega \to \infty$.

Using our model we have tested the effects of β on $\underline{Z}_{\alpha^*;\beta}$ while keeping other parameters constant. The example of Cole plot based on equation (47) showed that for values of $\beta \neq 0$ is not circular arc as in Cole model, Figure 2.

If we assume that the change of conducting properties of electrical circuits is only in the *CPE*, then *CPE* can be replaced and we can write.

$$\underline{Z}_{CPE_\alpha}(\omega) = \frac{1}{C_\alpha \cdot (j \cdot \omega)^\alpha} \to \underline{Z}_{GCPE\alpha^*;\beta}(\omega)$$
$$= \frac{1}{C_{\alpha^*;\beta}(\omega) \cdot (j \cdot \omega)^{\alpha^* + \beta \cdot \log\left(j \cdot \omega \cdot \tau_{\alpha^*;\beta}\right)}} \tag{49}$$

Cole model uses linear scaling as seen from the above equation. For generalized CPE (GCPE) we are scaling frequency for dispersion of impedance in non-linear manner. This can be seen from equation (49) i.e. scaling with function not only with one constant α value. Therefore, we have non linearized and non-constant scaling with two parameters $\alpha *$ and β. This is important advantages for modeling and describing natural complex systems. It should be noted that in Cole model linear scaling is valid and this is only one special case of our generalized model where $\beta = 0$ and non-generalized CPE.

The function $C_{\alpha^*;\beta}(\omega)$ is such a function of frequency where the non-negative constant $\tau_{\alpha^*;\beta}$ can be described by the following self-consistent equation.

$$\tau_{\alpha^*;\beta} = \left(\left(R_0^* - R_\infty^*\right) \cdot C_{\alpha^*;\beta}(\omega)\right)^{\overline{\frac{1}{\alpha^* + \beta \cdot \log\left(j \cdot \omega \cdot \tau_{\alpha^*;\beta}\right)}}} \tag{50}$$

$$C_{\alpha^*;\beta}(\omega) = \frac{\tau_{\alpha^*;\beta}^{\alpha^* + \beta \cdot \log\left(j \cdot \omega \cdot \tau_{\alpha^*;\beta}\right)}}{R_0 - R_\infty}$$

The Equation (49) introduces the element $\underline{Z}_{GCPE\alpha^*;\beta}(\omega)$, and provides generalization of *CPE* (*GCPE*). The electrical circuit comprising of the element R_∞^* in serial connection with the elements $R_0^* - R_\infty^*$ and $\underline{Z}_{GCPE\alpha^*;\beta}(\omega)$, which are themselves in a special parallel connection, is described by the Equation (47) and schematically presented in Figure 3. This means that $\underline{Z}_{GCPE\alpha^*;\beta}(\omega)$ is a function depending on the element $R_0^* - R_\infty^*$. Part of the electric circuit shown in Fig. 3, which includes parallel connection of *GCPE* element with $R_0^* - R_\infty^*$ element, we name a

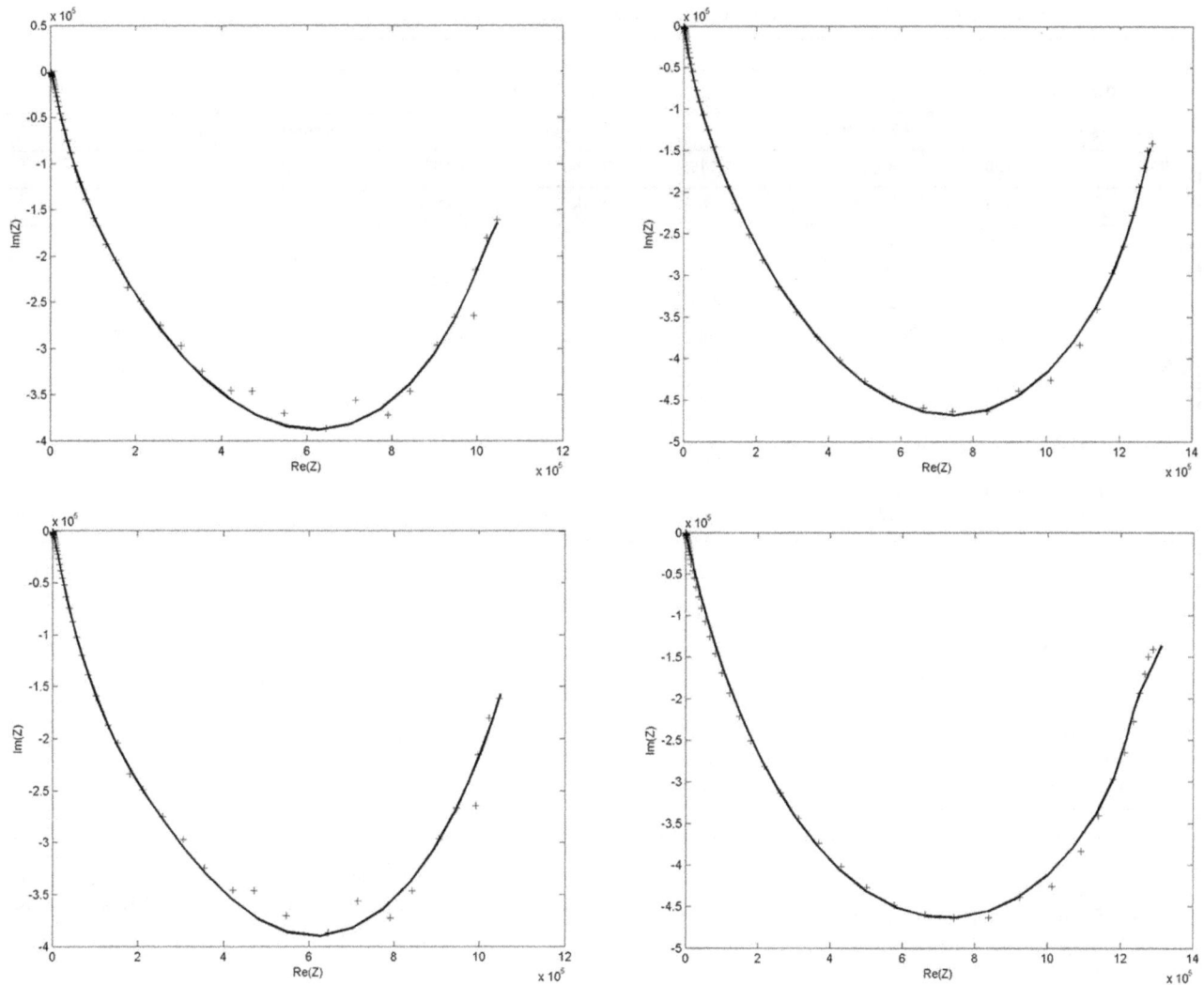

Figure 6. Cole Plot for Experimental Data and LM data Fitting. *Top Left*: Crosses represent experimental data of measurements using electrode $d = 2$ cm for $V_0 = 1.0$ V. Solid line represents LM fit for GGC1 model. *Top Right*: Crosses represent experimental data of measurements using electrode $d = 2.0$ cm for $V_0 = 1.0$ V. Solid line represents LM fit for GGC1 model. *Bottom Left*: Crosses represent experimental data of measurements using electrode $d = 0.25$ cm for $V_0 = 1.0$ V. Solid line represents LM fit for GC2 model. *Bottom Right*: Crosses represent experimental data of measurements using electrode $d = 2.0$ cm for $V_0 = 1.0$ V. Solid line represents LM fit for C1GC1 model.

reduced generalized Cole element. It should be noted that the following functional dependence of impedance $\underline{Z}(\omega) = K \cdot (j \cdot \omega)^{\alpha_1 + j \cdot \alpha_2}, \alpha_1, \alpha_2 \in \mathbb{R}, \alpha_2 \neq 0, K > 0$, described by Nigmatulin and Mehaute [28], is only the special case of the element $\underline{Z}_{GCPE\alpha^*;\beta}(\omega)$, whereas our equation describes the generalized case. Nigmatulin's equation [15] naturally explains temporal irreversibility phenomena which can appear in linear systems with remnant memory, while our model defines more complex behavior compared to the mentioned power law scenario. For the operator $(\tau_{\alpha^*;\beta} \cdot K)^{\alpha^*;\beta}$, described in the Equation (46), limitation values for α^* and β do not have to be strict as in the case for $\alpha(\alpha \in (0,1])$. Mathematically formulated our model is introducing a new parameter (β) and four corrected parameters $(R_0, R_\infty, \alpha, \tau_\alpha)$.

Using fractional calculus approach we have derived a new class of models for electrical impedance by generalizing Weyl fractional derivative operator, Cole model and constant phase element. These generalizations were described by the novel equation which presented parameter β which can adopt positive and negative

values. Therefore, for value of parameter $\beta = 0$ Cole model is a special case of our generalized model. Our generalized equation (47), containing non-integer integrals and derivatives either with real or complex power-law exponents, naturally explains temporal irreversibility phenomena. According to Nigmatullin and Trujillo these partial irreversibility phenomena can be declared as "remnant" memory of the complex system [15]. It would mean that in complex systems containing many entities only one part of their microscopic states will be conserved on the following level of intermediate scales and expressed in the form of the fractional integral.

We have derived a new primary model (equation 47) using fractional calculus approach for generalization of CPE and Cole equation together with introduction of a new parameter β which according to fractional calculus and Nigmatullin and Trujillo [15] we relate and interpret as remnant memory of the system. For values of parameter $\beta \neq 0$ our model defines the system with remnant memory and is represented in an impedance plot not as a circle arc but as a ellipsoid like arc type. For values of $\beta = 0$ the

Table 1. LM Fitted Parameters for Impedance Models C2, GC2, C1GC1.

1.0 V, d = 0.25 cm				1.0 V, d = 2.0 cm			
Parameters C2		**Parameters GC2**		**Parameters C2**		**Parameters C1GC1**	
R_0 (MΩ)	1.120	R^*_0 (MΩ)	1.119	R_0 (MΩ)	1.353	R'_0 (MΩ)	1.354
R_∞ (kΩ)	1.911	R^*_∞ (kΩ)	0.680	R_∞ (kΩ)	1.718	R'_∞ (kΩ)	2.963
α_1	0.743	α^*_1	0.819	α_1	0.784	α^{**}_1	0.532
τ_1 (s)	0.266	τ^*_1 (s)	0.240	τ_1 (s)	0.110	τ^{**}_1 (s)	1.900
$p(\alpha_1)$	0.267	$p(\alpha^*_1, \beta_1)$	0.422	$p(\alpha_1)$	0.200	$p(\alpha^{**}_1, \beta'_1)$	1.596
α_2	0.851	α^*_2	0.805	α_2	0.831	α_2	0.961
τ_2 (s)	1.414	τ^*_2 (s)	2.883	τ_2 (s)	0.687	τ_2 (s)	2.704
$p(\alpha_2)$	0.733	$p(\alpha^*_2, \beta_2)$	0.655	$p(\alpha_2)$	0.800	$p(\alpha_2)$	0.356
β_1	0.0	β_1	−0.015	β_1	0.0	β'_1	0.024
β_2	0.0	β_2	0.074	β_2	0.0	β'_2	0.0
Mean square errors (·10^7)							
4.19		3.90		3.64		2.76	

model does not take in account eventually existing remnant memory. According to the equation (47) our model is then deduced to the special case of Cole model, represented as a circle arc in an impedance plot, describing the system without taking in account remnant memory.

Our new parameter β adopting non restricted positive and negative values is introducing remnant memory. This is a new parameter different to α which can adopt only values between 0 and 1and is usually interpreted as a distribution of relaxation times [17,19,29].

Different mathematical approaches for studying electrical properties of human skin were attempted using impedance models based on serially connected reduced (R_∞ excluded) Cole elements (C1) (Yamamoto & Yamamoto [29]). They can be summarized in the following equation (Barsoukov and Macdonald [30]) for $n > 0$

$$\underline{Z}_{SC}(\omega) = R_\infty + (R_0 - R_\infty) \sum_{i=1}^{n} \frac{p(\alpha_i)}{1 + (j \cdot \omega \cdot \tau_{\alpha_i})^{\alpha_i}},$$

$$\sum_{i=1}^{n} p(\alpha_i) = 1, p(\alpha_i) \geq 0, \alpha_i \in (0,1] \tag{51}$$

Here we develop a second model based on serial connection of reduced generalized Cole elements (Equation (52) and electric circuit, Figure 4). This new type of a serial model is based on the generalized bioimpedance Equation (47).

Table 2. LM Fitted Parameters for Impedance Models GC2, C1GC1, GGC1.

1.0 V, d = 0.25 cm				1.0 V, d = 2.0 cm			
Parameters GC2		**Parameters GGC1**		**Parameters C1GC1**		**Parameters GGC1**	
R^*_0 (MΩ)	1.119	R^{**}_0 (MΩ)	0.642	R'_0 (MΩ)	1.354	R^{**}_0 (MΩ)	1.461
R^*_∞ (kΩ)	0.680	R^{**}_∞ (kΩ)	1.060	R'_∞ (kΩ)	2.963	R^{**}_∞ (kΩ)	2.530
α^*_1	0.819	α^{**}	0.940	α^{**}_1	0.532	α^{**}	0.707
τ^*_1 (s)	0.240	τ^* (s)	0.281	τ^{**}_1 (s)	1.900	τ^* (s)	0.604
$p(\alpha^*_1, \beta_1)$	0.422	β^*	−0.043233	$p(\alpha^{**}_1, \beta'_1)$	1.596	β^*	0.010132
α^*_2	0.805	γ	−0.131216	α_2	0.961	γ	−0.004308
τ^*_2 (s)	2.883	δ	−0.054793	τ_2 (s)	2.704	δ	−0.007025
$p(\alpha^*_2, \beta_2)$	0.655	ε	−0.009557	$p(\alpha_2)$	0.356	ε	−0.002536
β_1	−0.015	ζ	−0.000652	β'_1	0.024	ζ	−0.000206
β_2	0.074						
Mean square errors (·10^7)							
3.90		3.90		2.76		0.94	

$$\underline{Z}_{SGC}(\omega) = R_\infty^* + \left(R_0^* - R_\infty^*\right) \cdot$$

$$\sum_{i=1}^n \frac{p\left(\alpha_i^*; \beta_i\right)}{1 + \left(j \cdot \omega \cdot \tau_{\alpha_i^*; \beta_i}\right)^{\alpha_i^* + \beta_i \cdot \log\left(j \cdot \omega \cdot \tau_{\alpha_i^*; \beta_i}\right)}}, \quad (52)$$

$$p\left(\alpha_i^*; \beta_i\right) \geq 0, \ \lim_{\beta_i \to 0} p\left(\alpha_i^*; \beta_i\right) = p(\alpha_i)$$

The Equation (52) is direct generalization of Equation (51).

The Equation (53) denotes our combinatorial bioimpedance model using all permutations of serial connections between reduced Cole elements (Equation (51)) and our reduced generalized Cole elements described in Equations (52).

$$Z_{SGCC}(\omega) = R_\infty' + \left(R_0' - R_\infty'\right) \cdot$$

$$\left(\sum_{i=1}^{n_2} \frac{p(\alpha_i')}{1 + \left(j \cdot \omega \cdot \tau_{\alpha_i'}'\right)^{\alpha_i'}} + \sum_{l=1}^{n_1} \frac{p\left(\alpha_l^{**}; \beta_l'\right)}{1 + \left(j \cdot \omega \cdot \tau_{\alpha_l^{**}; \beta_l'}'\right)^{\alpha_l^{**} + \beta_l' \cdot \log\left(j \cdot \omega \cdot \tau_{\alpha_l^{**}; \beta_l'}'\right)}} \right) (53)$$

$$p(\alpha_i'), p\left(\alpha_l^{**}; \beta_l'\right) \geq 0, \alpha_i', \alpha_l^{**} > 0$$

A particular permutation case for our third model is using serial connection of reduced Cole elements and reduced generalized

Figure 7. Comparison of model quality evaluated by mean square error. Bioimpedance data fitting with all of our models, either single elements or their serial combination, by Levenberg-Marquardt (LM) nonlinear least squares algorithm for electrode size d = 0.25 and d = 2 cm.

Cole element and is based on Equation (53) and electric circuit shown in Figure 5.

For the purpose to set Equation (47) in more general form we have extended previously introduced members $\exp\left(-v \cdot \log(s) - \gamma \cdot (\log(s))^2\right)$ from Nigmatullin's Equation [15] with three more members $\exp\left(-\vartheta \cdot (\log(s))^3 - \ldots\right)$

$$\exp\left(-v \cdot \log(s) - \gamma \cdot (\log(s))^2\right) \rightarrow$$
$$\exp\left(-v \cdot \log(s) - \gamma \cdot (\log(s))^2 - \vartheta \cdot (\log(s))^3 - \ldots\right) \quad (54)$$

Formally from the equation (47) it follows

$$\underline{Z}(\omega) = R_\infty^{**} + \frac{R_0^{**} - R_\infty^{**}}{1 + (j\omega\tau)^{\alpha^{**} + \beta^{**} \cdot \log(j\omega\tau) + \gamma \cdot \log^2(j\omega\tau) + \ldots}} \quad (55)$$

The equation (55) represents further generalization of our generalized Cole (GGC1) including additional parameters. We have introduced partial sum of Maclaurin series determined by parameters $(\beta^*, \gamma, \delta \ldots)$. Cole equation is a special case of our generalized class of models for $\beta^* = \gamma = \delta = \ldots = 0$.

The purpose of this further generalization was to cover the broader frequency dispersion range using one element instead of using serially connected elements with larger number of parameters and inferior fitting.

Materials and Methods

1. Bioimpedance measurements

Bioimpedance skin measurements were performed at University of Belgrade on upper arm of human volunteers with Solartron *1255* Frequency Response Analyzer in combination with Solartron *1286* Pstat/Gstat. Experiments were done in shielded Faraday caged room. The linearity of all measurements with both electrode sizes was confirmed by testing the system with Solartron Schlumberger 12861 test module. The electrodes were made of stainless steel. We have used electrodes with diameter of 0.25 cm and 2.0 cm. The distance between outer edges of two electrodes was 5 cm. The electrode was completely covered with minimal amount of highly conductive cream (3.3 S/m) Grass EC33 obtained from Grass technologies. This cream is specifically designed for skin resistive and conductive measurements. Cream covered electrodes were gently placed on skin in order to avoid putting excess pressure to skin. Total required time for the frequency sweep measurement was about 10 minutes at 22°C and 50% relative humidity, thus, minimalizing artifacts production during measurements due to long cream exposure or cream penetration to skin, as well as sweating. Error of measurements was <0.1%. Twenty series of measurements were taken at each of the 61 different frequencies ranging between 0.1 Hz and 100.0 KHz. The applied voltage of alternating current was of 1.0 V amplitude. Total required time for the frequency sweep measurement was about 10 minutes.

1) Bioimpedance skin measurements were performed at University of Belgrade,

2) University of Belgrade Review Board (IRB) approved specifically use of oral consent for bioimpedance skin measurements on healthy human volunteers for this study.

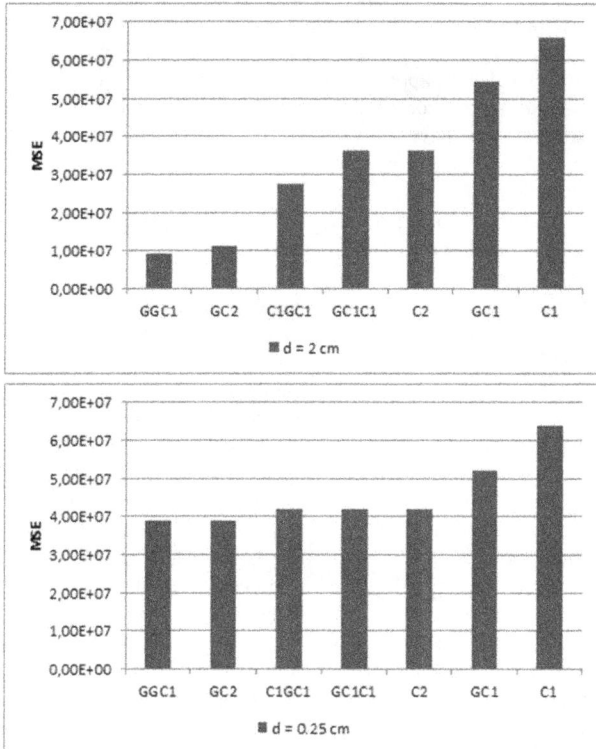

Please note that bioimpedance skin measurements are noninvasive and have been commonly used in public fitness centers for measurements of body fat content without medical or ethic commission authorizations and are thus generally deemed unnecessary for further ethics commission approval.

3) Each volunteers provided oral consent for bioimpedance skin measurements analyzed and reported in our study.

2. Experimental data fitting

Levenberg-Marquardt (LM) nonlinear least squares algorithm L2 (L_2-norm) [31], [32], with Levmar in Octave programming environment was used for fitting experimental data with different models [33]. Without complications this fitting calculation could use maximally ten parameters. This restriction encourages the implementation of LAPACK libraries in *C/C++*.

Discussion

1. Experimental measurements of bioimpedance on human skin

Bioimpedance measurements on human skin were performed under the conditions described in the experimental methods section. Obtained data are presented in Fig. 6 as Cole-Cole plot. In the *left* part of the Figure 6. we show electrical measurements results acquired using electrodes with 0.25 cm diameter and in the *right* part of Figure 6. results acquired using electrodes with 2.0 cm diameter. For both electrode sizes we have obtained arc type behavior when Z imaginary (*Im (Z)*) was plotted versus Z real (*Re (Z)*). There is a difference in the arc shape for two different electrode sizes. The maximal measured value of *Re (Z)* is greater for larger electrode surface then the maximal value of *Re (Z)* obtained for the smaller electrode surface (Fig. 6). The minimal measured value of *Im (Z)* is smaller for the larger electrode surface then the minimal value of *Im (Z)* obtained for the smaller electrode surface (Fig. 6). Therefore, one can observe that bioimpedance depends from the electrode surface area.

2. Experimental data fitting with derived mathematical models

We have used Levenberg-Marquardt (LM) nonlinear least squares algorithm for experimental data fitting with following models: (1) comprised of one Cole element (Equation (41)); (2) based on two serially linked reduced Cole elements (C2) (Equation (51) for n = 2); (3) based on three serially linked reduced Cole elements (C3) (Equation (51) for n = 3); (4) comprised of one generalized Cole element (GC1) (Equation (47)); (5) based on two serially linked reduced generalized Cole elements (GC2) (Equation (52) for n = 2); (6) based on one reduced Cole element serially linked to one reduced generalized Cole element (GC1C1) (Equation (53), $n_1 = 1$ and $n_2 = 1$, first permutation); (7) based on one reduced generalized Cole element serially linked to one reduced Cole element (C1GC1) (Equation (53), $n_1 = 1$ and $n_2 = 1$, second permutation). Although all models were analyzed using LM data fitting, we present in Table 1 and 2 only results obtained by three best performing models for human skin as the test system. These models are designated according to the abbreviations of corresponding of electric circuit elements. They are: C2, GC2 (schema shown in Fig. 4 for n = 2), and C1GC1 (schema shown in Fig. 5 for $n_1 = 1$ and $n_2 = 1$). Future experimental electrical biompedance measurements will demonstrate whether the presented seven model types could provide specific description for different organs, tissues and materials.

We have started with LM fitting of experimental data using Cole model, GC1, C2 and C3 models in order to obtain initial values for parameter: $R_0, R_\infty, \alpha 1, \tau 1, p(\alpha 1), \alpha 2, \tau 2$ and $p(\alpha 2)$, Table 1 and 2. The initial value of β for GC1 was 0.01. Among GC1, C2 and C3 models the C2 model provided the best fitting results (Table 1 and 2). In the second step, we have used the initial values of parameters obtained with C2 model in order to proceed with LM fitting of experimental data using GC2 and C1GC1 and GC1C1 models. The initial values for the parameters β_1 and β_2 were 0.01. The best fitting results were obtained for GGC1 followed by GC2 and C1GC1 models using the parameter values given in Table 1 and 2 and Figure 7. Bioimpedance data measurements, obtained with two different electrode sizes (0.25 cm and 2 cm in diameter), and fitting data, using GC2 and C1GC1 models, are shown in Fig. 6 and 7. The GC2 model provided the best fitting results for measurements with large electrodes, whereas GC2 and C1GC1 model provided the best fit for measurements with small electrodes, Figure 6. and 7. We have calculated mean squared errors for Cole model and our generalized models derived by fractional calculus. The results showed that our GGC1 model have 85% better experimental impedance data fitting obtained with electrodes of 2 cm diameter and 40% for 0.25 cm electrode diameter Figure 6 and 7. Model types are sorted in the qualitatively same ascending way according to increasing mean square error for both electrode size (d = 0.25 and d = 2 cm) (Fig. 7). These results showed that quality of model is invariant of electrode diameter.

Additional parameters introduced in GGC1 are correction of parameter β improving modeling of remnant memory. Using this particular model type we have further improved bioimpedance data fitting by Levenberg-Marquardt (LM) nonlinear least squares algorithm because this model gave best mean square error values as shown in Figure 7 and Table 1 and 2. In summary our fractional calculus approach adds to better understanding and description of the complex system and their electrical behavior. Therefore, we have presented quantitative evidence about improved precision of our models for description of human skin bioimpedance.

Three serially linked reduced Cole elements had two order of magnitudes higher mean square error and thus is the most inferior quality model.

Taking also in account the Equation (50), it can be concluded that τ is a function of β. Therefore, it seems that the parameter β in our impedance model is related to relaxation phenomena of electrical behavior of complex system such as human skin. Since fractional calculus is a mathematical approach dealing with derivatives and integrals of arbitrary and complex orders it adds a new dimension to understand and describe basic nature and behavior of complex processes, such as electrical properties of biological tissues, in an improved way. More precisely it contains in time information about the function at earlier points, thus it possesses a memory effect, and includes non-local spatial effects. Without the use of fractional calculus approach it would not be possible to make our new generalized type of superior and more precise class of models where Cole is a special case.

We have introduced fraction calculus generalization approach in order to be able to cope with complex multi-layered systems with unknown structures which also include simpler structures such as necessary gels and electrodes as additional element influencing and complicating interpretation and analyses of the experimental data.

Our GGC1 (one element) and GC2 (two elements) models provide significantly better fitting of the experimental data then Cole model which is actually, as previously explained, a special case for $\beta* = 0$ and $\beta = 0$ respectively. We have used human skin as one of the examples for complex system and our model is not limited only to such biological material, rather it is a generalized for any complex system including even mixture of biological and not biological systems such as gels and electrodes.

Our generalized *CPE*, generalized Cole model and serially connected reduced generalized Cole elements represent a valuable basis for further development of mathematical models for bio-systems and/or any kind of material using the fractional calculus approaches. We propose that this type of powerful modeling tools shall be further applied for noninvasive analysis of complexity of bio-systems and/or any kind of material.

Author Contributions

Conceived and designed the experiments: JBSK ZBV. Performed the experiments: JBSK ZBV. Analyzed the data: ZBV GML GNM JBSK. Contributed reagents/materials/analysis tools: ZBV GML JBSK. Wrote the paper: GNM ZBV JBSK.

References

1. Glade N, Demongeot J, Tabony J (2002) Numerical simulations of microtubule self-organisation by reaction-diffusion. Acta. Biotheoretica. 50: 4, 239–268.
2. Desmeulles G, Dupré D, Rodin V, Misery L (2003) Multi-agents computer simulation: towards a new way of research in allergy and immunodermatology. Journal of Investigative Dermatology 121: 1570.
3. Samko SG, Kilbas AA, Marichev OI (1993) Fractional Integrals and Derivatives: Theory and Application. New York: Gordon and Breach, 1006 p.
4. Podlubny I (1999) Fractional Differential Equations. San Diego-Boston-New York-London-Tokyo-Toronto: Academic Press, 368 p.
5. Hilfer R, (2000) Applications of Fractional Calculus in Physics. Singapore: World Scientific Pub. Co., 463 p.
6. Voet D, Voet JG (2011) Biochemistry. San Francisco: John Wiley and Sons, 1520 p.
7. Magin RL (2004) Fractional calculus in bioengineering 1. Critic. Rev. in Biomed. Eng. 32: 1–104.
8. Magin RL (2004) Fractional calculus in bioengineering 2. Critic. Rev. in Biomed. Eng. 32: 105–194.
9. Butzer PL, Nessel RJ (1971) Fourier Analysis and Approximation. Vol. 1: One-dimensional theory. Birkhauser, Basel, and Academic Press, New York, 553 p.
10. Popovic JK, Dolicanin D, Rapaic MR., Popovic SL, Pilipovic S, et al. (2011) A nonlinear two compartmental fractional derivative model. Eur. J. Drug. Metab. Pharmacokinet. 36(4): 189–96.
11. Lundstrom BN, Higgs M, William MH, Spain WJ, Fairhall AL (2008) Fractional differentiation by neocortical pyramidal neurons. Nat. Neuorosci. 11: 1335–1342.
12. Freed AD, Diethelm K (2006) Fractional calculus in biomechanics: a 3D viscoelastic model using regularized fractional derivative kernels with application to the human calcaneal fat pad. Biomechan. Model. Mechanobiol. 5: 203–215.
13. Sebaa N, Fellah A, Lauriks W, Depollier C (2006) Application of fractional calculus to ultrasonic wave propagation in human cancellous bone. Signal Processing 86: 2668–2677.
14. El-Sayed AMA, Rida SZ, Arafa AAM (2009) Exact Solutions of Fractional-Order Biological Population Model. Commun. Theor. Phys. 52: 992–996.
15. Nigmatullin RR, Trujillo JJ (2007) Mesoscopic Fractional Kinetic Equations Versus A Riemann-Liuoville Integral Type. In: Sabatier J, Agrawal OP, Machado JAT editors. Advances in Fractional Calculus. Dodrecht: Springer. 155–167.
16. Montagna W, Parakkal P F (1974) The Structure and Function of Skin. London, New York: Academic Press, 454 p.
17. Grimnes S, Martinsen ØG (2008) Bioimpedance and Bioelectricity Basics. Oxford: Elsevier, 471p.
18. Weinberg AM, Householder AS (1941) Statistical distribution of impedance elements in biological systems. Bull. Math. Biophys. 3: 129–135.
19. Yamamoto Y, Yamamoto T, Ozawa T (1986) Characteristics of skin admittance for dry electrodes and the measurement of skin moisturisation. Med. Biol. Eng. Comput. 24: 71–77.
20. Cole KS (1940) Permeability and Impermeability of Cell Membranes for Ions. Cold Spring Harbor Symp. Quant. Biol. 8: 110–122.
21. Cole KS, Cole RH (1941) Dispersion and Absorption in Dielectrics – I Alternating Current Characteristics. J. Chem. Phys. 9: 341–52.
22. El-Lakkani A (2001) Dielectric Response of Some Biological Tissues. Bioelectromagnetics 22: 272–279.
23. Dissado LA, Alison JM, Hill R M, McRae DA, Esrick MA (1995) Dynamic scaling in the dielectric response of excised EMT-6 tumours undergoing hyperthermia. Phys. Med. Biol. 40: 1067–1084.
24. Butzer PL, Westphal U (1975) An Access to Fractional Differentiation via Fractional Diference Quotients. In: Ross B editor. Fractional Calculus and Its Applications. Berlin: Springer-Verlag. 116–146.
25. Fricke H (1932) Theory of electrolytic polarization. Phil. Mag. 14: 310–318.
26. Atanackovic TM (2002) A modified Zener model of a viscoelastic body. Continuum Mechanics and Thermodynamics 14: 137–148.
27. Bohannan GW (2000) Interpretation of complex permittivity in pure and mixed crystals. AIP Conf. Proc. 535: 250–258.
28. Nigmatullin RR, LeMehaute A (2004) To the Nature of Irreversibility in Linear Systems. Magn. Reson. Solids. 6 No. 1: 165–179.
29. Yamamoto T, Yamamoto Y (1976) Electrical properties of the epidermal stratum corneum. Med. Biol. Eng.14 (2): 151–158.
30. Barsoukov E, Macdonald JR (2005) Impedance Spectroscopy Theory, Experiment and Applications. New Jersey: John Wiley & Sons, 595 p.
31. Richard CR, Aster C, Borchers B, Thurber CH (2005) Parameter Estimation and Inverse problems. Burlington: Elsevier, Academic Press 303 p.
32. Christian K, Nobuo Y, Masao F (2004) Levenberg-Marquardt methods with strong local convergence properties for solving nonlinear equations with convex constraints. J. Comput. Appl. Math. 6 Issue 2: 375–397.
33. ics website. Available: http://www.ics.forth.gr/lourakis/levmar/Accessed 2013 Feb 25.

16

Hydrogen Bonding Penalty upon Ligand Binding

Hongtao Zhao, Danzhi Huang*

Department of Biochemistry, University of Zurich, Zurich, Switzerland

Abstract

Ligand binding involves breakage of hydrogen bonds with water molecules and formation of new hydrogen bonds between protein and ligand. In this work, the change of hydrogen bonding energy in the binding process, namely hydrogen bonding penalty, is evaluated with a new method. The hydrogen bonding penalty can not only be used to filter unrealistic poses in docking, but also improve the accuracy of binding energy calculation. A new model integrated with hydrogen bonding penalty for free energy calculation gives a root mean square error of 0.7 kcal/mol on 74 inhibitors in the training set and of 1.1 kcal/mol on 64 inhibitors in the test set. Moreover, an application of hydrogen bonding penalty into a high throughput docking campaign for EphB4 inhibitors is presented, and remarkably, three novel scaffolds are discovered out of seven tested. The binding affinity and ligand efficiency of the most potent compound is about 300 nM and 0.35 kcal/mol per non-hydrogen atom, respectively.

Editor: Peter Butko, Anne Arundel Community College, United States of America

Funding: This work was supported by a grant (31003A_122442) of the Swiss National Science Foundation (www.snf.ch) to D.H. The funders had no role in study design, data collection and analysis, decision to publish, or preparation of the manuscript.

Competing Interests: The authors have declared that no competing interests exist.

* E-mail: dhuang@bioc.uzh.ch

Introduction

Hydrogen bonding is an exchange reaction whereby the hydrogen bond donors and acceptors of the free protein and ligand break their hydrogen bonds with water and form new ones in the protein-ligand complex [1,2,3]. About thirty years ago, Wilkinson and coworkers found mutation of Cys-35 in Tyrosyl-tRNA synthetase to Ser-35 causes poorer ATP binding and catalysis although the hydroxyl group of serine forms far stronger hydrogen bonds than does the thiol group of cysteine [1]. Analysis of the hydrogen bonding geometry revealed that a hydrogen bond of Ser-35 is at least 0.5 Å longer than the optimum. Accordingly, Ser-35 would have to lose a good hydrogen bond with a bound water molecule to form this weak hydrogen bond with ATP in the enzyme-substrate complex, and thus the mutant shows poorer binding and catalysis. Therefore, enthalpic loss in hydrogen bonding could take place upon ligand binding if not compensated by formation of good hydrogen bonds between the protein and ligand.

Virtual screening has emerged as an efficient tool in drug discovery from lead identification to optimization and beyond [4,5]. However, scoring functions that model the solvent environment as a continuum [6,7] are still grossly inaccurate [8]. The role of individual waters can be critical in predication of binding affinities, and continuum models often provide poor results in treating bound waters in a confined cavity [9]. Glide docks explicit waters into the binding site and measures the exposure of polar/charged groups to the explicit waters. When a polar/charged ligand or protein group is judged to be inadequately solvated, a desolvation penalty is assessed [9,10]. By contrast, most other scoring functions [11] do not properly take into account the enthalpic loss of hydrogen bonding upon ligand binding. Incorporation of bound water molecules into molecular docking was suggested for improvement of accuracy [12]. On the

other hand, in high-throughput molecular docking campaigns a significant part of binding poses are rather unrealistic, e.g. burial of polar atoms in hydrophobic sites, and thus discarding them at an early stage is desirable. Filters such as van der Waals efficiency based on arbitrary cutoff are often used to remove poses that unlikely bind [13]. However, it seems lack of a reliable and efficient filter with transferable cutoff among different proteins.

Protein kinases play an important role in cell-signaling pathways regulating a variety of cellular functions. Dysregulation of kinase activity has been implicated in pathological conditions ranging from neuronal disorders to cellular transformation in leukemia [14]. The tyrosine kinase erythropoietin producing human hepatocellular carcinoma receptor B4 (EphB4) is involved in cancer related angiogenesis [15]. So far, two high-throughput virtual screening campaigns have been reported, with two scaffolds identified in the low micromolar range [13,16]. Highly potent EphB4 inhibitors have been developed via chemical synthesis [17,18,19]. The marketed drug dasatinib, with Abl1 and Src as primary targets, also shows a very high affinity to Eph kinases [20].

Here, we report a new approach to calculate hydrogen bonding penalty (HBP) associated with ligand binding. HBP is further integrated into a binding energy calculation, and the fitted parameter of 1.7 kcal/mol is consistent with the estimate of contribution by formation of one neutral hydrogen bond ranging from 0.5 to 1.5 kcal/mol [21]. Moreover, statistics of HBP in kinase crystal structures and an application in a high-throughput docking campaign is presented.

Methods

Binding of a ligand to a protein involves the breakage of hydrogen bonds with water molecules and formation of new hydrogen bonds between the protein and ligand, which can be described by the following equation [21] by using one pair of donor (D) and acceptor (A):

$$D \cdots OH_2 + A \cdots HOH \leftrightarrow D \cdots A + HOH \cdots OH_2 \quad (1)$$

Based on hydrogen bonding being an exchange reaction [1,21,22], its energy can be described using normalized weights:

$$E_{HB-unbound} = [(w_D + w_{O-H_2O}) + (w_A + w_{H-H_2O})] * E_{HB} \quad (2)$$

$$E_{HB-bound} = [f_{hb} * (w_D + w_A) + (w_{O-H_2O} + w_{H-H_2O})] * E_{HB} \quad (3)$$

wherein, w_D and w_A is the hydrogen bonding weight of a donor or acceptor, respectively, f_{hb} stands for the fraction of hydrogen bonding relative to that of an optimum geometry, and E_{HB} is unit hydrogen bonding energy. Hydrogen bonds with water are assumed to be in the optimum geometry. HBP (p_{HB}) associates with ligand binding is then described as

$$p_{HB} = (1 - f_{hb}) * (w_D + w_A) \quad (4)$$

Probing hydrogen bonding status

Oxygen and nitrogen atoms in double or triple bonds are regarded as hydrogen bond acceptors, and hydrogen atoms bonded to oxygen, nitrogen or sulfur atoms are regarded as hydrogen bond donors. The existence of C—H...O hydrogen bonds has been confirmed by neutron diffraction data on organic compounds [23]. Analysis of 100 kinase crystal structures complexed with small molecule inhibitors at a resolution of at least 2.5 Å gives 64 short C—H...O interactions, showing typical hydrogen bonding features (Figure S1).

Each hydrogen bond donor or acceptor at the binding interface is firstly checked whether it forms hydrogen bond with water molecules. For this purpose, an optimum solvation radius (r_{sol}) is defined for each donor/acceptor and if a water molecule can be placed within 0.15 Å of the r_{sol} no penalty is applied. Here, 2.8 and 2.9 Å are used as r_{sol} for any oxygen and nitrogen, respectively, which were derived from an analysis of 397 crystal structures with X-ray resolutions below 1.0 Å (Figure S2). The r_{sol} of polar hydrogen is 1.9 Å (except 2.15 Å for H bonded to sulfur), which is the difference between the r_{sol} of nitrogen and the bond length [24] of N—H. The r_{sol} of other atom types are listed in Figure 1 and the values are mainly adapted based on the van der Waals radii of Bondi [25]. Details of probing hydrogen bonds with water were described in File S1. In case of not forming hydrogen bonds with water, the possibility of forming hydrogen bonds between the protein and ligand (including intra-molecular hydrogen bonds) is further checked and penalty (p_{HB}) is then calculated.

Fraction of hydrogen bonding

Similar to the strategy of evaluating hydrogen bonding energy in LUDI [26], the following equations are used to calculate the fraction of hydrogen bonding (f_{hb}) to that of an optimum geometry.

$$f_{hb}(r, \theta) = f(r) \cdot f(\theta) \quad (5)$$

	Atom types	Hydrogen bonding weight	r_{sol} (Å)
O	=O, −O⁻	1.0	2.8
	−OH	0.5	
	−O−	0.2	
N	=N−, ≡N	1.0	2.9
H	⟍|−NH⁺	1.0	1.88
	−OH, −NH, −NH₃⁺, −NH₂⁺	0.5	
	−NH₂	0.3	
	CH=N, HC=O	0	
	−SH	0.2	2.15
S	=S, −S−	-	3.2
C	Unsaturated	-	3.0
	Saturated	-	3.2
H	Apolar	-	2.28
F	-	-	2.99
Cl	-	-	3.27
Br	-	-	3.37
I	-	-	3.50

Figure 1. Hydrogen bonding weights and solvation radii of different atom types.

$$f(r) = \begin{cases} 1 & if \quad r \le 2.0 \\ 1 - 0.5 * (r - 2.0) & if \quad 2.0 < r \le 2.8 \quad (6) \\ 0 & if \quad r > 2.8 \end{cases}$$

$$f(\theta) = \begin{cases} 1 & if \quad \theta \ge 150^o \\ 4.0 * (-\cos\theta)^{0.25} - 2.86 & if \quad 110^o \le \theta < 150^o \,(7) \\ 0 & if \quad \theta < 110^o \end{cases}$$

wherein, r is the distance between the hydrogen atom and the acceptor and θ is the angle centered at hydrogen among donor, hydrogen and acceptor. The equation to calculate $f(r)$ and $f(\theta)$ as well as the upper and lower limit in r and θ are derived from the calculation using density functional theory [27]. In case of one hydrogen atom is shared by two acceptors or one acceptor interacting with two donors, the f_{hb} for the corresponding donor/acceptor is additive but with 1 as the upper limit.

Hydrogen bonding penalty

The HBP at the protein-ligand interface is summarized over each donor/acceptor as

$$P_{HB} = \sum_{pro,lig} w * (1 - \sum f_{hb}) \quad (8)$$

However, no penalty is applied for protein atoms which are not water accessible before ligand binding or participate in intramolecular hydrogen bonds. Initial guess of hydrogen bonding weights (w) is based on chemical intuition by considering atomic partial charge and water solubility of a few small molecules (Table S1). Empirical weights as proof-of-principle are then optimized with a trial-and-error procedure according to the fitted parameter in the binding free energy calibration.

Evaluation of binding free energy

The equation used for fitting the calculated energies to the experimental free energies of binding ($\Delta G = RT\ln(K_d)$) is a three-parameter model

$$\Delta G = \alpha \Delta E_{ff} + \beta P_{HB} + \gamma \quad (9)$$

where, ΔE_{ff} is the interaction energy between the ligand and the protein calculated by the CHARMm force filed [28] and P_{HB} stands for HBP. Three parameters α, β, and γ are generated with fitting. ΔE_{ff} is calculated by the following equation:

$$\Delta E_{ff} = \Delta E_{vdW} + \Delta E_{coul} + \Delta G_{solv} + \Delta E_{strain} \quad (10)$$

where, ΔE_{vdW} is the intermolecular van der Waals energy, ΔE_{coul} is the intermolecular Coulombic energy in vacuo, ΔE_{strain} is the strain energy of ligand upon binding, and ΔG_{sol} is the change in solvation energy of ligand and protein upon binding.

The van der Waals and Coulombic interaction energy are calculated by subtracting the values of the isolated components from the energy of the complex with CHARMM [29] and the CHARMm22 force filed [28]. The van der Waals energy is calculated using the default nonbonding cutoff of 14 Å. Coulombic energy is calculated using infinite cutoff and a dielectric constant of 2.0. The electrostatic solvation energy was calculated by the finite-difference Poisson approach (FDP) [30] using PBEQ module [31] in CHARMM and a focusing procedure with a final grid spacing of 0.25 Å. The size of the initial grid is determined by considering a layer of at least 12.5 Å around the solute. The dielectric discontinuity surface was delimited by the van der Waals surface. The ionic strength is set to zero and the temperature to 300 K. Two finite-difference Poisson calculations are performed for each of the three systems (protein, ligand, and protein/ligand complex). The exterior dielectric constant was set to 78.5 and 2.0 for the first and second calculation, respectively, while the solute dielectric constant is 2.0 to take polar fluctuations into account. The solvation energy is the difference between the two calculations. The strain energy of the ligand is the energy difference between the bound and global minimum. Here, the global minimum is the one showing the lowest $E_{vdW}+E_{coul}+E_{bonded}+G_{sol}$ among all the poses that have been minimized outside of the protein.

Twenty-three inhibitors [32] of CDK2 (1H0V), 24 inhibitors [18] (**8** to **32**, excluding **30**) of EphB4 (2VWX), and 27 uncharged inhibitors [33] of p38 alpha MAP kinase (3GC7) are used as the training set. Thirty type II inhibitors [34] of Braf (3II5), 14 charged inhibitors [33] of p38 alpha and another 20 p38 alpha inhibitors [35] (1YWR) are used as the test set. Protein structures were taken from the X-ray structure (PDB code indicated in the brackets) and prepared as described below. Some key physio-chemical properties of inhibitors are summarized in Figure S3.

Version 4 of AutoDock [36] was used to generate the binding poses over the conformational search space using the Lamarckian genetic algorithm. The binding site was determined by 4.0 Å away from any atom of the ligand complexed in the respective protein structure. The number of energy evaluations was 2,750,000 and the number of poses was 50. Poses were further clustered using all atom RMSD cutoff of 0.3 Å to remove redundancy and in average 20 cluster representatives were kept. All other parameters were set as default. A few poses for each inhibitor were also generated by manual modification of the scaffold present in the respective crystal structure. All poses were further minimized by CHARMM in the respective proteins. The protein structure was kept rigid in all steps.

Preparation of protein-ligand complexes

One hundred kinase crystal structures (including 15 different classes, File S2) complexed with small molecule inhibitors at a resolution of at least 2.5 Å were downloaded from Protein Data Bank for analysis of HBP. Hydrogen atoms were added according to the protonation states of chemical groups at pH 7. Partial charges were then assigned using MPEOE method [37,38]. The added hydrogen atoms were minimized by the conjugate gradient algorithm to a RMS of the energy gradient of 0.01 kcal - mol^{-1} $Å^{-1}$. During minimization, the electrostatic energy term was screened by a distance-dependent dielectric of $4r$ to prevent artificial deviations due to vacuum effects, and the default nonbonding cutoff of 14 Å was used. Furthermore, the positions of all heavy atoms were fixed.

Preparation of the compounds library for virtual screening

The compounds were selected from Zinc library [39]. Preparation included the assignment of CHARMm atom types, force field parameters [28], and partial charges [37,38], and energy minimization with a distance dependent dielectric function using the program CHARMM [29].

Enzymatic assay

In vitro kinase activity was measured using the Panvera Z'lyte Tyr2 kinase assay PV3191 (Invitrogen) according to the manufacturer's instructions. The reaction assay (10 μL) contained 7.5 ng of EphB4 kinase (Proqinase, Germany), 30 μM ATP, and 5% DMSO. The reaction was performed at room temperature for 1 h.

Results and Discussion

Statistics of hydrogen bonding penalty in kinase complexes

Small HBPs can be observed for the binding modes of inhibitors in the X-ray structures. One example is c-Kit tyrosine kinase with its apo and holo form in complex with Imatinib (PDB codes 1T45 and 1T46). In the apo conformation, donors/acceptors at the ATP binding site form hydrogen bonds with bound water molecules. While upon ligand binding, as shown in the holo conformation, some water molecules are displaced by Imatinib. HBP on the protein part is close to zero because new hydrogen bonds to the protein are formed to compensate for the replacement of the water molecules. However, one nitrogen atom of the Imatinib pyrimidine ring (N1 of Figure S4) becomes water inaccessible and does not form a new hydrogen bond, leading to a penalty of 1. By contrast, the other nitrogen atom (N2 of Figure S4) remains hydrogen bonding with a nearby bound water molecule and thus has no penalty.

To check the distribution of HBP values in crystal structures, 100 kinase-ligand complexes are investigated. In this data set, all the small molecule inhibitors have molecular weights from 200 to 700 g/mol and number of donors or acceptors from 2 to 11 (File S2). The HBP has been calculated for each of them and the values are in general small, with 62% smaller than 1 and 36% and 2% in the range from 1 to 2 and 2.0 to 2.1, respectively (Figure 2 and File S2). It has also been observed that larger HBPs appear in some X-ray structures, e.g., the structures of PDB code 3KVX and 1JSV, and the large values actually originate from poor fitting of small molecules to the density, a common problem in crystallography [40] which can be manifested by clash of atoms.

Distribution of HBPs for docked poses of small-molecule inhibitors is also evaluated. Here, the 138 molecules used in the binding free

energy calibration are docked into the corresponding protein binding sites with AutoDock. For each molecule, about 20 poses in average are generated. Then the HBPs and binding energies are calculated for all the poses. Firstly, the binding pose with the most favorable binding energy for each molecule (Figure S5) is selected and the distribution of HBPs is plotted. As can be observed from **B** of Figure 2, the distribution is similar to that of the 100 kinase complex structures (**A**). On the other hand, the distribution of all poses (**C**) spreads more widely with the largest HBP being 6.5. Compared with the HBPs in the crystal structures (**A**), 2 is a reasonable threshold, and about 50% of poses with unrealistic binding modes can be filtered out from further evaluations.

Hydrogen bonding penalty improves the accuracy of binding energies calculation

Binding energies can be calculated using equation 9 with the parameters obtained by least-squares fitting on the training data

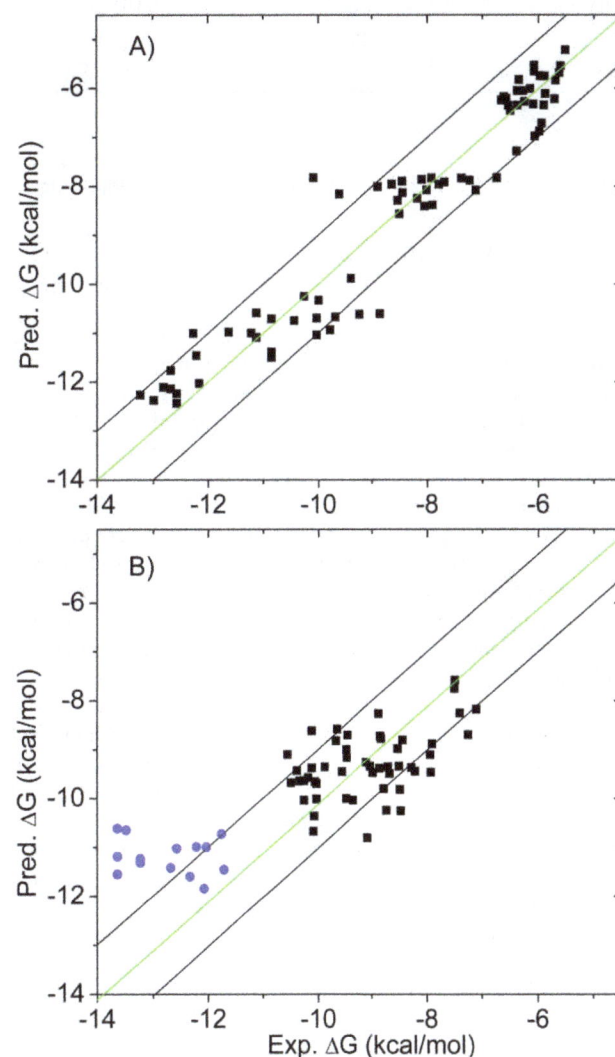

Figure 3. Comparison of the calculated versus experimental binding energies. A) Training set of 74 inhibitors. $R^2 = 0.92$ and RMS error = 0.69 kcal/mol; B) Validation set of 64 inhibitors. RMS error = 1.12 kcal/mol. The blue dots indicated the 14 p38α inhibitors with one formal charge. The green diagonal line is the ideal line of perfect prediction. The black diagonals delimit the 1 kcal/mol error region.

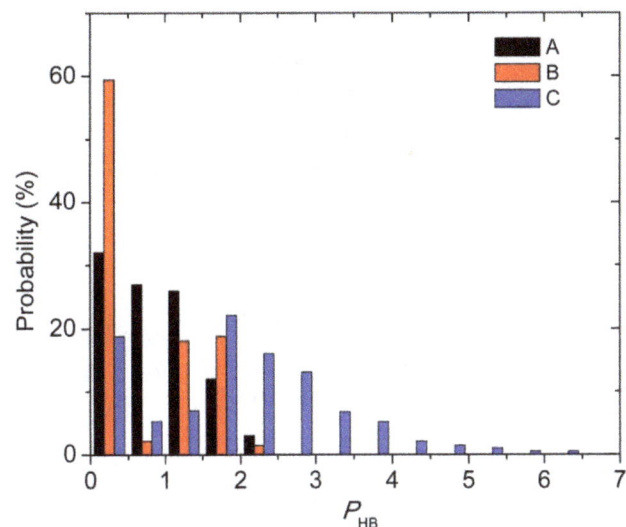

Figure 2. Distribution of hydrogen bonding penalties for: A) the binding modes in crystal structures of the 100 kinase complexes; B) poses with the most favorable calculated binding energies of the 138 molecules used in binding free energy calibration; C) all poses of the 138 molecules.

Table 1. Further validation of the three-parameter model with kinases and aspartic protease.

Protein	PDB code	ΔE_{ff} (kcal/mol)	P_{HB}	ΔG_{pred} (kcal/mol)	ΔG_{exp} (kcal/mol)
Abl	1OPJ	−64.80	1.24	−12.45	−10.81
Braf	1UWH	−57.61	1.27	−10.91	−10.45
JAK2	3E63	−30.18	0.00	−7.41	−7.91
Lck	2OFV	−59.13	0.53	−12.51	−13.23
JNK3	1PMV	−30.16	0.17	−7.12	−9.31
Ret	2X2L	−26.67	0.07	−6.58	−7.20
EGFR	1XKK	−66.60	2.30	−11.00	−10.91
CSrc	3G5D	−52.34	1.64	−9.19	−12.82
HIV-1 protease	1HIH	−65.71	1.49	−12.21	−11.01
	1HPX	−65.44	1.43	−12.26	−12.46
	1HXB	−61.24	0.95	−12.21	−13.49
	1HXW	−72.66	1.41	−13.78	−14.71
BACE-1	2QMF	−73.62	1.56	−13.72	−11.63
	2QP8	−68.76	0.47	−14.59	−11.05
	2XFI	−71.10	2.36	−11.83	−10.67

set of the 74 CDK2, EphB4, and p38α inhibitors as following:

$$\Delta G = 0.207 * \Delta E_{ff} + 1.72 * P_{HB} - 1.17 \qquad (11)$$

The calculated binding energies show high correlation with the experimental values (R-square of 0.92) and a small RMS error of 0.69 kcal/mol (Figure 3A). Here, the parameter β corresponds to the unit hydrogen bonding energy. Notably, the fitted value 1.72 kcal/mol is in agreement with the experimental value, e.g., breakage of a neutral hydrogen bond resulting in loss of energy from 0.5 to 1.5 kcal/mol [21]. Moreover, a charged primary amine or carboxyl group has a hydrogen bonding weight of 1.5 or 2.0, which can lead to a maximal penalty of 2.58 or 3.44 kcal/mol upon loss of the hydrogen bond/salt bridge. This value also agrees well with the experimental data (up to 4 kcal/mol) [21]. Hydrogen bonding weights were further used to rank the strength of individual hydrogen bonds in DNA base pairs, exhibiting good compatibility with the previously reported results (File S3).

The fitted model has been validated on a test set including 14 charged p38α inhibitors and 30 type II Braf inhibitors, with an RMS error of 1.12 kcal/mol (Figure 3B). Moreover, validation with different kinases shows general transferability of this model (Table 1). Transferability can be also seen for aspartic protease, e.g., HIV-1 protease and β-secretase, although a shift of 2.0 kcal/mol can be observed for the latter. Previously, we reported a two-parameter LIECE model for kinase inhibitors [13], which is not transferable for type II kinase inhibitors, HIV-protease or β-secretase inhibitors. The binding affinities predicted by the two-parameter LIECE on the 24 type I EphB4 inhibitors show about −5.0 kcal/mol shift compared with the experimental values (Table S2). Clearly, the incorporation of HBP into the scoring function improves the general transferability besides the role of ligand reorganization energy [41].

The derived model includes calculation of solvation energy by FDP which requires about 6 min on a single Intel 2.8 GHz CPU. Replacing the FDP approach with a distance-dependent dielectric model for solvation energy calculation gives similar accuracy for the neutral inhibitors at a much fast speed (10 seconds). However, distance-dependent dielectric model can only apply for non-charged compounds due to inaccurate treatment of the solvation effect, and also more false positives in a high-throughput virtual screening are observed. This comparison indicates that accurate calculation of solvation energies in prediction of binding affinities is necessary.

Virtual screening for EphB4 inhibitors

In a recent high throughput docking study for EphB4 inhibitors, ZINC "leads-now" library of about 20 million compounds

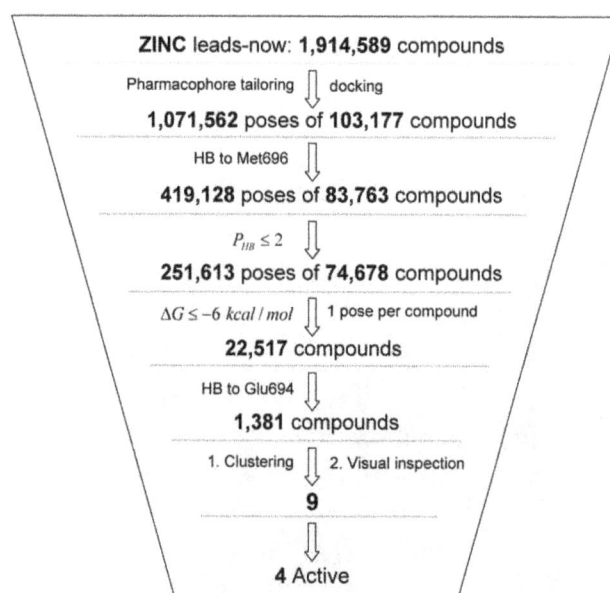

Figure 4. Schematic picture of the high throughput docking approach. HB stands for hydrogen bond. Met696 and Glu694 are the two key residues of the hinge loop (see also Figure 6).

Compound	Mw (g/mol)	P_{HB}	Pred. ΔG (kcal/mol)	IC$_{50}$ (μM)[a]
1	321	0.22	-7.9	0.38
2	333	0.74	-8.0	0.30
3	292	1.06	-6.6	15.7
4	270	0.61	-6.4	42% at 20 μM

Figure 5. Identified EphB4 inhibitors by high throughput docking. [a] All IC$_{50}$ values are means of two to four dose-response measurements.

(Mw$\leq$350 and cLogP$\leq$3.5) was first tailored by a pharmacophore model to generate a focused library of 103,177 compounds. This pharmacophore model was specifically designed for EphB4 type I inhibitors, consisting of a bi-dentate hydrogen bonding pattern and a conjugate hydrophobic group to be located in the deep ATP back pocket as well as geometric constraints thereof (H. Zhao, unpublished results). To our best knowledge, all known type I EphB4 inhibitors [13,16,17,18] can fulfill this model.

The focused library was docked by AutoDock 4 and about 1 million poses were generated by clustering with a RMSD cutoff of 1.0 Å. The cluster representatives which do not form a hydrogen bond to NH of Met696 were further filtered out. The HBP ($\leq$2) was then used to remove unrealistic poses (about 40%). The remaining poses were further ranked by the predicted binding energy, and the top about 30% compounds (22,517) with calculated binding energy smaller than -6 kcal/mol ($\sim$50 μM) were kept. Among them, 1381 compounds forming a hydrogen bond to Glu694 were selected and can be classified into 80 structural scaffolds. Finally, 7 scaffolds (9 compounds) of them were purchased for experimental measurements based on visual inspection of the binding modes, commercial availability and structural novelty. The procedures used in the virtual screening are shown in Figure 4. Comparison of the performances between the proposed and AutoDock 4 scoring function is shown in Figure S6.

Notably, 4 of the 9 tested compounds show inhibitory activity at micro-molar to high nano-molar range, with the most active compound showing IC$_{50}$ at 300 nM (Figure 5). Interestingly, the two compound also show a high ligand efficiency [42] of -0.35 kcal/mol per non-hydrogen atom. The predicted binding mode of compound **3** (Figure 6) is further confirmed by the preliminary X-ray crystallography (J. Dong, unpublished results).

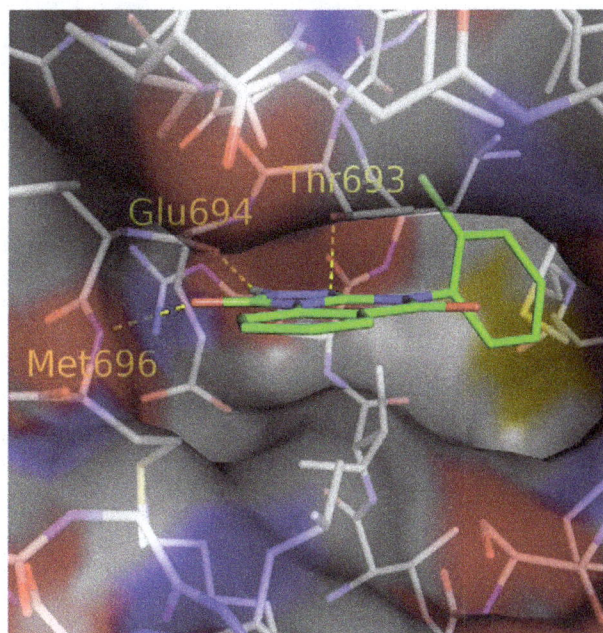

Figure 6. Binding mode of compound 3 (carbon atoms in green) predicted by docking. The intermolecular hydrogen bonds to the residues at the hinge loop (Glu694 and Met696) and the gatekeeper (Thr693) are shown by yellow dashed lines. The protein surface is colored based on atom types with carbon in white, oxygen in red, and nitrogen in blue. This figure was prepared using PyMOL (Delano Scientific, San Carlos, CA).

Conclusion

Hydrogen bonding in biological system is a complex phenomenon as water competes with ligand for the hydrogen bonding sites. Removal of a group that forms a hydrogen bond in unfavorable geometry actually improves binding [21]. In view of hydrogen bonding being an exchange reaction [1,21,22], a new approach is proposed to evaluate the HBP upon ligand binding. Analysis of the 100 crystal structures indicates the penalty in general is low, predominantly smaller than 2 for inhibitors. A high throughput docking case shows HBP can function as an efficient filter to remove poses that unlikely bind. Incorporation of HBP into binding free energy calculation can significantly improve the predictive accuracy and transferability. The fitted parameter of 1.72 kcal/mol means loss of a neutral hydrogen bond would result in a penalty of from 0.34 to 1.72 kcal/mol in binding energy, consistent with the experimental data from 0.5 to 1.5 kcal/mol [21]. Four inhibitors of three scaffolds were discovered out of nine tested, and the binding affinity and ligand efficiency of the most potent compound is about 300 nM and 0.35 kcal/mol per non-hydrogen atom, respectively.

Supporting Information

Figure S1   Scatter plot of C—H...O angles against H...O distances in short C—H...O interactions between ligands and proteins.

Figure S2   Distribution of distances between crystal water oxygen and oxygen or nitrogen atoms of proteins.

Figure S3   Distribution of some key properties of the inhibitors used in the training and test set.

Figure S4   2D plot of the binding mode of Imatinib. Upon ligand binding, one nitrogen atom of the Imatinib pyrimidine ring (N1) becomes water inaccessible and does not form a new hydrogen bond, leading to a penalty of 1. By contrast, the other nitrogen atom (N2) remains hydrogen bonding with a nearby bound water molecule and thus has no penalty.

Figure S5   Poses with the most favorable binding energy of inhibitors of CDK2 (A), EphB4 (B), p38 α (C), Braf (D)

and another set of p38α inhibitors (E). The molecules with bonds in red are the binding modes of the corresponding scaffolds in the crystal structures.

Figure S6   Distribution of predicted binding affinities by Autodock4 (black) and the proposed scoring function (red) on 74,678 compounds passing the first two filters (HB to Met696 and $P_{HB} \leq 2$ kcal/mol). Bin size: 0.1 kcal/mol.

Table S1   MPEOE partial charge and water solubility of model small molecules used to generate initial guess of hydrogen bonding weights.

Table S2   Two-parameter LIECE energy and hydrogen bonding penalty on the 24 EphB4 inhibitors.

File S1   Probing hydrogen bonds formed with implicit water.

File S2   Hydrogen bonding penalty of the 100 kinase complex structures.

File S3   Ranking the strength of individual hydrogen bonds in DNA base pairs.

Acknowledgments

We thank Dr. Amedeo Caflisch for useful discussions and comments on the manuscript. We thank Dr. Jing Dong for the preliminary X-ray structure. We are grateful to Armin Widmer (Novartis Basel) for continuous support with the program WITNOTP, which was used for visual analysis. Calculations were performed on the Schroedinger cluster at the Informatikdienste, University of Zurich.

Author Contributions

Conceived and designed the experiments: DH HZ. Performed the experiments: HZ. Analyzed the data: HZ DH. Contributed reagents/materials/analysis tools: HZ. Wrote the paper: HZ DH.

References

1. Wilkinson AJ, Fersht AR, Blow DM, Winter G (1983) Site-directed mutagenesis as a probe of enzyme structure and catalysis: tyrosyl-tRNA synthetase cysteine-35 to glycine-35 mutation. Biochemistry 22: 3581–3586.
2. Winter G, Fersht AR, Wilkinson AJ, Zoller M, Smith M (1982) Redesigning enzyme structure by site-directed mutagenesis: tyrosyl tRNA synthetase and ATP binding. Nature 299: 756–758.
3. Wilkinson AJ, Fersht AR, Blow DM, Carter P, Winter G (1984) A large increase in enzyme-substrate affinity by protein engineering. Nature 307: 187–188.
4. Bajorath J (2002) Integration of virtual and high-throughput screening. Nat Rev Drug Discov 1: 882–894.
5. Langer T, Hoffmann RD (2001) Virtual screening: an effective tool for lead structure discovery? Curr Pharm Des 7: 509–527.
6. Honig B, Nicholls A (1995) Classical electrostatics in biology and chemistry. Science 268: 1144–1149.
7. Feig M, Onufriev A, Lee MS, Im W, Case DA, et al. (2004) Performance comparison of generalized born and Poisson methods in the calculation of electrostatic solvation energies for protein structures. J Comput Chem 25: 265–284.
8. Schneider G (2010) Virtual screening: an endless staircase? Nat Rev Drug Discov 9: 273–276.
9. Friesner RA, Murphy RB, Repasky MP, Frye LL, Greenwood JR, et al. (2006) Extra precision glide: docking and scoring incorporating a model of hydrophobic enclosure for protein-ligand complexes. J Med Chem 49: 6177–6196.
10. Friesner RA, Banks JL, Murphy RB, Halgren TA, Klicic JJ, et al. (2004) Glide: a new approach for rapid, accurate docking and scoring. 1. Method and assessment of docking accuracy. J Med Chem 47: 1739–1749.
11. Wang R, Lu Y, Wang S (2003) Comparative evaluation of 11 scoring functions for molecular docking. J Med Chem 46: 2287–2303.
12. Thilagavathi R, Mancera RL (2010) Ligand-protein cross-docking with water molecules. J Chem Inf Model 50: 415–421.
13. Kolb P, Huang D, Dey F, Caflisch A (2008) Discovery of kinase inhibitors by high-throughput docking and scoring based on a transferable linear interaction energy model. J Med Chem 51: 1179–1188.
14. Hunter T (1998) The role of tyrosine phosphorylation in cell growth and disease. Harvey Lect 94: 81–119.
15. Pennisi A, Ling W, Li X, Khan S, Shaughnessy JD, Jr., et al. (2009) The ephrinB2/EphB4 axis is dysregulated in osteoprogenitors from myeloma patients and its activation affects myeloma bone disease and tumor growth. Blood 114: 1803–1812.
16. Zhou T, Caflisch A (2010) High-throughput virtual screening using quantum mechanical probes: discovery of selective kinase inhibitors. ChemMedChem 5: 1007–1014.
17. Miyazaki Y, Nakano M, Sato H, Truesdale AT, Stuart JD, et al. (2007) Design and effective synthesis of novel templates, 3,7-diphenyl-4-amino-thieno and furo-[3,2-c]pyridines as protein kinase inhibitors and in vitro

evaluation targeting angiogenetic kinases. Bioorg Med Chem Lett 17: 250–254.

18. Bardelle C, Cross D, Davenport S, Kettle JG, Ko EJ, et al. (2008) Inhibitors of the tyrosine kinase EphB4. Part 1: Structure-based design and optimization of a series of 2,4-bis-anilinopyrimidines. Bioorg Med Chem Lett 18: 2776–2780.

19. Lafleur K, Huang D, Zhou T, Caflisch A, Nevado C (2009) Structure-based optimization of potent and selective inhibitors of the tyrosine kinase erythropoietin producing human hepatocellular carcinoma receptor B4 (EphB4). J Med Chem 52: 6433–6446.

20. Karaman MW, Herrgard S, Treiber DK, Gallant P, Atteridge CE, et al. (2008) A quantitative analysis of kinase inhibitor selectivity. Nat Biotechnol 26: 127–132.

21. Fersht AR, Shi JP, Knill-Jones J, Lowe DM, Wilkinson AJ, et al. (1985) Hydrogen bonding and biological specificity analysed by protein engineering. Nature 314: 235–238.

22. Hine J (1972) Structural Effects on Rates and Equilibria .15. Hydrogen-Bonded Intermediates and Stepwise Mechanisms for Proton-Exchange Reactions between Oxygen-Atoms in Hydroxylic Solvents. J Am Chem Soc 94: 5766–&.

23. Steiner T, Saenger W (1993) Role of C-H...O Hydrogen-Bonds in the Coordination of Water-Molecules - Analysis of Neutron-Diffraction Data. J Am Chem Soc 115: 4540–4547.

24. Cordero B, Gomez V, Platero-Prats AE, Reves M, Echeverria J, et al. (2008) Covalent radii revisited. Dalton Trans. pp 2832–2838.

25. Bondi A (1964) Van Der Waals Volumes+Radii. Journal of Physical Chemistry 68: 441–&.

26. Bohm HJ (1994) The development of a simple empirical scoring function to estimate the binding constant for a protein-ligand complex of known three-dimensional structure. J Comput Aided Mol Des 8: 243–256.

27. Morozov AV, Kortemme T, Tsemekhman K, Baker D (2004) Close agreement between the orientation dependence of hydrogen bonds observed in protein structures and quantum mechanical calculations. Proc Natl Acad Sci U S A 101: 6946–6951.

28. Momany FA, Rone R (1992) Validation of the General-Purpose Quanta(R)3.2/Charmm(R) Force-Field. J Comput Chem 13: 888–900.

29. Brooks BR, Bruccoleri RE, Olafson BD, States DJ, Swaminathan S, et al. (1983) Charmm - a Program for Macromolecular Energy, Minimization, and Dynamics Calculations. J Comput Chem 4: 187–217.

30. Warwicker J, Watson HC (1982) Calculation of the electric potential in the active site cleft due to alpha-helix dipoles. J Mol Biol 157: 671–679.

31. Im W, Beglov D, Roux B (1998) Continuum Solvation Model: computation of electrostatic forces from numerical solutions to the Poisson-Boltzmann equation. Computer Physics Communications 111: 59–75.

32. Gibson AE, Arris CE, Bentley J, Boyle FT, Curtin NJ, et al. (2002) Probing the ATP ribose-binding domain of cyclin-dependent kinases 1 and 2 with O(6)-substituted guanine derivatives. J Med Chem 45: 3381–3393.

33. Stelmach JE, Liu L, Patel SB, Pivnichny JV, Scapin G, et al. (2003) Design and synthesis of potent, orally bioavailable dihydroquinazolinone inhibitors of p38 MAP kinase. Bioorg Med Chem Lett 13: 277–280.

34. Berger DM, Torres N, Dutia M, Powell D, Ciszewski G, et al. (2009) Non-hinge-binding pyrazolo[1,5-a]pyrimidines as potent B-Raf kinase inhibitors. Bioorg Med Chem Lett 19: 6519–6523.

35. Koch P, Jahns H, Schattel V, Goettert M, Laufer S (2010) Pyridinylquinoxalines and pyridinylpyridopyrazines as lead compounds for novel p38 alpha mitogen-activated protein kinase inhibitors. J Med Chem 53: 1128–1137.

36. Goodsell DS, Olson AJ (1990) Automated docking of substrates to proteins by simulated annealing. Proteins 8: 195–202.

37. No KT, Grant JA, Scheraga HA (1990) Determination of Net Atomic Charges Using a Modified Partial Equalization of Orbital Electronegativity Method .1. Application to Neutral Molecules as Models for Polypeptides. Journal of Physical Chemistry 94: 4732–4739.

38. No KT, Grant JA, Jhon MS, Scheraga HA (1990) Determination of Net Atomic Charges Using a Modified Partial Equalization of Orbital Electronegativity Method .2. Application to Ionic and Aromatic-Molecules as Models for Polypeptides. Journal of Physical Chemistry 94: 4740–4746.

39. Irwin JJ, Shoichet BK (2005) ZINC–a free database of commercially available compounds for virtual screening. J Chem Inf Model 45: 177–182.

40. Hawkins PC, Warren GL, Skillman AG, Nicholls A (2008) How to do an evaluation: pitfalls and traps. J Comput Aided Mol Des 22: 179–190.

41. Yang CY, Sun H, Chen J, Nikolovska-Coleska Z, Wang S (2009) Importance of ligand reorganization free energy in protein-ligand binding-affinity prediction. J Am Chem Soc 131: 13709–13721.

42. Hopkins AL, Groom CR, Alex A (2004) Ligand efficiency: a useful metric for lead selection. Drug Discov Today 9: 430–431.

17

Nanopore Fabrication by Controlled Dielectric Breakdown

Harold Kwok◊, Kyle Briggs◊, Vincent Tabard-Cossa*

Department of Physics, University of Ottawa, Ottawa, Ontario, Canada

Abstract

Nanofabrication techniques for achieving dimensional control at the nanometer scale are generally equipment-intensive and time-consuming. The use of energetic beams of electrons or ions has placed the fabrication of nanopores in thin solid-state membranes within reach of some academic laboratories, yet these tools are not accessible to many researchers and are poorly suited for mass-production. Here we describe a fast and simple approach for fabricating a single nanopore down to 2-nm in size with sub-nm precision, directly in solution, by controlling dielectric breakdown at the nanoscale. The method relies on applying a voltage across an insulating membrane to generate a high electric field, while monitoring the induced leakage current. We show that nanopores fabricated by this method produce clear electrical signals from translocating DNA molecules. Considering the tremendous reduction in complexity and cost, we envision this fabrication strategy would not only benefit researchers from the physical and life sciences interested in gaining reliable access to solid-state nanopores, but may provide a path towards manufacturing of nanopore-based biotechnologies.

Editor: Adam Hall, Wake Forest University School of Medicine, United States of America

Funding: This work was supported by the Natural Sciences and Engineering Research Council of Canada, the Canada Foundation for Innovation, and Ontario Network of Excellence. The funders had no role in study design, data collection and analysis, decision to publish, or preparation of the manuscript.

Competing Interests: A patent application was filed on the content of the work presented. "Fabrication of nanopores using high electric fields", with publication number: WO2013167955 A1, and application number: PCT/IB2013/000891). Since this study, the authors have been awarded a research grant from NSERC in collaboration with Abbott Laboratories, but that funding was not used to support the work presented here. There are currently no further patents, products in development or marketed products to declare.

* E-mail: tcossa@uOttawa.ca

◊ These authors contributed equally to this work.

Introduction

Nanopore sensing relies on the electrophoretically driven translocation of biomolecules through nanometer-scale holes embedded in thin insulating membranes to confine, detect and characterize the properties or the activity of individual biomolecules electrically, by monitoring transient changes in ionic current [1–4]. The field was initially shaped by the ability of researchers to exploit biological channels to translocate single molecules [5,6]. It rapidly expanded when new techniques to fabricate individual molecular-sized holes in thin solid-state materials were developed over the last decade [7–12]. These techniques, based on beams of high-energy particles, either produced by a dedicated ion beam machine (i.e ion-beam sculpting) or a transmission electron microscope (i.e TEM drilling), allowed researchers to control the nanopore size at the sub-10-nm length scale with single nanometer precision, thus greatly diversifying the breadth of applications. Since then, a host of applications for DNA, RNA and proteins analysis using both biological and solid-state nanopores have been demonstrated [4,13,14]. Compared to their organic counterparts, solid-state nanopores were expected to emerge as an essential component of any practical nanopore-based instrumentation due to the size control, increased robustness of the membrane, and their natural propensity for integration with wafer-scale technologies, including CMOS and microfluidics [15,16]. Yet, this prospect is significantly hindered due to the constraints and limitations imposed by ion beam sculpting and transmission

electron microscopy-based drilling, which, to this date, remain the only viable tools for achieving nanopores fabrication with dimensional control at the 1-nm scale. The complexity, low-throughput, and high-cost associated with these techniques restrict accessibility of the field to many researchers, greatly limit the productivity of the community, and prevent mass production of nanopore-based devices. Alternative nanofabrication strategies are therefore needed for the field to continue to thrive, and for the promised health-related applications to be successfully commercialized (including single-molecule DNA sequencing). Here, we introduce a fabrication technique based on the use of high electric fields to control dielectric breakdown in solution. The method is automated, simple, and low-cost, allowing nanopores to be created directly in aqueous solution with sub-nm precision, greatly facilitating use and improving yield of functional solid-state nanopore devices. We envision this fabrication strategy will not only provide a path towards nanomanufacturing of nanopore-based devices for a wide range of biotechnology applications, but will democratize the use of solid-state nanopores, while offering researchers new strategies for designing nanofluidics devices, as well as integrating nanopores with CMOS and microfluidics technologies.

Results and Discussion

We fabricate individual nanopores on thin insulating solid-state membranes directly in solution. A thin silicon nitride (SiN_x)

membrane, supported by a silicon frame, is mounted in a liquid cell and separates two reservoirs containing an aqueous solution of 1M KCl. Ag/AgCl electrodes immersed on both sides of the membrane are connected to a custom-built resistive feedback current amplifier, which allow trans-membrane potentials of up to ± 20 V to be applied. The setup shown in Figure 1 is otherwise identical to what is commonly used for biomolecular detection [17], which greatly facilitates the transition to sensing experiments, eliminating further handling of membranes. See Material and Methods section and Section S1 for more detail.

A single nanopore is fabricated by applying a constant potential difference, ΔV, across a t = 10-nm or 30-nm thick SiN_x membrane, to produce an electric field, $E = \Delta V/t$ in the dielectric membrane in the range of 0.4-1 V/nm (Figure 2a). At these high field strengths, a sustainable leakage current, $I_{leakage}$, is observed through the membrane, which remains otherwise insulating at low fields. $I_{leakage}$ rapidly increases with electric field strength, but is typically tens of nanoamperes for our operating conditions. We attribute the dominant conduction mechanism to a form of trap-assisted tunneling of electrons, supplied by ions in solution [18–21] (Figure 2b and 2c), since the membrane is too thick for significant direct tunnelling [18], and migration of impurities cannot produce lasting currents [22]. Direct migration of electrolyte ions is also unlikely, or negligible, since for a given electric field strength, a higher $I_{leakage}$ is observed in thicker membranes (Figure 2e). A larger current is observed on thicker membranes since the number of charge traps (defects) per unit area is greater, as their number in the material increases with volume. We provide additional discussion on the characteristics of the leakage current in Section S2.

We observe the creation of a single nanopore (i.e. fluidic channel spanning the membrane) by a sudden irreversible increase in $I_{leakage}$, which is attributed to the onset of ionic current (Figure 2d and 2f) due to a discrete dielectric breakdown event. As the current continues to increase, the nanopore further enlarges (Figure 2g). We use a feedback control mechanism to rapidly terminate the trans-membrane potential when the current exceeds a pre-determined threshold, I_{cutoff}. A threshold, set as $I_{cutoff}/I_{leakage} < 1.2$, which is generally sufficient to terminate ΔV within ~ 0.1 s of the breakdown event, can produce nanopores on the order of 2-nm in diameter as shown by the I-V curves in Figure 2 h (see Section S3 for additional results). In addition, following the nanopore fabrication event, we can continue to enlarge its size with sub-nm precision by applying moderate AC electric field square pulses in the range of ± 0.2-0.3 V/nm, similar to Beamish et al. [23,24]. This allows the nanopore size to be precisely tuned, for a particular sensing application, directly in neutral KCl solution.

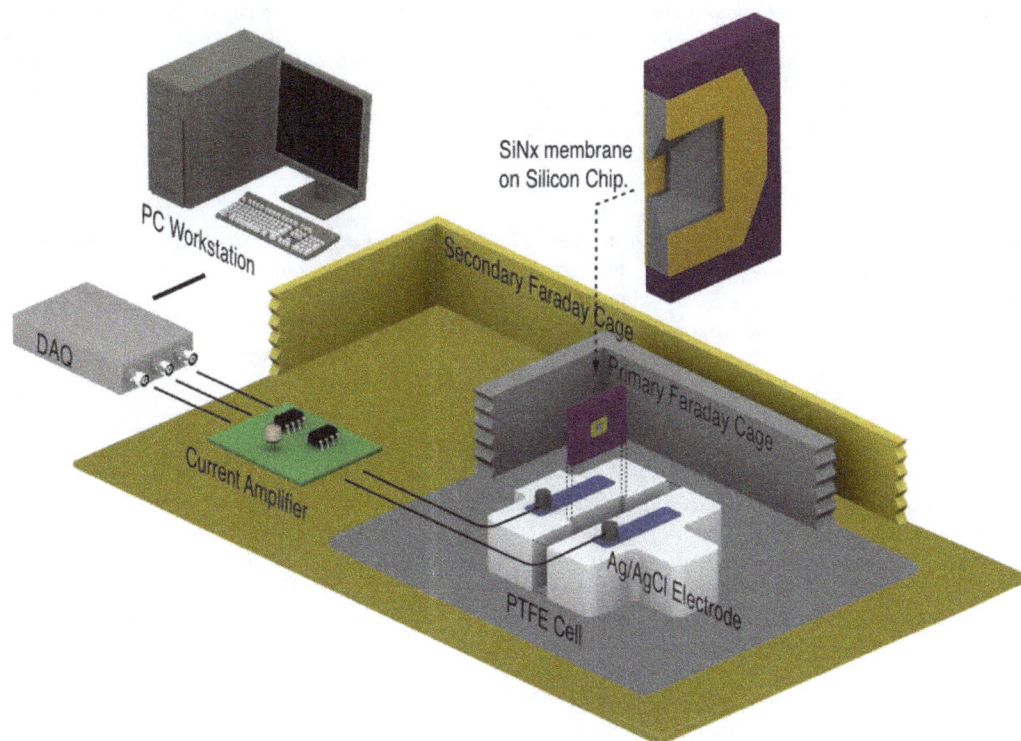

Figure 1. Schematic of the fabrication setup. A computer-controlled custom current amplifier is used to apply voltages up to ± 20 V and measure the current with sub-nA sensitivity from one of the two Ag/AgCl electrodes positioned on either sides of the membrane. It is noteworthy to realize that this experimental setup is identical (with the exception of the custom current amplifier replacing the commonly used Axopatch 200B) to the instrumentation used to study DNA or proteins translocation through nanopores.

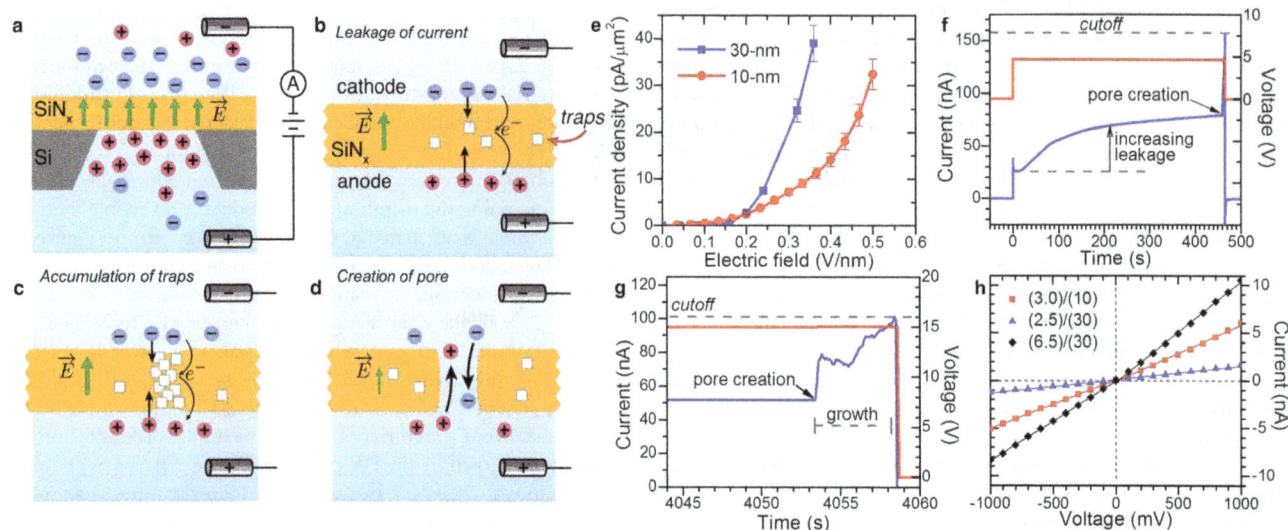

Figure 2. Nanopore formation by dielectric breakdown. a) Application of a trans-membrane potential generates an electric field inside the SiN$_x$, and charges the interfaces with opposite ions. b) Leakage current through the membrane follows a trap-assisted tunneling mechanism. Free charges (electrons or holes) can be produced by redox reactions at the surface or by field ionization of incorporated ions. The number of available charged traps (structural defects) sets the magnitude of the observed leakage current. c) Accumulation of charge traps produced by electric field-induced bond breakage or energetic charges carries leads to a highly localized conductive path, and a discrete dielectric breakdown event. d) A nanopore is formed following removal of the defects. e) Leakage current density for SiN$_x$ membranes (50-μm × 50-μm). The leakage current is fully reversible and stable, unless high fields are sustained, see Section S2 f) Leakage current at 5 V, on a 10-nm-thick SiN$_x$ membrane, in 1 M KCl at pH13.5. Pore created is ~5-nm (18 nS). The slowly increase leakage current, following the capacitive spike, is a result of the accumulation of traps in the membrane. g) Experiment performed at 15 V, on a 30-nm-thick SiN$_x$ membrane, in 1 M KCl pH10. The nanopore is allowed to grow until a pre-determined threshold current is reached, at which point the voltage is turned off. The observed current fluctuations at the onset of pore formation are attributed to significant low-frequency noise at this voltage. Pore created is ~3-nm (2.9 nS). h) Current-to-voltage curves for 3 nanopores fabricated on different membranes. The legend indicates the (pore diameter)/(membrane thickness) in nm. Measurements performed in 1 M KCl pH8, with an Axopatch 200B.

I-V Characteristics and Noise

To infer the nanopore size upon fabrication, we measure its ionic conductance, G, and relate it to an effective diameter, d, assuming a cylindrical geometry and accounting for access resistance [25,26], using $G = \sigma \left[\dfrac{4t}{\pi d^2} + \dfrac{1}{d} \right]^{-1}$, where σ is the bulk conductivity of the solution. This method, practical for nanopores fabricated in liquids, provides a reasonable first order estimate of the pore size [26,27] as confirmed by DNA translocations, and compares well with actual dimensions obtained from TEM images (see Sections S4 and S8). I-V curves are performed in a ±1 V window, where the leakage current can safely be ignored. Figure 2 h reveals an ohmic electric response in 1 M KCl. The majority of our nanopores exhibit linear I-V curves upon fabrication. The remaining nanopores that show signs of self-gating or rectification can be conditioned, by applying moderate electric field pulses [23], to slightly enlarge them until an ohmic behaviour is attained in high salt. Such I-V characteristics imply a relatively symmetric internal electric potential pore profile [28] which supports the symmetrical geometry with a uniform surface charge distribution assumed by our pore conductance model. Otherwise, one would expect significant rectification from multiple ≤1-nm fluidic paths or from a single narrow nano-crack of similar conductance, due to strong electrostatic double layer overlap. To further characterize the nanopores, we examined the noise in the ionic current by performing power spectral density measurements. Our fabrication method consistently produces nanopores with low-$1/f$ noise levels, comparable to fully wetted TEM-drilled nanopores (see Section S5)[29,30]. This may be attributed to the fact that nanopores are created directly in liquid rather than in

vacuum. Thus far, we have successfully fabricated hundreds of individual nanopores ranging from 1 to 25-nm in size with comparable electrical characteristics that are stable for days, 66 of which are included in Figure 2. The success rate for fabricating a nanopore under the experimental parameters presented here is estimated at >99%.

Dielectric Breakdown Mechanism

In order that a single, well-defined nanopore be created each time, we postulate that the leakage current must be highly localized on the insulating membrane, since for conductive substrates (semiconductors or metals) anodic oxidation leads to an array of nanopores[31–33]. The leakage spot(s) must also modify the membrane at the nanoscale since an aqueous KCl solution at neutral pH is not an active etchant of SiN$_x$. To elucidate the mechanism leading to the formation of a nanopore, we investigate the fabrication process as a function of applied voltage, membrane thickness, electrolyte composition, concentration, and pH. Figure 3a shows the time-to-pore creation, τ, as a function of the trans-membrane potential for 30-nm-thick membranes, in 1 M KCl buffered at various pHs. Interestingly, τ scales exponentially with the applied voltage irrespective of other conditions, and can be as short as a few seconds. For a given voltage, pH has a strong effect. As seen in Figure 3b, τ can be reduced by 1,000-fold when lowering the pH from 7 to 2. We have also observed that lower salt concentrations increase the fabrication time (see Section S6). Overall, for a given fabrication condition τ is relatively consistent, though variations by a factor of 4 are common, and is uncorrelated with the size of the fabricated pore. Figure 3c shows τ for 10-nm-thick SiN$_x$ membranes, buffered at pH10 in various 1 M Cl-based aqueous solutions. The fabrication time in these thinner membranes also decreases exponentially with potential, but the value required for forming a nanopore is now reduced by ~1/3 compared to 30-nm-thick membranes, irrespective of the different cations (K^+, Na^+, Li^+) tested. This observation indicates that the applied electric field in the membrane is the main driving force for initiating the fabrication of a single nanopore. Fields in the range of 0.4-1 V/nm are close to the dielectric breakdown strength of low-stress SiN_x films[19], and are key for intensifying the leakage current, which is thought to ultimately cause breakdown in thin insulating layers[34]. The exponential dependence of τ on potential implies the same electric field dependency, which is reminiscent of the time-to-dielectric breakdown in gate dielectrics[34]. According to the current understanding, dielectric breakdown mechanisms proceed as follows[34–36]: (i) probabilistic accumulation of charge traps (i.e. structural defects) by electric field-induced bond breakage or generated by charge injection from the anode or cathode, (ii) increasing up to a critical density forming a highly localized conductive path, and (iii) causing physical damage due to substantial power dissipation and the resultant heating. We propose that the process by which we fabricate a nanopore in solution is similar, though we control the damage to the nanoscale by limiting the localized leakage current, at the onset of the first, discrete breakdown event. As indicated by Figure 3, the likelihood of defect formation within the silicon nitride membrane increases with the applied voltage and the strength of the electric field. At low values, the accumulation of charge traps is accomplished with relatively low efficiency compared to the amount of charge carriers traversing the membrane, since a leakage current of tens of nanoamperes can be sustained for hours or days. Given the stochastic nature of the pore creation process, multiple simultaneous nanoscale breakdown events are unlikely. Termination of the applied voltage following the occurrence of the first breakdown event, observed by the sudden irreversible increase in $I_{leakage}$, ensures that ultimately a single nanopore is created. Moreover, the fabrication of a single nanopore may result from the fact that the formation path of a nanopore experiences increased electric field strength during growth, which locally reinforces the rate of defect generation. The process by which material is removed from the membrane remains unclear, but broken bonds could be chemically dissolved by the electrolyte or following a conversion to oxides/hydrides [37,38]. Another possibility is shearing due to localized plasticity of the membrane as a result of heating at the breakdown spot, but the efficiency of heat dissipation at the nanoscale, resulting from high surface-area-to-volume ratios, makes this less likely [22]. We explain the pH dependency on the fabrication time, for 30-nm thick SiN_x membranes, by the fact that breakdown at low pH is amplified by impact ionization producing an avalanche, due to the increased likelihood of hole injection or H^+ incorporation from the anode (see Section S6 for more detail). To support the general character of nanofabrication by dielectric breakdown, we created nanopores in a different material (silicon dioxide) and present the data in Section S7.

DNA Translocations

We performed DNA translocation experiments to demonstrate that these nanopores can be leveraged for the benefit of single-molecule detection. Electrophoretically driven passage of a DNA molecule across a membrane is expected to transiently block the flow of ions in a manner that reflects the molecule length, size, charge and shape. The results using a ~6.4-nm-diameter pore, as estimated from conductance measurements, in a 10-nm thick SiN_x membrane are shown in Figure 4. The scatter plot shows event

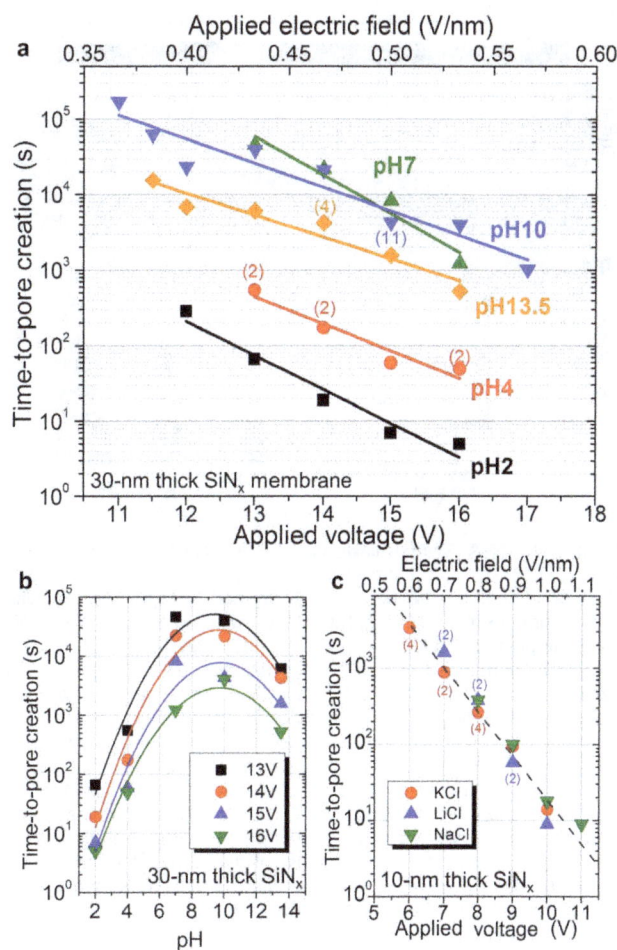

Figure 3. Time-to-pore creation as a function of experimental conditions. a) Semi-log plot of fabrication time of individual nanopores created in 30-nm-thick SiN_x membranes in 1 M KCl buffered as indicated, versus the applied voltage and the calculated applied electric field. The number of separate nanopores each data point is averaged over is indicated in parentheses. The vast majority of nanopores plotted are sub-5-nm in size (i.e. <7 nS). b) Semi-log plot of fabrication time versus pH for the data plotted in a). c) Semi-log plot of fabrication time of individual nanopores created in 10-nm-thick SiN_x membranes in 1 M Cl-based electrolyte buffered at pH 10 for different cationic species versus the applied voltage and the calculated applied electric field. All 66 nanopores plotted are sub-5-nm in size (i.e. <20 nS).

duration and average current blockage of over 2,400 single-molecule translocations events of 5-kb dsDNA. The characteristic shape of the events is indistinguishable to data obtained on TEM-drilled nanopores [26,39–41]. The observed quantized current blockades strongly support the presence of a single nanopore spanning the membrane. Using dsDNA (~2.2 nm in diameter) as a molecular-sized ruler, the value of the single-level blockage events, $\Delta G = 7.4 \pm 0.9$ nS, provides an effective pore diameter of 6.0 ± 0.5-nm consistent with the size extracted from the pore conductance model [26]. This result also suggests that the membrane thickness at the vicinity of the nanopores has not been significantly altered. We observed similar DNA translocation signatures from most nanopores tested (revealing >80% success rate in detecting DNA for N = 19 nanopores tested), and provide further discussion and additional translocation data in Section S8.

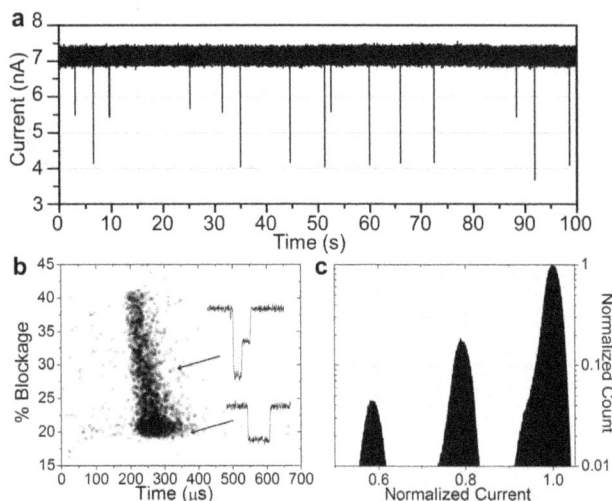

Figure 4. DNA Translocations. a) Ionic current trace showing multiple DNA translocation events through a ~6.4-nm pore in a 10-nm-thick SiN_x membrane. Experiments performed with 10µg/mL of 5-kb DNA fragments in 3.6 M LiCl pH8, at 200 mV using an Axopatch 200B. Data sampled at 250 kHz and low-pass filtered at 100 kHz. b) Scatter plot of the normalized average current blockade (0% presenting a fully opened pore, and 100% a fully blocked pore) versus the total translocation time of a single-molecule event. Each data point represents a single DNA translocation event. The majority of the events are unfolded. There are very few anomalously long events, indicating weak DNA-pore interactions. The inset shows ionic current signatures of two single-molecule translocation events, passing in a linear and partially folded conformation. c) Histogram of the current level revealing the expected quantization of the amplitude of current blockades. Quantized levels corresponding to zero, one, two dsDNA strands in the nanopore are clearly observed.

Outlook

Nanopore fabrication by controlled dielectric breakdown in solution represents a major reduction in complexity and cost over current fabrication methods, which will greatly facilitate accessibility to the field to many researchers, and provides a path to commercialize nanopore-based technologies. While we attribute the nanopore creation process to an intrinsic property of the dielectric membrane, such that the nanopore can form anywhere on the surface, our current understanding strongly suggests that the position of the pore can be determined by locally controlling the electric field strength or the material dielectric strength. This could be achieved, for instance, by nanopatterning or locally thinning the membrane, by positioning of a nanoelectrode, or by confining the field to specific areas on the membrane via micro- or nanofluidic channel encapsulation (see Section S9). The latter would also allow for the simple integration of independently addressable nanopores in an array format on a single chip.

Materials and Methods

Dielectric Membranes

Silicon Nitride (SiN_x) membranes used in our experiments are commercially available as transmission electron microscope (TEM) windows (Norcada product # NT005X and NT005Z). Each membrane is made of 10-nm or 30-nm thick low-stress (<250 MPa) SiN_x, deposited on 200-µm thick lightly doped silicon (Si) substrate by low-pressure chemical vapour deposition

(LPCVD). A 50-µm × 50-µm window on the backside of the Si substrate is opened by a KOH anisotropic chemical etch. Prior to mounting into liquids, SiN_x membranes can be cleaned in oxygen plasma for 30 s at 30 W to facilitate wetting of the membrane surface, though this is not a requirement. All solutions used were filtered and degassed prior to use. The absence of pre-existing structural damages (e.g. pinholes, nano-cracks) is inferred by the fact that no current (<pA) is measured across a membrane at low voltages (<±1 V) prior to nanopore fabrication. The membrane side opposing the Si etch pit is the reference point for all applied voltages in this article. Silicon dioxide membranes were also purchased from TEMWindows (product# SO100-A20Q33). Note that we have also successfully fabricated nanopores on SiN_x membranes purchased from TEMWindows, and on custom fabricated SiN_x membranes.

Instrumentation and Data Acquisition

A schematic of the experimental setup is shown in Figure 1. A silicon chip with an intact silicon nitride membrane is sandwiched between two silicone gaskets (shown in purple on the figure). It is then positioned between the two electrolyte reservoirs in a PTFE (polytetrafluoroethylene) or a PEEK (polyether ether ketone) fluidic cell. The two reservoirs filled with liquid electrolyte are electrically connected to a current amplifier by two Ag/AgCl electrodes. The entire system is encapsulated in a grounded faraday cage to isolate from electromagnetic interference. Data acquisition and measurement automation were performed using custom-designed LabVIEW software controlling a National Instruments USB-6351 or PXIe-6366 DAQ card. The value of the trans-membrane potential is set by the DAQ card. Leakage current is digitized at 250 kHz and the signal is filtered at 10 Hz. When a current exceed a pre-set threshold, the voltage bias is immediately ceased by the software (response time is ~100 ms). I-V measurements and ionic current signal during DNA translocations are recorded using an Axopatch 200B with a 4-pole Bessel filter set at 100 kHz, with at 250 kHz sampling rate. Data analysis was carried out using custom-designed LabVIEW software to measure the duration and depth of each current blockade events.

DNA Studies

We performed DNA translocation studies, using dsDNA fragments of 100 bp, 5 kbp, 10 kbp purchased from Fermantas (NoLimits products) in 1 M KCl pH8 or in 3.6 M LiCl pH8 at a final concentration of 10µg/mL. Lambda DNA (48.5 kbp) purchased from NewEngland BioLabs was also used.

Supporting Information

Section S1   Section S1 contains additional information regarding the experimental setup.

Section S2   Section S2 provides further discussion on the leakage current.

Section S3   Section S3 shows I-V curves of 8 nanopores <2.5-nm in diameter.

Section S4   Section S4 shows TEM images of nanopores fabricated by controlled dielectric breakdown.

Section S5 Section S5 presents the ionic current noise characteristics of 4 nanopores.

Section S6 Section S6 presents additional data of the time-to-pore creation versus trans-membrane potential and electric field strength.

Section S7 Section S7 demonstrates fabrication of a nanopore on a silicon dioxide membrane.

Section S8 Section S8 presents additional DNA translocation data and their analysis.

Section S9 Section S9 discusses strategies for localizing nanopores on a membrane.

Acknowledgments

The authors would like to thank Y. Liu for aid in TEM imaging and L. Andrzejewski for valuable technical support.

Author Contributions

Conceived and designed the experiments: VTC HK. Performed the experiments: HK KB. Analyzed the data: KB HK VTC. Contributed reagents/materials/analysis tools: HK. Wrote the paper: VTC. Discovered the fabrication process: HK.

References

1. Venkatesan BM, Bashir R (2011) Nanopore sensors for nucleic acid analysis. Nature nanotechnology 6: 615–624. doi:10.1038/NNANO. 2011.129.
2. Dekker C (2007) Solid-state nanopores. Nature nanotechnology 2: 209–215. doi:10.1038/nnano.2007.27.
3. Branton D, Deamer DW, Marziali A, Bayley H, Benner S a, et al. (2008) The potential and challenges of nanopore sequencing. Nature biotechnology 26: 1146–1153. doi:10.1038/nbt.1495.
4. Kasianowicz JJ, Robertson JWF, Chan ER, Reiner JE, Stanford VM (2008) Nanoscopic porous sensors. Annual review of analytical chemistry (Palo Alto, Calif) 1: 737–766. doi:10.1146/annurev.anchem.1.031207.112818.
5. Bezrukov SM, Vodyanoy I, Parsegian VA (1994) Counting polymers moving through a single ion channel. Nature 370: 279–281.
6. Kasianowicz JJ, Brandin E, Branton D, Deamer DW (1996) Characterization of individual polynucleotide molecules using a membrane channel. Proceedings of the National Academy of Sciences of the United States of America 93: 13770–13773.
7. Li J, Stein D, McMullan C, Branton D, Aziz MJ, et al. (2001) Ion-beam sculpting at nanometre length scales. Nature 412: 166–169. doi:10.1038/35084037.
8. Storm AJ, Chen JH, Ling XS, Zandbergen HW, Dekker C (2003) Fabrication of solid-state nanopores with single-nanometre precision. Nature materials 2: 537–540. doi:10.1038/nmat941.
9. Storm a J, Chen JH, Ling XS, Zandbergen HW, Dekker C (2005) Electron-beam-induced deformations of SiO[sub 2] nanostructures. Journal of Applied Physics 98: 014307. doi:10.1063/1.1947391.
10. Kuan AT, Golovchenko JA (2012) Nanometer-thin solid-state nanopores by cold ion beam sculpting. Applied physics letters 100: 213104–2131044. doi:10.1063/1.4719679.
11. Russo CJ, Golovchenko JA (2012) Atom-by-atom nucleation and growth of graphene nanopores. Proceedings of the National Academy of Sciences of the United States of America 109: 5953–5957. doi:10.1073/pnas.1119827109.
12. Yang J, Ferranti DC, Stern L a, Sanford C a, Huang J, et al. (2011) Rapid and precise scanning helium ion microscope milling of solid-state nanopores for biomolecule detection. Nanotechnology 22: 285310. doi:10.1088/0957-4484/22/28/285310.
13. Miles BN, Ivanov AP, Wilson KA, Doğan F, Japrung D, et al. (2013) Single molecule sensing with solid-state nanopores: novel materials, methods, and applications. Chemical Society reviews 42: 15–28. doi:10.1039/c2cs35286a.
14. Oukhaled A, Bacri L, Pastoriza-Gallego M, Betton J-M, Pelta J (2012) Sensing Proteins through Nanopores: Fundamental to Applications. ACS chemical biology 7: 1935–1949. doi:10.1021/cb300449t.
15. Rosenstein JK, Wanunu M, Merchant CA, Drndic M, Shepard KL (2012) Integrated nanopore sensing platform with sub-microsecond temporal resolution. Nature methods 9: 487–492. doi:10.1038/nmeth.1932.
16. Jain T, Guerrero RJS, Aguilar CA, Karnik R (2013) Integration of solid-state nanopores in microfluidic networks via transfer printing of suspended membranes. Analytical chemistry 85: 3871–3878. doi:10.1021/ac302972c.
17. Tabard-Cossa V (2013) Instrumentation for Low-Noise High-Bandwidth Nanopore Recording. In: Edel J, Albrecht T, editors. Engineered Nanopores for Bioanalytical Applications. Elsevier. pp. 59–88.
18. Frenkel J (1938) On Pre-Breakdown Phenomena in Insulators and Electronic Semi-Conductors. Physical Review 54: 647–648. doi:10.1103/PhysRev.54.647.
19. Habermehl S, Apodaca RT, Kaplar RJ (2009) On dielectric breakdown in silicon-rich silicon nitride thin films. Applied Physics Letters 94: 012905. doi:10.1063/1.3065477.
20. Jeong DS, Hwang CS (2005) Tunneling-assisted Poole-Frenkel conduction mechanism in HfO[sub 2] thin films. Journal of Applied Physics 98: 113701. doi:10.1063/1.2135895.
21. Kimura M, Ohmi T (1996) Conduction mechanism and origin of stress-induced leakage current in thin silicon dioxide films. Journal of Applied Physics 80: 6360. doi:10.1063/1.363655.
22. Lee S, An R, Hunt AJ (2010) Liquid glass electrodes for nanofluidics. Nature nanotechnology 5: 412–416. doi:10.1038/nnano.2010.81.
23. Beamish E, Kwok H, Tabard-Cossa V, Godin M (2012) Precise control of the size and noise of solid-state nanopores using high electric fields. Nanotechnology 23: 405301. doi:10.1088/0957-4484/23/40/405301.
24. Beamish E, Kwok H, Tabard-Cossa V, Godin M (2013) Fine-tuning the Size and Minimizing the Noise of Solid-state Nanopores. Journal of visualized experiments: JoVE: e51081. doi:10.3791/51081.
25. Vodyanoy I, Bezrukov SM (1992) Sizing of an ion pore by access resistance measurements. Biophysical journal 62: 10–11. doi:10.1016/S0006-3495(92)81762-9.
26. Kowalczyk SW, Grosberg AY, Rabin Y, Dekker C (2011) Modeling the conductance and DNA blockade of solid-state nanopores. Nanotechnology 22: 315101. doi:10.1088/0957-4484/22/31/315101.
27. Frament CM, Dwyer JR (2012) Conductance-Based Determination of Solid-State Nanopore Size and Shape: An Exploration of Performance Limits. The Journal of Physical Chemistry C 116: 23315–23321. doi:10.1021/jp305381j.
28. Kosińska ID (2006) How the asymmetry of internal potential influences the shape of I-V characteristic of nanochannels. The Journal of chemical physics 124: 244707. doi:10.1063/1.2212394.
29. Tabard-Cossa V, Trivedi D, Wiggin M, Jetha NN, Marziali A (2007) Noise analysis and reduction in solid-state nanopores. Nanotechnology 18: 305505. doi:10.1088/0957-4484/18/30/305505.
30. Smeets RMM, Keyser UF, Dekker NH, Dekker C (2008) Noise in solid-state nanopores. Proceedings of the National Academy of Sciences of the United States of America 105: 417–421. doi:10.1073/pnas.0705349105.
31. Thompson GE, Wood GC (1981) Porous anodic film formation on aluminium. Nature 290: 230–232. doi:10.1038/290230a0.
32. Létant SE, Hart BR, Van Buuren AW, Terminello LJ (2003) Functionalized silicon membranes for selective bio-organism capture. Nature materials 2: 391–395. doi:10.1038/nmat888.
33. Tseng AA, Notargiacomo A, Chen TP (2005) Nanofabrication by scanning probe microscope lithography: A review. Journal of Vacuum Science & Technology B: Microelectronics and Nanometer Structures 23: 877. doi:10.1116/1.1926293.
34. Lombardo S, Stathis JH, Linder BP, Pey KL, Palumbo F, et al. (2005) Dielectric breakdown mechanisms in gate oxides. Journal of Applied Physics 98: 121301. doi:10.1063/1.2147714.
35. McPherson JW, Mogul HC (1998) Underlying physics of the thermochemical E model in describing low-field time-dependent dielectric breakdown in SiO[sub 2] thin films. Journal of Applied Physics 84: 1513. doi:10.1063/1.368217.
36. DiMaria DJ, Cartier E, Arnold D (1993) Impact ionization, trap creation, degradation, and breakdown in silicon dioxide films on silicon. Journal of Applied Physics 73: 3367. doi:10.1063/1.352936.
37. Liu H, Steigerwald ML, Nuckolls C (2009) Electrical double layer catalyzed wet-etching of silicon dioxide. Journal of the American Chemical Society 131: 17034–17035. doi:10.1021/ja903333s.
38. Jamasb S, Collins S, Smith RL (1998) A physical model for drift in pH ISFETs. Sensors and Actuators B: Chemical 49: 146–155.
39. Chen P, Gu J, Brandin E, Kim Y-R, Wang Q, et al. (2004) Probing Single DNA Molecule Transport Using Fabricated Nanopores. Nano Letters 4: 2293–2298. doi:10.1021/nl048654j.
40. Fologea D, Brandin E, Uplinger J, Branton D, Li J (2007) DNA conformation and base number simultaneously determined in a nanopore. Electrophoresis 28: 3186–3192.
41. Li J, Gershow M, Stein D, Brandin E, Golovchenko J a (2003) DNA molecules and configurations in a solid-state nanopore microscope. Nature materials 2: 611–615. doi:10.1038/nmat965.

18

Design of Miniaturized Double-Negative Material for Specific Absorption Rate Reduction in Human Head

Mohammad Rashed Iqbal Faruque*, Mohammad Tariqul Islam

Centre for Space Science (ANGKASA), Research Centre Building, Universiti Kebangsaan Malaysia, UKM, Bangi, Selangor D. E., Malaysia

Abstract

In this study, a double-negative triangular metamaterial (TMM) structure, which exhibits a resounding electric response at microwave frequency, was developed by etching two concentric triangular rings of conducting materials. A finite-difference time-domain method in conjunction with the lossy-Drude model was used in this study. Simulations were performed using the CST Microwave Studio. The specific absorption rate (SAR) reduction technique is discussed, and the effects of the position of attachment, the distance, and the size of the metamaterials on the SAR reduction are explored. The performance of the double-negative TMMs in cellular phones was also measured in the cheek and the tilted positions using the COMOSAR system. The TMMs achieved a 52.28% reduction for the 10 g SAR. These results provide a guideline to determine the triangular design of metamaterials with the maximum SAR reducing effect for a mobile phone.

Editor: Jeongmin Hong, University of California, Berkeley, United States of America

Funding: This work was supported by the Universiti Kebangsaan Malaysia, under grants Dana Lonjakan Penerbitan (UKM- DLP-2014). The funders had no role in study design, data collection and analysis, decision to publish, or preparation of the manuscript.

Competing Interests: The authors have declared that no competing interests exist.

* Email: rashedgen@yahoo.com

Introduction

Portable communication devices are widely used. Because the use of such mobile devices increases every year, an extensive study on the health risk from hazardous electromagnetic fields is currently in progress. The specific absorption rate (SAR) is the parameter used to evaluate power absorption in the human head. Radio frequency (RF) safety guidelines have been issued to prevent exposure to excessive electromagnetic fields in terms of the SAR [1]. The exposure of the human head to the near-field of a cellular mobile phone can be appraised by measuring the SAR in a human-head phantom or by calculating the exposure using a human-head numerical model [2].

Mobile phone radiation of the body's cells, brain or immune system has been suggested to elevate the threat of developing diseases, ranging from cancer to Alzheimer's disease. Laboratory tests on cockroaches have shown that radiation from mobile phones can have an adverse effect on overall health [3–5]. Note that research effort has also been dedicated to people suffering from headaches, fatigue and a loss of concentration after using their cellular mobile phones.

The study by Jensen and Rahmat-Samii considered a monopole, a side mounted Planar-Inverted F Antenna (PIFA), a top mounted curved inverted F-antenna and a back-mounted PIFA that utilized the FDTD method. They aimed to understand the effects of the tissue position and the corporal model on the antenna performance [6]. Gandhi et al. studied cellular telephones operating at 835 and 1900 MHz using a $\lambda/4$ and a $3\lambda/4$ monopole antenna. They observed that the homogeneous model overestimated the SAR for a $\lambda/4$ antenna above a handset at 835 MHz with a radiation power of 600 mW [7].

Ref. [3] investigated the antenna efficiency, bandwidth, and SAR as a function of a mobile phone's armature-associated parameters, such as the length, thickness, width, and partition from the phantom. This statistical study established that the SAR increased while the radiation efficiency decreased when the resonant frequency of the armature equaled the resonant frequency of the antenna.

The effects of attaching conductive materials to mobile phones for SAR reduction were reported in ref. [8–10]. These studies showed that the position of the shielding material is an important factor for the effectiveness of SAR reduction. The spatial peak SAR needs to be reduced at the design stage of the material because the possibility of a spatial peak SAR exceeding the recommended exposure limit cannot be completely ruled out. The experiment described in [9–14] involved a perfect electric conductor (PEC) reflector positioned between a head model and folded loop antenna driver. The results of the experiment demonstrated that the radiation efficiency may be improved, and a decreased peak SAR value can be obtained. Such an antenna structure sacrifices the availability of signals received from all directions to the phone model. Ferrite sheet attachment reduced the SAR due to the suppression of surface currents on the front side of the phone model [10]. However, the relationship between the maximum SAR reducing effect and the parameters, such as the attaching location, size and material properties of the ferrite sheet, remains unknown. Furthermore, a bottom position is preferred to reduce the SAR. Moreover, SAR also depends on the type of handset or radiator used in a handset [15–18].

The SAR value due to a dipole antenna that is placed next to a plane phantom (flat phantom) was analyzed in ref. [19]. The authors demonstrated that the spatial peak of the SAR was directly related to the antenna's current distribution for frequencies above

300 MHz in the near field. Two relationships were discovered: (i) flanked by the SAR and the magnetic field and (ii) along with the SAR and the antenna's feed point current.

In [20], the authors analyzed the dependence of the exposure of the head on the antenna's radiation patterns using FDTD computations and compared the SAR using various head models. The authors demonstrated that the shaped-head model filled with a homogeneous liquid absorbed the most power and hence resulted in the highest SAR. The spherical model also exhibited a higher SAR than the anatomically correct model.

The main breakthrough of metamaterials is their ability to efficiently guide and control electromagnetic waves via the engineering of their material properties [21–24]. In recent years, significant research has been performed worldwide in order to study, develop and design metamaterials and their applications, particularly in electromagnetics. Metamaterials exhibit negative electrical permittivity and or negative permeability. When both the permittivity and permeability are negative, a metamaterial exhibits a negative refractive index, i.e. it is a left-handed material.

Different methods have been proposed over the last 20 years to reduce the SAR produced by emissions from handset antennas to levels below the current maximum exposure levels of the international standards, including auxiliary antenna elements, ferrite loading, EBG/AMC surfaces and low SAR handset antenna techniques. New results or ideas that promise useful devices for the future are being proposed each year. A number of detailed reviews and books have been published recently [25]. Among these results and ideas, the investigation of metamaterials is currently one of the most active topics in engineering and physics. Metamaterials are an emerging technology that has the potential to significantly change everyday life in the near future [26–28].

In particular, the problems to be solved in EM absorption reduction require an accurate representation of the mobile phone, anatomical representation of the head, alignment of the phone and the head, and an appropriate design of the metamaterials. Metamaterial techniques seem promising options to reduce the SAR in terms of low cost and ease-of implementation in mobile phones. This paper focuses on an antenna design that utilizes new metamaterial developments for SAR reduction.

In this paper, we emphasize artificial structures to design double negative TMMs that are attached to the PCB of mobile phones in order to reduce the SAR in the human head. This paper is structured as follows. Section 2 describes the numerical analysis of the handset in conjunction with the SAM phantom head. The modeling and analysis of the FDTD method coupled with the lossy-Drude model for SAR reduction are also discussed in Section 2. The impacts of the attachment of triangular split ring resonators (TSRRs) on SAR based on double negative TMMs are analyzed in Section 3. The design methodology of a TSRRs structure, the design simulation, and the fabrication of TMMs are explained in Section 4. Section 5 describes the experimental validation of the measurement results, and Section 6 concludes the paper.

Model and Method

A simulation model that includes a handset with a PIFA-type antenna and the SAM phantom head provided by CST Microwave Studio (CST MWS) was used in this study. The numerical simulations were carried out using the finite integration technique (FIT) package in CST Microwave Studio. In these simulations, we added the losses that are typically associated with the resonant behavior of metamaterials, and the dielectric materials were considered perfect electric conductors. The relative permittivity and conductivity of the individual components were set to comply with industrial standards. In addition, the definitions in [13] were adopted for the material parameters involved in the SAM phantom head. To precisely characterize the performance over a broad frequency range, dispersive models for all dielectrics were used in the simulation [29–31].

The electrical properties of the materials used for the simulation are listed in Table 1. In the simulation model, a helical PIFA-type antenna was used, which is used for Global System for Mobile (GSM) 900 MHz applications. A high-quality geometrical approximation can be obtained for this structure from the meshing scheme of the FDTD method. The use of this meshing scheme in turn led to challenges in obtaining convergent results within a short simulation time.

In this paper, the FDTD method with lossy-Drude polarization and magnetization models is used to simulate the DNG medium. Therefore, the permittivity and permeability are described as follows in the frequency domain:

$$\epsilon(\omega) = \epsilon_0 \left(1 - \frac{\omega_{pe}^2}{\omega(\omega - j\,\Gamma_e)} \right) \tag{1}$$

$$\mu(\omega) = \mu_0 \left(1 - \frac{\omega_{pm}^2}{\omega(\omega - j\,\Gamma_m)} \right) \tag{2}$$

where ω_p and Γ denote the corresponding plasma and damping frequencies, respectively. Here, the Drude model is better suited for the FDTD simulations for both the permeability and permittivity functions. This approach provides a much wider bandwidth over which the negative values of the permittivity and permeability can be achieved. This option was only selected for numerical convenience and does not alter any conclusions derived from these simulations. Either case shows negative. However, choosing the Drude model for the FDTD simulation also allows significantly shorter simulation times, especially for low-loss media. In other words, the FDTD simulation will take longer to reach a steady state in the Lorentz model counterpart because the resonance region where the permittivity and permeability acquire their negative values would be very narrow in said model. With this method, we can treat the metamaterials as homogeneous materials with frequency-dispersive material parameters.

The handset featuring an antenna at 900 MHz considered in this study is shown in Fig. 1. This handset was modeled as a quarter-wavelength PIFA antenna positioned onto a rectangular conducting box that was 10 cm tall, 4 cm wide, and 1.5 cm thick. The PIFA antenna was placed on the top surface of the conducting box. A non-uniform meshing scheme was adopted to enable the major computational power to be dedicated to the regions along the inhomogeneous boundaries for rapid and flawless analysis. The minimum and maximum mesh sizes were 0.3 mm and 1.0 mm, respectively. The SAM head model used in this research consisted of approximately 2,097,152 cubical cells of 1 mm resolution. A time-step of 0.1 nanoseconds was used in the simulation, and the simulation duration was approximately eight sinusoidal cycles to ensure steady state conditions. The second-order Mur absorbing boundaries acting on the electric fields were implemented to absorb the outgoing scattered waves. An antenna excitation was introduced by specifying a sinusoidal voltage diagonally at the one-cell gap between the helix and the top surface of the conducting box.

Table 1. Electrical properties of the materials considered in the simulation.

Phone Materials	ε_r	σ (S/m)
Circuit Board	4.4	0.05
Housing Plastic	2.5	0.005
LCD Display	3.0	0.02
Rubber	2.5	0.005
SAM Phantom Head		
Shell	3.7	0.0016
Liquid @ 900 MHz	40	1.42

The effectiveness of the SAR reduction and the antenna performance in different arrangements, sizes, and material properties of the materials and the metamaterials determined from the simulations are presented below. The head models used in this study were based on a MRI-based head model from the website named "whole brain Atlas". Six types of tissues, bone, brain, muscle, eyeball, fat, and skin, were used in this model [12]. Numerical simulations of the SAR value were performed using the FDTD method. The parameters for the FDTD computation were defined as follows. The simulation domain was $128 \times 128 \times 128$ cells. The cell sizes were set to $\Delta x = \Delta y = \Delta z = 1$ mm. The computational domain was terminated with 8 cells of a perfect matched layer (PML). A PIFA antenna was modeled using the thin-wire approximation.

Impact of the TSRRs Attachment on the SAR

In this section, the designed TSRRs were placed between a human head of a phantom and the antenna. This arrangement reduced the SAR value. To study this reduction at the GSM 900 band, different positions, sizes, and metamaterials were also analyzed by using the FDTD method to implement the detailed human head model.

The dispersive models for all dielectrics were used for the simulation to accurately analyze the TSRRs. The antenna was aligned to be parallel to the head axis. The distance between the antenna and the head was varied from 5 mm to 20 mm. Finally, a distance of 20 mm was chosen to compare the different metamaterials. The output power of the mobile phone model was set before the SAR was simulated. In this paper, the output power of the cellular phone was set to 600 mW at an operating frequency of 900 MHz. In the real world, the output power of the mobile phone will not exceed 250 mW for normal use, while the maximum output power can reach to 1 W or 2 W when the base station is far away from the mobile station (cellular phone). The SAR simulation results were compared with the results in [12], [30], and [31] for validation. When the phone model was placed 20 mm away from the human head model without a metamaterial, the simulated SAR 1 g peak value was 2.002 W/kg, and the SAR 10 g value was 1.293 W/kg. The level of SAR reduction was higher with metamaterial attachments. The results reported in [30] and [31] are 2.17 W/kg and 2.28 W/kg, respectively, for a SAR value of 1 g. These reported SAR levels are due to the mobile antenna position, i.e. the antenna structures were not properly aligned, and the antennae themselves also differed. These SAR values are significantly better than the result reported in [12],

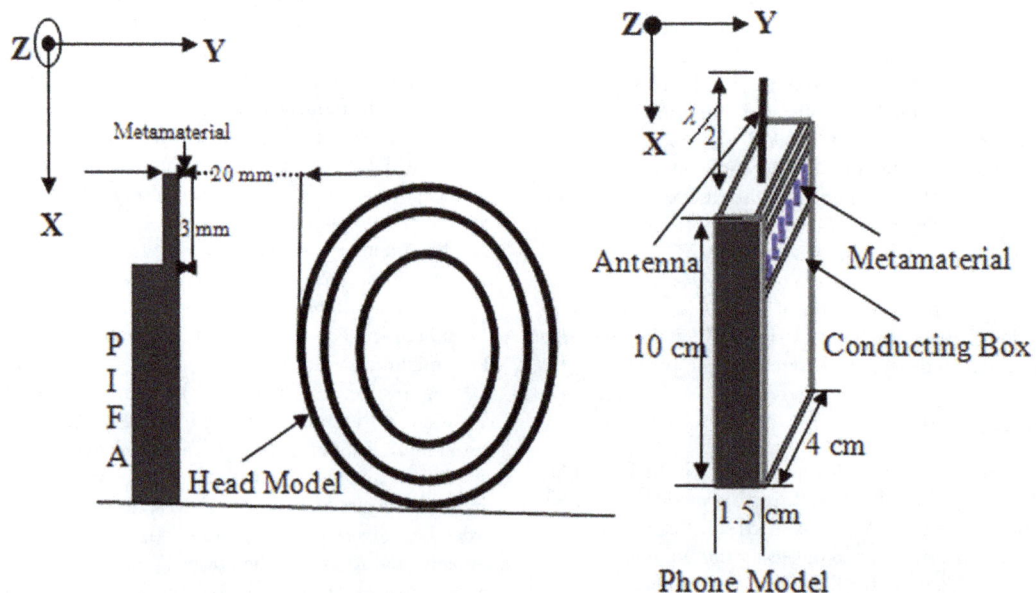

Figure 1. The head and antenna model for SAR calculation.

which was 2.43 W/kg for a SAR value of 1 g. This reported SAR value was achieved via different radiating powers and antenna designs.

The distance from the antenna feeding point to the edge of the metamaterial was A = 3 mm. The size of the metamaterial in the x–z plane was 62 mm×38 mm, and the thickness was 6 mm. The SAR value and the antenna performance with the metamaterial were studied. To calculate the antenna's radiated power, the source impedance (Z_S) was assumed to be the complex conjugate of the free space radiation impedance ($Z_S = 105.18+j81.97\Omega$). The source voltage ($V_S$) was chosen such that the radiated power in free space is equal to 600 mW ($V_S = \sqrt{0.6.8.R_{R0}}$). The source impedance and source voltage were held constant at the Z_S and V_S values when analyzing the effect of the metamaterials and the human head on the antenna performance.

The power radiated from the antenna was calculated by comparing the radiation impedance in this situation ($Z_R = R_R+ jX_R$) with that using the following [14] equation:

$$P_R = \frac{1}{2}V_S^2 \frac{R_R}{|Z_R+Z_S|^2} \quad (3)$$

The total power absorbed in the head was calculated using the following:

$$P_{abs} = \frac{1}{2}\int_V \sigma|E|^2 dv \quad (4)$$

Different negative medium parameters were investigated to study the efficiency of the SAR reduction effectiveness. We positioned the negative permittivity media between the antenna and the human head to intercept the electromagnetic waves. Initially, the plasma frequencies of the media were set to $\omega_{pe} = 9.239\times10^9$ rad/s, which resulted in media with $\mu=1$ and $\varepsilon=-3$ at 900 MHz. Media with a larger negative permittivity, i.e., $\mu=1$, and $\varepsilon=-5$ or $\mu=1$ and $\varepsilon=-7$, were also analyzed. We set $\Gamma_e = 1.22\times10^8$ rad/s, indicating that the media experience losses. The SAR 1 g peak level reduced to 1.0724 W/kg for media with $\mu=1$ and $\varepsilon=-3$, and metamaterials also affected impedance. Compared to the control experiment without metamaterials, the radiated power was reduced by 14.55%, while the SAR was reduced by 46.44%. When traditional media are used, the SAR reduction effectiveness decreased compared to that of the metamaterials. In addition, the radiated power from the antenna was nearly unaffected when the metamaterials were used. Comparisons of the SAR reduction effectiveness for different positions and sizes of the metamaterials were performed.

The radiation pattern of the PIFA antenna combined with $\mu=1$ and $\varepsilon=-3$ metamaterials were analyzed to further examine whether the metamaterials affect the antenna performance. The radiation patterns were obtained from the near- and far-field transformation of the Kirchhoff surface integral representation (KSIR) [23]. All radiation patterns were normalized to the maximum gain attained without any added materials.

Methods and Analysis of SAR for the proposed TSRRs

The SAR was reduced by placing the triangular metamaterials (TMMs) between the antenna and the head. Note that the TMMs were smaller than the operating wavelength and that the structures resonated due to their internal capacitance and inductance. The

stop band can be tuned to the operational bands of cellular phone radiation. The TMMs were designed on a printed circuit board to allow them to be easily integrated into cellular phones. The overall dimensions of the TMMs were determined by a periodic arrangement of the sub-wavelength resonators.

1. Construction and Design of the TSRRs

Using FDTD analysis, this research establishes that TMMs can reduce the 1 g SAR and 10 g SAR peak levels in the head. In this section, the TMMs are evaluated in a cellular phone 900 MHz band. Periodically arranged TSRRs can act as TMMs. The structure of TSRRs consists of two conductive material concentric triangular rings. Both triangular rings have a gap, and each ring is placed opposite to the gap of the other ring. The schematic of the structure of TSRRs used in this study is shown in Fig. 2.

To construct the TMMs for SAR reduction, the TSRRs were used as the resonator model, as shown in Fig. 2. The resonators operated in the 900 MHz band. The TSRRs consisted of two triangular rings, each with gaps on the opposite sides [9]. Note that the SRRs were introduced by Pendry et al. (1999) [22] and subsequently used by Smith et al. (2000) to synthesize the first left-handed artificial medium [21]. The metamaterials in this work were designed with periodic TSRRs arrangements to reduce the SAR value. By properly designing the TSRRs structure parameters, a negative effective medium parameter can be achieved for both the 900 MHz band.

The splits in the rings allow the SRR unit to resonate at wavelengths much larger than the diameter of the rings, i.e., the splits preclude the half-wavelength requirement for resonance incurred by closed rings. The purpose of the second split ring, which is split inside and whose split is oriented opposite to the first, is to generate a large capacitance in the small gap region between the rings, which considerably lowers the resonant frequency and concentrates the electric field. By combining the split ring resonators into a periodic medium to ensure a strong (magnetic) coupling between the resonators, unique properties emerge from the composite. In particular, because these resonators respond to the incident magnetic field, the medium can be viewed as having an effective permeability, $\mu_{eff}(\omega)$. However, we can use a physical approach and alter the dielectric function of the surrounding medium, which creates scattering properties that can distinguish whether the band gaps are due to either the $\mu_{eff}(\omega)$ or $\varepsilon_{eff}(\omega)$ of the SRR being negative.

Combining the SRR medium that has a frequency band gap due to a negative permeability with a thin wire medium produces a left-handed material in the region where both $\mu_{eff}(\omega)$ or $\varepsilon_{eff}(\omega)$ have negative values.

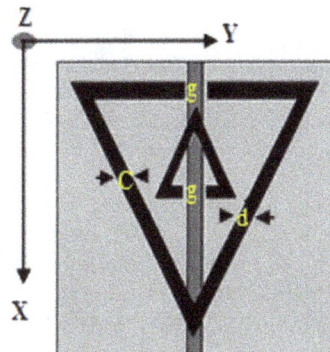

Figure 2. The structure of the TSRRs.

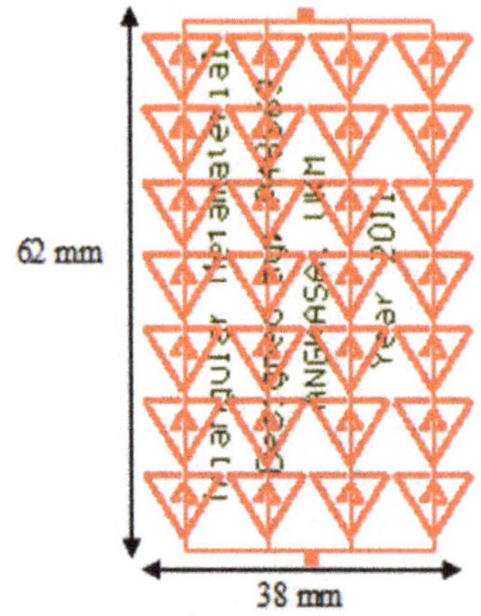

Figure 3. Arrays of TSRRs.

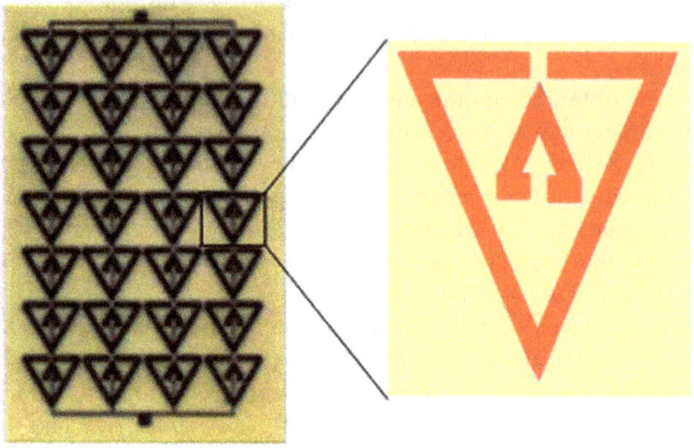

Figure 4. Fabricated TSRRs structure for the SAR measurement.

In Fig. 2, the resonator structures are defined by the following structure parameters: the triangular ring thickness, c, the triangular ring gap, d, the triangular ring size, l, and the split gap, g. Here, c_0 is the speed of light in free space, and r is the radius of the inner ring. Fig. 3 presents the TSRRs arrays used for the SAR calculations in this work. These arrays were divided into four columns and seven rows. The dimensions of the TSRRs arrays considered in this work were 62 mm×38 mm. Fig. 4 shows the fabricated TSRRs used for the SAR measurement.

2. Results and Discussion

Numerical simulations that predict the transmission properties depend on the system's various structure parameters. Such complex simulations are typically performed by the FDTD method. Periodic boundary conditions can reduce the computational domain, while an absorbing boundary condition can be used to represent the propagation regions. The total-field/scatter-field formulation excites the plane wave. The regions inside the computational domain and those outside the TSRRs were set to be vacuums.

The permittivity and permeability are the two parameters that determine the metamaterial response to electromagnetic waves. At the resonant frequency of a double negative TMM, both the permeability and permittivity are negative, as shown in Fig. 5, which results in a negative refractive index. The permittivity and permeability can be determined by measuring the complex EM wave reflection and transmission coefficients of a material sample.

The figure shows that a region in the stop band exists where the permittivity and permeability is negative. Polarizing the magnetic field along the split ring axes, $H\|$, will generate a magnetic field that might either oppose or enhance the incident field. A bulky capacitance in the region among the rings will be generated, and the electric field will be strongly concerted. Strong field coupling was evident among TSRRs, and the permeability medium was negative at the stop band. In contrast, the magnetic field was parallel to the TSRRs plane for $H^\perp$.

Here, the author assumed small magnetic effects and a small, positive, and slowly varying permeability. In the $H^\perp$ condition, these structures can be viewed as sporadically arranging the metallic wires. The unremitting wires behave like a high-pass filter, which means that the permittivity can be negative underneath the plasma frequency.

The dissimilarity among the two stop bands in the $H\|$ and $H^\perp$ cases illustrates the difference among the magnetic and electric responses of the TSRRs. Theoretical investigations have shown

that the $H\|$ band gap is due to negative permeability, and the $H^\perp$ band gap is due to negative permittivity. This study shows that both of the two incident polarizations can produce a stop band. In addition, a region in the stop band exists where the permittivity and permeability are negative. Polarizing the magnetic field alongside the split ring axes results in a magnetic field that may either resist or augment the incident field.

To verify our FDTD simulation, the structure parameters of the TSRRs were set to those defined in [24]: $d = g = c = 0.33$ mm and $l = 3$ mm. The thickness and dielectric constant of the circuit board were 4.4 mm and 0.45, respectively. Twenty-eight unit elements were used in the propagation direction. Periodic boundary conditions were implemented normal to the direction of propagation. Fig. 6 illustrates the simulated and measured transmission spectra of the TSRRs studied. In [24], the measured results indicate that the SRRs exhibit a stop band extending from 8.1 to 9.5 MHz, whereas in this work, the simulation and the measured results indicate that the TSRRs exhibit a stop band extending from 8.3 to 9.4 MHz. This difference is due to the use of the square ring SRRs in [24] and the modified TSRRs with a novel TMMs shape used in this research.

The stop bands of the TSRRs were designed to be at 900 MHz and 1800 MHz. The periodicity along the x-, y-, and z-axes were $L_x = 62$ mm, $L_y = 1.5$ mm, and $L_z = 62$ mm, respectively. To obtain a stop band at 1800 MHz, the TSRRs parameters were chosen to be $c = 1.8$ mm, $d = 0.6$ mm, $g = 0.6$ mm, and $r = 12.7$ mm. The periodicity along the x-, y-, and z-axes were $L_x = 50$ mm, $L_y = 1.5$ mm, and $L_z = 50$ mm, respectively. The thickness and dielectric constant of the circuit boards for both the 900 MHz and 1800 MHz bands were 0.508 mm and 3.38, respectively. Once the geometric parameters were properly chosen, the TSRRs medium could exhibit a stop band at approximately 900 MHz and 1800 MHz.

Experimental Validation

The SAR measurement was performed using the COMOSAR measurement system. The system uses a robot to position the SAR probe inside the head phantom. The head phantom is filled with a liquid with dielectric properties selected based on IEEE standard 1528, which are $\varepsilon_r = 41.5$ and $\sigma = 0.97$ S/m for 900 MHz and $\varepsilon_r = 40$ and $\sigma = 1.4$ S/m for 1800 MHz. The measured and simulated SAR values without the inclusion of TMMs are shown in Fig. 7, and the distance between the head and phone model was 20 mm.

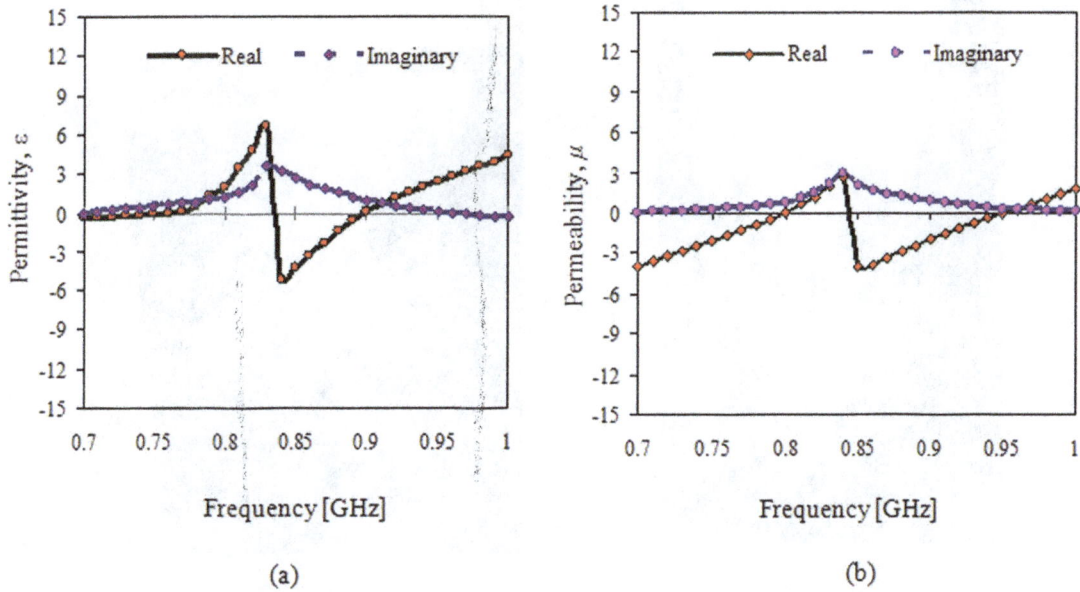

Figure 5. Materials measured at the resonant frequency are double negative: a) negative permittivity and b) negative permeability.

Fig. 7 indicates that the simulated SAR value is greater than 3.29% for the SAR value of 1 g and 3.82% for the SAR 10 g value because the distance between the head and phone model was not correctly calibrated for the measurement stage.

In addition, the distance between the source and the internal surface of the phantom position affects the SAR. For a 5 mm distance, a positioning uncertainty of ± 0.5 mm would produce a SAR uncertainty of $\pm 20\%$. Therefore, accurate device positioning is essential for accurate SAR measurements. In the measurement, the antennas with the TMMs are in contact with the SAM phantom head. During the measurement, the radiation power has been set to the maximum for the phone being tested (as required by the standard) of 33 dBm for GSM 900.

Moreover, peak SAR 1 g values of 2.002 W/kg and peak SAR 10 g values 1.293 W/kg were reached without the attachment of TMMS. When the TMMs were attached, the peak SAR 1 g value was 1.017 W/kg and the peak SAR 10 g value was 0.617 W/kg.

Figs. 8 and 9 illustrate how to setup the apparatus for the cheek position and the tilted position measurements to determine the SAR value. Figs. 8 (a) and (b) show the cheek position of the measurement without metamaterials and with the attachment of TMMs. Fig. 9 (a) and Fig. 9 (b) show the 15° tilted position without metamaterials and with TMMS attachment.

Figs. 10 and 11 show the simulated and measured SAR values when the antenna with the TMMs attached was in the cheek position, which resulted in SAR 1 g values of 1.132 W/kg and 1.035 W/kg, respectively. The simulated and measured SAR differed by 9.70% for the SAR 1 g value. The higher SAR values measured for the cheek position than for the tilted position can be attributed to the influence of the ground plane on the distribution

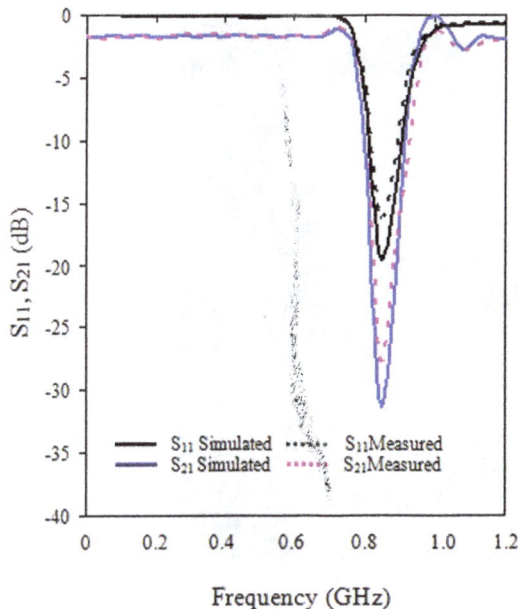

Figure 6. Modeled transmission spectra of the TSRRs plane in the *yz* plane.

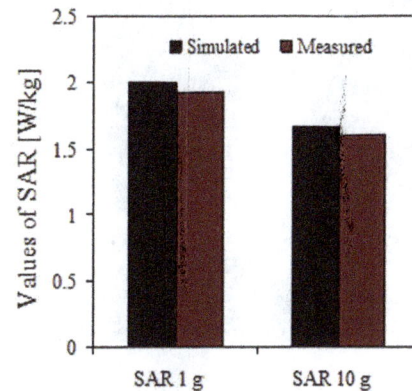

Figure 7. Comparison on SAR simulation and measurement results without the inclusion of TMMs.

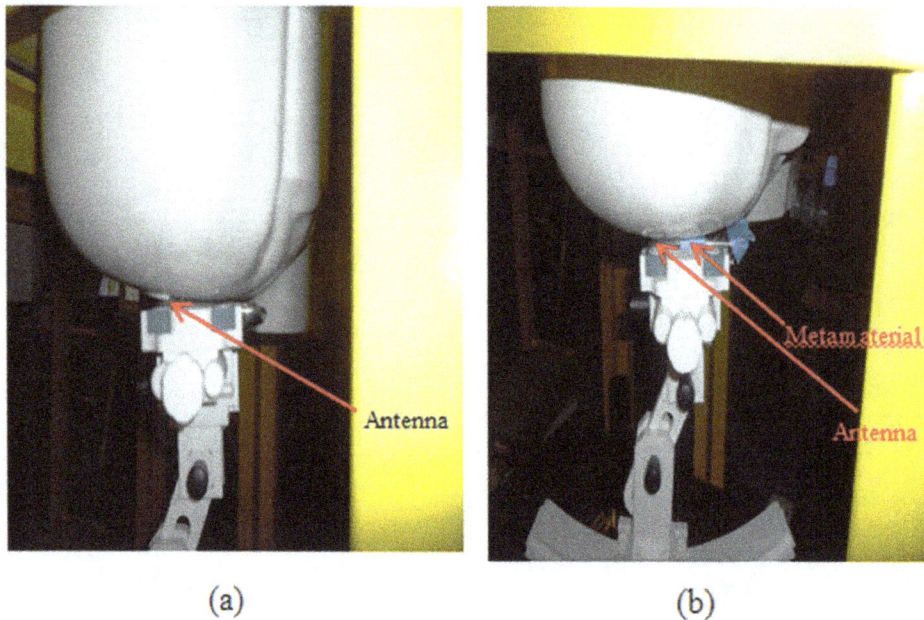

Figure 8. Measurement of the SAR values at the cheek position a) without a metamaterial and b) with a metamaterial.

of the surface currents; consequently, the power deposited inside the head is higher at the GSM frequency band.

The simulated and measured SAR values were obtained using the tilted position for the TMMs with the antenna, which resulted in simulated and measured SAR 1 g values of 1.0963 W/kg and 1.017 W/kg, respectively, using the antenna with TMMs. The simulated and measured SAR values differed by 7.23% for the SAR 1 g value. Regarding the difference in the absolute values of peak SAR, the phone model casing for the simulation differed from the case used for the measurement. In addition, Scotch tape was used to attach the TMMs and the antenna during the

measurement stages. The simulated and measured results also differ because the parameters for the measurement system change with the water evaporation and temperature. Furthermore, the measurement system contains several parameters (e.g. source, network emulator, probe, and electronic evaluation procedures) that affect the SAR calculation but are not included in the simulation.

The SARs of the hot-spot positions without and with the attachment of the TMMs metamaterial for different positions are shown in Fig. 12 (a), (b), (c) and (d). The hot spots can be observed in the brain, especially in the region close to the eyeballs and at the

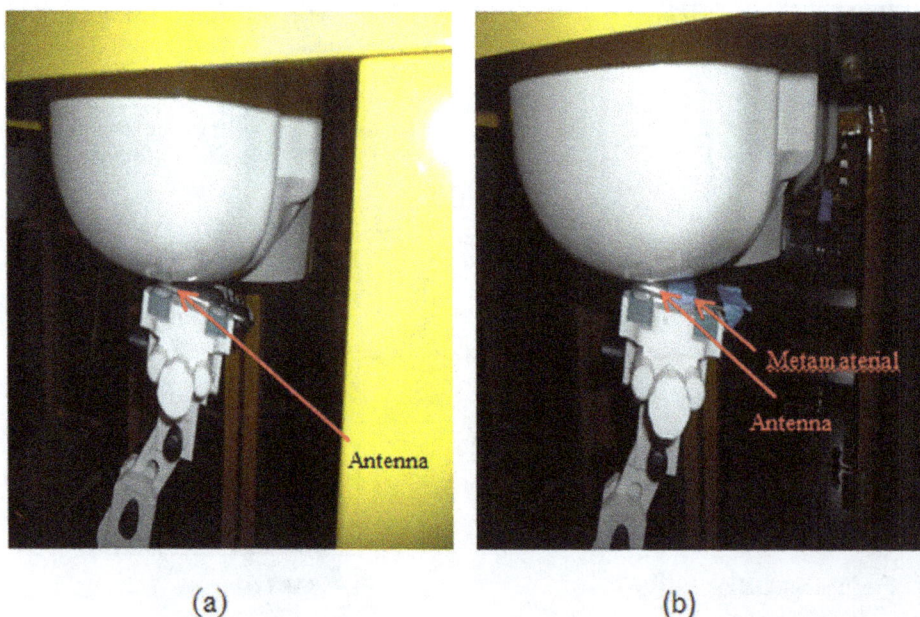

Figure 9. Measurement of the SAR values at the tilt position a) without a metamaterial and b) with a metamaterial.

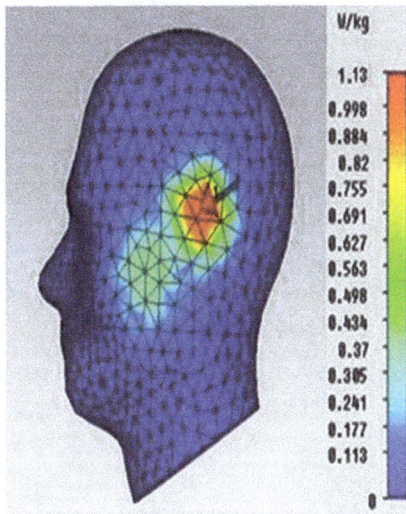

Figure 10. Simulated SAR value of the antenna with the TMMs attachment in the cheek position.

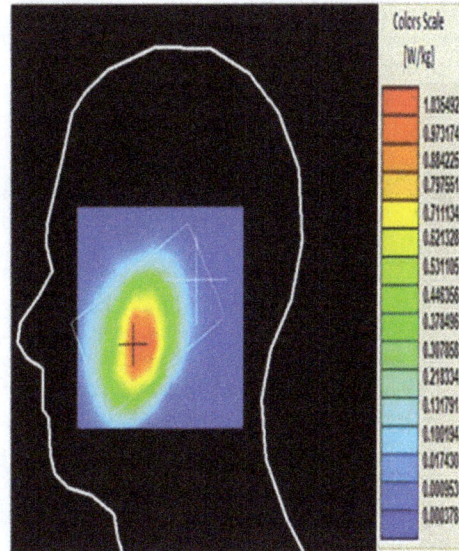

Figure 11. Measured SAR of the antenna with the TMMs attached in the cheek position.

muscle. Generally, hot spots are likely to occur at tissue interfaces with high dielectric contrast (i.e. fat and muscle tissue).

Limitations of the Study, Research Questions and Future Work

The differences in the EM interactions between an antenna and the head caused by the double negative metamaterials have been discussed in this study. The developed TMMs parameters, such as the length, width, gaps and ring size, have been examined, and the frequency range of the negative ε and μ was analyzed. In addition, the simulated and experimentally measured SAR values were presented for the design of metamaterials.

A negative permittivity medium can also be constructed by periodically arranging the metallic thin wires. However, we found that when thin wires operated at 900 MHz are too large for practical application. Because the SRR structures resonate due to internal capacitance and inductance, they are smaller than the wavelength of radiation. Moreover, this study failed to reduce the SAR value for antennae that operate at 1800 MHz.

Some of the main issues that we intend to focus in this study include the following:

1. Is the existing material sufficient to reduce the SAR in the human body?
2. Are the proposed SRRs capable of effectively reducing the SAR?
3. Will the use of designed SRRs shrink the existing printed circuit board devices in mobile phones?

The field of metamaterials is relatively new, which translates into a very large scope for the further development of several factors presented in this study.

1. The development of metamaterials for wideband applications.
2. The use of PCB techniques to design the negative permittivity medium so that it can be directly implemented into portable devices.

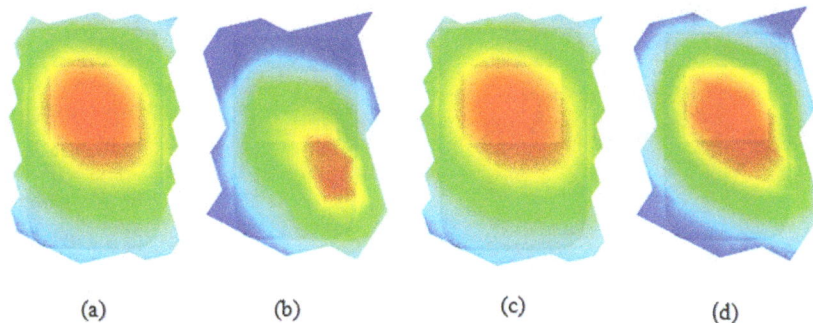

| (a) | (b) | (c) | (d) |

Hot spot position (for the cheek position): a) without metamaterial and b) with metamaterial.

Hot spot position (for the tilted position): c) without metamaterial and d) with metamaterial.

Figure 12. SAR comparisons at the different positions between an antenna without TSRRs and an antenna with TSRRs.

3. The experimental realization of far-field sub-diffraction imaging at microwave wavelengths and the development of algorithms for complex sub wavelength structures.

Conclusion

The electromagnetic energy absorption reduction between an antenna and the human head has been discussed. The proposed double-negative metamaterials in the phone model achieved a 10 g SAR value of 0.617 W/kg and a 1 g SAR value of 1.0175 W/kg. Based on the FDTD method with the lossy-Drude model, both the 1 g SAR and 10 g SAR peak values of the head can be abridged by placing metamaterials between the antenna and the human head. The use of all dielectric metamaterials tuned to the exact ε and μ values required that the electromagnetic energy be diverted from the cellular phone user's head, which are

promising options to improve the SAR reductions. The achieved results provide constructive information for the design of communication equipment that complies with the safety requirements.

Acknowledgments

This work was supported by the Universiti Kebangsaan Malaysia, under grants Dana Lonjakan Penerbitan (UKM- DLP-2014).

Author Contributions

Conceived and designed the experiments: MRIF MTI. Performed the experiments: MRIF MTI. Analyzed the data: MRIF MTI. Contributed reagents/materials/analysis tools: MRIF MTI. Wrote the paper: MRIF MTI.

References

1. International Commission on Non-Ionizing Radiation (ICNIRP) (1988) Guidelines for limiting exposure to time-varying electric, magnetic, and electromagnetic fields (up to 300 GHz). Health Physics 74: 494–522.
2. Mochizuki S, Wakayanagi H, Hamada T, Watanabe S (2007) Effects of ear shape and head size on simulated head exposure to a cellular phone. IEEE Trans on Electromagnetic Compatibility 49: 512–518.
3. Kiveka's O, Ollikainen J, Lehtiniemi T, Vainikainen P (2004) Bandwidth, SAR, and efficiency of internal mobile phone antennas. IEEE Trans on Electromagnetic Compatibility 46: 71–86.
4. Lin JC (2000) Specific absorption rates (SARs) induced in head tissues by microwave radiation from cell phones. IEEE Trans on Antennas and Propagation Magazine 42: 138–140.
5. Watanabe S, Taki M, Nojima T, Fujiwara O (1996) Characteristics of the SAR distributions in a head exposed to electromagnetic fields radiated by a hand-held portable radio. IEEE Trans on Microwave Theory & Technique 44: 1874–1883.
6. Jensen MA, Rahmat-Samii Y (1995) EM Interaction of handset antennas and a human in personal communications. Proceedings of the IEEE 83: 7–17.
7. Gandhi OP, Lazzi G, Furse CM (1996) Electromagnetic absorption in the human head and neck for mobile telephones at 835 and 1900 MHz. IEEE Trans on Microwave Theory & Technique 44: 1884–1897.
8. Fung LC, Leung SW, Chan KH (2002) An investigation of the SAR reduction methods in mobile phone application. IEEE International Symposium on Electromagnetic Compatibility. Minneapolis, MN, USA 2, 656–660.
9. Faruque MRI, Islam MT, Misran N (2012) Design analysis of new metamaterial for EM absorption reduction. Progress In Electromagnetics Research 124: 119–135.
10. Tay RYS, Balzano Q, Kuster N (1998) Dipole configuration with strongly improved radiation efficiency for hand-held transceivers. IEEE Trans on Antennas & Propagation 46: 798–806.
11. Li CH, Chavannes N, Kuster N (2009) Effects of hand phantom on mobile phone antenna performance. IEEE Trans on Antennas and Propagation 57: 2763–2770.
12. Hawang JN, Chen FC (2006) Reduction of the peak SAR in the human head with metamaterials. IEEE Trans on Antennas & propagation 54: 3763–3770.
13. Manapati MB, Kshetrimayum RS (2009) SAR reduction in human head from mobile phone radiation using single negative metamaterials. Journal of Electromagnetic Waves and Applications 23: 1385–1395.
14. Faruque MRI, Islam MT, Misran N (2011) Analysis of electromagnetic absorption in the mobile phones using metamaterials. Electromagnetics Journal 31: 215–232.
15. Cabedo A, Anguera J, Picher C, Ribó M, Puente C (2009) Multi-Band handset antenna combining a PIFA, slots, and ground plane modes. IEEE Trans on Antennas and Propagation 57: 2526–2533.
16. Anguera J, Andújar A, Huynh MC, Orlenius C, Picher C, et al. (2013) Advances in antenna technology for wireless handheld devices. International Journal on Antennas and Propagation Article ID 838364: 1–25.
17. Risco S, Anguera J, Andújar A, Picher C, Pajares J (2012) Comparison of a monopole and a PIFA handset antenna in the presence of the human head. Microwave and Optical Technology Letters 54: 454–459.
18. Rowell C, Lam EY (2012) Mobile-phone antenna design. IEEE Antennas and Propagation Magazine 54: 14–34.
19. Kuster N, Balzano Q (1992) Energy absorption mechanism by biological bodies in the near field of dipole antennas above 300 MHz. IEEE Trans on Vehicular Technology 41: 17–23.
20. Okoniewski M, Stuchly MA (1996) A study of the handset antenna and human body interaction. IEEE Trans on Microwave Theory & Technique 44: 855–1864.
21. Smith DR, Padilla WJ, Vier DC, Nemat-Nasser SC, Schultz S (2000) Composite medium with simultaneously negative permeability and permittivity. Physical Review Letters 84: 4184–4187.
22. Pendry JB, Holen AJ, Robbins DJ, Stewart WJ (1999) Magnetism from conductors and enhanced nonlinear phenomena. IEEE Trans on Microwave Theory & Technique 47: 2075–2084.
23. Sievenpiper D (1999) High-impedance electromagnetic surfaces with a forbidden frequency band. IEEE Trans on Microwave Theory & Technique 47: 2059–2074.
24. Bayindir M, Aydin K, Ozbay E, Markos P, Soukoulis CM (2002) Transmission properties of composite metamaterials in free space. Applied Physics Letters 81: 120–122.
25. Caloz C, Itoh T (2005) Electromagnetic metamaterials, transmission line theory and microwave applications. Wiley-IEEE Press. 376p.
26. Ziolkowski RW (2003) Design, fabrication, and testing of double negative metamaterials. IEEE Trans on Antennas & Propagation 51: 1516–1529.
27. Zhang J, Xiao S, Jeppesen C, Kristensen A, Mortensen NA (2010) Electromagnetically induced transparency in metamaterials at near-infrared frequency. Optics Express 18: 17187–92.
28. Xu HX, Wang GM, Qi MQ, Cai T, Cui TJ (2013) Compact dual-band circular polarizer using twisted Hilbert-shaped chiral metamaterial. Optics Express 21: 24912–21.
29. Islam MT, Faruque MRI, Misran N (2009) Design analysis of ferrite sheet attachment for SAR reduction in human head. Progress In Electromagnetics Research 98: 191–205.
30. Wang J, Fujiwara O (1997) FDTD computation of temperature rise in the human head for portable telephones. IEEE Trans on Microwave Theory & Technique 47: 1528–1534.
31. Kuo CM, Kuo CW (2003) SAR distribution and temperature increase in the human head for mobile communication. IEEE Antennas and Propagation Society International Symposium, Columbus, OH, USA 2, 1025–1028.

19

Comprehensive Analysis of Human Cells Motion under an Irrotational AC Electric Field in an Electro-Microfluidic Chip

Clarisse Vaillier[1,2,◊]**, Thibault Honegger**[1,2,3,◊]**, Frédérique Kermarrec**[4]**, Xavier Gidrol**[4]**, David Peyrade**[1,2]*

1 Univ. Grenoble Alpes, LTM, Grenoble, France, **2** CNRS, LTM, Grenoble, France, **3** Department of Electrical Engineering and Computer Science, Massachusetts Institute of Technology, Cambridge, Massachusetts, United States of Amercia, **4** CEA, Institut de Recherches en Technologies et Sciences pour le Vivant, Grenoble, France

Abstract

AC electrokinetics is a versatile tool for contact-less manipulation or characterization of cells and has been widely used for separation based on genotype translation to electrical phenotypes. Cells responses to an AC electric field result in a complex combination of electrokinetic phenomena, mainly dielectrophoresis and electrohydrodynamic forces. Human cells behaviors to AC electrokinetics remain unclear over a large frequency spectrum as illustrated by the self-rotation effect observed recently. We here report and analyze human cells behaviors in different conditions of medium conductivity, electric field frequency and magnitude. We also observe the self-rotation of human cells, in the absence of a rotational electric field. Based on an analytical competitive model of electrokinetic forces, we propose an explanation of the cell self-rotation. These experimental results, coupled with our model, lead to the exploitation of the cell behaviors to measure the intrinsic dielectric properties of JURKAT, HEK and PC3 human cell lines.

Editor: Aristides Docoslis, Queen's University at Kingston, Canada

Funding: This work is supported by French National Agency ANR through Nanoscience and Nanotechnology Program Project NANOSHARK No. ANR-11-NANO-0001. The funders had no role in study design, data collection and analysis, decision to publish, or preparation of the manuscript.

Competing Interests: The authors have declared that no competing interests exist.

* E-mail: david.peyrade@cea.fr

◊ These authors contributed equally to this work.

Introduction

AC electrokinetic forces have been used in numbers of methods ranging from particle/cell characterization [1,2], separation [3,4] or manipulation [5,6] and can be applied to biosensors, cell therapeutics, drug discovery, medical diagnostics, microfluidic and particle filtration [7] thanks to various designs of electrodes and/or microchannels. These forces induce both liquid and micro-scaled objects motions, namely electro-hydrodynamic (EHD) and dielectrophoretic (DEP) forces. EHD is coupling both linear and non-linear electrokinetic phenomenon that have been discovered and studied in microfluidic channels during the past decade, respectively electrothermal effect (ETE) and AC/induced charged electroosmosis (ACEO/ICEO)[8,9]. EHD forces create motion of liquid that drags micro-objects along streamlines. Those forces are specific to the electric properties of the suspension media and are difficult to tune in microsystems. On the contrary, DEP has been discovered by Pohl [10] in the 1950's. DEP is a contactless induced force that polarizes micro-objects and induces their motion relatively to the electrodes, providing a non-uniform distribution of the electric field. What is significantly interesting in using DEP to manipulate micro-objects is that its magnitude and direction of the force are directly linked to the frequency and voltage of the applied electric field, which makes the applied force and thus the movement of the object tunable by the electric field properties. There is however a competition between EHD and DEP forces in microsystems [11,12], which results in a variety of behaviors of objects relatively to the electrodes. Besides understanding the physics of this competition, there has been couple of studies describing the observed motions of micro- and nanoparticles in such microsystems [13,14]. However, cells are fundamentally different than colloidal particles, either by size, shape, deformability and electrical properties, which results in very different behaviors than the ones previously reported with commercial or engineered particles. For example, cells can present different polarizabilities if alive or dead [15] when applying the same AC fields. Moreover, recent work has reported self-rotation under non rotating fields and the origin of this observation is still unclear [16]. Here, we present a qualitative and quantitative analysis of the induced motion of human cells by non-uniform AC electric fields. Based on the state-of-art comprehensive analysis of colloidal particles motion under such fields, we first report and analyze the motion of three human cells lines when tuning the parameters of the applied electric field. We then suggest possible mechanisms that could lead to those behaviors. We finally exploit those motions to measure the values of the electrical properties of such cells.

Theory

Castellanos et al. presented a model [12] based on a scaling law approach that described the motion of colloidal particles between planar electrodes. This model described the comprehension of the competition between DEP and EHD forces in the assumption that the electric field distribution is semi-circular and $E = V/\pi r \vec{u}_\theta$

where V is the amplitude of the applied voltage and r is the distance to the center of the gap. Here, we adapt their model to human cells to provide a better understanding of the competition of forces applied on cells and to explain their motions.

Dielectrophoresis

Non-uniform electric fields can be used to induce motion of cells. When a cell is suspended in a viable dielectric medium, the applied AC electric field causes the cell to polarize, giving rise to a net dipole moment in the cell. If the electric field is non-uniform, the cell will experience a force. This force is referred to as Dielectrophoresis. By adjusting the experimental conditions, it is possible to move cells towards (positive dielectrophoresis) or away from high field regions (negative dielectrophoresis).The dielectrophoretic force is given in equation (1) [17].

$$\langle F_{DEP} \rangle = \pi \varepsilon_m a^3 \text{Re}[CMF(\omega)] \nabla |E^2| \qquad (1)$$

where $\nabla |E^2|$ is the gradient of the square of the RMS electric field E, ω is the angular velocity of the electric field, a is the cell radius, Re[] indicates the real part and $CMF(\omega)$ is the Clausius-Mossotti factor (CMF) that translates the relative polarizability of the cell to the medium at a given frequency. The CMF depends on the complex permittivities of the cell and of the medium (permittivity ε_m, conductivity σ_m). In the single shell model of a human cell [18], as illustrated in Figure 1, the dielectric properties of a cell are generally expressed with the membrane capacitance C_{mem} and conductance G_{mem}. The membrane of mammalian cells is generally poorly conductive and G_{mem} is usually negligible compared to C_{mem} [19].

Assuming the membrane thickness $\delta << a/2$, the general form of the CMF of cells is given in equation (2) [20,21].

$$\text{Re}[CMF(\omega)] = \text{Re} \left[\frac{w^2 (\tau_m \tau_1 - \tau_m^* \tau_3) - 1 + iw(\tau_m^* - \tau_1 - \tau_m)}{2 - w^2 (\tau_m^* \tau_3 - 2\tau_1 \tau_m) + iw(\tau_m^* + 2\tau_1 + \tau_m)} \right] \quad (2)$$

Where $\quad C_{mem} = \frac{\varepsilon_0 \varepsilon_2}{\delta}, G_{mem} = \frac{\sigma_2}{\delta}, \tau_1 = \frac{\varepsilon_1}{\sigma_1}, \tau_3 = \frac{\varepsilon_3}{\sigma_3}, \tau_m = \frac{aC_{mem}}{\sigma_3},$

$\tau_m^* = \frac{aC_{mem}}{\sigma_1}.$ The main assumption of this model is that the cells are spherical in shape. The model must be modified for other geometries, like spheroids or ellipsoids, as reviewed in [22] or in [23].

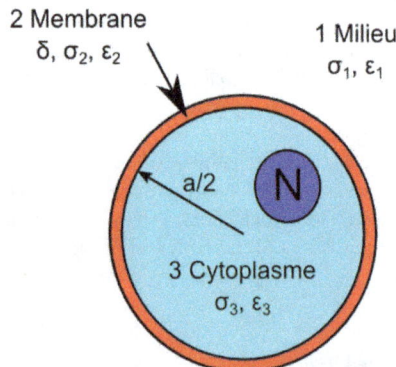

Figure 1. Single shell model of a mamalian cell with dielectric parameters annotated.

This model also assumes that the cell is much smaller than the characteristic length of the electrode separation (i.e. to make the dipole approximation). This assumption basically states that the field gradient length scale is much larger than the cell length scale. In most cases when using cells in electro-microsystems, this assumption may not be valid because the electrode gap is in the same order of magnitude than the cell diameter. In this case, the use of a multipole model boosts the magnitude of the polarization factor, but have essentially no influence on the cell crossover frequency and behavior when varying frequencies ore medium conductivity [24]. We therefore proceed with the original core-shell model as a first order approach to describe the cell behaviors. Adapting Castellanos model to the single shell model of human cells, the DEP displacement expresses as shown in equation (3)

$$u_{DEP} = \frac{\varepsilon_m a^2}{6\pi^2 \eta r^3} \text{Re}[CMF(\omega)](cV)^2 t \qquad (3)$$

where η is the viscosity of the medium, $c = \frac{\Omega}{\sqrt{1 + \Omega^2}}$ a factor that takes into account the reduction of the voltage in the medium due to electrode polarization (Ω is defined hereafter), V the applied voltage peak to peak, r the distance and t the duration at which the velocity is evaluated.

Figure 2.a1 plots the evolution of the theoretical values of the CMF of 10 μm cells when increasing the medium conductivity. DEP is mostly negative at cell culture medium conductivities (typically PBS or DMEM whose conductivities are around 1 S/m). The classical shape presents a crossover frequency f_{x0}, where DEP becomes null and changes direction.

Our recent works [25,26] have used a method to experimentally evaluate the CMF of any polarizable particle, allowing a direct measurement of their dielectrophoretic properties.

Electrohydrodynamical forces (EHD)

Whereas DEP induces motion of the cell itself, the application of an AC electric field inside a fluid creates two major EHD phenomena, namely ACEO and ETE motions. Those effects will alter the DEP manipulation of cells.

First, AC electroosmosis (ACEO) refers to the flow motion created on the electrodes' surfaces when AC signals are applied. The motion of charges in the electrical double layer induced by the dissymetry of the tangential field will create convection rolls at the electrode interface.

Therefore, there exists an optimal AC frequency at which the product of the electric field and the interface potential, referred as the zeta potential, reaches a maximum. The frequency dependency of fluid motion in co-planar electrodes can be estimated by introducing a non-dimensional frequency Ω given in equation (4) [12,28].

$$\Omega = \Lambda \frac{w \varepsilon_m \pi r}{2 \sigma_m \lambda_D}, \text{ where } \Lambda = \frac{C_S}{C_S + C_D}, C_D = \frac{\varepsilon_m}{\lambda_D}, \lambda_D = \sqrt{D \frac{\varepsilon_m}{\sigma_m}} \quad (4)$$

Where D is the diffusion coefficient of the medium and λ_D the Debye layer distance. The value of the Stern layer capacitance C_s ~ 0.007 F.m^{-2} has been used from previous experiments in literature [12]. In the semi-circular electric field approach, the resulting mean velocity induced by ACEO is given in equation (5).

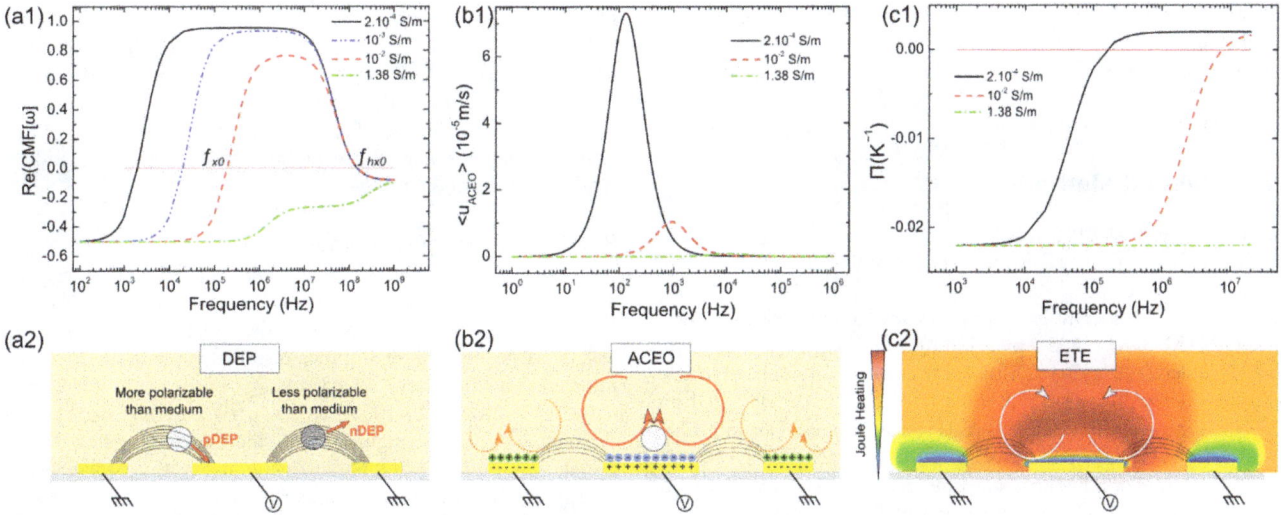

Figure 2. Plots of critical parameters and streamlines induced by AC electrokinetic forces. (a.1) Plot of the real part of the Clausius-Mossotti factor for human cells with single shell models (parameters extracted respectively from [27] for HeLa-60) and $\sigma_m = 2.10-2$ S/m, **(b.1)** Plot of the ACEO mean velocities of the fluid near the electrodes (x = 1 μm) for several conductivities of the fluidic medium for water ($\varepsilon_f = 78$) and **(c.1)** Plot of the Π factor as a function of frequency for several conductivities of the medium. Review of predominant forces in presence of AC electric field and the induced motion of liquid and cells: **(a.2)** Dielectrophoresis (DEP) induces attraction (p-DEP) or repelling (n-DEP) of cells from high field region (in co-planar cases, electrode edges), **(b.2)** AC electroosmosis (ACEO) are electrohydrodynamic forces that create convective rolls over the electrodes edges and drag cell with them and **(c.2)** electrothermal effects (ETE).

$$u_{ACEO} = 0.1 \Lambda \frac{\varepsilon_m V^2}{\eta r} \frac{\Omega^2}{\left(1+\Omega^2\right)^2} t \qquad (5)$$

Figure 2.b1 shows the velocities of 10 μm cells when increasing the medium conductivity, which makes ACEO significantly unobservable at cell culture medium conductivities.

Second, the electrothermal effect (ETE) is observed when a non-uniform electric field is applied over a fluid, Joule heating is produced inside the volume of the fluid, which leads to temperature gradient ∇T in the fluid. This variation produces spatial gradients in the local permittivity and conductivity of the fluid, given as α and β respectively (equation (6)) [11].

$$\alpha = \frac{1}{\varepsilon_m} \frac{\varepsilon_m}{\nabla T} = -0.4\% K^{-1} \qquad \beta = \frac{1}{\sigma_m} \frac{\sigma_m}{\nabla T} = 2\% K^{-1} \qquad (6)$$

Furthermore, those gradients generate mobile space charges ρ, in the bulk fluid, following $\rho = \nabla(\varepsilon_m E) = \nabla \varepsilon_m E + \varepsilon_m \nabla E$ and $\frac{d\rho}{dt} + \nabla(\varepsilon_m E) = 0$ in AC fields. The time average of the electric force that acts on the fluid through viscosity and leads to fluid transport is given in equation (7).

$$\langle F_{ETE} \rangle = 0.5 \varepsilon_m \nabla T \, E^2 \Pi(\omega) \text{ with } \Pi(\omega) = \left(\frac{\alpha - \beta}{1 + \left(\omega \frac{\varepsilon_m}{\sigma_m}\right)^2} - \frac{\alpha}{2} \right) \quad (7)$$

where the factor Π plays a significant role in the magnitude and direction of the force as shown in Figure 2.c1 for different medium conductivities. When the left term on equation is greater than the right one, Π is positive and the fluid flows from the edge to the center of the electrode. For negative values of Π, the flow pattern is in the opposite direction, as shown in Figure 2.c2. Finally, in the semi-circular approach, the induced displacement of the ETE is given in equation (8).

$$u_{ETE} = \frac{\varepsilon_m \sigma_m}{3a\pi k \eta (\pi r)^3} \Pi(\omega)(cV)^4 t \qquad (8)$$

where k is the thermal conductivity of the medium.

Whereas ACEO is a dominant force at small frequencies of the AC field (typically below 10 kHz) and decreases in magnitude when raising the conductivity of the medium, ETE will remain constant at all frequencies and will strongly increase with the conductivity of the medium. On the contrary, the DEP force can change direction in the $f < 10^6$ Hz range frequencies and when the conductivity of the medium is low ($\sigma_m < 10^{-2}$ S/m) but will remain negative at higher conductivities.

In the conductivity range of biological medium ($\sigma_m \sim 1$ S/m), it has been shown that EHD forces become dominant when rising the voltage [12] but there exists a window of operation in which DEP is still active (V < ~3 V_{pp}) and where biological sample can be manipulated without being injured by heat [29].

The large differences between inorganic particles and cells lead to distinct overall behaviors when immersed in non-uniform AC electric fields. First, size has a cubic square dependency on the DEP force and generally human cells are one or two folds much bigger in diameter than colloidal particles. Second, human cells are mostly composed of water whose permittivity is similar to the one of their suspension media. Moreover, cells exhibit a membrane that separates the interior of the cell, which usually has specific conductivities with all the proteins and the cell apparatus, from the external environment, whose conductivity is buffered. This membrane is permeable to ions and other charged molecules and act as a barrier to the polarization effect of the interior of the cell, until the electric breakdown point where the membrane becomes transparent to the electric field. This complex polarization process gives rise to an opposite behavior than the one observed for uniform dielectric inorganic particles, whose motion under electric fields have been fully supplied by the community.

Material and Methods

Cell culture and preparation

PC3, JURKAT and HEK lines were commercially available cell lines and were purchased at the American Type Culture Collection, respectively http://www.lgcstandards-atcc.org/Products/All/CRL-1435.aspx, http://www.lgcstandards-atcc.org/Products/All/TIB-152.aspx, http://www.lgcstandards-atcc.org/Products/All/CRL-1573.aspx. These cells were cultivated in 25 cm^2 tissue-culture treated flasks (Product 353109, Corning Life Sciences), at 37°C, in 5% carbon dioxide. Both JURKAT and PC3 lines grew in RPMI standard medium, and HEK in DMEM standard medium, all supplemented with 5% of fetal calf serum and penicillin-streptomycin mix (1%). For anchorage-dependent cells, they were collected when confluence reaches 80%, using trypsin-EDTA complex (0.25%, Sigma-Aldrich) during 5 minutes at 37°C. JURKAT cells were collected when clusters reached about fifteen cells, by dynamic pipetting.

Cells were finally suspended in sucrose-dextrose medium (8.5%/0.3%) in deionized water respectively, because of its very low conductivity. DMEM medium was added to adjust the medium conductivity up to the wanted value.

Cell viability

The cell viability in the sucrose-dextrose medium was average (half of cells die in 5 hours), but the very low conductivity of this medium and its ability to conserve the osmotic pressure viable for cells makes it suitable to perform electrokinetic handling during a limiting amount of time (~1 hour). The experiment being conducted in less than an hour, the cell viability was not observed to decrease significantly. Based on our experiments, we observed that dielectrophoresis had a damaging effect on cells only under highly stressful conditions, ($|E| > 0.150$ V/μm) or at low frequencies ($f \leq 1$ kHz). Those conditions were avoided during our experiments.

Microfabrication of chips

The fabrication of the chips was based on glass-electrode and soft lithography technology. Briefly, a glass slide was deposited with a bi-layer Ti:Au 15 nm:135 nm by metal evaporation. The slides were then patterned with AZ1512HS, exposed through a mask with a mask aligner (MJB4, Suss), and etched by Ion Beam Etching (Plassys MU400). Resist was stripped in acetone. For soft lithography, PolyDiMethylSiloxane base was mixed with curing agent in a 10:1 ratio (Sylgard 184, Dow Corning), degassed for 15 minutes to remove air bubbles and cured 5 minutes at 110°C on a prefabricated mold with SU-8.

Observation of cell motion

The cells were placed on top of 50 μm width interdigitated electrodes spaced by 50 μm in a suspending medium whose conductivity had previously been measured. Motion was recorded through a camera (GiGE, AVT Manta, G-201C) mounted on a modified Leika microscope (INM 20) controlled by a home-made Labview software (National Instrument, Labview 8.0, NI-IMAQ Vision v4.5). The recording of the movie was launched before the electric field was turned on. Measurements were conducted on a minimum of 3 cells and when possible on more cells (up to 15 cells), depending on the effective number of cells ongoing rotation.

Fits

Fits were performed with Origin8 (OriginLab). The Nonlinear Multiple Variables Fitting tool of Origin was used on the experimental datas and fitting variables were initialized with typical known values (e.g. C_{mem} are initialized at 5 mF/m). Once fitted, the values obtained for the fitting parameters were used to plot the corresponding curve.

Rotation speed measurements

The tested frequency and voltage were applied on the electrode. The recorded video was opened with Labview (National Instruments). A new template was defined (generally the entire cell, or part of the cell membrane), directly by taking the "region of interest" on the frame. Depending on the adjustable score (between 1 and 1000, 1000 being a perfect correlation between the template and the original video), the template was tracked on all selected frames. A text file was generated with template center coordinates for each frame, and the angle since the precedent frame. The speed was calculated from the coordinates, as the average rotation from angle datas, during a given number of frames (related to the time by the frame rate). The magnitude of the electric field was calculated as $|E| = V_{p-p}/e$ where V_{p-p} was the peak to peak voltage applied on the electrodes and e was the gap between the signal and ground electrodes.

Results and Discussion

Cells were placed on a microfluidic chip on which Au electrodes were activated. As soon as the field was applied, several induced motions were observed depending on the parameters of the applied voltage (frequency, voltage) and on the medium conductivity. Here, for human cells, we report three different regimes, which have been summarized in Fig. 3. We observed cell destruction (regime 1), cell dielectrophoresis (regime 2) or cell self-rotation (regime 3). On this figure we also have plotted the computed velocities of cells induced by DEP and EHD motion based on the model described here before. For the two studied conductivities, DEP determines the overall displacement of cells (regime 2). However, around the inflection point created by the crossover frequency, ETE overpowers DEP and influences cells motion described by regimes 1 and 3. We have also supplied a movie spotting each behavior in Movie S1. The study of cell destruction and dielectrophoresis is conducted on one cell type (HEK epithelial cells), while three different cell types (HEK, Jurkat T-cells and PC3) were studied for the self-rotation phenomenon.

Regime 1: Cell destruction (ACEO dominant and ETE)

At low frequencies ($f < 5$ kHz) and at $\sigma_m = 2.10^{-4}$ S/m, cells experienced mostly ACEO and were generally attracted to (and maintained above) the edge of the activated electrode. When increasing the magnitude of the electric field (higher voltage or smaller inter-electrodes distance), ETE becomes predominant over other forces and cells are carried away in big convective rolls. In this later case, cell destruction is commonly observed in a few seconds (~4 s) as shown on Figure 4, either due to the presence of reactive oxygen species or a temperature elevation that has both been observed to lead to cell destruction [30].

Regime 2: Electrode edge collection or repulsion of cells (DEP dominant)

When spanning frequency and voltage, cells are attracted to or repelled from the electrode edges (Figure 5), which is a typical behavior of a dominant DEP regime. Due to the large size of human cells (a ~> 10 μm) and to the cubic relation of DEP force

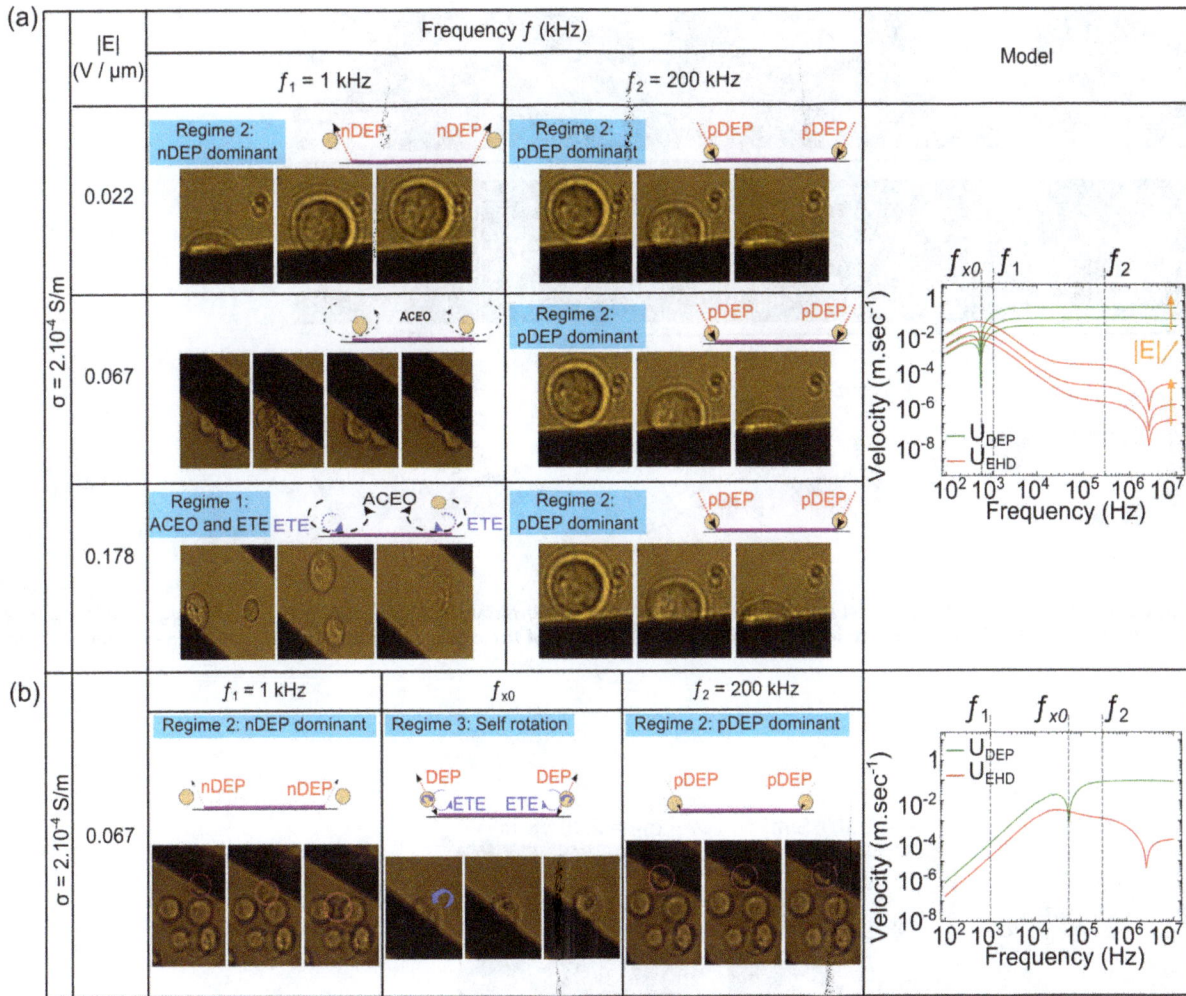

Figure 3. Summary of cells behaviors at (a) $\sigma_m = 2.10^{-4}$ S/m and (b) $\sigma_m = 2.10^{-2}$ S/m. Photographs and schemes illustrate the cell motion for typical frequencies and magnitudes with corresponding graphs of the DEP (U_{DEP}, green line) and EHD (U_{EHD}, in red) velocities. The velocities were calculated according to the theoretical model presented in the first paragraph and position of the field was taken for x = 1 µm. The boxed text refers to the related paragraph.

to the cell radius, cells often experience DEP, even if they are not very close to the electrodes (i.e. $x > 2a$ where x is the distance from the electrode edges to the cell), which is not necessarily the case for colloidal particles[29].

First, cells are repelled from the electrode edges both in the electrode plane and in the z-axis. Repulsion of cells is foreseen when $Re[CMF(\omega)] < 0$ (negative DEP), which corresponds to a range of low frequencies ($f < f_{x0}$) and very high frequencies ($f > f_{hx0}$), as shown on Figure 2.a1. Whereas f_{hx0} is hardly observable because of difficulties to conduct very high frequencies in microsystems [21], f_{x0} is commonly observed and depends on cell type, size and medium conductivity [31]. For example, at $\sigma_m = 2.10^{-4}$ S/m, we observe $f_{x0} \sim 50$ kHz for human cells, and at $\sigma_m = 10^{-2}$ S/m, we observe $f_{x0} \sim 100$ kHz. When approaching the crossover frequency, the velocity of cells reduces because the DEP force tends to vanish.

Second, cells are attracted to the electrode edges at higher frequencies ($f_{x0} < f < f_{hx0}$), namely the positive DEP regime when $Re[CMF(\omega)] > 0$. Once on the top of the electrode edges, cells do not move anymore and are able to attach on the surface of the glass slide. Moreover, at those frequencies, cells can chain together

and create organized assembly as reported previously with human liver cells [32] or 3T3 mouse cells [33]. This pearl chain formation results from the distortion of the electric field distribution by the first attracted cell, creating a high strength distribution at its own edges and therefore attracting another cell to it, and so on.

The influence of the medium conductivity is quantified by measuring cells velocities for both n-DEP and p-DEP regimes. As shown on Figure 5, the magnitude of the DEP force, which is observed *via* the cell motion, is vanishing when increasing the medium conductivity.

Regime 3: Self-rotation in non-rotating field (ETE)

Around the crossover frequency ($f = f_{x0} \pm 10$ kHz) and at $\sigma_m \leq 10^{-2}$ S/m, a self-rotation phenomenon of the cells is observed. Cells rotate counterclockwise above the edges of electrodes, with a y-axis of rotation as shown on Figure 6.a1. We observed the self-rotation for human cells lines JURKAT, HEK and PC3, as reported in Figure 6.b and 6.c, and the rotation speeds were maximal at the first crossover frequencies f_{x0}. This phenomenon is particularly surprising since the electric field is non-rotational compared to electro-rotation experiments (ROT). We here report

(a)

(b)

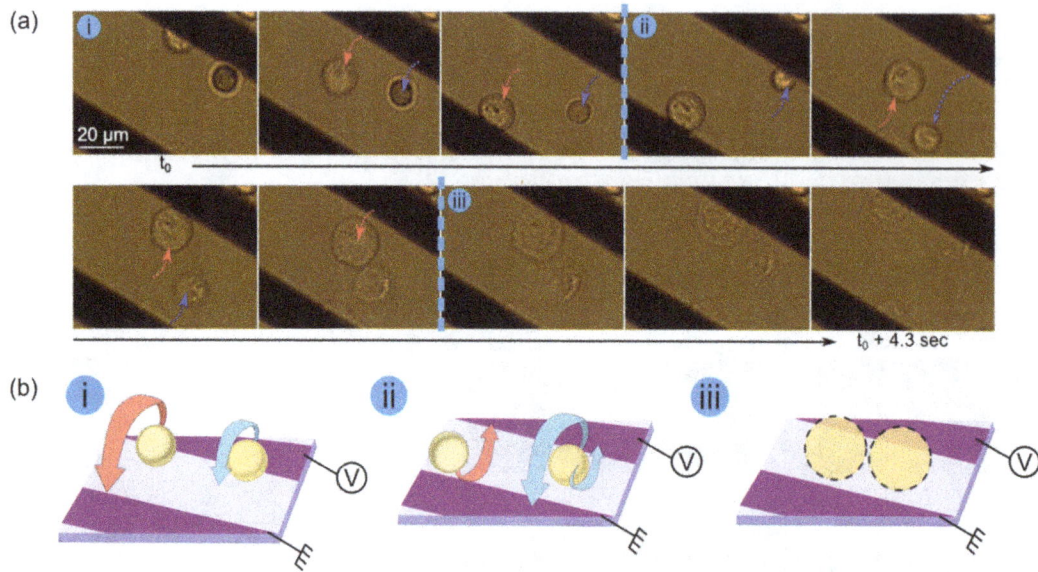

Figure 4. Time-lapse sequence images of cell destruction in conditions of dominant ACEO and ETE (f = 1 kHz, V_{pp} = 8 V, σ_m = 2 10^{-4} S/m) (a) picture of HEK cells taken in a microfluidi chip and (b) schema of the observed motions. Cells are dragged in bulk rolls and the membrane rapidly breaks.

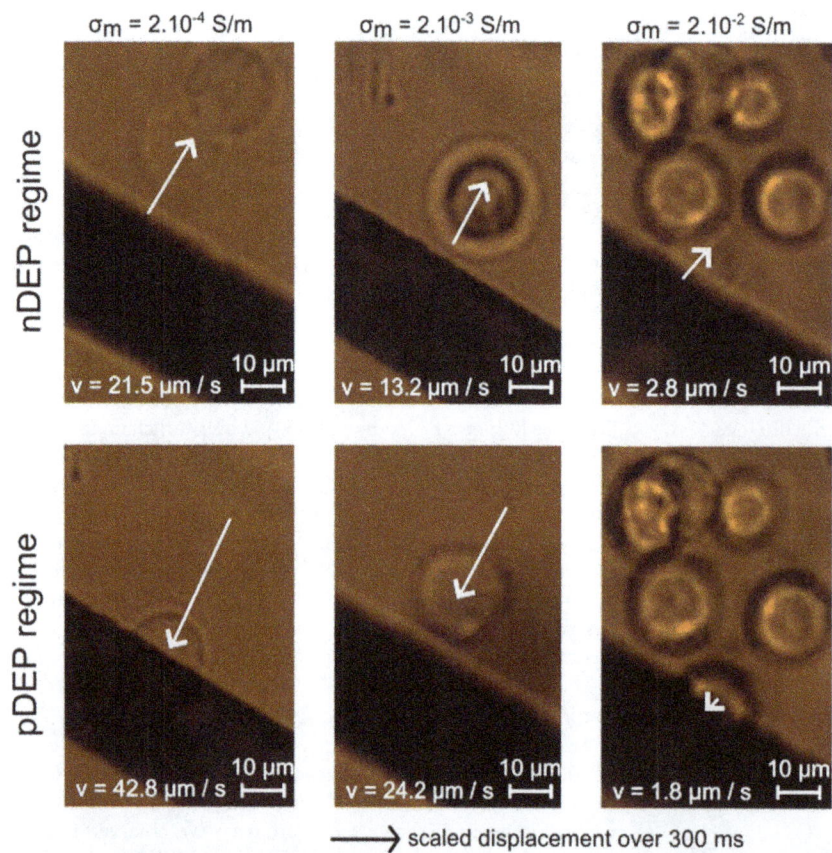

scaled displacement over 300 ms

Figure 5. Response of HEK cells to dielectrophoresis for increasing medium conductivities. nDEP and pDEP are applied at f = 1 kHz and f = 200 kHz, respectively. The arrow represents cell motion during 5 frames (300 ms), the picture being the last image. DEP is stronger at low conductivities compared to EHD forces so cells experience larger displacement at higher velocities at low conductivities.

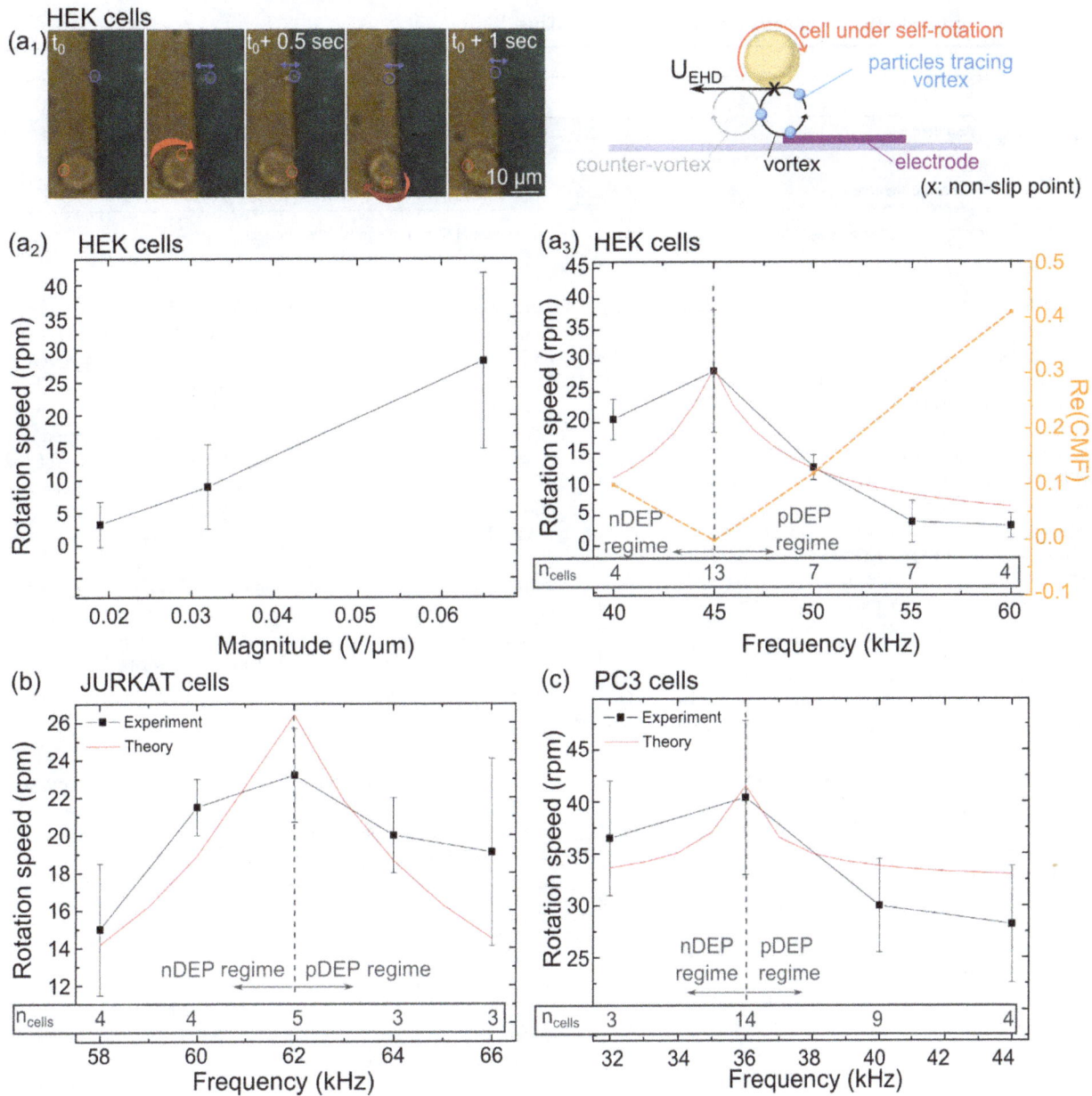

Figure 6. Rotation study of three human cell lines. (a1) Time-lapse sequence images of the rotation of HEK cells in the z-axis, in presence of 1 μm polystyrene colloid (highlighted in blue circles). Particles were added to observe medium stream lines. The red circle pinpoints a visible organelle. Rotation is studied at $\sigma_m = 2.10^{-2}$ S/m when varying **(a2)** magnitude of the electric field at f = 45 kHz or **(a3)** frequency at magnitude 0.065 V/μm (V = 10 Vp-p). The dashed line plots the values of |Re[CMF(ω)]| at the same frequencies, bringing out the relation between DEP effect and ETE. Rotation studies of **(b)** of JURKAT cells and **(c)** PC3 cells (electric field magnitude is 0.089 V/μm (V = 4Vp-p) and $\sigma_m = 2~10^{-2}$ S/m.). The inset on the lower part of the graphs shows the number of cells used for each mean value.

the same behavior than the ones recently observed by Chau *et al.* [16] with melan-a cells or lymphocytes and by Ouyang *et al.* [34] with melanin pigmented cells. In both cases, the origin of such phenomenon was uncertain, as explained by the authors.

Chau *et al.* have hypothesized that the rotational effect of cells is due to the uneven distribution of mass within the cells, thus creating a dipole moment that may drive the cell to rotate continuously. We believe that in that case, the electric field would not penetrate the cell membrane around the crossover frequency f_{x0}, and prevents a dipole creation inside the cell itself. Ouyang *et al.*, have observed the self-rotation of melan-A cells at very high

frequencies (f > 10 MHz) and do not report any peak in the rotation speed. They hypothesized the existence of a tangential force in the lower part of the cell that induces a torque and self-rotation. Since their experiments were performed at very high frequencies, there is a low chance to actually have the presence of a crossover frequency (f_{x0} or f_{hx0}). However, their experiment shows a quadratic dependency of the rotation speed according to the voltage they used. We believe they have observed a competition of force between positive DEP that attracts the cells towards their active electrodes, and ETE that induces local

Table 1. Table of the dielectric parameters for three human cell lines.

Line	Cell Type	σ_3 (S/m)	ε_3	C_{mem} (mF/m²)	Cell size (µm)
HEK	Adherent cell	0.50 ± 0.1	60	3.27 ± 0.05	15 ± 1.0
JURKAT	Circulant cell	0.65 ± 0.12	60	2.38 ± 0.04	15 ± 1.0
PC3	Cancerous adherent cell	0.9 ± 0.15	60	3.44 ± 0.02	18 ± 1.0

The cytoplasm conductivity σ_3 and membrane capacitance C_{mem} are calculated from experimental fit to the competitive model.

vortexes at the edges of the electrodes and thus induces self-rotation of cells.

Instead, we suggest that the self-rotation effect is the result of a competition of forces between DEP and the ETE: Around the crossover frequency, cells DEP regimes are being overpowered by ETE forces and at the exact crossover frequency, the DEP force vanishes, letting ETE forces inducing vortexes like induced motion of liquid above the electrode edges, thus dragging the cells into a self-rotation motion.

To support our hypothesis, streamlines of fluid motion were visualized by mixing 1 µm polystyrene colloids to the cells, at the crossover frequency f_{x0} of cells, as it can be better seen in Movie S1. We observed the rotation of particles just above the edge of the activated electrode, in the z-axis. As overlapped on Figure 6a$_1$, the particles were first pushed up away, then attracted to the edge of the electrode ($t_0 + 0.5$ sec), and finally ended the cycle by being pushed away again in the other direction ($t_0 + 1$ sec). Those observations translate a convective-roll like motion at the edge of electrode. This type of motion was described to be induced by the ETE [13,29] in co-planar or bi-planar electrodes configuration. Evolution of $|Re[CMF(\omega)]|$ values compared to the rotation speed is inverted: the faster the rotation, the less $Re[CMF(\omega)]$ as shown on the Figure 6a$_3$.

Since ETE forces are liquid induced motions, their influence on the global cell motion is quenched by the magnitude of the DEP force that dominates the cell behavior (attracted or repelled to the electrode edges). However, when approaching the DEP crossover frequency, the DEP force vanishes and the ETE behavior dominates. At this particular frequency, the dependency on voltage raises as shown on Figure 6.a$_2$. The cell is much bigger than the convective rolls which induces its self-rotation by slip-free rolling. On the contrary, smaller particles are dragged by the fluid flow and follow the stream in convective rolls motion, as drawn in Figure 6.a$_1$.

Following this explanation, the cell rotation Ω_{fx0} at the crossover frequency f_{x0}, is expressed with a slip-free rolling condition with the global electrohydrodynamical u_{EHD} induced velocity on the cell, corrected by a compensation factor α_{lim} revealing the force competition between ETE and DEP, as shown in equation (9).

$$\Omega_{fx_0} = \frac{\pi}{30} \frac{u_{EHD}}{a} \alpha_{lim} \text{ where } u_{EHD} = u_{ACEO} + u_{ETE} \text{ and}$$

$$\alpha_{lim} = \frac{|u_{EHD}|}{2|u_{EHD} + u_{DEP}|} \tag{9}$$

We have fitted the experimental data observed on the three cell lines and this model. Respective fits are shown on Figure 6. One

can observe that the trends are well described by the model, emphasizing the competitiveness competitively between EHD and DEP with a maximum rotation value at the crossover frequency. The fits have been performed when varying σ_3 and C_{mem} whose values for the three cell lines are reported in Table 1.

We address the robustness of the rotation method by confronting the measured values of crossover frequencies of the three cell lines determined by the rotation and the more classical observation method, as shown in Movie S1.

We do not observe any changes in cell behavior compared to $\sigma_m = 10^{-2}$ S/m when rising the conductivity of the medium ($\sigma_m > 10^{-2}$ S/m). Indeed, the potential drops within the double layer become null thus the ACEO force fails, DEP is the dominant force over ETE so that DEP is overall the main force acting on the cells unless around the crossover frequency as described in the last paragraphs.

Conclusion

AC electrokinetic forces have been reported to identify and sort cells according to their dielectric differences and we have presented here the behaviors of three human cells lines when placed in non-uniform electric field. We have analyzed and reported the influence of key parameters of the field that radically change cell motions and shown how to take advantage of those behaviors to characterize and/or handle human cells by AC fields in microfluidic channels. Our analysis gives a sense to the observations reported by several other works related to human cells. We believe that our results will help the community to better understand their experimental observations when handling human cells, to design enhanced experiments when using AC fields to characterize or handle cells and finally to allow the establishment of an accurate database of human cell dielectric properties for accurate cell detection and sorting.

Supporting Information

Movie S1 The movie shows the different behaviors of JURKAT cells reported in the article. Videos are displayed in real time.

Author Contributions

Conceived and designed the experiments: CV TH. Performed the experiments: CV TH FK XG. Analyzed the data: CV TH. Contributed reagents/materials/analysis tools: CV TH FK XG. Wrote the paper: CV TH FK XG DP.

References

1. Minerick AR, Zhou R, Takhistov P, Chang H-C (2003) Manipulation and characterization of red blood cells with alternating current fields in microdevices. Electrophoresis 24: 3703–3717.
2. Vahey MD, Voldman J (2009) High-throughput cell and particle characterization using isodielectric separation. Anal Chem 81: 2446–2455.
3. Gagnon Z, Mazur J, Chang H-C (2010) Integrated AC electrokinetic cell separation in a closed-loop device. Lab Chip 10: 718–726. Available: http://www.ncbi.nlm.nih.gov/pubmed/20221559.
4. Lenshof A, Laurell T (2010) Continuous separation of cells and particles in microfluidic systems. Chem Soc Rev 39: 1203–1217. Available: http://www.ncbi.nlm.nih.gov/pubmed/20179832.
5. Kua CH, Lam YC, Rodriguez I, Yang C, Youcef-Toumi K (2007) Dynamic cell fractionation and transportation using moving dielectrophoresis. Anal Chem 79: 6975–6987. Available: http://www.ncbi.nlm.nih.gov/pubmed/17702529.
6. Honegger T, Peyrade D (2013) Moving pulsed dielectrophoresis. Lab Chip 13: 1538–1545. Available: http://dx.doi.org/10.1039/C3LC41298A.
7. Pethig R, Menachery A, Pells S, De Sousa P (2010) Dielectrophoresis: A review of applications for stem cell research. J Biomed Biotechnol 2010.
8. González A, Ramos A, Green NG, Castellanos A, Morgan H (2000) Fluid flow induced by nonuniform ac electric fields in electrolytes on microelectrodes. II. A linear double-layer analysis. Phys Rev E 61: 4019–4028. doi:10.1103/PhysRevE.61.4019.
9. Squires TM, Bazant MZ (2004) Induced-charge electro-osmosis. J Fluid Mech 509: 217–252. Available: http://dx.doi.org/10.1017/S0022112004009309.
10. Pohl HA (1950) The motion and precipitation of suspensoids in divergent electric fields. J App Phy 22: 869. Available: http://jap.aip.org/jap/copyright.jsp.
11. Ramos A, Morgan H, Green NG, Castellanos A (1998) Ac electrokinetics: A review of forces in microelectrode structures. J Phys D Appl Phys 31: 2338–2353.
12. Castellanos A, Ramos A, Lez AG, Green NG, Morgan H (2003) Electrohydrodynamics and dielectrophoresis in microsystems: Scaling laws. J Phys D Appl Phys 36: 2584–2597.
13. Morgan H, Holmes D, Green NG (2003) 3D focusing of nanoparticles in microfluidic channels. IEEE Proc-Nanobiotechnology 150,2: 76–81. doi:10.1049/ip-nbt:20031090.
14. Oh J, Hart R, Capurro J, Noh H (2009) Comprehensive analysis of particle motion under non-uniform AC electric fields in a microchannel. Lab Chip 9: 62–78. Available: http://dx.doi.org/10.1039/B801594E.
15. Shafiee H, Sano MB, Henslee EA, Caldwell JL, Davalos R V (2010) Selective isolation of live/dead cells using contactless dielectrophoresis (cDEP). Lab Chip 10: 438–445.
16. Chau L-H, Liang W, Cheung FWK, Liu WK, Li WJ, et al. (2013) Self-Rotation of Cells in an Irrotational AC E-Field in an Opto-Electrokinetics Chip. PLoS One 8: e51577.
17. Jones TB (2003) Basic theory of dielectrophoresis and electrorotation. Eng Med Biol Mag IEEE 22: 33–42.
18. Kaler K V, Jones TB (1990) Dielectrophoretic spectra of single cells determined by feedback-controlled levitation. Biophys J 57: 173–182. Available: http://linkinghub.elsevier.com/retrieve/pii/S0006349590825200.
19. Gascoyne PRC, Vykoukal JV, Schwartz JA, Anderson TJ, Vykoukal DM, et al. (2004) Dielectrophoresis-based programmable fluidic processors. Lab Chip 4: 299–309.
20. Pethig R, Jakubek LM, Sanger RH, Heart E, Corson ED, et al. (2005) Electrokinetic measurements of membrane capacitance and conductance for pancreatic I^2-cells. IEE Proc Nanobiotechnology 152: 189–193.
21. Chung C, Waterfall M, Pells S, Menachery A, Smith S, et al. (2011) Dielectrophoretic Characterisation of Mammalian Cells above 100 MHz. J Electr Bioimp 2: 64–71.
22. Gagnon ZR (2011) Cellular dielectrophoresis: applications to the characterization, manipulation, separation and patterning of cells. Electrophoresis 32: 2466–2487. Available: http://www.ncbi.nlm.nih.gov/pubmed/21922493. Accessed 2014 January 22.
23. Honegger T, Scott MA, Yanik MF, Voldman J (2013) Electrokinetic confinement of axonal growth for dynamically configurable neural networks. Lab Chip 13: 589–598. Available: http://dx.doi.org/10.1039/C2LC41000A.
24. Jones TB, Washizu M (1996) Multipolar dielectrophoretic and electrorotation theory. J Electrostat 37: 121–134. Available: http://www.sciencedirect.com/science/article/pii/030438869600006X. Accessed 2014 February 11.
25. Honegger T, Berton K, Picard E, Peyrade D (2011) Determination of Clausius-Mossotti factors and surface capacitances for colloidal particles. Appl Phys Lett 98: 181906. Available: http://link.aip.org/link/?APL/98/181906/1.
26. Honegger T, Peyrade D (2012) Dielectrophoretic properties of engineered protein patterned colloidal particles. Biomicrofluidics 6: 44112–44115. Available: http://dx.doi.org/10.1063/1.4771544.
27. Huang Y, Wang XB, Becker FF, Gascoyne PR (1997) Introducing dielectrophoresis as a new force field for field-flow fractionation. Biophys J 73: 1118–1129. Available: http://linkinghub.elsevier.com/retrieve/pii/S000634959778144X.
28. Ramos A, Morgan H, Green NG, Castellanos A (1999) The role of electrohydrodynamic forces in the dielectrophoretic manipulation and separation of particles. J Electrostat 47: 71–81.
29. Honegger T, Peyrade D (2013) Comprehensive analysis of alternating current electrokinetics induced motion of colloidal particles in a three-dimensional microfluidic chip. J Appl Phys 113: 194702. Available: http://dx.doi.org/10.1063/1.4804304.
30. Desai SP, Voldman J (2011) Cell-based sensors for quantifying the physiological impact of microsystems. Integr Biol 3: 48–56. Available: http://dx.doi.org/10.1039/C0IB00067A.
31. Su H-W, Prieto JL, Voldman J (2013) Rapid dielectrophoretic characterization of single cells using the dielectrophoretic spring. Lab Chip 13: 4109–4117. Available: http://dx.doi.org/10.1039/C3LC50392E.
32. Ho C-T, Lin R-Z, Chang W-Y, Chang H-Y, Liu C-H (2006) Rapid heterogeneous liver-cell on-chip patterning via the enhanced field-induced dielectrophoresis trap. Lab Chip 6: 724–734. Available: http://dx.doi.org/10.1039/B602036D.
33. Gupta S, Alargova RG, Kilpatrick PK, Velev OD (2010) On-chip dielectrophoretic coassembly of live cells and particles into responsive biomaterials. Langmuir 26: 3441–3452. Available: http://www.ncbi.nlm.nih.gov/pubmed/19957941.
34. Ouyang M, Cheung WK, Liang W, Mai JD, Liu WK, et al. (2013) Inducing self-rotation of cells with natural and artificial melanin in a linearly polarized alternating current electric field. Biomicrofluidics 7: 54112. Available: http://dx.doi.org/10.1063/1.4821169.

20

Implanted Miniaturized Antenna for Brain Computer Interface Applications: Analysis and Design

Yujuan Zhao[1], Robert L. Rennaker[2], Chris Hutchens[3], Tamer S. Ibrahim[1,4]*

1 Department of Bioengineering, University of Pittsburgh, Pittsburgh, Pennsylvania, United States of America, **2** Behavioral and Brain Sciences, Erik Jonsson School of Engineering, University of Texas Dallas, Richardson, Texas, United States of America, **3** School of Electrical and Computer Engineering, Oklahoma State University, Stillwater, Oklahoma, United States of America, **4** Department of Radiology, University of Pittsburgh, Pittsburgh, Pennsylvania, United States of America

Abstract

Implantable Brain Computer Interfaces (BCIs) are designed to provide real-time control signals for prosthetic devices, study brain function, and/or restore sensory information lost as a result of injury or disease. Using Radio Frequency (RF) to wirelessly power a BCI could widely extend the number of applications and increase chronic in-vivo viability. However, due to the limited size and the electromagnetic loss of human brain tissues, implanted miniaturized antennas suffer low radiation efficiency. This work presents simulations, analysis and designs of implanted antennas for a wireless implantable RF-powered brain computer interface application. The results show that thin (on the order of 100 micrometers thickness) biocompatible insulating layers can significantly impact the antenna performance. The proper selection of the dielectric properties of the biocompatible insulating layers and the implantation position inside human brain tissues can facilitate efficient RF power reception by the implanted antenna. While the results show that the effects of the human head shape on implanted antenna performance is somewhat negligible, the constitutive properties of the brain tissues surrounding the implanted antenna can significantly impact the electrical characteristics (input impedance, and operational frequency) of the implanted antenna. Three miniaturized antenna designs are simulated and demonstrate that maximum RF power of up to 1.8 milli-Watts can be received at 2 GHz when the antenna implanted around the dura, without violating the Specific Absorption Rate (SAR) limits.

Editor: Masaya Yamamoto, Institute for Frontier Medical Sciences, Kyoto University, Japan

Funding: This work was supported by NIH R01NS062065 (http://www.nih.gov/). The funders had no role in study design, data collection and analysis, decision to publish, or preparation of the manuscript.

Competing Interests: The authors have declared that no competing interests exist.

* Email: tibrahim@pitt.edu

Introduction

Brain Computer Interfaces (BCIs) are devices designed to establish a communication link between the human brain and neuroprosthetic devices to assist individuals with neurological conditions. However, because of the limitation of the power supply, most BCIs require a direct power connection with the external devices. The BCIs could be only implanted inside the subjects' brain for a very limited time, which limit BCIs' functionality and therefore limit the applications of clinical practice.

Battery can be used as BCI power supply units [1–3]. However, batteries present significant challenges due to the size, mass, toxic composition, and finite lifetime. There are several research groups using the inductive coupling method to transfer the power wirelessly [4–6]. The coupling coils have been typically designed to operate at 10 MHz or below (quasi-static conditions). The drawback of the inductive coupling is that its transmission mainly depends on the changing of magnetic field flux, which requires a relatively large (diameter of several centimeters) implanted coil precisely aligned with an external coil. The distance between two coupling coils is limited to approximately one centimeter in order to maintain the effective coupling results [7].

There are some groups studying implanted antennas to transmit data wirelessly into the human body. Most of these implanted

antennas have been designed to operate at the medical implant communication service (MICS) band of 402–405 MHz. The implantable small profile patch antennas' characteristics and their radiation were evaluated [8,9]. The transmission and reflection of microstrip antennas affected by different superstrates and substrates were studied [10], through numerical analysis and measurements. The effects of different inner insulating layers and external insulating layers and power loss were discussed [11] analytically, using a spherical model. Besides the radiation efficiency impacts of insulating layers were presented [12]. For GHz and above operating frequencies, the impact of the coating on antenna performance was studied by an implanted antenna radiation measurement setup [13]. A pair of microstrip antennas working at microwave frequencies (1.45 GHz and 2.45 GHz) established a data telemetry link for a dual-unit retinal prosthesis [14].

Recent research reveals that the electromagnetic field penetration depth inside the tissue can be asymptotically independent of frequency at high frequencies, and the optimal frequency for the millimeter sized implanted antennas is in the gigahertz range. [15] An implanted antenna operating in the gigahertz range could be designed into a very small profile and also solve the difficulties in designing efficient high data rate [16]. Therefore, an implanted antenna (operating in the gigahertz range) provides a promising approach to accomplish a long term implantation of BCI in users as well as transmits power effectively.

Most of the abovementioned works are assuming that the implanted antennas are connected with 50 Ohm transmission lines. It is noted however, that the ratio between received RF power and tissue absorption depends on the input impedance of the receive antenna [15]. To realize the conjugate matching (i.e. optimal performance), the antenna loads including connected wires and implanted chips could be designed to other values rather than being restricted to 50 Ohms. For example the optimal choice was a 5.6 Ohms load in Poon's study [15]. In our work, we simulate and characterize the input impedance of the implanted BCI RF power receiving antenna operating at an RF above 1 GHz. The input impedance and efficiency of wireless implanted antenna is evaluated for different 1) thickness of insulating layers 2) dielectric properties of insulating layers 3) location of implants, and 4) tissue compositions. Lastly, three miniaturized implanted antenna designs are compared and the maximum received power under the SAR regulations are calculated based on the FDTD simulation results.

Materials and Methods

FDTD simulation and the transmission line feed model

The input impedance of an antenna of the classic structure could be calculated analytically when the antenna is placed in the free space, buried in materials [17], or even when insulated antenna is embedded inside a homogeneous lossy material [18]. However, it is extremely challenging to analytically calculate the impedance of an insulated antenna with arbitrary structures embedded in the human brain, which integrates many different lossy tissue materials.

The FDTD method has great advantages for simulating interactions of electromagnetic waves with biological tissues [19]. In this work, a one dimensional transmission line feed model [2,20] is implemented into our in-house three dimensional (3D) FDTD method package in order to study the input impedance of the implanted antenna. This simulation package has been widely utilized and verified in many papers [21–24]. The perfectly matched layers (PML) are used as the absorbing boundary conditions and the power radiated from the antenna in the FDTD model propagates similarly as it does in the lossless/lossy medium of infinite extent. The material of the antenna is simulated as the perfect electric conductor (PEC) to model very good conducting materials. To get the accurate computational results, the integration contour of the currents is shifted one cell from the antenna drive point to avoid the electric fringing field in the gap [2]. To analyze the ultra-thin (micrometers) insulating layers effects on the antennas performance, thin material sheets are modeled using a three dimensional sub-cell modeling formula in FDTD [25]. This efficient sub-cell modeling method removes the limitation that spatial information should be much larger than the cell grid and therefore greatly reduce the computer storage requirement and computational time.

At the feeding location, the antenna is excited by the virtual transmission line [26], which is injected with a differentiated Gaussian pulse with sufficient frequency content around the intended operational frequency. The differentiated Gaussian pulse is:

$$G(t) =$$

$$\frac{1}{T \times 10^{-12}}(t - S \times T \times 10^{-9}) \exp\left(-\left(\frac{(t - S \times T \times 10^{-9})}{T \times 10^{-9}}\right)^2\right) \quad (1)$$

The parameter T affects the pulse-width and the time delay of the pulse. S is a temporal delay parameter. A set of suitable parameters for S (5.8) and T (0.1) have been chosen for a wideband spectrum of frequencies ranging from 1 GHz to 4 GHz according to the geometries of the antennas to be simulated.

Antenna geometry and antenna performance parameters

The antenna reciprocity theorem [27] guarantees that a good transmitting antenna is also a good receiving antenna. The transmission/radiation efficiency is in part proportional to the radiation resistance [12,27]. Generally for one specific antenna design, the radiation resistance of the antenna increases when the antenna size is larger [28]. In addition, the chip circuitry (attached to the implanted antenna) typically possesses high input impedance values (~80–200 Ohms). Therefore, for efficient operation (minimal mismatch), it is highly favorable to have the input impedance of the implanted antenna in the same range (~80–200 Ohms). The input impedance of a folded dipole antenna is approximately four times of the impedance of a dipole antenna when the length of the folded dipole equals to half wavelength [27], which is on the order of about 300 Ohm in the free space. As a result, a modified folded dipole antenna (rectangular antenna) was chosen for the following analysis.

Due to the inhomogeneous and lossy environment (human head), the relation between power reception and the implantation depth of the antenna does not strictly follow the Friis transmission formula as it is not a far field RF problem. Therefore the radiation pattern is not used to study the antennas' performance in this work. Since the RF power is absorbed by the body and can result in tissue heating, the major concern about the wireless powering the BCI devices is mainly related to this safety issue. As a result, the main performance parameter of the BCI implanted antennas mainly depends on power reception in relation to tissue absorption i.e. SAR rise. Thus any geometry/feeding design of the antenna will aim at achieving maximum power reception for a given local SAR. Furthermore, from circuit theory, a maximum transfer of power from a given voltage source to a load occurs when the load impedance is the complex conjugate of the source impedance. Therefore, the input impedance of the implanted antenna is studied as the major power transmission indicator. The antennas can be used at any frequency where they exhibit enough power receptivity for a given local SAR. The input impedance and the received power of the implanted antenna are calculated through voltage and current information from the transmission line feed model [2,20] used in this study.

Human Head model

Antennas are implanted inside a 3D 19 materials head model which is developed from 1.5 tesla MRI images [29]. The tissue properties are defined [24] based on the study [30]. In order to compare the different effects of phantoms and the head model, two phantoms (different shapes) with the same single tissue material are also implemented, which are shown in Figure 1. The size of the head model/phantom is 182 mm×187 mm×230 mm. The implantable electrode arrays are normally implanted inside the cortex and the processing chip is between the dura and the grey

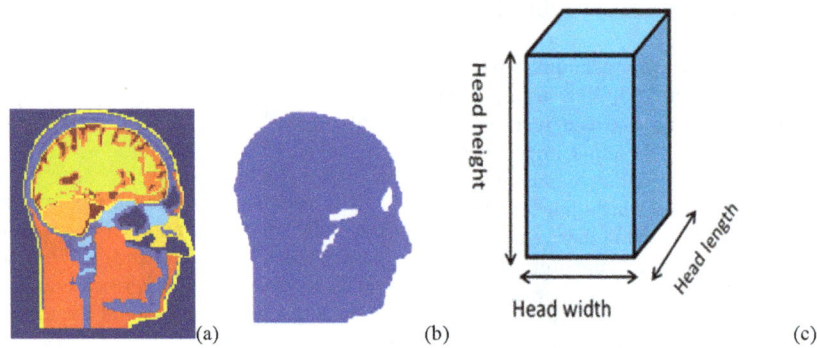

Figure 1. Phantoms and head models. a) Sagittal cross sections of the multi tissues head model at the middle slice; b) Sagittal cross section of the homogenous head shape phantom model at the middle slice; c) Homogeneous rectangular shape phantom model.

matter [1]. Therefore, the dielectric properties of these two single-tissue head phantoms are calculated from the average of properties of the dura and the grey matter [30] (relative permittivity of 46 and conductivity of $\sigma = 1.6$ S/m).

Results

Effects of ultra-thin insulating layers on the input impedance of the implanted antennas

Biocompatible insulating materials are used to surround implanted antennas in order to prevent metallic oxidation and avoid the short circuit effect from the high conductive human head tissues. These biocompatible insulating layers could even the electromagnetic wave transition between the source and the head model and reduce the coupling with the lossy human tissues [11]. From the antenna miniaturization techniques aspect, the dielectric loading (biocompatible insulating material) has also been shown to be a very effective way of reducing the dimensions of the antenna [31]. Furthermore, the tissue model in the area immediately surrounding the implant affects the antenna performance considerably [9]. In this work, the impacts from the micrometer scale insulating layers are studied.

Description of the rectangular with a length of 13 mm and width of 3 mm (the thickness and width of the wire of this implanted antenna is negligible) surrounded by the insulating layer is shown in Figure 2a. In the Figure 2a, the dark rectangular line is the antenna wire and the grey part is the biocompatible insulating material mesh. The excitation is located at one of the longer parallel wires. The antenna surrounded by the insulating layer is numerically implanted into the center of the brain of the 3D anatomically detailed human head model (Figure 2b).

The simulation spatial resolution is set to 1 mm in this study. The thicknesses of the insulating layers are changing from 25 um to 330 um (thin material sheets are modeled using the three dimensional sub-cell modeling formula in FDTD [25]). Since the biocompatible materials are usually polymers and ceramics which are low conductive materials, the relative permittivity of the insulating layers is simulated as 2.1 (polycarbonate) in this simulation and the conductivity is approximately zero [32,33].

The results in Figure 3 demonstrate that the thickness of insulating layers significantly impacts antenna's resonance frequency and input impedance, which in turn will affect antenna's radiation efficiency. The results could be explained: when an antenna is implanted inside the human head model, the dielectric constant of insulating layers (2.1 in this case), is much smaller than that of the head tissues. The velocity of the electromagnetic wave is higher in the small dielectric constant material thus yielding longer

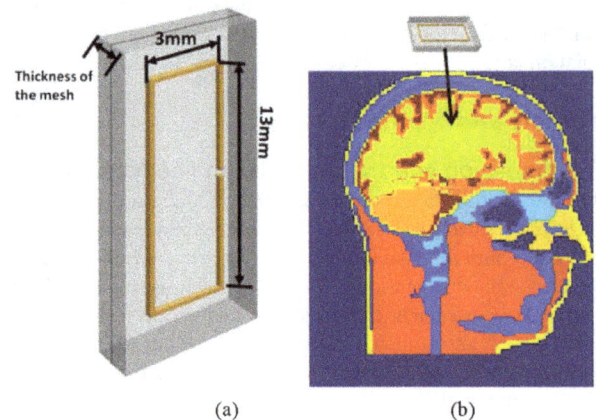

Figure 2. Simulated antenna geometry and its location. a) Geometry of the implanted rectangular antenna; b) Antenna position inside the head model (sagital view of the head model is shown), the color bar scale represents the relative permittivity values.

operating wavelength. Therefore the resonant frequency of the same length antenna will shift to higher frequency when compared to non-insulating cases. This effect increases when the insulating layer becoming thicker (from 25 um to 330 um). The real part of the input impedance also increases because of the decreased average dielectric constant of the whole surrounding volume of the implanted antenna, including the insulating material and the brain tissues. In other words, the lossy human tissue material is moved away from the near field of the implanted antenna with a micrometer insulating layer which will lead to higher radiation efficiency. For example, the 330 um insulating layer antenna real part of the input impedance (which is 420 Ohm) more than doubles that obtained with the 25 um insulating layer antenna (which is 180 Ohm) as shown in Figure 3.

From the simulation results plot of the frequency and input impedance in Figure 3, the input impedance values don't change dramatically for insulating layers with different thickness if the operating frequency is larger than the resonant frequency (1.7 GHz−4 GHz). Therefore, for this implanted antenna, if operational frequency is chosen at this frequency band, the mismatch from the thicknesses changing will be minimal.

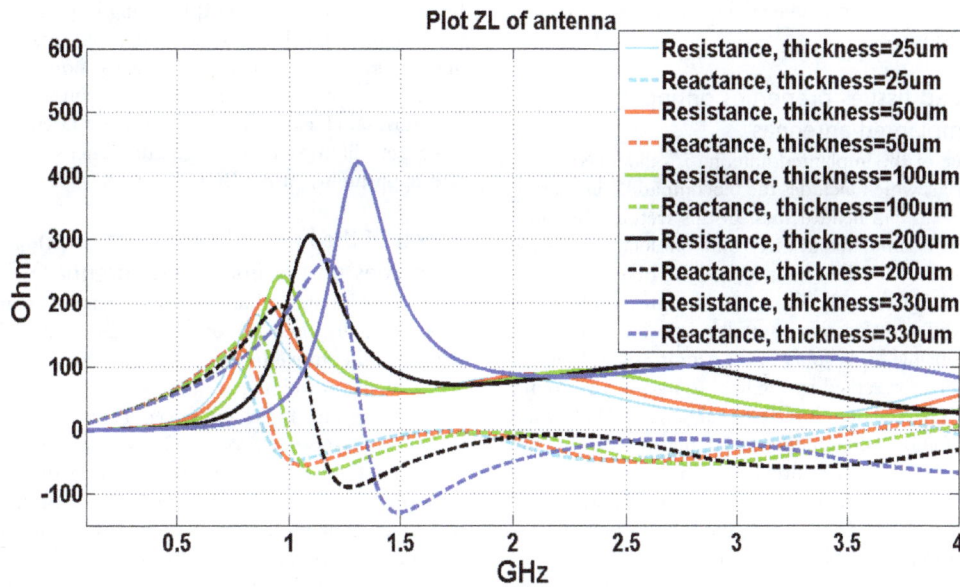

Figure 3. Effects of thin insulating layers on the input impedance of the implanted antenna inside the head model.

Effects of the insulating layer dielectric properties on the input impedance of the antennas

The same geometry of the rectangular implanted antenna shown in Figure 2a is simulated with two different biocompatible insulating layers (the simulated insulating layers have the same thickness of 0.33 mm in the two simulations) inside the human head model. The simulation results are shown in Figure 4.

The simulation results in this section show not only that the thickness of the insulating material affects antenna performance, but also the dielectric property of the insulating materials influence the performance of the implanted antenna inside the human brain. The results reveal that the antenna resonant frequency shifts to a lower frequency when the antenna is embedded inside a high

dielectric constant insulating layer. Figure 4 also shows that the first resonant frequency is around 1.4 GHz if the relative permittivity is 2.1. If the antenna is embedded in the material with relative permittivity of 21, the center resonant frequency will be around t 0.9 GHz. Higher averaged dielectric constant of the media surrounding the antenna reduces the wavelength of the electromagnetic waves inside the media. As the length of the antenna depends on the wavelength of the antenna's operational frequency; High dielectric constant insulating layer consequently facilitates the reduction of antennas geometric dimensions. However, high dielectric constant insulating layer may reduce the real part of the input impedance of the antenna which in turn may hamper the radiation efficiency. Therefore, a balance design

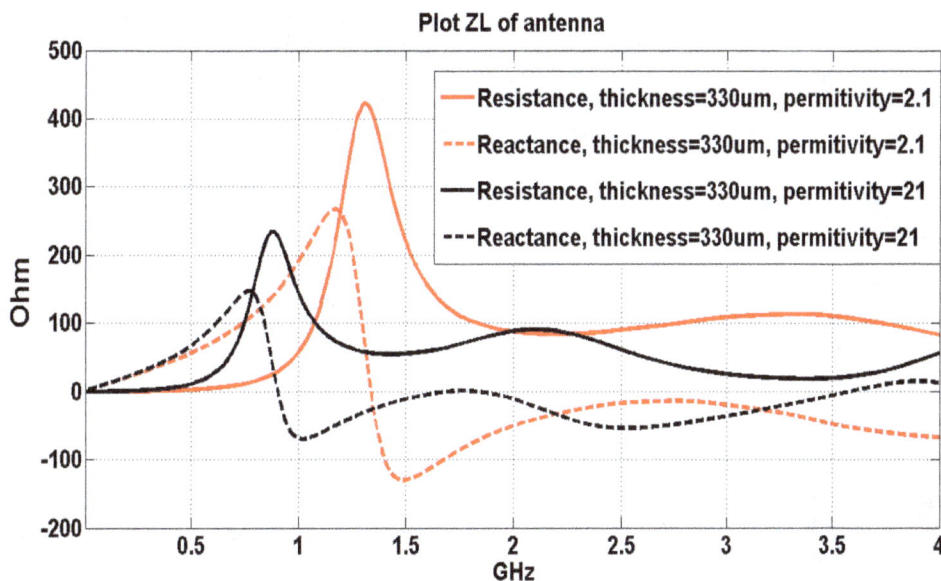

Figure 4. Simulation results of antennas surrounded with insulating layers with the same thickness but the different dielectric properties.

of high radiation efficiency and smaller dimensions is crucial to achieve optimal performance.

Effects of the head tissues properties on input impedance of implanted antennas

The performance of the implanted antenna is influenced by all surrounding materials which includes the biocompatible insulating layers and the lossy human head tissues. In this section, the same rectangular antenna is simulated at three different locations inside the human brain model. For the clinical usage, the BCI devices are normally implanted between the dura and the grey matter [1]. Hence, the three different locations are all proposed around the dura which is responsible for keeping in the cerebrospinal fluid. In Figure 5, the dura is represented by the light orange color around the brain cortex. Above the dura is the cortical bone and below the dura are the combination tissues of the dura and grey matter in the head model. Their constitutive properties and the simulated antenna positions in this head model are listed in Table 1. The same insulating layer (thickness of 1 mm and relative permittivity of 2.1) is used for three different simulation cases.

Table 1 shows that at 2.4 GHz the conductivity and relative permittivity of grey matter (1.773 S/m and 48.994 respectively) are similar with the dura's dielectric property (1.639 S/m and 42.099 respectively) and different for that of the bone(0.385 S/m and 11.4) [30]. These similarities and differences hold true for all other frequencies of interest. Figure 6 displays input impedance of the implanted antenna at the three different implanted positions inside the human brain shown in Figure 5.

Since these three implantation positions are adjacent to each other, we assume that any performance differences of the antenna are not caused by the implantation depth. The results show that the implanted antenna performs differently in bone and in the dura while the same antenna performs relative similar when the antenna is implanted in the dura and directly under the dura. In addition, the brain tissues are separated from the implanted antenna by the biocompatible insulating layers. The frequency shifts and the impedance varieties caused by the tissues properties changes are not as significant as the biocompatible insulating layers' impacts.

The input impedance of the antenna implanted above the dura, where cortical bone is present, is larger than the other two cases. Therefore, the antenna implanted in low conductivity tissues (e.g. cortical bone) may facilitate the antenna radiation efficiency. In addition, the antenna frequency could be altered with time caused

by the saline absorption [13] resulting in instability in the antenna performance. The brain tissues with properties are stable over time and less saline content (i.e. the cortical bones) may be preferable for antenna implantations from the considerations of antenna transmission efficiency as well as the RF circuit stabilization. This of course will impact the design and dimensions of the micro wires and applicability of the BCI.

Effects of the human head phantom shape and dielectric properties on the implanted antennas

A head shape phantom with single liquid mixture was experimentally used by other groups to test the human head effects on the implanted antenna. For example, in [34] the return loss and transmission parameters were measured using a head shape phantom by Schmidt & Partner Engineering for the dosimetric assessment system. To answer whether a multi tissues head phantom is necessary for measuring the implanted antenna performance accurately, and whether a head shape phantom with one homogeneous material could be used to test implanted antenna performance (frequency bandwidth and input impedance), the antenna performance is studied inside three different 3D phantom models. We utilized a multi-tissue head model, a homogenous head model, and a rectangular phantom model, all of which have the same head height, length, and width (see Figure 1.) As mentioned, the relative permittivity is $\varepsilon = 46$ and conductivity is $\sigma = 1.6$ S/m for the rectangular phantom model and the homogenous head model.

The 3 mm by 12 mm rectangular antenna with 1 mm insulating layer is implanted 19 mm under the top of the multi tissues head model (Figure 1a) (the spatial resolution of the simulation is 1 mm), which is just under the dura of this head model. It is centered at the coronal and axial directions. The same insulated rectangular antenna is implanted at the exactly same physical positions inside the homogenous head shape phantom and the rectangular shape phantom model respectively.

The simulation results are presented in Figure 7 and it demonstrates that the performances of the implanted antenna are highly similar inside the three head/phantom models, although the shapes of the head phantoms are different. Especially, the results are identical when the antenna is implanted inside the homo-head model and when it is inside the homo-phantom model. This verifies that the phantom model shape is not necessary to assess the implanted antenna's performances (input impedance and resonance frequency) for this application. Rectangular homogenous phantom could be used instead of a more complex head shape phantom to assess the BCI implanted antenna's specific characteristics (frequency band and input impedance).

While homogenous rectangular head-sized phantom could be used to study the implanted antenna's bandwidth and input impedance, the head shape as well as the presence of different types of tissues is necessary to study heating/SAR/power transmission. This is because SAR as well as the power will change when RF waves go through different tissue, therefore the rectangular homogenous phantom may not be accurate to advise such information.

Designs of the implanted antennas

Around 2.4 GHz, the minimum wavelength (15 mm) shows up in high water content material Cerebra Spinal Fluid (CSF) in human head tissues. Results of the one-cell-gap-feeding models show convergence to the true value if using fine grids [2,35], so spatial resolution of 0.165 mm ($\lambda_{min}/\Delta x = 90$) is implemented for the following miniaturized antenna designs. The time resolution of

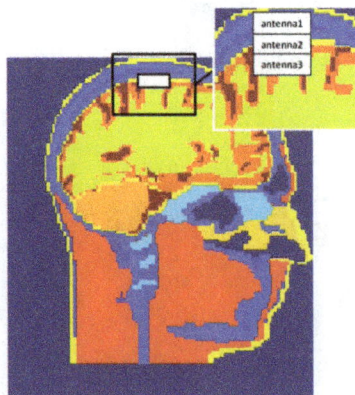

Figure 5. Implanted antenna at three different locations inside the human head model.

Table 1. Dielectric property of three adjacent major tissues at three different locations inside the human head (Fig. 6) at 2.4 GHz.

Tissue	Conductivity(S/m)	Relative permittivity
Bone Cortical (1.6 cm from the surface)	0.385	11.410
Dura (1.9 cm from the surface)	1.639	42.099
Brain Grey Matter (2.24 cm from the surface)	1.773	48.994

FDTD is calculated based on the stability conditions to satisfy the stability criterion.

Three implanted antenna designs are simulated and compared in this study. The same insulating material is used for these implanted antenna simulations (the thickness is 0.33 mm). The thickness of 0.33 mm is chosen because it is a feasible thickness to manufacture and assemble. The surrounding biocompatible material is peek [11] polymer (the relative permittivity is 3.2) which has excellent mechanical properties (stiffness, toughness and durability).

The first antenna design considered is a rectangular antenna. The detailed geometry is shown in Figure 8a. Its input impedance as a function of frequency was calculated using the FDTD model and is shown in Figure 8b. The first resonant frequency (when the imaginary part of the input impedance is zero) is around 1.6 GHz. In order to reduce the circuit mismatching effect, the frequency bandwidth could be chosen between 2 GHz and 4 GHz (because the impedance of the antenna is relative stable in this frequency band).

The second implanted antenna design considered is a serpentine antenna or a meander line antenna [36] which substantially has the greater length in a specific surface area. The geometry detail of the implanted serpentine antenna is shown in Figure 9a. The size of the implanted serpentine antenna (length of 13.695 mm and width of 3.96 mm) is almost the same as the length of the implanted rectangular antenna (length of 13.695 mm and width of 4.29 mm), but has a much longer physical wire length (55.935 mm for the serpentine antenna and 31.35 mm for the rectangular antenna). From the simulation results of the input impedance and

frequency in Figure 9b, the first resonant frequency is around 1.38 GH, which is 220 MHz lower than the first resonant frequency of the implanted rectangular antenna. The frequency bandwidth could be chosen between 1 GHz and 2 GHz (the impedance of the antenna is relative stable in this frequency band). The real part of the input impedance of the serpentine antenna is almost one fifth of that associated with the rectangular antenna at their respective bandwidths (stable resistance slope as a function of frequency); 18 Ohm around 1.5 GHz for the serpentine antenna and 100 Ohm around 2.4 GHz for the rectangular antenna.

The third implanted antenna design considered is a dipole antenna. The geometry detail of the implanted dipole antenna is shown in Figure 10. The first resonant frequency is around 5.2 GHz, which shows that the dipole antenna is electrically shorter than the other two antennas. Since the 5.2 GHz falls out of our accurate simulated range (1 GHz to 4 GHz), the impedance and frequency plot is not shown here. The real part of the input impedance around 2 GHz is around 14 Ohm.

Maximum power reception without SAR violations

The SAR safety regulations regarding RF power deposition in the head varies for different applications. In this work, the power receptions of the implanted antennas are analyzed based on the IEEE RF safety Standard developed by the International Committee on Electromagnetic Safety (ICES) [37] (IEEE, 2005) and the International Commission on Non-ionizing Radiation Protection (ICNIRP) safety regulations [38] with respect to human exposure to radiofrequency electromagnetic fields up to 300 GHz. With respect to SAR limits, the frequency is from 100 kHz to

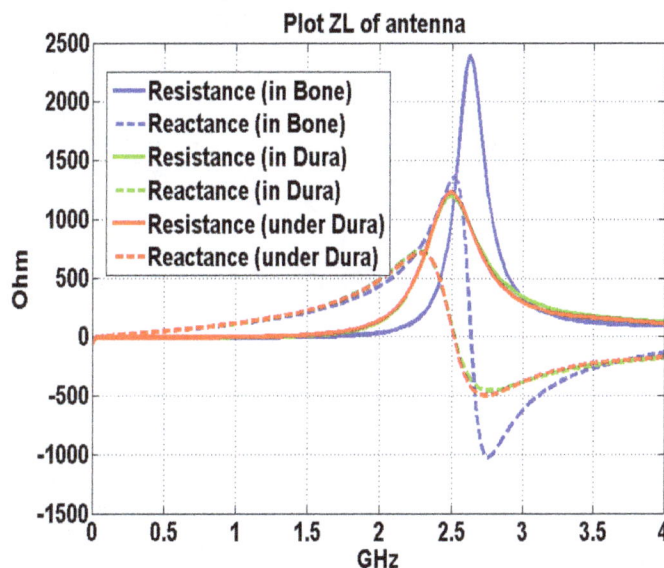

Figure 6. Input Impedance of the implanted rectangular antenna at three different locations inside the human head (Figure 5).

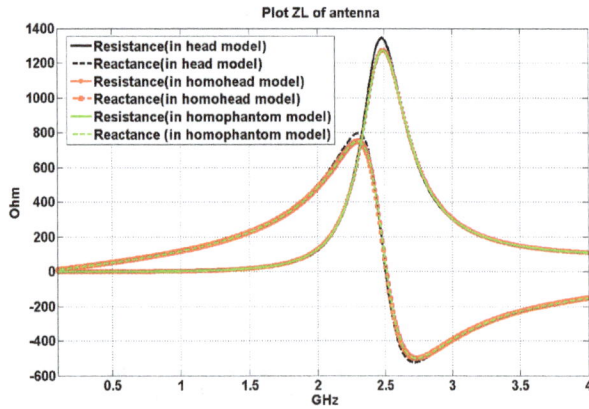

Figure 7. Input Impedance of the antenna when implanted 19 mm inside the multi-tissue head model, the head shape homogenous phantom model and the rectangular homogenous phantom model.

3 GHz in IEEE regulation and 100 kHz–10 GHz in ICNIRP regulation. According these two SAR regulations, the local SAR peak averaged over any 10 g of tissue in the head must be less than or equal to 2 W/kg.

In order to calculate the maximum power reception under the SAR limitations, a dipole antenna is chosen as the external transmitting antenna and the three different implanted antennas are simulated as the receiving antennas. Based on the analysis of these three designed antennas (especially the rectangular antenna and the serpentine antenna), the common preferred frequency band is around 2 GHz. Therefore, the length of the external antenna is defined as75 mm (with negligible thickness). Its resonant frequency is around 2 GHz (simulated and analyzed when the head model existing in the environment near the antenna). The distance between the transmit and receive antennas is about 30 mm; the inner antenna is just under the dura and the outside antenna is about 10 mm away from the surface of the head. Their excitation positions of transmit and receive antennas are vertically centered and placed at the same plane. The multi tissues head model is used to study the maximum received power from the implanted receiving antenna without violating the SAR limits.

Considering the implanted rectangular/serpentine/dipole antennas' input impedance characteristics, the simulated load of implanted chip and circuits (virtual transmission line connected to the antenna ports) are modified to match with the real part of input impedance of the implanted receiving antenna at frequency 2.0 GHz. Considering there are also reactive parts, it is not a perfect match. Hence the calculated (in this work) maximum available power will represent a less optimized scenario: while the real part of impedance is identical for both the implanted receiving

(a)

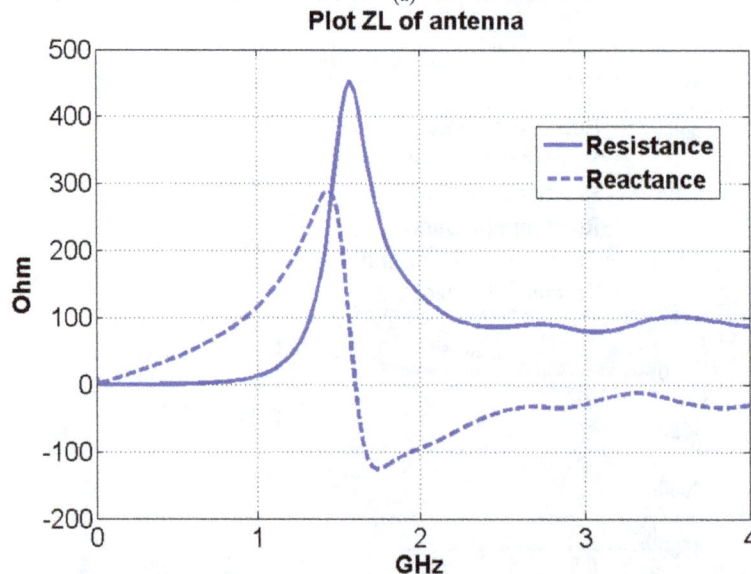

(b)

Figure 8. Implanted rectangular antenna, a) geometry and b) input impedance.

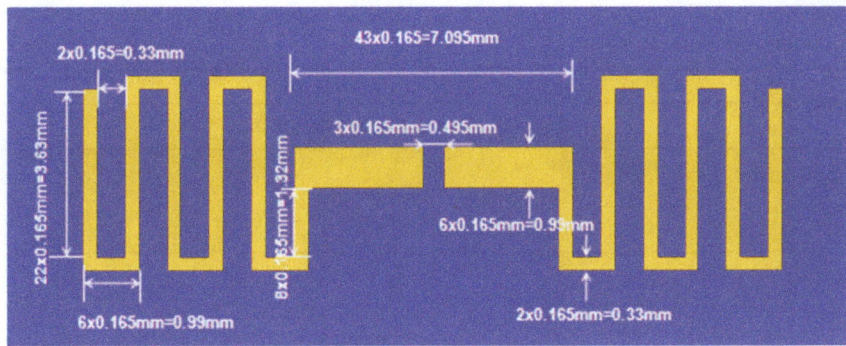

(a)

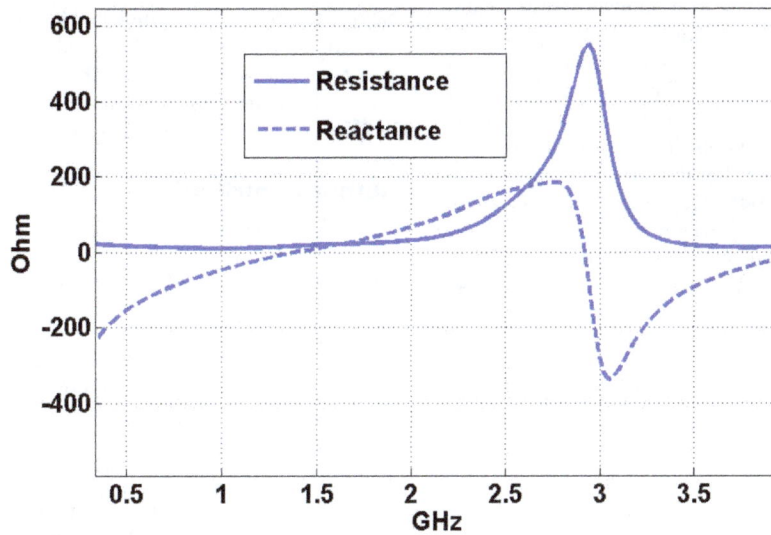

(b)

Figure 9. Implanted serpentine antenna, a) geometry and b) input impedance.

antenna and the chip circuitry/transmission lines, no matching circuit is utilized to compensate for the mismatch in the imaginary part. The calculated maximum power received by the three antenna designs at the SAR limit is shown in the Table 2. The results could be changed from the calculated results in this work (more power can be received potentially) once the source matched to the load perfectly.

Table 2 shows the serpentine antenna allows for more power reception at the SAR limit than the rectangular antenna: the maximum received power is 1.8 mW at the SAR limit when the serpentine antenna is implanted around the dura. While the results

show the superiority of the serpentine antenna in terms of power reception, the higher input impedance of the rectangular antenna allows for better interfacing with the typically expected high input impedance of the chip circuitry (less impedance mismatch).

Furthermore, the maximum power reception has also been investigated when the rectangular antenna implanted inside the cortical bone. The calculated result shows that the rectangular antenna implanted at the bone could receive about 2.5 times more RF power at the SAR limit than that obtained when the antenna is implanted at dura.

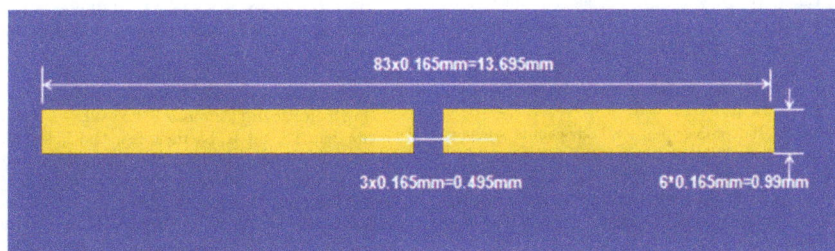

Figure 10. Geometry of the implanted dipole antenna.

Table 2. Maximum power reception under IEEE and ICNIRP SAR limit (2 Watts perKg per 10 gm) at 2 GHz when the implanted antenna is placed right under the dura.

Antenna	Maximum power reception (mW)
Rectangular antenna	1.3
Serpentine antenna	1.8
Dipole antenna	0.58

Conclusion

Miniaturized antennas designs for the BCI application were simulated and analyzed in this work. The simulation results show that the micrometer thickness insulating layer can significantly impact implanted antenna performance. The proper selection of the dielectric properties of the biocompatible insulating layers and the implantation position inside head brain tissues would facilitate the RF power transmission/reception. The shape of the head model may be not a critical factor, but the dielectric properties of surrounding tissues can impact the implanted antennas' input impedance and its operational frequency bandwidth.

Based on three miniaturized antenna designs' simulation results, the maximum power of 1.8 mW could be received by an implanted serpentine antenna when it is implanted inside the dura at the IEEE and ICNIRP SAR limit. Assuming a 25% RF/DC conversion efficiency (due to the switching nature of the harvester circuits), the implantable BCI device can consume 450 uW or less based on the results in this work. Our current designs of simple implantable chip consume about 35 uW [39] which means the designed miniaturized antenna could provide sufficient power to this available chip design if placed in the dura.

Author Contributions

Conceived and designed the experiments: YZ RR CH TI. Performed the experiments: YZ TI. Analyzed the data: YZ TI. Contributed reagents/materials/analysis tools: YZ TI. Wrote the paper: YZ RR TI.

References

1. Chestek CA, Gilja V, Nuyujukian P, Kier RJ, Solzbacher F, et al. (2009) HermesC: low-power wireless neural recording system for freely moving primates. IEEE transactions on neural systems and rehabilitation engineering : a publication of the IEEE Engineering in Medicine and Biology Society 17: 330–338.

2. Hertel TW, Smith GS (2003) On the convergence of common FDTD feed models for antennas. Ieee Transactions on Antennas and Propagation 51: 1771–1779.

3. Farshchi S, Pesterev A, Nuyujukian P, Guenterberg E, Mody I, et al. (2010) Embedded Neural Recording With TinyOS-Based Wireless-Enabled Processor Modules. Ieee Transactions on Neural Systems and Rehabilitation Engineering 18: 134–141.

4. Song YK, Borton DA, Park S, Patterson WR, Bull CW, et al. (2009) Active Microelectronic Neurosensor Arrays for Implantable Brain Communication Interfaces. Ieee Transactions on Neural Systems and Rehabilitation Engineering 17: 339–345.

5. Harrison RR, Kier RJ, Chestek CA, Gilja V, Nuyujukian P, et al. (2009) Wireless Neural Recording With Single Low-Power Integrated Circuit. Ieee Transactions on Neural Systems and Rehabilitation Engineering 17: 322–329.

6. Irazoqui PP, Mody I, Judy JW (2005) Recording brain activity wirelessly. Ieee Engineering in Medicine and Biology Magazine 24: 48–54.

7. Kim S, Zoschke K, Klein M, Black D, Buschick K, et al. (2007) Switchable polymer-based thin film coils as a power module for wireless neural interfaces. Sensors and Actuators a-Physical 136: 467–474.

8. Kim J, Rahmat-Samii Y (2004) Implanted antennas inside a human body: Simulations, designs, and characterizations. Ieee Transactions on Microwave Theory and Techniques 52: 1934–1943.

9. Kiourti A, Nikita KS (2013) Numerical assessment of the performance of a scalp-implantable antenna: effects of head anatomy and dielectric parameters. Bioelectromagnetics 34: 167–179.

10. Soontornpipit P, Furse CM, Chung YC (2004) Design of implantable microstrip antenna for communication with medical implants. Ieee Transactions on Microwave Theory and Techniques 52: 1944–1951.

11. Merli F, Fuchs B, Mosig JR, Skrivervik AK (2011) The Effect of Insulating Layers on the Performance of Implanted Antennas. Ieee Transactions on Antennas and Propagation 59: 21–31.

12. Hall PS, Hao Y (2006) Antennas and propagation for body-centric wireless communications. Boston: Artech House. 1 online resource (xiii, 291 s.) p.

13. Warty R, Tofighi MR, Kawoos U, Rosen A (2008) Characterization of Implantable Antennas for Intracranial Pressure Monitoring: Reflection by and Transmission Through a Scalp Phantom. Ieee Transactions on Microwave Theory and Techniques 56: 2366–2376.

14. Gosalia K, Lazzi G, Humayun M (2004) Investigation of a microwave data telemetry link for a retinal prosthesis. Ieee Transactions on Microwave Theory and Techniques 52: 1925–1933.

15. Poon ASY, O'Driscoll S, Meng TH (2010) Optimal Frequency for Wireless Power Transmission Into Dispersive Tissue. Ieee Transactions on Antennas and Propagation 58: 1739–1750.

16. Yakovlev A, Kim S, Poon A (2012) Implantable Biomedical Devices: Wireless Powering and Communication. Ieee Communications Magazine 50: 152–159.

17. King RWP, Smith GS (1981) Antennas in Matter: Fundamentals, Theory, and Applications. Cambridge, MA: The MIT Press.

18. Fenwick RC, Weeks WI (1963) Sumberged antenna characteristics. Ieee Transactions on Antennas and Propagation 11: 296–305.

19. Ibrahim TS, Hue YK, Tang L (2009) Understanding and manipulating the RF fields at high field MRI. Nmr in Biomedicine 22: 927–936.

20. Taflove A, Hagness SC (2005) Computational electrodynamics : the finite-difference time-domain method. Boston: Artech House. xxii, 1006 p., [1008] f. de pl. en coul. p.

21. Krishnamurthy N, Zhao T, Ibrahim TS (2013) Effects of receive-only inserts on specific absorption rate, B field, and Tx coil performance. J Magn Reson Imaging.

22. Tang L, Hue YK, Ibrahim TS (2011) Studies of RF Shimming Techniques with Minimization of RF Power Deposition and Their Associated Temperature Changes. Concepts Magn Reson Part B Magn Reson Eng 39B: 11–25.

23. Ibrahim TS, Tang L, Kangarlu A, Abraham R (2007) Electromagnetic and modeling analyses of an implanted device at 3 and 7 Tesla. Journal of Magnetic Resonance Imaging 26: 1362–1367.

24. Zhao Y, Tang L, Rennaker R, Hutchens C, Ibrahim TS (2013) Studies in RF Power Communication, SAR, and Temperature Elevation in Wireless Implantable Neural Interfaces. PLoS One 8: e77759.

25. Maloney JG, Smith GS (1992) The Efficient Modeling of Thin Material Sheets in the Finite-Difference Time-Domain (Fdtd) Method. Ieee Transactions on Antennas and Propagation 40: 323–330.

26. Maloney JG, Shlager KL, Smith GS (1994) A Simple FDTD Model for Transient Excitation of Antennas by Transmission Lines. IEEE TRANSAC-TIONS ON ANTENNAS AND PROPAGATION 42: 289–292.

27. Balanis CA (2005) Antenna theory : analysis and design. Hoboken, NJ: John Wiley. xvii, 1117 p. p.

28. Endo T, Sunahara Y, Satoh S, Katagi T (2000) Resonant frequency and radiation efficiency of meander line antennas. Electronics and Communications in Japan Part Ii-Electronics 83: 52–58.

29. Ibrahim TS, Lee R, Baertlein BA, Abduljalil AM, Zhu H, et al. (2001) Effect of RF coil excitation on field inhomogeneity at ultra high fields: A field optimized TEM resonator. Magnetic Resonance Imaging 19: 1339–1347.

30. Andreuccetti D, Fossi R, Petrucci C (1997) An Internet resource for the calculation of the dielectric properties of body tissues in the frequency range 10 Hz–100 GHz. Website at http://niremf.ifac.cnr.it/tissprop/, IFAC-CNR, Florence (Italy). pp. Based on data published by C.Gabriel et al. in 1996.

31. Skrivervik AK, Zurcher JF, Staub O, Mosig JR (2001) PCS antenna design: The challenge of miniaturization. Ieee Antennas and Propagation Magazine 43: 12–26.

32. Alberti G, Casciola M, Massinelli L, Bauer B (2001) Polymeric proton conducting membranes for medium temperature fuel cells (110–160°C). Journal of Membrane Science 185: 73–81.
33. Kobayashi T, Rikukawa M, Sanui K, Ogata N (1998) Proton-conducting polymers derived from poly(ether-etherketone) and poly(4-phenoxybenzoyl-1,4-phenylene). Solid State Ionics 106: 219–225.
34. Chen ZN, Liu GC, See TSP (2009) Transmission of RF Signals Between MICS Loop Antennas in Free Space and Implanted in the Human Head. Ieee Transactions on Antennas and Propagation 57: 1850–1854.
35. Zhao HP, Shen ZX (2009) Weighted Laguerre Polynomials-Finite Difference Method for Time-Domain Modeling of Thin Wire Antennas in a Loaded Cavity. Ieee Antennas and Wireless Propagation Letters 8: 1131–1134.
36. Nakano H, Tagami H, Yoshizawa A, Yamauchi J (1984) Shortening Ratios of Modified Dipole Antennas. Ieee Transactions on Antennas and Propagation 32: 385–386.
37. IEEE (2005) IEEE Standard for Safety Levels with Respect to Human Exposure to Radio Frequency Electromagnetic Fields, 3 kHz to 300 GHz. IEEE Standard for Safety Levels with Respect to Human Exposure to Radio Frequency Electromagnetic Fields, 3 kHz to 300 GHz.
38. ICNIRP (1998) Guidelines for limiting exposure to time-varying electric, magnetic, and electromagnetic fields (up to 300 GHz). International Commission on Non-Ionizing Radiation Protection. Health Phys. 494–522.
39. Hutchens C, Rennaker RL 2nd, Venkataraman S, Ahmed R, Liao R, et al. (2011) Implantable radio frequency identification sensors: wireless power and communication. Conference proceedings : Annual International Conference of the IEEE Engineering in Medicine and Biology Society IEEE Engineering in Medicine and Biology Society Conference 2011: 2886–2892.

21

Dielectric Barrier Discharge Ionization in Characterization of Organic Compounds Separated on Thin-Layer Chromatography Plates

Michał Cegłowski[1]*, Marek Smoluch[2], Michał Babij[3], Teodor Gotszalk[3], Jerzy Silberring[2,4], Grzegorz Schroeder[1]

1 Department of Supramolecular Chemistry, Faculty of Chemistry, Adam Mickiewicz University in Poznan, Poznań, Poland, 2 Department of Biochemistry and Neurobiology, Faculty of Materials Science and Ceramics, AGH-University of Science and Technology, Krakow, Poland, 3 Faculty of Microsystem Electronics and Photonics, Wroclaw University of Technology, Wroclaw, Poland, 4 Center for Polymer and Carbon Materials, Polish Academy of Sciences, Zabrze, Poland

Abstract

A new method for on-spot detection and characterization of organic compounds resolved on thin layer chromatography (TLC) plates has been proposed. This method combines TLC with dielectric barrier discharge ionization (DBDI), which produces stable low-temperature plasma. At first, the compounds were separated on TLC plates and then their mass spectra were directly obtained with no additional sample preparation. To obtain good quality spectra the center of a particular TLC spot was heated from the bottom to increase volatility of the compound. MS/MS analyses were also performed to additionally characterize all analytes. The detection limit of proposed method was estimated to be 100 ng/spot of compound.

Editor: Andrew C. Gill, University of Edinburgh, United Kingdom

Funding: This work was supported by the Polish National Science Center (NCN: www.ncn.gov.pl; grant no. 2011/03/B/ST5/01573). GS received the funding. The funders had no role in study design, data collection and analysis, decision to publish, or preparation of the manuscript.

Competing Interests: The authors have declared that no competing interests exist.

* Email: ceglowski.m@gmail.com

Introduction

Thin-layer chromatography (TLC) is a very simple, cost-effective and fast chromatographic technique allowing separation of most chemical mixtures [1]. Detection of separated compounds relies mostly on their optical visualization using UV light or appropriate reagents, such as Dragendorff reagent. The information obtained by these means is, however very limited. For definite identification of the compound, which has been visualized, it must be compared with a standard that has been simultaneously eluted on TLC. Coupling TLC and mass spectrometry (MS), which is a very useful tool for structural analysis of organic compounds, would make a simple and easy to operate technique. That is why the application of mass spectrometry in characterization of compounds separated on TLC plates has been a subject of interest for many scientists [2,3] and recently some reviews on this topic have been published [4,5,6,7]. The TLC-MS methods can be divided into indirect sampling TLC-MS and direct sampling TLC-MS [5].

Indirect sampling TLC-MS is of less importance because it requires time-consuming processes, such as scratching of the spot containing particular compound, followed by solvent extraction of the adsorbed substance [8]. The direct sampling TLC-MS techniques allow for mass spectrometric analysis of compounds directly from the surface of TLC plates. These methods can be further divided into vacuum-based and ambient TLC-MS approaches [5].

Desorption/ionization methods that operate under vacuum have been used to obtain mass spectra of compounds directly from the surface of TLC plates. Several techniques such as fast atom bombardment (FAB) [9,10], secondary ion mass spectrometry (SIMS) [11], laser desorption (LDI) [12], matrix-assisted laser desorption/ionization (MALDI) [13,14], surface-assisted laser desorption/ionization (SALDI) [15] have been applied to characterize compounds separated by TLC. These techniques however, generate several problems: obtaining high vacuum in the ion source considerably increases the time of analysis, volatile compounds have poor sensitivity, MALDI matrices produce interferences with mass spectra, poor reproducibility in quantitative analysis, diffusion of analytes on TLC plates after applying MALDI matrix solution [16].

Some of the presented problems were solved by coupling ambient MS techniques with TLC. Hence, the TLC plates do not need to be placed in a vacuum chamber for ionization, and the analysis is much faster without the risk that volatile compounds will desorb from the plate before MS analysis. Moreover, the TLC plate size is not limited by the dimensions of a vacuum chamber. Techniques, such as electrospray ionization (ESI) [17,18,19], electrospray-assisted laser desorption ionization (ELDI) [20], desorption electrospray ionization (DESI) [21], laser desorption/ atmospheric pressure chemical ionization (LD-APCI) [22], atmospheric pressure matrix-assisted laser desorption/ionization (AP-MALDI) [23,24] and direct analysis in real time (DART) [25,26] have been adopted to generate mass spectra of compounds directly from TLC plates.

Dielectric barrier discharge ionization (DBDI) produces low-temperature plasma by dielectric barrier discharges (DBD) [27]. DBD are obtained at ambient conditions and are formed between two electrodes with a dielectric layer that separates them. The presence of a dielectric layer limits the average current in the gas space, which causes formation of the low-temperature plasma containing a large number of high energy electrons [28,29]. DBD plasma was used to produce mass spectra of compounds desorbed from different surfaces [30], to detect nonvolatile chemicals directly on various surfaces [31], utilized as an ion source for liquid chromatography/mass spectrometry [32], and coupled *on line* with TLC for mercury speciation [33].

Herein we report the coupling of DBD plasma source with TLC. DBDI is capable of providing very fast and selective ionization of each particular compound separated on TLC plate. The ionized compounds can be characterized using MS and MS/MS modes. Combination of these approaches can find many applications, particularly in synthetic organic laboratories and forensic sciences.

Materials and Methods

Synthesis

Synthesis of compounds **1–6**, whose structures are presented in Figure 1, as well as information about reagents used are given in the **Information S1**.

Instrumental Design

The DBD plasma source consists of quartz capillaries (O.D. 1.5 mm and I.D. 0.8 mm). The electrodes of 2 mm width were made of copper rings surrounding the capillary tube and the gap between the inner edges was 5 mm. The grounded electrode was placed 6 mm apart from the capillary end. The plasma was operated with helium (99.996% purity) flow of 1 L/min, by applying a voltage of 8 kV. Additional data describing DBD ion source are given in details elsewhere [34]. A Bruker Esquire 3000

quadrupole ion trap mass spectrometer (Bruker Daltonics, Bremen, Germany) was used for all measurements. The typical ESI-MS source settings were found to be optimal also for the DBDI source, with the exception of the mass spectrometer entrance glass capillary voltage, where lower potential (1 kV compared to the standard ESI setting (4.5 kV) was used. The temperature of the glass capillary was set to 200°C, the drying gas flow was maintained at 3 L/min., and the nebulizer gas (N_2) was not applied. The scan range was set from 80 to 500 m/z. For MS/MS experiments the isolation width was set to 2 m/z and the fragmentation amplitude was in the range of 0.5 to 0.8 unit.

TLC preparation and detection

Compounds **1–6** were dissolved in dichloromethane to obtain final concentration of 10 mg/mL. One µL sample solutions were spotted on a TLC plate (Merck Millipore TLC silica gel 60 F_{254} aluminium sheets). The compounds were spotted in ester-alcohol pairs on a single TLC plate. The plates with compounds **1** and **2** were developed with CH_2Cl_2/Et_2O (1:1, v/v) solution, those with **3** and **4** were developed with CH_2Cl_2/Et_2O (5:1, v/v), those with **5** and **6** were developed with CH_2Cl_2/Et_2O (10:1, v/v). The plate containing all six compounds was developed with CH_2Cl_2/Et_2O (10:1, v/v). For MS analysis, the spots were visualized on TLC with UV light and marked with a pencil. The TLC plate was then cut through the centers of all spots, resulting as a narrow strip, and a heating element, of a diameter similar to the size of a spot, was attached to the bottom of a particular spot. The strip was then heated from the bottom and held manually so that the plasma released from DBD ion source would emerge just above the marked spot. The heating element consisted of resistance wire which temperature was constantly increased in the range 25°C–400°C using manual control until the signal of particular analyte was visible on mass spectra. Each analysis has been completed in less than one minute. After each m/z measurement the heating element was allowed to cool down. It was then attached to the bottom of the next spot and the measurements were continued. Figure 2 displays a schematic illustration of coupling DBDI to TLC, whereas Figure 3 shows a photograph of experimental setup used for TLC-MS analysis.

Results and Discussion

To examine the performance of DBDI in the on-spot detection of compounds separated on TLC plates, solutions of six compounds were deposited onto plates. The compounds were applied in ester-alcohol pairs on a single TLC plate, because these pairs are resolved properly on TLC and alcohols are products of ester reduction. The spectra obtained from three TLC plates with the six compounds are presented in Figure 4. Each spectrum represents the major ion corresponding to the protonated molecule with little or no background ions. The most intense signals belonging to contaminants are observed at m/z 279.1 and 205.1 and derive from di-n-butyl phthalate (DBP) [35], which is commonly used as a plasticizer in plastics, from which DBD ion source has been manufactured. Particularly, the tube used to transport helium from gas cylinder to DBD ion source, is made of polyvinyl chloride (PVC) for which DBP is used as a plasticizer. We therefore, believe that the possible source of contamination is desorption of DBP by helium stream. An advantage of DBDI technique is its ability to produce intact (usually protonated) species with little or no fragmentation what substantially simplifies identification of the compounds separated on TLC.

To show that good quality mass spectra can be generated even when compounds are poorly resolved on TLC, all six compounds

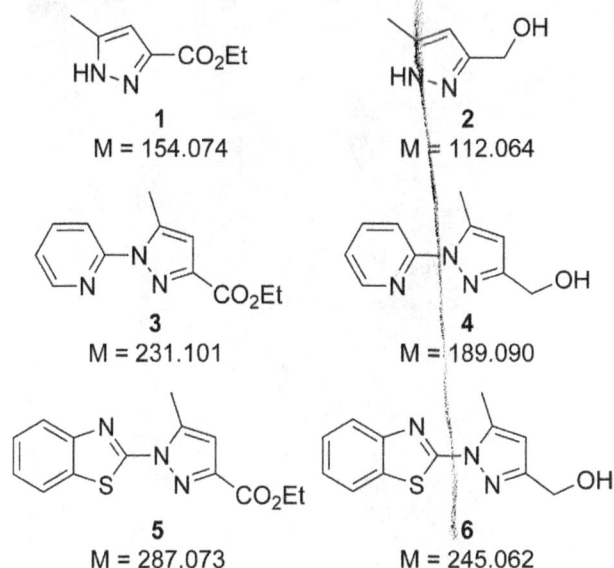

1
M = 154.074

2
M = 112.064

3
M = 231.101

4
M = 189.090

5
M = 287.073

6
M = 245.062

Figure 1. Structures of compounds separated on TLC plates.

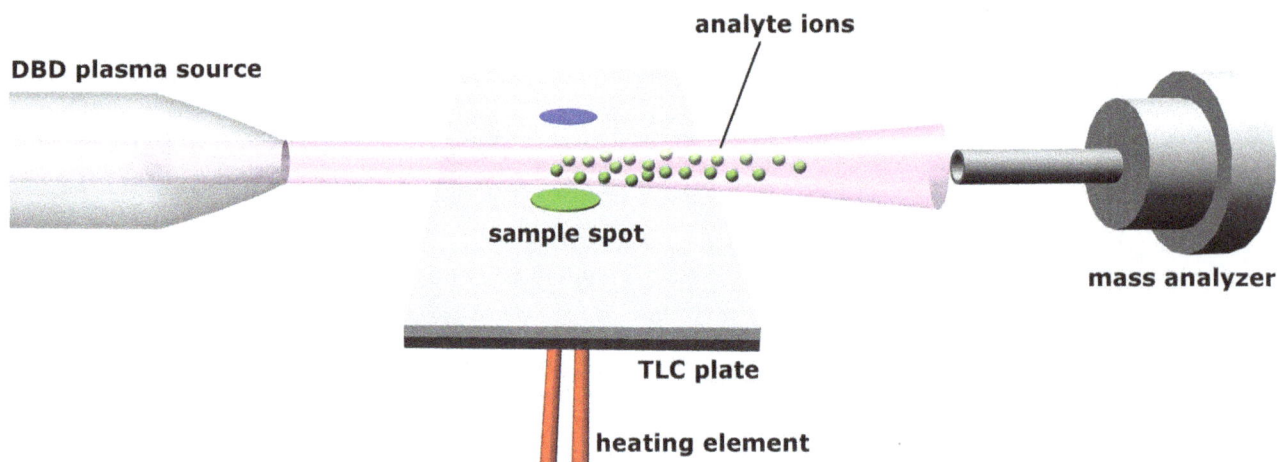

Figure 2. Schematic illustration of DBDI coupled to TLC.

were deposited and separated on a single TLC plate. The resulting mass spectra of all six compounds are presented in the **Information S1**. Each spectrum obtained represents the major ion corresponding to the protonated molecule of a particular compound with only small traces of other compounds resolved on TLC.

The structures of the analytes were confirmed by MS/MS analysis of all compounds. The exemplary MS/MS spectrum of

the signal at m/z 190.1 (assigned to protonated compound **4**) is presented in Figure 5. Only one daughter ion ($[\mathbf{4}+\mathrm{H} - \mathrm{H_2O}]^+ = m/z$ 172.1) appeared in the mass spectrum. MS/MS spectra of all compounds examined are presented in the **Information S1**.

The detection limits of compounds deposited on TLC plates analyzed by DBD ion source were estimated using the solutions containing different concentrations of compounds **1–6**. Each solution was deposited onto TLC plate and after air-drying the

Figure 3. Photograph of experimental setup used for TLC-MS analysis.

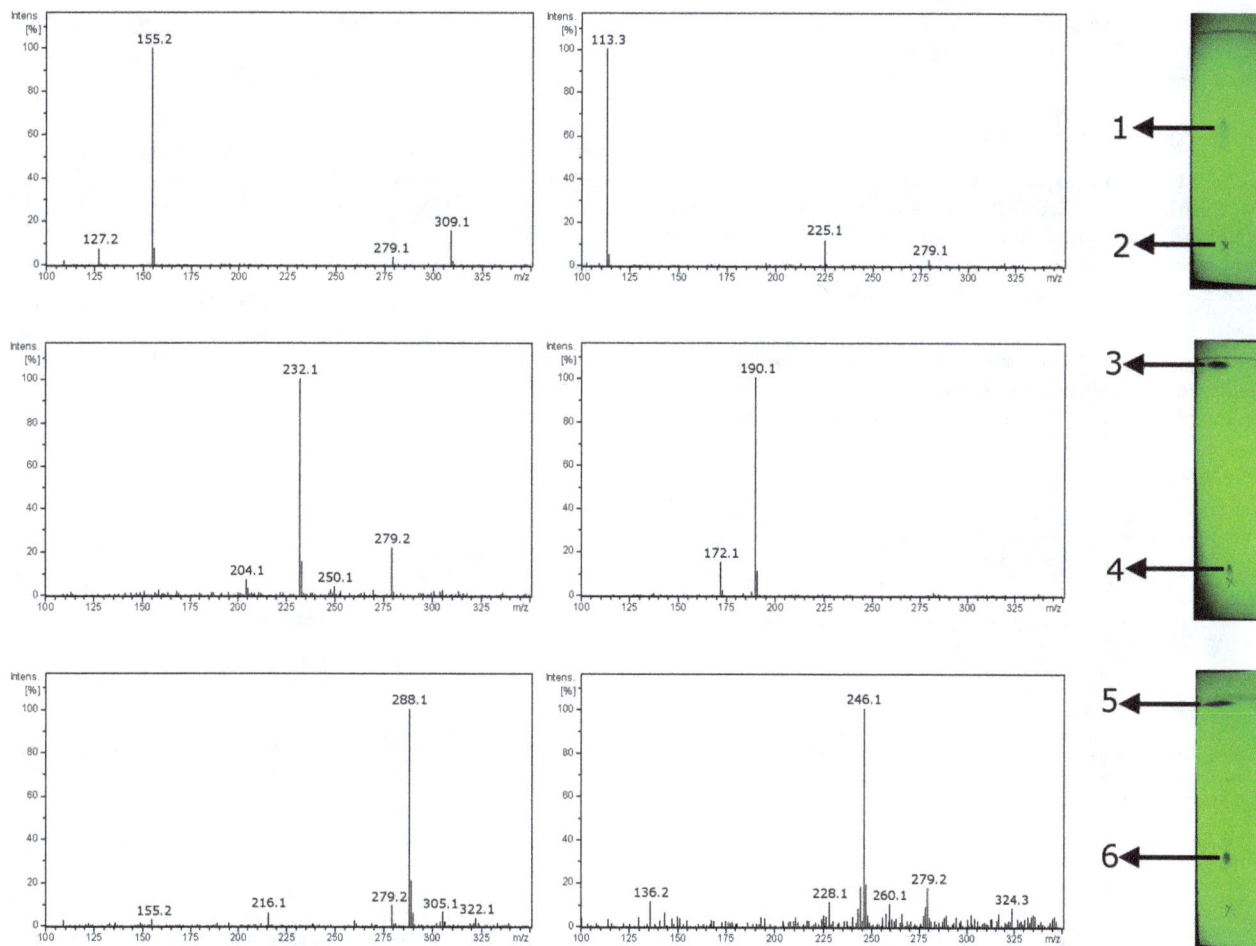

Figure 4. Photographs of TLC plates (visualized in UV light) with corresponding mass spectra obtained from particular spots. The spots were assigned to the label of particular compound. Left panel represents the mass spectra of compounds with higher R_f value (esters).

plate was heated as described in the Methods section and introduced into the gas stream. The best result was obtained for compound **3** where 100 ng/spot was detected. However, in the spectrum obtained (presented in Figure 6) the most intense signal comes from DBP contamination. Nevertheless, this result is sufficient for TLC-MS analysis because the amount of 100 ng/spot of compound **3** could not be seen under the UV light, therefore this amount is far less than needed for common TLC analysis. It is worth noting, that this is only semi-quantitative analysis because the ion intensity of an analyte desorbed from TLC plate varies depending on the thickness and composition of silica gel, temperature, diffusion coefficient of the analyte on the

TLC plate during mass analysis and positioning of TLC plate in the gas stream.

Conclusions

In this paper we demonstrated the use of DBDI in conjunction with TLC separations. This technique appeared to be effective for direct analysis of compounds that can be separated by thin layer chromatography. The coupling of DBDI and TLC has many advantages, such as ionization under ambient conditions, low-cost of a DBD ion source, easy and fast analysis of volatile and semi-volatile compounds, unlimited dimensions of TLC plate and fast sample switching. These advantages make DBDI a robust and

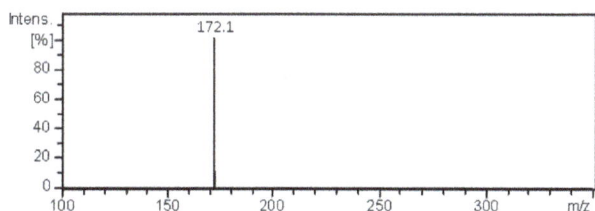

Figure 5. MS/MS spectrum of the signal at *m/z* **190.1 derived from compound 4.**

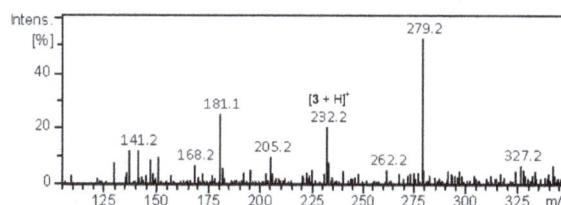

Figure 6. Mass spectrum of compound 3 (100 ng/spot) recorded directly from TLC spot.

convenient mass spectrometric technique, which allows for fast screening of TLC plates that are run every day in many laboratories.

Supporting Information

Figure S1 Photograph of TLC plate (visualized under UV light), on which all six compounds have been separated, combined with mass spectra obtained from respective spots. The spots and the mass spectra were assigned to the number labeling of particular compound.

Figure S2 MS/MS spectra of ions at: a) m/z 155.2 (assigned to protonated compound 1); b) m/z 113.3 (assigned to protonated compound 2); c) m/z 232.1 (assigned to protonated compound 3); d) m/z 288.1 (assigned to protonated compound 5); e) m/z 246.1 (assigned to protonated compound 6).

Information S1 Information about reagents used and synthesis of compounds 1–6.

Author Contributions

Conceived and designed the experiments: MC MS JS GS. Performed the experiments: MC MS. Analyzed the data: MC MS JS GS. Contributed reagents/materials/analysis tools: MC MS MB TG JS GS. Contributed to the writing of the manuscript: MC JS GS.

References

1. Sherma J (2002) Planar Chromatography. Anal Chem 74: 2653–2662.
2. Busch KL (1995) Mass spectrometric detectors for samples separated by planar electrophoresis. J Chromatogr, A 692: 275–290.
3. Wilson ID, Morden W (1996) Advances and applications in the use of HPTLC-MS-MS. J Planar Chromatogr – Mod TLC 9: 84–91.
4. Fuchs B, Süß R, Nimptsch A, Schiller J (2009) MALDI-TOF-MS Directly Combined with TLC: A Review of the Current State. Chromatographia 69: 95–105.
5. Cheng S-C, Huang M-Z, Shiea J (2011) Thin layer chromatography/mass spectrometry. J Chromatogr, A 1218: 2700–2711.
6. Pasilis SP, Van Berkel GJ (2010) Atmospheric pressure surface sampling/ionization techniques for direct coupling of planar separations with mass spectrometry. J Chromatogr, A 1217: 3955–3965.
7. Morlock G, Schwack W (2010) Coupling of planar chromatography to mass spectrometry. Trends Anal Chem 29: 1157–1171.
8. Ghosh P, Krishna Reddy MM, Ramteke VB, Sashidhar Rao B (2004) Analysis and quantitation of diazepam in cream biscuits by high-performance thin-layer chromatography and its confirmation by mass spectrometry. Anal Chim Acta 508: 31–35.
9. Chang TT, Lay JO, Francel RJ (1984) Direct analysis of thin-layer chromatography spots by fast atom bombardment mass spectrometry. Anal Chem 56: 109–111.
10. Tames F, Watson ID, Morden W, Wilson ID (1999) Detection and identification of morphine in urine extracts using thin-layer chromatography and tandem mass spectrometry. J Chromatogr B 729: 341–346.
11. Oriňák A, Arlinghaus HF, Vering G, Oriňáková R (2004) Modified chromatographic thin layer surface as an interface to couple thin layer chromatography with ToF-SIMS. Surf Interface Anal 36: 1122–1125.
12. Kubis AJ, Somayajula KV, Sharkey AG, Hercules DM (1989) Laser mass spectrometric analysis of compounds separated by thin-layer chromatography. Anal Chem 61: 2516–2523.
13. Gusev AI, Vasseur OJ, Proctor A, Sharkey AG, Hercules DM (1995) Imaging of thin-layer chromatograms using matrix-assisted laser desorption/ionization mass spectrometry. Anal Chem 67: 4565–4570.
14. Gusev AI, Proctor A, Rabinovich YI, Hercules DM (1995) Thin-Layer Chromatography Combined with Matrix-Assisted Laser Desorption/Ionization Mass Spectrometry. Anal Chem 67: 1805–1814.
15. Chen Y-C, Shiea J, Sunner J (1998) Thin-layer chromatography–mass spectrometry using activated carbon, surface-assisted laser desorption/ionization. J Chromatogr, A 826: 77–86.
16. Cheng S-C, Huang M-Z, Shiea J (2009) Thin-Layer Chromatography/Laser-Induced Acoustic Desorption/Electrospray Ionization Mass Spectrometry. Anal Chem 81: 9274–9281.
17. Van Berkel GJ, Sanchez AD, Quirke JME (2002) Thin-Layer Chromatography and Electrospray Mass Spectrometry Coupled Using a Surface Sampling Probe. Anal Chem 74: 6216–6223.
18. Hsu F-L, Chen C-H, Yuan C-H, Shiea J (2003) Interfaces To Connect Thin-Layer Chromatography with Electrospray Ionization Mass Spectrometry. Anal Chem 75: 2493–2498.
19. Chai W, Leteux C, Lawson AM, Stoll MS (2003) On-Line Overpressure Thin-Layer Chromatographic Separation and Electrospray Mass Spectrometric Detection of Glycolipids. Anal Chem 75: 118–125.
20. Lin S-Y, Huang M-Z, Chang H-C, Shiea J (2007) Using Electrospray-Assisted Laser Desorption/Ionization Mass Spectrometry To Characterize Organic Compounds Separated on Thin-Layer Chromatography Plates. Anal Chem 79: 8789–8795.
21. Van Berkel GJ, Ford MJ, Deibel MA (2005) Thin-Layer Chromatography and Mass Spectrometry Coupled Using Desorption Electrospray Ionization. Anal Chem 77: 1207–1215.
22. Peng S, Ahlmann N, Kunze K, Nigge W, Edler M, et al. (2004) Thin-layer chromatography combined with diode laser desorption/atmospheric pressure chemical ionization mass spectrometry. Rapid Commun Mass Spectrom 18: 1803–1808.
23. Salo P, Salomies H, Harju K, Ketola R, Kotiaho T, et al. (2005) Analysis of small molecules by ultra thin-layer chromatography-atmospheric pressure matrix-assisted laser desorption/ionization mass spectrometry. J Am Soc Mass Spectrom 16: 906–915.
24. Salo PK, Vilmunen S, Salomies H, Ketola RA, Kostiainen R (2007) Two-Dimensional Ultra-Thin-Layer Chromatography and Atmospheric Pressure Matrix-Assisted Laser Desorption/Ionization Mass Spectrometry in Bioanalysis. Anal Chem 79: 2101–2108.
25. Morlock G, Schwack W (2006) Determination of isopropylthioxanthone (ITX) in milk, yoghurt and fat by HPTLC-FLD, HPTLC-ESI/MS and HPTLC-DART/MS. Anal Bioanal Chem 385: 586–595.
26. Smith NJ, Domin MA, Scott LT (2008) HRMS Directly From TLC Slides. A Powerful Tool for Rapid Analysis of Organic Mixtures. Org Lett 10: 3493–3496.
27. Na N, Zhao M, Zhang S, Yang C, Zhang X (2007) Development of a dielectric barrier discharge ion source for ambient mass spectrometry. J Am Soc Mass Spectrom 18: 1859–1862.
28. Kogelschatz U (2003) Dielectric-Barrier Discharges: Their History, Discharge Physics, and Industrial Applications. Plasma Chem Plasma Process 23: 1–46.
29. Kogelschatz U (2002) Filamentary, patterned, and diffuse barrier discharges. IEEE Trans Plasma Sci 30: 1400–1408.
30. Harper JD, Charipar NA, Mulligan CC, Zhang X, Cooks RG, et al. (2008) Low-Temperature Plasma Probe for Ambient Desorption Ionization. Anal Chem 80: 9097–9104.
31. Gilbert-López B, Schilling M, Ahlmann N, Michels A, Hayen H, et al. (2013) Ambient Diode Laser Desorption Dielectric Barrier Discharge Ionization Mass Spectrometry of Nonvolatile Chemicals. Anal Chem 85: 3174–3182.
32. Hayen H, Michels A, Franzke J (2009) Dielectric Barrier Discharge Ionization for Liquid Chromatography/Mass Spectrometry. Anal Chem 81: 10239–10245.
33. Liu Z, Zhu Z, Zheng H, Hu S (2012) Plasma Jet Desorption Atomization-Atomic Fluorescence Spectrometry and Its Application to Mercury Speciation by Coupling with Thin Layer Chromatography. Anal Chem 84: 10170–10174.
34. Babij M, Gotszalk T, Kowalski ZW, Nitsch K, Silberring J, et al. (2013) Miniature plasma jet for mass spectrometry. pp.8902081–8902086.
35. Xu D, Deng X, Fang E, Zheng X, Zhou Y, et al. (2014) Determination of 23 phthalic acid esters in food by liquid chromatography tandem mass spectrometry. J Chromatogr, A 1324: 49–56.

22

Salt-Bridge Energetics in Halophilic Proteins

Arnab Nayek[1], Parth Sarthi Sen Gupta[1], Shyamashree Banerjee[1], Buddhadev Mondal[2], Amal K. Bandyopadhyay[1]*

1 The Department of Biotechnology, The University of Burdwan, Burdwan, West Bengal, India, 2 Department of Zoology, Burdwan Raj College, The University of Burdwan, Burdwan, West Bengal, India

Abstract

Halophilic proteins have greater abundance of acidic over basic and very low bulky hydrophobic residues. Classical electrostatic stabilization was suggested as the key determinant for halophilic adaptation of protein. However, contribution of specific electrostatic interactions (i.e. salt-bridges) to overall stability of halophilic proteins is yet to be understood. To understand this, we use Adaptive-Poison-Boltzmann-Solver Methods along with our home-built automation to workout net as well as associated component energy terms such as desolvation energy, bridge energy and background energy for 275 salt-bridges from 20 extremely halophilic proteins. We then perform extensive statistical analysis on general and energetic attributes on these salt-bridges. On average, 8 salt-bridges per 150 residues protein were observed which is almost twice than earlier report. Overall contributions of salt-bridges are -3.0 kcal mol^{-1}. Majority (78%) of salt-bridges in our dataset are stable and conserved in nature. Although, average contributions of component energy terms are equal, their individual details vary greatly from one another indicating their sensitivity to local micro-environment. Notably, 35% of salt-bridges in our database are buried and stable. Greater desolvation penalty of these buried salt-bridges are counteracted by stable network salt-bridges apart from favorable equal contributions of bridge and background terms. Recruitment of extensive network salt-bridges (46%) with a net contribution of -5.0 kcal mol^{-1} per salt-bridge, seems to be a halophilic design wherein favorable average contribution of background term (-10 kcal mol^{-1}) exceeds than that of bridge term (-7 kcal mol^{-1}). Interiors of proteins from halophiles are seen to possess relatively higher abundance of charge and polar side chains than that of mesophiles which seems to be satisfied by cooperative network salt-bridges. Overall, our theoretical analyses provide insight into halophilic signature in its specific electrostatic interactions which we hope would help in protein engineering and bioinformatics studies.

Editor: Eugene A. Permyakov, Russian Academy of Sciences, Institute for Biological Instrumentation, Russian Federation

Funding: No current funding sources for this study.

Competing Interests: The authors have declared that no competing interests exist.

* E-mail: akbanerjee@biotech.buruniv.ac.in

Introduction

The family halobactereaceae or halophiles are archaea that thrive in natural habitat of saturated brine [1] and pH optima in neutral range. Intracellular salt concentration is similar to that in the environment outside the cell. Thus, the entire protein machinery of halophiles is dependent on high salt concentration for function and stability [2,3]. In general, high concentration of salt is detrimental to mesophilic proteins. It enhances aggregation and collapse of 3D structure of proteins. It also interferes with electrostatic interactions due to charge screening and reduces natural hydration of proteins [1]. In contrast to its mesophilic counterpart, halophilic proteins maintain structural and functional integrity only in saturated salt solution, withdrawal of which causes gradual loss of tertiary structure. At low salt condition such unfolding is caused both by non-specific electrostatic and hydrophobic destabilization [1,4,5].

How then halophilic proteins remain stable in high salt environment? Genome, proteome-wide analyses as well as studies on specific halophilic proteins showed a number of compositional biases in their sequences. In general, higher abundance of acidic over basic residues, low content of bulky hydrophobic residues over less bulky ones in sequences of halophilic proteins are

observed [1,6–10]. In 3D structures, majority of these acidic residues are found specifically positioning on the surface of proteins which were proposed to facilitate excess protein hydration thereby making the surface less hydrophobic, more flexible and thus help to overcome deleterious effect of salt by promoting non-specific electrostatic interactions with salts in solution [10,11]. Detailed studies on malate dehydrogenase from *Haloarcula marismortui* confirmed interactions among surface acidic residues with hydrated salt ions which were argued to prevent aggregation [12] and also help to achieve functional state of protein. Similarly clusters of acidic residues are observed on the surface of atomic structure of dihydrofolate reductase, proliferating cell nuclear antigen (PCNA) from *Haloferax volcanii* [13,14] and glucose dehydrogenase from *H. mediterranei* [15]. Further, reduction of hydrophobic surface is achieved by another novel strategy by making surface of these proteins deficit of lysine residue [7,9,15].

Physical Chemistry and native state of halophilic proteins are made by conventional weak interactions of which non-specific classical electrostatic interactions have received the major focus [1,13,16,17]. Salt and pH dependent spectroscopic studies on halophilic ferredoxin from *Halobacterium salinarum* showed high salt not only contribute to classical electrostatic stability but also play role in solvent mediated stabilization [18]. It has been argued that classical electrostatic stability is saturated at a salt concentration of ~0.1M and thus high salt (4M) is needed for maintaining hydrophobic interactions [1,18]. Crystal structure analyses of Malate dehydrogenase from *Haloarcula marismortui* [19] showed greater number of salt-bridges than its mesophilic counterpart which enhanced enzyme stability at high salt concentrations. In pH dependent urea induced kinetic studies of the native state of ferredoxin from *Halobacterium salinarum* showed optimal kinetic stability at neutral pH which becomes unstable at either acidic or alkaline pH indicated presence of stabilizing salt-bridge in this protein [20].

Computational studies based on PBE, a pioneer theoretical analysis was carried out on crystal structure of ferredoxin and malate dehydrogenase from *Haloarcula maismortu* for understanding contribution of salt and pH on classical electrostatic stability. Further, in comparative analyses of halophilic and non-halophilic proteins, it has been demonstrated that the former gain stability with increment of salt concentration or decrease of pH in low salt [21].

Ion-pair or salt-bridge is one major contributor for the stability of proteins in general [22,23]. It is more so in case of proteins adapted in extreme of environment such as at high salt or temperature [19,20,24,25]. Either experimental or theoretical determination of interaction energy of salt-bridges shows that they could either be stabilizing [26–29] or destabilizing [30–32]. Energy of salt-bridge can be partitioned into three component terms such as columbic attraction of opposite charges, their desolvation and background interactions. The first term is always contributing and other two terms could either be contributing or costly. The favorable charge-charge attraction within a salt-bridge is often opposed by the unfavorable desolvation of charges and is further modulated by charge-dipole interactions as well as by the ionization behavior of nearby charge groups [31]. Thus the net energy of a given salt-bridge could either be stabilizing [26–29] or destabilizing [30–32] or insignificant [33] and the same are entertained both in experimental or theoretical scenario. As far as calculation of net energy of salt-bridge is concerned, computational procedure is advantageous over experimental Pka and double mutation cycle methods, in that separation of direct and indirect terms as well as pH and ionic strength variation are possible for a given set of parameters and structural model of proteins [29,31,34]. Poisson-Boltzmann Equation (PBE) is an ideal continuum electrostatic descriptor for bimolecular system and thus its solvers methods such as Delphi [24,25,28,29,31,35] and APBS [36] are most popularly used.

Here we present results of systematic and extensive analysis of involvement of acidic and basic residues in the formation of monomeric and network salt-bridges and their contribution to overall stability in the native state of proteins from a dataset of 20 high resolution (≤ 1.4 Å) crystal structures from halophilic domain possessing a total of 275 non-equivalent salt-bridges. Frequencies of these salt-bridges for each of six pairing partners (such as Arg-Asp, Arg-Glu, Lys-Asp, Lys-Glu, His-Asp and His-Glu) and their presence in secondary structures are also been worked out. We also report computation of different component energy terms using APBS methods for: (a) determination of net contribution of salt-bridges in halostability, (b) binary classification of sat-bridges into stable or unstable exposed or buried, isolated or networked and H-bonded or non-H-bonded categories and (c) establishing correlation of average accessibility of salt-bridges with these energy terms. Our present study shows details of salt-bridge energetic of halophilic proteins, knowledge of which has potential implication in comparative bioinformatics and protein engineering.

Results and Discussion

General characteristics of salt-bridges and its partners in halophilic proteins

Halophilic proteins are reported to possess excess of negative charges over basic residues that contribute to the overall stability by non-specific electrostatic interactions [1,3,7,9,10,12,14,21]. However, such electrostatic interactions, which are saturated at around 0.1M NaCl, was reported to make less contribution to the overall stability of halophilic proteins [1]. Again lower content of bulky hydrophobic residues [1,9] that are also present under low water activity situation in saturated brine solution [37], hydrophobic force seems to have lower contribution to halostability. Thus, arguably specific electrostatic interactions which are less affected by the presence of multimolar salts [22] seem to have major contribution to the stability of halophilic proteins. To the best of our knowledge, details of involvement and contributions of salt-bridges and their energetics using computational approach involving crystallographic structures are absent. We therefore used 20 high resolution halophilic protein structures to understand salt-bridge energetics using PBE solver methods [38].

Table 1 show acidic and basic residues extracted from 20 unique chains (i.e. A chain) of 20 crystal structures and their participation in salt-bridge formation. Following points are noteworthy from the table. Firstly, although normalized composition of acidic residue (19.76%) exceeds than that of basic residues (11.09%), lower fraction of the former (27.58%) than the later (40.3%) participates in salt-bridge formation. It is worth noting here that acidic residues participate both in non-specific electrostatic as well as salt-bridge stability. However, these unequal normalized frequencies of acidic and basic residues participating in salt-bridge formation also indicate presence of network salt-bridges. Secondly, among basic residues, Arg has higher abundance and also contribute greater fraction of it for salt-bridge formation. In turn Glu which has lower abundance than Asp, contributes higher fraction in salt-bridge formation. The preferences of basic and acidic residues for salt-bridge formation as seen above might have relation with their side chain structure, length, relation to the stability of secondary structures and accessibility.

To check salt bridging partner's positional distribution in protein sequences, we have made grouped frequency distribution plot of intervening residue distances for a class interval of 5 and the same is plotted in figure 1. In this plot we have consider only first 100 residues for each of 20 proteins. Thus, the grouped frequency for each class interval represents their value for the overlapping region of 20 protein sequences. The plot shows that salt-bridging acidic and basic residues when present closer in sequence tend to form more number of salt-bridges. The frequency decreases with increase in intervening residues between the bridging partners. Our observation in context of halophilic proteins obtains the similar pattern as was entertained earlier in the context of mesophilic proteins [29]. This kind of coding pattern in relation to specific electrostatic interactions which is seen in halophilic (present study) and mesophilic cases [29] are reminiscent of hierarchal protein folding [39,40].

In general, protein structure is more conserved than their primary sequences. Secondary structure [41] which determines protein topology is thus more conserved in evolution. Earlier studies with salt-bridges showed that major fraction of these residues are distributed in secondary structure region [29]. Are salt-bridges conserved in halophiles? The question is justified as halophilic proteins to adapt in extreme of salt solution, have to pass through critical transition of evolution in relation to salt with reference to its mesophilic counterparts. In this context we were

Table 1. Absolute and normalized frequency of total and salt-bridge forming acidic and basic residues as obtained from 20 halophilic proteins (see Materials and Methods).

Residues/class	f_t^d	$f_t^d\,in\%$	f_{sb}^d	$f_{sb}^d\,in\%$	$f_{sb}^R\,in\%$
Arg	279	5.61	127	2.55	45.45
Lys	144	2.89	57	1.14	39.44
His	129	2.59	39	0.78	30.11
Basic	552	11.09	223	4.47	**40.30**
Asp	501	10.07	126	2.53	25.12
Glu	482	9.69	145	2.92	30.13
Acidic	983	19.76	271	5.45	**27.58**
Total	1535	30.85	494	9.92	32.16

f_t^d: Residues absolute frequency in 20 protein chains in the database. $f_t^d\,in\%$: Percent frequency w.r.t. total residues (4975). f_{sb}^d: Residues salt-bridge absolute frequency. $f_{sb}^d\,in\%$: Percent salt-bridge frequency w.r.t total residues. $f_{sb}^R\,in\%$: Percent salt-bridge frequency w.r.t. corresponding salt-bridge forming residues. Total residues in 20 unique chains 4975.

interested to observe the distribution of salt-bridging candidates in secondary structures. Main chain dihedral angles of salt-bridge partners were extracted from energy minimized structures for all 20 halophilic proteins and plotted in figure 2. The plot shows the distribution of dihedral angles (φ,ψ) for salt-bridge partners occupying mostly the defined right handed α-helix (Rα), beeta sheet (β) regions. About 78% of these residues fall in these conserved regions of Ramachandran plot. Such secondary structure specific distribution of salt-bridges not only indicates that they are conserved in halophilic evolution but also highlights their specific nature of interactions.

Table 2 shows frequency of all possible salt-bridge pairs such as Lys-Asp, Lys-Glu, Arg-Asp, Arg-Glu, His-Asp and His-Glu. Arg has highest pairing frequencies with both the acidic residues than Lys and His. Again, in all cases of Asp and Glu, the later is used

more in number for formation of salt-bridges by basic residues. As far as distribution of Arg mediated salt-bridges in the core and on the surface is concerned, this residue is favored over Lys. The difference in the proportions of Glu and Asp that are buried in the protein interior is not so large. These facts are in line with earlier observations [40].

Observations of all 275 salt-bridges of halophilic proteins in our dataset

Salt-bridge is specific electrostatic interaction that contributes to the overall stability of native state of proteins. The fact that halophilic proteins are devoid of bulky hydrophobic residues [1] and non-specific electrostatic interactions cause marginal stability; specific electrostatic interactions which are less affected by the presence of multi molar salt concentrations [22] are expected to

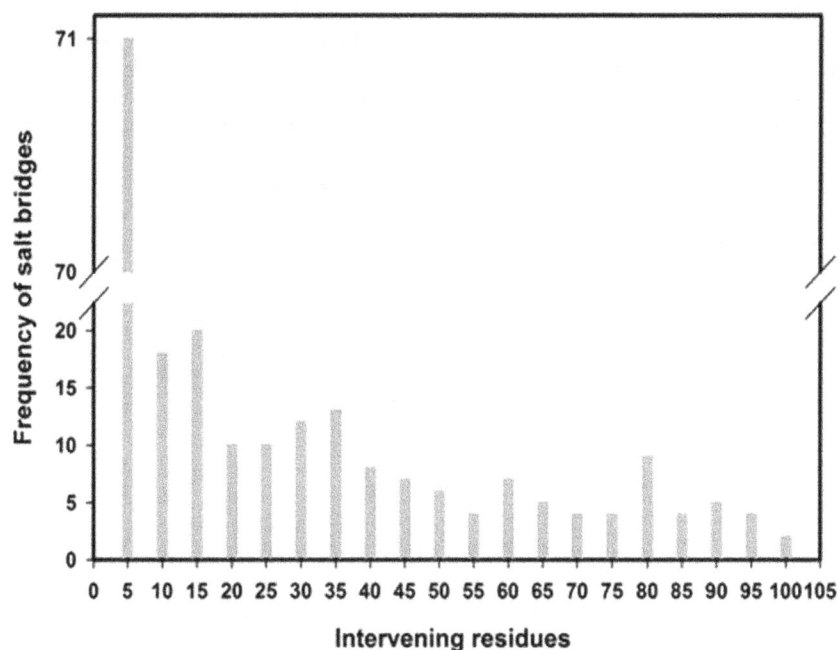

Figure 1. Histogram showing the frequency of salt-bridges against intervening residues for first 100 residues from N-terminal region of 20 halophilic proteins. This length is common for all proteins in our database.

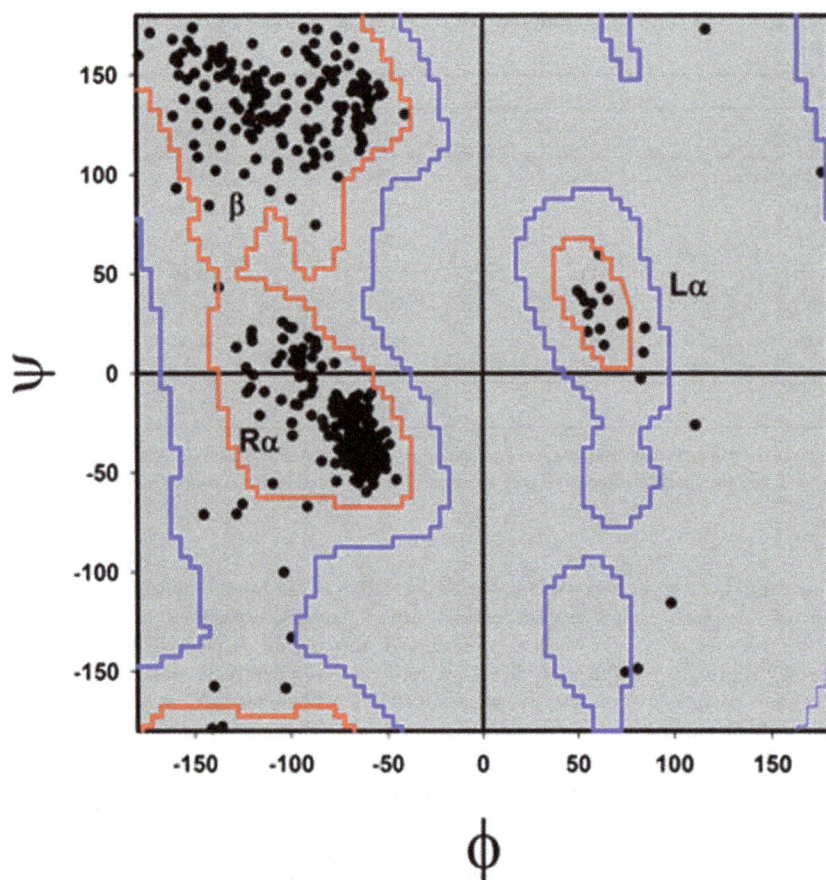

Figure 2. Ramachandran plot of salt-bridge forming residues (275×2). In this plot core region is outlined in blue and allowed region in red. These boundary values are taken from Lovell et. al., 2003 [52].

make effective contributions to the stabilization. We use APBS methods along with our in-house automation for calculation of all three associated energy terms (such as $\mathbf{\Delta\Delta G_{dslv}}$: desolvation, $\mathbf{\Delta\Delta G_{brd}}$: bridge and $\mathbf{\Delta\Delta G_{prt}}$: background) for finding net salt-bridge energy ($\mathbf{\Delta\Delta G_{tot}}$ i.e. sum of the above three component terms). $\mathbf{\Delta\Delta G_{dslv}}$ and $\mathbf{\Delta\Delta G_{prt}}$ are indirect interaction terms of which the former is an unfavorable term that originates due to desolvation of charges during folding and the later is due to interaction of charges could either be favorable or unfavorable. $\mathbf{\Delta\Delta G_{brd}}$ is a direct term that always causes favorable contribution due to interactions of charges in the folded state [29].

Heterogeneity in databases might affect statistical generality [42], we therefore involved homogenous dataset of 275 salt-bridges obtained strictly from extremely halophilic proteins. Our observations suggest that halophilic proteins utilize more of specific electrostatic interactions than mesophilic ones (that includes prokaryotic and eukaryotic proteins) studied earlier [29] in that in the former 275 salt-bridges from 20 proteins making on average 13.8 salt-bridges per protein and in the later 222 salt-bridges from 36 proteins making on average 6.2 salt-bridges per protein.

Net salt-bridge energy, on average, is -3.0 (± 4.0) kcal mol^{-1} which is contributed almost equally by bridge ($\mathbf{\Delta\Delta G_{brd}} = -6.9 \pm 4.0$ kcal mol^{-1}) and background ($\mathbf{\Delta\Delta G_{prt}} = -6.7 \pm 6.0$ kcal mol^{-1}) energy terms and unfavored by desolvation ($\mathbf{\Delta\Delta G_{dslv}} = 10.6 \pm 6$ kcal mol^{-1}) term (Table 3). The average equal favorable contribution of both bridge and background energy terms is contrasted by earlier observations that, on average, background term ($\mathbf{\Delta\Delta G_{prt}} = -3.9 \pm 4.0$ kcal mol^{-1}) contributes only about half the value of bridge term ($\mathbf{\Delta\Delta G_{brd}} = -6.3 \pm 4.0$ kcal mol^{-1}) [29]. Overall, desolvation cost is overbalanced by the sum of bridge and background terms.

To obtain more realistic view on energy contribution of 275 candidate salt brides, distribution of various energy terms are shown in the Figure 3 (A through E). Almost symmetric distribution of $\mathbf{\Delta\Delta G_{tot}}$ term containing both stable and unstable salt-bridges, majority of which are falling in the stabilizing zone ($\mathbf{\Delta\Delta G_{tot}} < 0$) with a maximum near zero from the negative side is observed (figure 3A). Quantitatively 78% (213 of 275) of candidate salt-bridges in our dataset contribute to stability and rest 22% (62 of 275) cause destabilization. Profiles of component energy terms shows that the distribution of bridging (Figure 3C) and desolvation

Table 2. Frequency salt-bridges formed by each of six possible pairs.

Residues	Asp	Glu
Lys	28	38
Arg	82	84
His	18	25

Table 3. Average energy terms in various salt-bridge categories.

Salt-bridge class	Energy terms in Kcal per Mol.			
	$\Delta\Delta G_{dslv}$	$\Delta\Delta G_{brd}$	$\Delta\Delta G_{prt}$	$\Delta\Delta G_{tot}$
1 All	+10.57±5.50	−6.88±4.10	−6.66±5.74	−2.96±4.06
2 Stable	+10.91±5.36	−7.54±3.92	−7.72±5.85	−4.35±3.47
Unstable	+9.41±5.81	−4.60±3.87	−2.99±3.34	+1.82±1.50
3 Buried	+14.87±6.07	−9.14±4.69	−9.85±6.46	−4.13±4.39
Exposed	+8.34±3.51	−5.70±3.17	−5.00±4.50	−2.35±3.74
4 Isolated	+9.22±5.33	−6.66±4.13	−3.84±4.06	−1.28±2.91
Networked	+12.20±5.27	−7.14±4.05	−10.3±5.63	−4.97±4.32
5 H-bonded	+11.24±5.32	−7.82±3.83	−6.79±5.66	−3.38±4.00
No H-bonds	+7.47±5.24	−2.50±1.81	−6.03±6.02	−1.04±3.74

All: whole dataset of 275 salt-bridges; Stable: 213 salt-bridges with $\Delta\Delta G_{tot}$<0 Kcal/mol; Unstable: 62 salt-bridges with $\Delta\Delta G_{tot}$>0 Kcal/mol; Buried: 94 salt-bridges with average ASA of ≤20%; Exposed: 181 salt-bridges with average ASA of >20%; Networked: 125 salt-bridges that participate in salt-bridge networks; Isolated: 150 salt-bridges that do not form part of salt-bridge networks; H-bonded: 226 salt-bridges containing at least one hydrogen bond between side-chain charged groups; and Non H-bonded: 49 salt-bridges that do not contain any hydrogen bond between their side-chain charged groups.

(Figure 3B) terms occupying stabilizing and destabilizing zone respectively while the background term (Figure 3D) possesses both stabilizing (major fraction) and destabilizing population of salt-bridges. This would mean that both bridge and background term act favorably and thus overcome unfavorable desolvation penalty to make net salt-bridge energy stabilizing. The observation that the background term i.e. $\Delta\Delta G_{prt}$, which originates due to charge-dipole interactions, contains both stabilizing (major fractions) and destabilizing populations indicate it is most sensitive to protein microenvironment among all component terms. An inspection of these individual terms shows that only 4% (11 of 275) are destabilizing with an average destabilization of 0.7±0.8 kcal-mol^{-1} which is much less (11.3% with an average destabilization of 0.9 kcal mol^{-1}) than earlier observations [29].

This fact implore us to make closer look on 62 destable salt-bridges in our dataset (see above) for all three component terms for their nature of contribution (favorable or unfavorable) to net stability. In this population both bridge and background terms are still negative, except the above 11 background terms which are positive. However, the magnitude of the negative values are far weaker than stable cases (213 of 275) such that their collaborative effect could not compensate the desolvation cost, indicating both these two terms are responsive to protein environment. In comparison to bridge term, background term seems to be more sensitive to protein microenvironment as some of these values (11 individuals of 62 destable cases) are positive. However our observation (mentioned above), unlike earlier one [29], that overall contributions of these two energy terms to net stability are seen to be almost equal (Table 3). As far as protein environment is concerned, halophilic proteins possess unique design in that in sequences polar as well as negatively charged residues dominate over bulky hydrophobic residues [1] and in structures majority of these residues are distributed on the surface [37]. In halophilic proteins, lower content of bulky hydrophobic residues might indicate mandatory more polar protein interior than their mesophilic counterparts. Thus the context and composition of such excessive charges and dipoles constitute local microenviron-

ment around individual salt-bridge partner which determine the magnitude of their contribution to net stability. In our observations 107 instances of 275 salt-bridges where favorable contribution of background term exceed than that of bridge term for overcoming desolvation penalty and thus making salt-bridges stabilizing: an unique phenomena for halophilic proteins than mesophilic ones [29].

To check relatedness among different energy terms with average accessible surface area (ASA$_{av}$), correlation plots are presented in figure 4 (A through F) for all 275 candidate salt-bridges similar to earlier studies [29,31]. $\Delta\Delta G_{dslv}$, which is energy change due to desolvation of salt-bridge partners due to folding, is linearly and negatively correlated with ASA$_{av}$ (Figure 4 B).

The linear best fit of log of $\Delta\Delta G_{dslv}$ against ASA$_{av}$ could be expressed by the equation:

$$\log(\Delta\Delta G_{dslv}) = -0.01 ASA_{av} + 1.3$$

The fitted line has correlation coefficient 0.71 and RMSD 0.41 (INSET of Figure 4B). The linear correlation between $\Delta\Delta G_{dslv}$ and ASA$_{av}$ with negative slope indicate transfer of salt-bridges into the core of protein involve greater desolvation penalty.

The dependence of observed data of $\Delta\Delta G_{brd}$ (bridge term), $\Delta\Delta G_{prt}$ (background term) and $\Delta\Delta G_{tot}$ (net energy term) on ASA$_{av}$ are shown in Figure 4: C, D and A respectively. Figure 4F shows such dependence by fitted lines only (without the observed data) for $\Delta\Delta G_{brd}$, $\Delta\Delta G_{prt}$, $\Delta\Delta G_{tot}$ and $\Delta\Delta G_{dslv}$ (INSET of Figure 4F). From the figure, it is apparent that $\Delta\Delta G_{dslv}$ is negatively correlated (correlation coefficient r = 0.71) whereas both bridge and background terms are positively correlated (r = 0.51 and r = 0.48 respectively) with ASA$_{av}$. Again, the curve for background term (blue line in Figure 4F) is seen to cross the bridge term (red line) around 22% of ASA$_{av}$, indicating its stronger effect upon burial. However, such gain in energy upon burial of salt-bridges which is seen in both background and bridge terms were argued to be due to local environmental effect caused by linear gradient of dielectric constant and charge screening on protein surface (dielectric constant = 80) and in the interior (dielectric constant = 4) [29]. The fact of lower linear dependence (low correlation coefficient as above) of these two energy terms on ASA$_{av}$; stronger background than bridge term upon burial ($\Delta\Delta G$ more negative of ASA$_{av}$<20%) and more importantly large spread of individual salt-bridge energy of these terms against ASA$_{av}$ (Figure 4 C and D) might indicate involvement of additional non-linear factors. As mentioned above, local microenvironment of a given salt-bridge could additionally and differentially be modulated by the presence of permanent dipoles (presence of peptide bond, helix, and polar groups) and non-specific side chain charges at neutral pH (not participating in salt-bridge) apart from the above uniform effect of dielectric constant mediated charge screening. In these aspects, the dependence of net energy term on ASA$_{av}$ is very low (the correlation coefficient 0.07 pink lines Figure 4F) might be due to combined effect of its component terms.

Stabilizing and destabilizing salt-bridges

Our database contains 275 salt-bridges obtained from 20 halophilic enzymes and proteins, making it suitable for understanding salt-bridge energetics. In table 3, salt-bridge energy classes (row: 1 to 5) are presented as buried or exposed (table 4) and stable or unstable (table 5) formats. In our dataset a total of 213 (78%) out of 275 salt-bridges possess $\Delta\Delta G_{tot}$<0, hence are stable. This observation, like Kumar and Nussinov (1999) and unlike Hendsch and Tidor (1994), shows majority of salt-bridges in

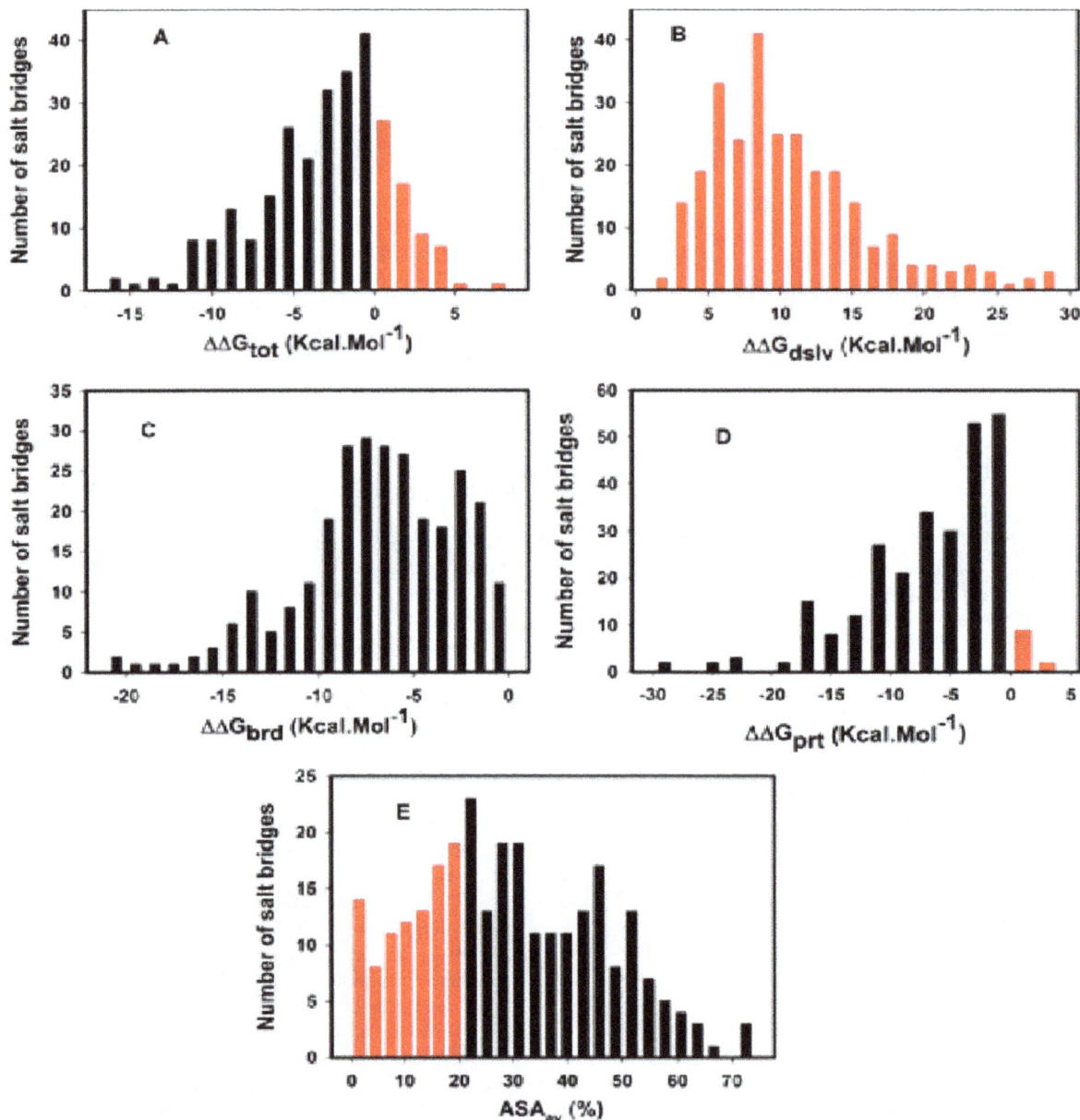

Figure 3. Histogram showing distribution of 275 salt-bridges from 20 halophilic proteins extracted using SBION [53] as a function of $\Delta\Delta G_{tot}$ Kcal Mol^{-1} (A), $\Delta\Delta G_{dslv}$ Kcal Mol^{-1} (B), $\Delta\Delta G_{brd}$ Kcal Mol^{-1} (C), $\Delta\Delta G_{prt}$ Kcal Mol^{-1} (D) and ASA$_{av}$ (%) (E). Figure (A) through (D), black bar indicates stabilizing (i.e. $\Delta\Delta G_{tot}<0$) and red bar indicates destabilizing (i.e. $\Delta\Delta G_{tot}>0$) salt-bridges. Figure (E) red bar (ASA$_{av}$%≤20) indicates salt-bridges present in the core.

our data set are stabilizing. The rest 62 (22%) salt-bridges are destabilizing type (Table 5, row 4). The instability in this population of salt-bridges are stemmed from the weaker contributions of background (which is weak by about 4.8 kcal mol^{-1}) and bridge energy terms (weak by about 3 kcal mol; Table 3) than that of stable cases. Our observation is in line with earlier studies involving mesophilic proteins [29].

How the buried and exposed salt-bridges contribute to the stability of halophilic proteins? Table 5 (row 3) shows out of 213 (78% of total 275) stable salt-bridges, 36% (76 of 213) are buried and remaining 64% are exposed. Similarly in unstable salt-bridges, 29% (18 of 62) are buried and remaining 71% are exposed. Interestingly, buried stabilizing population of salt-bridges is higher in our observations than observed in earlier study (present 36%;

earlier 29%) by Kumar and Nussinov (1999). Nevertheless, overall distributions of buried and exposed salt-bridges under stable and unstable categories remain the same as earlier [43]. With respect to total buried (94 in Table 4), stable buried salt-bridges constitutes 81% (76 in 94) and remaining 19% are destabilizing type. At this point it is worth noting that only a little fraction of total destabilizing salt-bridges (18 of 62) are buried. In other words, of all buried salt-bridges (94 in table 4) very little fractions are destabilizing type (18 of 94). Statistical analyses of 275 salt-bridges by APBS methods (present studies) and 222 salt-bridges from mesophilic proteins by DELPHI software packages [29] show identical observation that majority of buried salt-bridges are stabilizing; are not in parallel with studies that involve selected salt-bridges and model systems [31,44]. In this connection the contrast

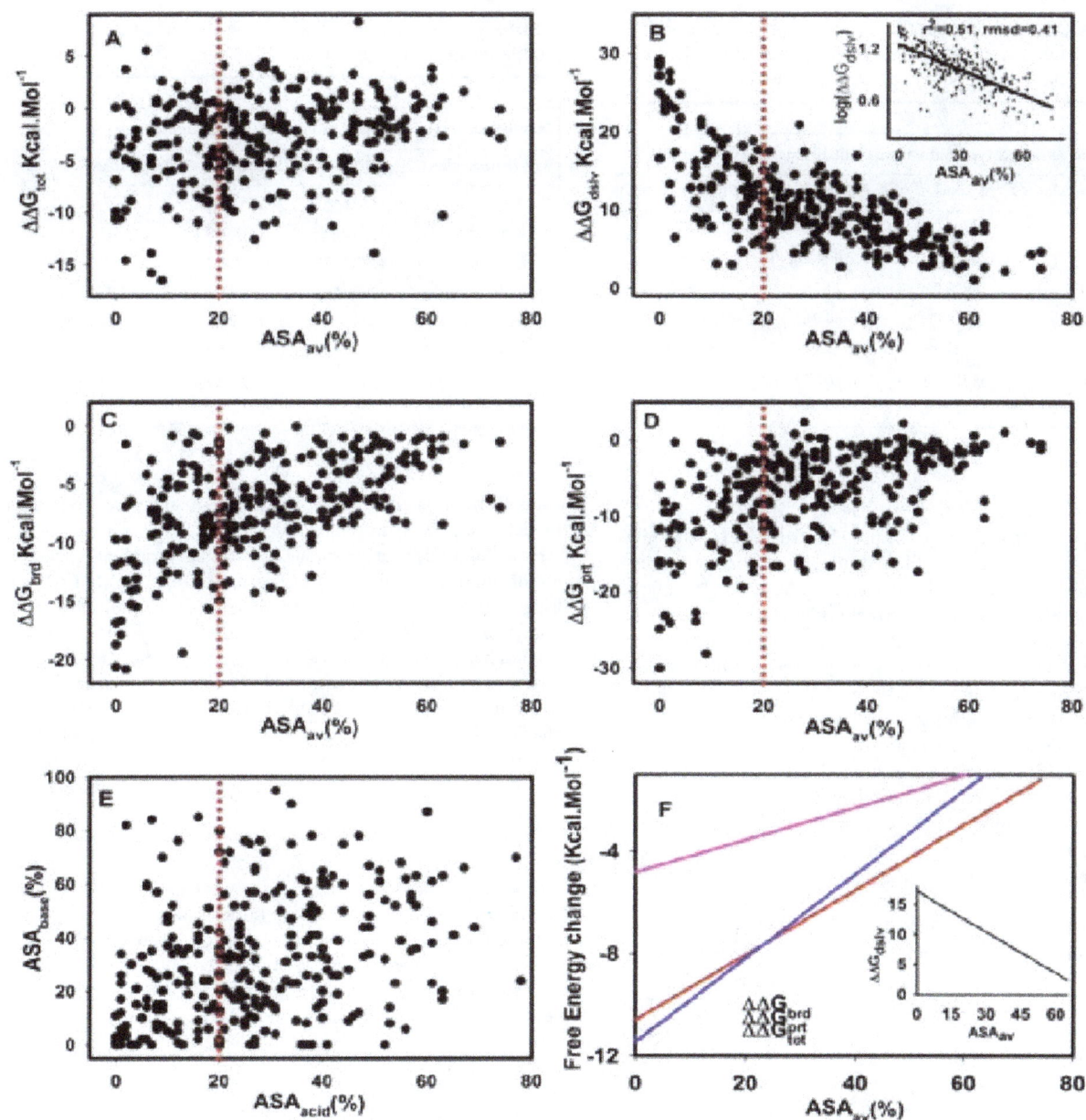

Figure 4. Analyses of correlation of ASA$_{av}$ (%) (X-axis) with ΔΔG$_{tot}$ (A), ΔΔG$_{dslv}$(B), ΔΔG$_{brd}$ (C) and ΔΔG$_{prt}$ (D) for 275 salt-bridges from 20 halophilic proteins are presented. Correlation of accessibilities of individual salt-bridge partners is shown in E. Linear fit of logarithm of ΔΔG$_{dslv}$ and ASAav (%) along with fitting parameters (correlation co-efficient and RMSD) are shown in the INSET of B. Plot F shows fitted lines for correlation of ΔΔGprt (blue), ΔΔGbrd (red) and ΔΔGtot (pink) with ASAav (%) along with that for ΔΔGdslv (in INSET of F).

was argued to be due to the use of selected dataset and inclusion of ion-pairs with centroid distances greater than 4 Å [29].

Desolvation cost which was considered to be the sole factor for making buried salt-bridges always unstable [31,44] may not be the only criteria for salt-bridge instability. In our study, a total of 44 out of 62 (Table 5) unstable salt-bridges, a greater proportion than burial ones (18 of 62), are seen to be exposed. How come destable salt-bridge populations are more under exposed than buried condition? Greater solvent mediated charge screening effect on salt-bridge charges (that affects both bridge and background terms) is apparent on the protein surface than in the solvent sequestered protein interior [29]. Further under halophilic situation, protein surface possess higher abundance of negative over positive

charges, polar uncharged side chains (i.e. dipoles) than protein interior and thus unfavorable electrostatic interactions between biased proximity of these charges with surface salt-bridges (affect **ΔΔG$_{brd}$** term); unfavorable charge-dipole interactions with surface salt-bridges (affect **ΔΔG$_{prt}$** term) might cause reduction of favorable contributions of bridge and background terms. Additional factor that might contribute to buried salt-bridges being more stable is network salt-bridges (see below). In contrast to mesophilic situation [29], our dataset shows 50 of 76 total stable buried salt-bridges are network type (table 4).

Table 4. Distribution of various salt-bridge classes under buried and exposed categories. Number outside indicate absolute count and percentage in parentheses.

Buried and exposed types in various salt bridge classes

Salt bridge classes	Buried	Exposed	Total
1 All	94 (34.2%)	181 (65.8%)	275 (100%)
2 Stable	76 (35.7%)	137 (64.3%)	213(77.5%)
Unstable	18 (29.0%)	44 (71.0%)	62(22.5%)
3 Networked	50 (40.0%)	75 (60.0%)	125(45.5%)
Isolated	44 (29.3%)	106 (70.7%)	150(54.5%)
4 H-bonded	82 (36.3%)	144 (63.7%)	226(82.2%)
No H-bonds	12 (24.5%)	37 (79.5%)	49(17.8%)

Table 5. Distribution of various salt-bridge classes under stable and unstable categories.

Stable and unstable types in various salt bridge classes

Salt bridge class	Stable	Unstable	Total
1 All	213 (77.5%)	62 (22.5%)	275 (100%)
2 Buried	76 (80.9%)	18 (19.1%)	94(34.2%)
Exposed	137 (75.7%)	44 (24.3%)	181(65.8%)
3 Networked	106 (84.8%)	19(15.2%)	125(45.5%)
Isolated	107 (71.3%)	43 (28.7%)	150(54.5%)
4 H-bonded	186 (82.3%)	40 (17.7%)	226(82.2%)
No H-bonds	27 (55.1%)	22 (44.9%)	49(17.8%)

Number outside indicate absolute count and percentage in parentheses.

Buried and exposed salt-bridges

The location of salt-bridges in protein structure is determined by average accessibility [45,46] of candidate salt-bridge partners, using a probe radius of 1.4 Å. An average ASA of $\leq 20\%$ indicates the salt-bridge is in the core of protein otherwise exposed.

As far as contribution of buried salt-bridges is concerned, overall experimental observations showed that buried salt-bridges could be stabilizing, indifferent or mostly destabilizing [27,32,33,47,48]. Theoretical studies on buried salt-bridge also were shown to contribute little to protein stability. Thus there arises an apparent conflict about the contribution of buried salt-bridges: a point we consider below based on our and others observations [29,43].

Computational analyses on salt-bridges in our dataset were classified into buried and exposed categories (table 4). 94 (34%) out of 275 salt-bridges are buried and rest 181 (66%) are exposed on proteins surface (Table 4). The unfavorable desolvation cost of these buried salt-bridges is indeed very high (i.e. 14.9 kcal mol^{-1}) which is outweighed by joint effect of favorable bridge and background energy terms (Table 3 row 3). Interestingly, the favorable contribution of these two terms under buried condition exceed by about 4 kcal mol^{-1} than that under exposed condition making buried salt-bridges, on average, more stable. The overall stability of buried salt-bridges is seen to be -4.1 ± 4.4 kcal mol^{-1} and that for exposed ones is -2.4 ± 3.7 kcal mol^{-1} (Table 3 row 6). A similar observation was obtained in theoretical analyses involving 222 salt-bridges from 36 monomeric proteins [29].

The apparent conflict of buried salt-bridges with higher average net stability in our dataset that involve large population of candidate salt-bridges could get reasonable resolution if we consider the case on per protein basis. In studies with 38 protein structures, it was concluded that, on average, there exist 5 salt-bridges per 150 residues protein of which only 1 is under buried condition [43]. Similarly studies based on 36 proteins, it was shown that a total of 4 (222*150/9271) salt-bridges per 150 residues protein of which only 1 (55*150/9271) is under buried condition. Applying the above scale, our study with 20 halophilic proteins shows that a total of 8 (275*150/4975) salt-bridges per 150 residues protein with 3 (94*150/4975) under buried condition. It is thus apparent that buried salt-bridges on per protein basis are rare [43] in general. At this juncture it is worth noting that under halophilic condition, specific electrostatic make dominant contribution (78% and stable type) in that more number of salt-bridges and grater fraction of it is present in the protein interior than mesophilic ones. Such additional contribution of salt-bridges

seems to be important under halophilic situation to compensate the deficit of low hydrophobic interactions [1].

However, unlike earlier studies [29] not all buried salt-bridges in our database are stable type. Out of 94, 76 (81%) buried salt-bridges are stable and that of 18 (19%) are unstable (table 4). In other words, on average, out of 3 buried salt-bridges in a 150 residues halophilic protein 1 is unstable and this figure is expected to be much narrower under mesophilic situation (see above). Thus it is apparent that finding stable buried salt-bridges depends on selection of i) type of protein and ii) candidate buried salt-bridge in that protein. Again, selection of buried salt-bridges is mostly limited due to the fact that it occurs in very low frequency in proteins in general (see above). Like earlier ones [29,43], our study on large database shows that obtaining stable buried salt-bridge is purely context dependent.

How the buried salt-bridges overcome desolvation penalty and gain higher net stability? Although buried salt-bridges suffers from large desolvation penalty, entropic cost of localizing salt-bridges partners in the protein interior is minimized [49] at the same time. Hence buried salt-bridges are enthalpically favored. Again, in our database out of 94 buried salt-bridges, 50 form networked (Table 4). Under mesophilic condition, out of 66 buried salt-bridges only 6 are networked [29] indicating extensive network salt-bridges under halophilic situation. The fact that network salt-bridges contribute greater net (Table 3) stability (-5 kcal mol^{-1}) and are mostly stable (85%; Table 5); halophilic situations seems to utilize the extensive networking to overcome desolvation penalty and thus to gain extra average net stability with buried salt-bridges. Apart from the above effects other important factor that might acts favorably, already noted above in the context of 275 salt-bridges, is the protein local microenvironments attributed with differential distribution of charges and dipoles in the vicinity of salt-bridges might modulate bridging and background terms to a varying degree. Overall the observations of highly stable buried salt-bridges indicate favorable contributions of these above factors.

Isolated and networked salt-bridges in halophilic proteins

A salt-bridge between two oppositely charged residues is considered to be networked if at least one of these charged residues forms additional salt-bridge(s) with the other one(s). Otherwise, the salt-bridge is considered to be isolated. In our dataset, 125 of 275 (~46%) salt-bridges are network type of which 40 triads, 12 tetrads, 1 pentad and 1 hexad are observed. The remaining 150 salt-bridges are isolated.

Salt-bridges mainly confer stability to tertiary structure of proteins [43]. However, their contribution to stability depends on their location, geometry and interactions of bridge partner's vicinity with other side chains in proteins [50]. While isolated salt-bridges provide marginal stabilization, network salt-bridges cause cooperative stabilization. In comparison to mesophilic proteins, in halophilic proteins higher level of network salt-bridges are observed. Halophiles like thermophiles are extremophiles. Our database shows 46% salt-bridges forming network and rest 54% are isolated type (Table 5). Such extensive network salt-bridges in halophilic proteins are not available with its mesophilic counterparts which form only 8% network salt-bridges [29] indicating extremophilic design in halophiles. Table 4 shows 50 (40%) of 125 network salt-bridges are buried which is greater than even isolated ones (29%). As far as stability is concerned, unlike mesophiles [29], 85% of network salt-bridges (106 out of 125, table 5) are stabilizing type and rest 15% is marginally unstable. In isolated case (Table 5), stable population constitutes 71% and that for unstable is 29%. Overall the net stability of network salt-bridges is about 4 times (Table 3) than that of isolated salt-bridges (network -4.97 kcal $\cdot$ mol^{-1} and Isolated -1.28 kcal mol^{-1}). Greater abundance and cooperative nature of network salt-bridges [51] seems to have greater significance under halophilic context. Sequence of halophilic proteins contains extra negative charges [1,9] and in structures these residues are largely present on protein surface [37]. However, a sizable fraction of these excess charges are also present in the protein core. Network and isolated salt-bridge formation in protein core would satisfy these solvent sequestered charge residues. The fact that network salt-bridges contribute to higher stability and promote cooperative interactions among charges, halophilic design seems to utilize advantage of these interactions especially in their protein core such that obligatory presences of extra charges are satisfied. However, unlike earlier observation [43], the presence of destable population of network salt-bridges (15%) is not fully clear (table 5 row 4). A closer look on this population shows that lion's share of desolvation cost is balanced by bridge and background terms while making $\Delta\Delta G_{tot}$ slightly positive. This phenomenon might indicate a trend towards maintenance of local flexibility a prerequisite for functionality, rather than stability and hence rigidity, is more critical.

Hydrogen bonded and non-hydrogen bonded salt-bridges in halophilic protein

Salt-bridges in the dataset are also characterized based on their association with hydrogen atom. A hydrogen-bonded salt-bridge is identified by the presence of at least one pair of side-chain charged group atoms, with opposite partial charges, within a 3.5 A [29] distance. 226 out of 275 (82%, table 5) salt-bridges in our dataset contain at least one side-chain to side-chain hydrogen bond.

A total of 226 out of 275 (82%) salt-bridges in our dataset contain at least one hydrogen bond between bridging partner side-chains. The remaining 49 salt-bridges are devoid of such bond (Table 5). On average, the hydrogen bonded salt-bridges are more stabilizing than non-hydrogen bonded salt-bridges in that net energy gains ($\Delta\Delta G_{tot}$) for the former is -3 kcal mol^{-1} that for the later is -1.0 kcal mol^{-1} (table 3). The greater stability of H-bonded population of salt-bridges indicates excess electrostatic stabilization on salt-bridge self one. Similar net stabilizing effects are also expected due to salt-bridge proximity to dipoles or charges with modulation of component energy terms to some varying degrees. Thus, hydrogen bonded salt-bridges (as well as charges and dipole bonded ones) provides direct evidence for salt-bridges sensitivity to its local environment. However, hydrogen bonded salt-bridges have stronger bridge energy term than that of non-

hydrogen bonded ones. The average $\Delta\Delta G_{brd}$ for the hydrogen bonded salt-bridges is -8 kcal mol^{-1}, while that for the non-hydrogen bonded salt-bridges is -3 kcal mol^{-1} (Table 3). Like network salt-bridges, H-bonded salt-bridges are distributed both in the core and on the surface of proteins. A total of 82 out of 226 (36%) of the H-bonded salt-bridges are buried and 144 (64%) are exposed (table 4). As far as stability is concerned, 82% (186 out of 226) H-bonded salt-bridges are stabilizing and remaining 18% are destabilizing. Overall, formation of halophilic protein interior which is lack of bulky hydrophobic residues are contributed favorably both by network salt-bridge formation (see above) and hydrogen bonded salt-bridges.

Conclusion

Our study with 275 salt-bridges from 20 halophilic proteins shows that about 80% of salt-bridges are conserved and contribute to halostability. In halophilic proteins, net salt-bridge energy is -3.0 kcal mol^{-1} and hence stable which is favored almost equally by bridge and background energy terms and disfavored by desolvation term. Both the former component terms are affected by factors: 1) dielectric constant of the medium and 2) local microenvironment of charges and dipoles in the vicinity of salt-bridges. While the former cause uniform modulation based on position of salt-bridge from the surface to the core; the later vary with context and composition of protein hence non-linear. Comparison between bridge and background terms for their energy contribution to 275 salt-bridges, the former contains 213 highly stable and 62 less stable energy populations, the later include 213 highly stable, 51 less stable and 11 unstable energy populations indicating the background term is more sensitive to protein local microenvironment. Specific polar nature of halophilic proteins over mesophilic ones seems to contribute to both background and bridge energy terms to overcome desolvation penalty and making net salt-bridge energy favorable. Halophiles recruits, on average, 8 salt-bridges per 150 residues protein which is almost double than that of mesophilic proteins. The fact that halophilic proteins have lower content of bulky hydrophobic residues and under saturated salt solution hydrophobic interactions but not the specific electrostatic ones are severely affected due to low water activity situation, additional salt-bridges compensates the deficit of hydrophobic force. These proteins possess higher proportion of buried salt-bridges than mesophiles and four-fifth of which are stable. Extensive networked salt-bridges have been another attribute of halophilic proteins of which 2/5 are found under buried condition. Halophilic protein interior is relatively more polar than their mesophilic counterpart, the cooperative networked salt-bridges in this protein core are crucial not only to overcome greater desolvation cost and thus making buried salt-bridges stable but also to satisfy isolated charges. Similar to networked salt-bridges hydrogen bonded salt-bridges also contribute to halostability. Under halophilic situation formation of protein core with relatively more polar residues, buried networked and hydrogen bonded salt-bridges play crucial role.

Materials and Methods

Dataset

We obtain 275 non-equivalent salt-bridges from 20 halophilic proteins following the definition of salt-bridges [46]. Every protein contains at least 50 residues. The three-dimensional structures of these proteins have been solved by X-ray crystallography, whose resolution is better than or equal to 1.4 Å and are available in the protein data bank (PDB) [44]. The PDB identity codes of these 20 halophilic proteins are **1DOI, 1ITK, 1MOG, 1MOJ, 1VDR, 2AZ3, 2IJQ, 2VWG, 2X98, 2ZUA, 3B73, 3CRJ, 3EEH, 3IFV, 3PUG, 3QTA, 3U1D, 4AF1, 4E19, 4JCO**.

Salt-bridge extraction and categorization

A pair of oppositely charged residues (Asp or Glu with Arg, Lys or His) forms ion pairs in native protein structures. An ion pair is defined as a salt-bridge if they meet the following criteria: (i) The centroids of the side chain charged groups in oppositely charged residues lie within 4.0 Å of each other [43] and (ii) at least one pair of Asp or Glu side-chain carboxyl oxygen atoms and side-chain nitrogen atoms of Arg, Lys or His are within a 4.0 Å distance [25,29]. The three-dimensional atomic coordinates of the charged atoms participating in ion pairs have been extracted from their respective PDB files. From the distances, ion pairs within 4 Å have selected for constructing our dataset of 275 salt-bridges. Of the observed salt-bridges, the percentage of individual basic and acidic residues involved in salt-bridge formation was calculated. The salt-bridge dataset is divided into several categories based on geometry i.e. hydrogen bonded or non-hydrogen bonded, location in the protein i.e. buried or solvent exposed and networking i.e. networked or isolated.

Computation of continuum electrostatic energy contributions by salt-bridges

Energy contribution due to electrostatic interactions in proteins is computed using continuum electrostatics using APBS methodologies. PBE is a continuum description of electrostatics for proteins. This method models the protein as a low dielectric medium in which the charges of ionizable groups and the partial charges of permanent dipoles are assigned to the corresponding atoms according to the three-dimensional structures of the protein. The solvent is represented as a high dielectric medium and mobile ions are taken into account through the ionic strength. PB solver such as APBS method [36] and Delphi software package [24,25,28,29,31,35] are popularly used for finding salt-bridge energies. Salt-bridge energy obtained by this *in silico* approach was found to be consistent with experimental observations [32,41]. We have followed the former procedure and model as devised by Hendsch & Tidor, (1994) along with our in-house computational automations for obtaining values of different energy terms associated with salt-bridges. The electrostatic energy contribution ($\Delta\Delta G_{tot}$) can be decomposed into three different energy terms: (i) $\Delta\Delta G_{dslv}$ is the energy difference caused by desolvation of charges. It is an unfavorable term. (ii) $\Delta\Delta G_{prt}$ is the energy difference due to background interactions of charge with permanent dipoles of the peptide backbone, of helices, or of non-ionizable polar side chains. (iii) $\Delta\Delta G_{brd}$ is the favorable bridge energy term that represents the electrostatic interaction between two charged residues side-chain groups in the folded state of the protein. First two terms are indirect and pH independent terms and the last one is direct and pH dependent term.

The total electrostatic energy contribution to the salt-bridge formation $\Delta\Delta G_{tot}$ is taken as sum of indirect and direct terms:

$$\Delta\Delta G_{tot} = \Delta\Delta G_{dslv} + \Delta\Delta G_{brd} + \Delta\Delta G_{prt}$$

The electrostatic energy contribution to salt-bridge formation is calculated relative to a mutation of its salt-bridging side-chains to their hydrophobic isosteres. Hydrophobic isosteres are identical with the charged residue side-chains, with the exception that their partial atomic charges are set to zero. The energy minimization of initial structures is carried out with 100 steps of steepest descent followed by 500 steps of conjugate gradient using CHARMM force field in the GROMACS software. This procedure improves the accuracy of the continuum electrostatic calculations [45]. In each case, all hydrogen atoms are added, the protonation state of all charged residues are defined at pH 7.0, atomic radii and charges are assigned according to CHARMM force field using the program PDB2PQR [46]. Continuum electrostatic calculations are performed with the APBS [38]. The linearised PBE is solved on a $97 \times 97 \times 97$ Å^3 cubic grid box with finer grid spacing (0.5 Å per grid step) using iterative finite-difference methods [42,43]. The solvent probe radius of 1.4 Å is used to define the molecular surface. The internal protein dielectric constant of 4.0 and the external solvent dielectric constant of 76 are used for each calculation. The ionic strength of 0.2M NaCl is used.

In each calculation, initially the molecule occupies 23% of the grid and the Debye-Huckel boundary conditions are applied. Results of this rough calculation are used as a boundary condition for a focused calculation in which the molecule occupies 92% of the grid. The results of the focused calculations are presented here. APBS outputs the energy values in units of κT, where κ is the Boltzmann constant and T is absolute temperature. These values are multiplied by a conversion factor of 0.592 to obtain the results in units of kilo calories per mole (Kcal mol^{-1}) at room temperature (25°C).

Database linking

PDB:**1DOI, 1ITK, 1MOG, 1MOJ, 1VDR, 2AZ3, 2IJQ, 2VWG, 2X98, 2ZUA, 3B73, 3CRJ, 3EEH, 3IFV, 3PUG, 3QTA, 3U1D, 4AF1, 4E19, 4JCO**.

Supporting Information

Table S1   Details of energetics of 275 salt-bridges from 20 extremely halophilic proteins. The table also shows different salt-bridge classes (as in Table 3).

Acknowledgments

We thankfully acknowledge the computational facility Laboratory of the Department of Biotechnology, The University of Burdwan.

Author Contributions

Conceived and designed the experiments: AKB. Performed the experiments: AN PSSG AKB. Analyzed the data: AN PSSG AKB SB BM. Contributed reagents/materials/analysis tools: AN PSSG AKB SB BM. Wrote the paper: AKB AN. Energy calculation: AN PSSG AKB.

References

1. Lanyi JK (1974) Salt-dependent properties of proteins from extremely halophilic bacteria. Bacteriol Rev 38: 272–290.
2. Ginzburg M, Sachs L, Ginzburg BZ (1970) Ion metabolism in a Halobacterium. I. Influence of age of culture on intracellular concentrations. J Gen Physiol 55: 187–207.
3. Eisenberg H (1995) Life in unusual environments: progress in understanding the structure and function of enzymes from extreme halophilic bacteria. Arch Biochem Biophys 318: 1–5.
4. Hecht K, Langer T, Wrba A, Jaenicke R (1990) Lactate dehydrogenase from the extreme halophilic archae bacterium Halobacterium marismortui. Biol Chem Hoppe-Seyler 371: 515–519.

5. Bandyopadhyay AK, Sonawat HM (2000) Salt Dependent Stability and Unfolding of [Fe2-S2] Ferredoxin of Halobacterium salinarum: Spectroscopic Investigations. Biophys J 79: 501–510.

6. Rao JKM, Argos P (1981) Structural stability of halophilic proteins. Biochemistry 20: 6536–6543.

7. Kennedy SP, Ng WV, Salzberg SL, Hood L, DasSarma S (2001) Understanding the adaptation of Halobacterium species NRC-1 to its extreme environment through computational analysis of its genome sequence. Genome Res 11: 1641–1650.

8. Bolhuis A, Kwan D, Thomas JR (2008) Halophilic adaptations of proteins. Protein adaptation in extremophiles. Nova Science Publishers Inc (USA) pp. 71–104.

9. Paul S, Bag SK, Das S, Harvill ET, Dutta C (2008) Molecular signature of hypersaline adaptation: insights from genome and proteome composition of halophilic prokaryotes. Genome Biol 9: R70.

10. Tadeo X, Lopez-Mendez B, Trigueros T, Lain A, Castano D, et al. (2009) Structural basis for the amino acid composition of proteins from halophilic archaea. PLoS Biol 7: e1000257.

11. Frolow F, Harel M, Sussman JL, Mevarech M, Shoham M (1996) Insights into protein adaptation to a saturated salt environment from the crystal structure of a halophilic 2Fe-2S ferredoxin. Nat Struct Biol 3: 452–458.

12. Mevarech M, Frolow F, Gloss LM (2000) Halophilic enzymes: proteins with a grain of salt. Biophys Chem 86: 155–164.

13. Pieper U, Kapadia G, Mevarech M, Herzberg O (1998) Structural features of halophilicity derived from the crystal structure of dihydrofolate reductase from the Dead Sea halophilic archaeon, Haloferax volcanii. Structure 6: 75–88.

14. Winter JA, Christofi P, Morroll S, Bunting KA (2009) The crystal structure of Haloferax volcanii proliferating cell nuclear antigen reveals unique surface charge characteristics due to halophilic adaptation. BMC Struct Biol 9: 55.

15. Britton K L, Baker PJ, Fisher M, Ruzheinikov S, Gilmour DJ, et al. (2006) Analysis of protein solvent interactions in glucose dehydrogenase from the extreme halophile Haloferax mediterranei. Proc Natl Acad Sci (USA) 103: 4846–4851.

16. Mevarech M, Eisenberg H, Neumann E (1977) Malate dehydrogenase isolated from extremely halophilic bacteria of the Dead Sea. 1. Purification and molecular characterization. Biochemistry 17: 3781–3785.

17. Bonete MJ, Pire C, Llorca FI, Camacho ML (1996) Glucose dehydrogenase from the halophilic archaeon Haloferax mediterranei: enzyme purification, characterization and N-terminal sequence. FEBS Lett 383: 227–229.

18. Bandyopadhyay AK, Krishnamoorthy G, Sonawat HM (2001) Structural stabilization of [2Fe-2S] ferredoxin from Halobacterium salinarum. Biochemistry 405: 1284–1292.

19. Dym O, Mevarech M, Sussman JL (1995) Structural features that stabilize halophilic malate dehydrogenase from an archaebacterium. Science 267: 1344–1346.

20. Bandyopadhyay AK, Krishnamoorthy G, Padhy LC, Sonawat HM (2007) Kinetics of salt-dependent unfolding of [2Fe-2S] ferredoxin of Halobacterium salinarum. Extremophiles 4: 615–625.

21. Elcock AH, McCammon JA (1998) Electrostatic contributions to the stability of halophilic proteins. J Mol Biol 280: 731–748.

22. Dill KA (1990) Dominant forces in protein folding. Biochemistry 29: 7133–7155.

23. Pace CN (1990) Conformational stability of globular proteins. Trends Biochem Sci 15: 14–17.

24. Kumar S, Tsai CJ, Nussinov R (2000) Factors enhancing protein thermostability. Protein Eng 13: 179–191.

25. Kumar S, Nussinov R (2001) How do thermophilic proteins deal with heat? Cell Mol Life Sci 58: 1216–1233.

26. Horovitz A, Fersht AR (1992) Co-operative interactions during protein folding. J Mol Biol 224: 733–740.

27. Marqusee S, Sauer RT (1994) Contribution of a hydrogen bond/salt-bridge network to the stability of secondary and tertiary structures in lambda repressor. Protein Sci 3: 2217–2225.

28. Lounnas V, Wade RC (1997) Exceptionally stable salt-bridges in cytochrome P450cam have functional roles. Biochemistry 36: 5402–5417.

29. Kumar S, Nussinov R (1999) Salt-bridge stability in monomeric proteins. J Mol Biol 293: 1241–1255.

30. Dao-pin S, Anderson DE, Baase WA, Dahlquist FW, Matthews BW (1991) Structural and thermodynamic consequences of burying a charged residue within the hydrophobic core of T4 lysozyme. Biochemistry 30: 11521–11529.

31. Hendsch ZS, Tidor B (1994) Do salt-bridges stabilize proteins? A continuum electrostatic analysis. Protein Sci 3: 211–226.

32. Waldburger CD, Schildbach JF, Sauer RT (1995) Are buried salt-bridges important for protein stability and conformational specificity? Nat Struct Biol 2: 122–128.

33. Barril X, Aleman C, Orozco M, Luque FJ (1998) Salt-bridge interactions: stability of ionic and neutral complexes in the gas phase, in solution and in proteins. Proteins: Struct Funct Genet 32: 67 79.

34. Dong F, Zhou HX (2002) Electrostatic contributions to T4 lysozyme stability: solvent-exposed charges versus semi-buried salt-bridges. Biophys J 83: 1341–1347.

35. Li L, Li C, Sarkar S, Zhang J, Witham S, et al. (2012) DelPhi: a comprehensive suite for DelPhi software and associated resources. BMC Biophys 4: 9.

36. Guest WC, Cashman NR, Plotkin S (2010) Electrostatics in the stability and misfolding of the prion protein. Biochem Cell Biol 88: 371–381.

37. Karan R, Capes MD, DasSarma S (2012) Function and biotechnology of extremophilic enzymes in low water activity. Aquatic Biosystems 8: 4.

38. Baker NA, Sept D, Joseph S, Holst MJ, McCammon JA (2001) Electrostatics of nanosystems: application to microtubules and the ribosome. Proc Natl Acad Sci (U.S.A) 98: 10037–10041.

39. Baldwin RL, Rose GD (1999) Is protein folding hierarchic? II. Folding intermediates and transition states. Trends Biochem Sci 24: 77–84

40. Tsai CJ, Lin SL, Wolfson HJ, Nussinov R (1997) Studies of protein-protein interfaces: a statistical analysis of the hydrophobic effect. Protein Sci 6: 53–64

41. Fleming PJ, Gong H, Rose GD (2006) Secondary structure determines protein topology. Protein Sci 15: 1829–1834.

42. Madigan D, Ryan PB, Schuemie M, Stang PE, Overhage JM, et al. (2013) Evaluating the impact of database heterogeneity on observational study results. Am J Epidemiol 178: 645–651

43. Barlow DJ, Thornton JM (1983) Ion-pairs in proteins. J Mol Biol 168: 867–885.

44. Hendsch ZS, Sindelar CV, Tidor B (1998) Parameter dependence in continuum electrostatic calculations: a study using protein salt-bridges. J Phys Chem ser B 102: 4404–4410.

45. Lee BK, Richards FM (1971) The interpretation of protein structures estimation of static accessibility. J Mol Biol 55: 379–400.

46. Tsai CJ, Nussinov R (1997) Hydrophobic folding units derived from dissimilar monomer structures and their interactions. Protein Sci 6: 24–42.

47. Lebbink JH, Consalvi V, Chiaraluce R, Berndt KD, Ladenstein R (2002) Structural and thermodynamic studies on a salt-bridge triad in the NADP-binding domain of glutamate dehydrogenase from Thermotoga maritima: cooperativity and electrostatic contribution to stability. Biochemistry 41: 15524–15535.

48. Marti DN, Bosshard HR (2003) Electrostatic interactions in leucine zippers: thermodynamic analysis of the contributions of Glu and His residues and the effect of mutating salt-bridges. J Mol Biol 330: 621–637.

49. Sun DP, Sauer U, Nicholson H, Matthews BW (1991) Contributions of engineered surface salt-bridges to the stability of T4 lysozyme determined by directed mutagenesis. Biochemistry 30:7142–7153.

50. Missimer JH, Steinmetz MO, Baron R, Winkler FK, Kammerer RA, et al. (2007) Configurational entropy elucidates the role of salt-bridge networks in protein thermostability. Protein Sci 16: 1349–1359.

51. Albeck S, Unger R, Schreiber G (2000) Evaluation of direct and cooperative contributions towards the strength of buried hydrogen bonds and salt-bridges. J Mol Biol 298: 503–520.

52. Lovell SC, Davis IW, Arendall WB 3rd, de Bakker PI, Word JM, et al. (2003) Structure validation by Calpha geometry: phi, psi and Cbeta deviation. Proteins 50(3): 437–450.

53. Gupta PSS, Mondal S, Mondal B, Ul Islam RN, Banerjee S, et al. (2014) SBION: A Program for Analyses of Salt-Bridges from Multiple Structure Files Bioinformation 10(3): 164–166

Permissions

All chapters in this book were first published in PLOS ONE, by The Public Library of Science; hereby published with permission under the Creative Commons Attribution License or equivalent. Every chapter published in this book has been scrutinized by our experts. Their significance has been extensively debated. The topics covered herein carry significant findings which will fuel the growth of the discipline. They may even be implemented as practical applications or may be referred to as a beginning point for another development.

The contributors of this book come from diverse backgrounds, making this book a truly international effort. This book will bring forth new frontiers with its revolutionizing research information and detailed analysis of the nascent developments around the world.

We would like to thank all the contributing authors for lending their expertise to make the book truly unique. They have played a crucial role in the development of this book. Without their invaluable contributions this book wouldn't have been possible. They have made vital efforts to compile up to date information on the varied aspects of this subject to make this book a valuable addition to the collection of many professionals and students.

This book was conceptualized with the vision of imparting up-to-date information and advanced data in this field. To ensure the same, a matchless editorial board was set up. Every individual on the board went through rigorous rounds of assessment to prove their worth. After which they invested a large part of their time researching and compiling the most relevant data for our readers.

The editorial board has been involved in producing this book since its inception. They have spent rigorous hours researching and exploring the diverse topics which have resulted in the successful publishing of this book. They have passed on their knowledge of decades through this book. To expedite this challenging task, the publisher supported the team at every step. A small team of assistant editors was also appointed to further simplify the editing procedure and attain best results for the readers.

Apart from the editorial board, the designing team has also invested a significant amount of their time in understanding the subject and creating the most relevant covers. They scrutinized every image to scout for the most suitable representation of the subject and create an appropriate cover for the book.

The publishing team has been an ardent support to the editorial, designing and production team. Their endless efforts to recruit the best for this project, has resulted in the accomplishment of this book. They are a veteran in the field of academics and their pool of knowledge is as vast as their experience in printing. Their expertise and guidance has proved useful at every step. Their uncompromising quality standards have made this book an exceptional effort. Their encouragement from time to time has been an inspiration for everyone.

The publisher and the editorial board hope that this book will prove to be a valuable piece of knowledge for researchers, students, practitioners and scholars across the globe.

List of Contributors

Helena W. Qi, Priyanka Nakka, Connie Chen, Mala L. Radhakrishnan
Department of Chemistry, Wellesley College, Wellesley, Massachusetts, United States of America

Wilfred R. Hagen
Department of Biotechnology, Delft University of Technology, Delft, The Netherlands

Chen Song
School of Biomedical, Biomolecular and Chemical Sciences, The University of Western Australia, Perth, Australia
Department of Theoretical and Computational Biophysics, Max Planck Institute for Biophysical Chemistry, Go¨ttingen, Germany

Ben Corry
School of Biomedical, Biomolecular and Chemical Sciences, The University of Western Australia, Perth, Australia

Long-Ho Chau and Shih-Chi Chen
Centre for Micro and Nano Systems, The Chinese University of Hong Kong, Hong Kong

Wenfeng Liang
State Key Laboratory of Robotics, Shenyang Institute of Automation, Chinese Academy of Sciences, Shenyang, China

Florence Wing Ki Cheung and Wing Keung Liu
School of Biomedical Sciences, Faculty of Medicine, The Chinese University of Hong Kong, Hong Kong

Wen Jung Li
Centre for Micro and Nano Systems, The Chinese University of Hong Kong, Hong Kong
State Key Laboratory of Robotics, Shenyang Institute of Automation, Chinese Academy of Sciences, Shenyang, China
Department of Mechanical and Biomedical Engineering, City University of Hong Kong, Hong Kong

Gwo-Bin Lee
Department of Power Mechanical Engineering, National Tsing Hua University, Hsinchu, Taiwan

Ming Lei, Ze Li, Shaohui Yan, Baoli Yao, Dan Dan, Yujiao Qi, Jia Qian, Yanlong Yang, Peng Gao, Tong Ye
State Key Laboratory of Transient Optics and Photonics, Xi'an Institute of Optics and Precision Mechanics, Chinese Academy of Sciences, Xi'an, China

Hiroshi Ishikita
Career-Path Promotion Unit for Young Life Scientists, Graduate School of Medicine, Kyoto University, Kyoto, Japan
Precursory Research for Embryonic Science and Technology, Japan Science and Technology Agency, Saitama, Japan

Simin Öz and Achim Breiling
Division of Epigenetics, DKFZ-ZMBH Alliance, German Cancer Research Center, Heidelberg, Germany

Christian Maercker
Mannheim University of Applied Sciences, Mannheim, Germany
Genomics and Proteomics Core Facilities, German Cancer Research Center, Heidelberg, Germany

Chao Wang
Institute of Mechanical Manufacturing Technology, China Academy of Engineering Physics, Mianyang, People's Republic of China
Department of Chemistry, Harbin Institute of Technology, Harbin, People's Republic of China

Surong Hu, Wen Huang and Lunfu Tian
Institute of Mechanical Manufacturing Technology, China Academy of Engineering Physics, Mianyang, People's Republic of China

Xijiang Han
Department of Chemistry, Harbin Institute of Technology, Harbin, People's Republic of China

Shuang Yu, Shujun Sun and Kai Zhang
Academy for Advanced Interdisciplinary Studies, Peking University, Beijing, China

Yongdong Liang
College of Engineering, Peking University, Beijing, China

Jing Fang and Jue Zhang
Academy for Advanced Interdisciplinary Studies, Peking University, Beijing, China
College of Engineering, Peking University, Beijing, China

Abdossamad Talebpour, Robert Maaskant, Aye Aye Khine and Tino Alavie
Qvella Incorporation, Richmond Hill, Ontario, Canada

Ray S. Kasevich
Stanley Laboratory of Electrical Physics, Great Barrington, Massachusetts, United States of America

David LaBerge
Department of Cognitive Sciences, University of California Irvine, Irvine, California, United States of America

Wenfeng Liang
State Key Laboratory of Robotics, Shenyang Institute of Automation, Chinese Academy of Sciences, Shenyang, China
University of Chinese Academy of Sciences, Beijing, China
Department of Mechanical and Biomedical Engineering, City University of Hong Kong, Kowloon, Hong Kong

Yuliang Zhao
Department of Mechanical and Biomedical Engineering, City University of Hong Kong, Kowloon, Hong Kong

Lianqing Liu, Yuechao Wang and Zaili Dong
State Key Laboratory of Robotics, Shenyang Institute of Automation, Chinese Academy of Sciences, Shenyang, China

Wen Jung Li
State Key Laboratory of Robotics, Shenyang Institute of Automation, Chinese Academy of Sciences, Shenyang, China
Department of Mechanical and Biomedical Engineering, City University of Hong Kong, Kowloon, Hong Kong

Gwo-Bin Lee
Department of Power Mechanical Engineering, National Tsing Hua University, Hsinchu, Taiwan

Xiubin Xiao and Weijing Zhang
Department of Lymphoma, Affiliated Hospital of Military Medical Academy of Sciences, Beijing, China

Sébastien Joucla, Pascal Branchereau, Daniel Cattaert and Blaise Yvert
Université Bordeaux, Institut des Neurosciences Cognitives et Intégratives d'Aquitaine, UMR5287, Bordeaux, Talence, France

CNRS, Institut des Neurosciences Cognitives et Intégratives d'Aquitaine, UMR5287, Bordeaux, Talence, France

Jesus Eduardo Lugo, Rafael Doti and Jocelyn Faubert
Visual Psychophysics and Perception Laboratory, School of Optometry, University of Montreal, Montreal, Quebec, Canada

Zoran B. Vosika and Jovana B. Simic-Krstic
Department of Biomedical Engineering, Faculty of Mechanical Engineering at University of Belgrade, Belgrade, Serbia

Goran M. Lazovic
Department of Mathematics, Faculty of Mechanical Engineering at University of Belgrade, Belgrade, Serbia

Gradimir N. Misevic
Department of Research, Gimmune GmbH, Zug, Switzerland

Hongtao Zhao and Danzhi Huang
Department of Biochemistry, University of Zurich, Zurich, Switzerland

Harold Kwok, Kyle Briggs and Vincent Tabard-Cossa
Department of Physics, University of Ottawa, Ottawa, Ontario, Canada

Mohammad Rashed Iqbal Faruque and Mohammad Tariqul Islam
Centre for Space Science (ANGKASA), Research Centre Building, Universiti Kebangsaan Malaysia, UKM, Bangi, Selangor D. E., Malaysia

Clarisse Vaillier and David Peyrade
Univ. Grenoble Alpes, LTM, Grenoble, France
CNRS, LTM, Grenoble, France

Thibault Honegger
Univ. Grenoble Alpes, LTM, Grenoble, France
CNRS, LTM, Grenoble, France
Department of Electrical Engineering and Computer Science, Massachusetts Institute of Technology, Cambridge, Massachusetts, United States of Amercia

Frédérique Kermarrec and Xavier Gidrol
CEA, Institut de Recherches en Technologies et Sciences pour le Vivant, Grenoble, France

Yujuan Zhao
Department of Bioengineering, University of Pittsburgh, Pittsburgh, Pennsylvania, United States of America

Robert L. Rennaker
Behavioral and Brain Sciences, Erik Jonsson School of Engineering, University of Texas Dallas, Richardson, Texas, United States of America

Chris Hutchens
School of Electrical and Computer Engineering, Oklahoma State University, Stillwater, Oklahoma, United States of America

Tamer S. Ibrahim
Department of Bioengineering, University of Pittsburgh, Pittsburgh, Pennsylvania, United States of America
Department of Radiology, University of Pittsburgh, Pittsburgh, Pennsylvania, United States of America

Michał Cegłowski and Grzegorz Schroeder
Department of Supramolecular Chemistry, Faculty of Chemistry, Adam Mickiewicz University in Poznan, Poznań, Poland

Marek Smoluch
Department of Biochemistry and Neurobiology, Faculty of Materials Science and Ceramics, AGH-University of Science and Technology, Krakow, Poland

Michał Babij and Teodor Gotszalk
Faculty of Microsystem Electronics and Photonics, Wroclaw University of Technology, Wroclaw, Poland

Jerzy Silberring
Center for Polymer and Carbon Materials, Polish Academy of Sciences, Zabrze, Poland

Arnab Nayek, Parth Sarthi Sen Gupta, Shyamashree Banerjee and Amal K. Bandyopadhyay
The Department of Biotechnology, The University of Burdwan, Burdwan, West Bengal, India

Buddhadev Mondal
Department of Zoology, Burdwan Raj College, The University of Burdwan, Burdwan, West Bengal, India

Index